Met. 0. 848

METEOROLOGICAL OFFICE

British Rainfall 1966

THE ONE HUNDRED AND SIXTH ANNUAL VOLUME

LONDON: HER MAJESTY'S STATIONERY OFFICE: 1973

U.D.C. 551.506.1 (41-4)

HER MAJESTY'S STATIONERY OFFICE

Government Bookshops

49 High Holborn, London WC1V 6HB
13A Castle Street, Edinburgh EH2 3AR
109 St Mary Street, Cardiff CF1 1JW
Brazennose Street, Manchester M60 8AS
50 Fairfax Street, Bristol BS1 3DE
258 Broad Street, Birmingham B1 2HE
80 Chichester Street, Belfast BT1 4JY

Government publications are also available through booksellers

SBN 11 400209 6*
Printed in England for Her Majesty's Stationery Office by Manor Farm Press, Alperton, Wembley, Middlesex

Contents

LISTS OF MAPS AND GRAPHS PARTS I AND II

LIST OF TABLES, PART II

Introduction

The presentation of data in this volume of *British Rainfall* follows closely that introduced with the 1961 volume. However, for the first time since 1961 the General Table of Rainfall is no longer subdivided into river board areas, the river boards having ceased to exist at the end of March 1965.

The data published in Section 3 of Part II have been extended slightly by including a graph showing the percentage departure of average rainfall (1916–50) with each graph showing the percentage departure of actual rainfall in 1966.

As in previous volumes, in order to publish as much useful data as possible, marginally incomplete records have been completed by estimating the missing data. Estimated data are printed in brackets throughout the volume.

BRITISH RAINFALL 1966

Part I General table of rainfall

AREAL GROUPING OF RAINFALL STATIONS

based on drainage areas

See page opposite for index

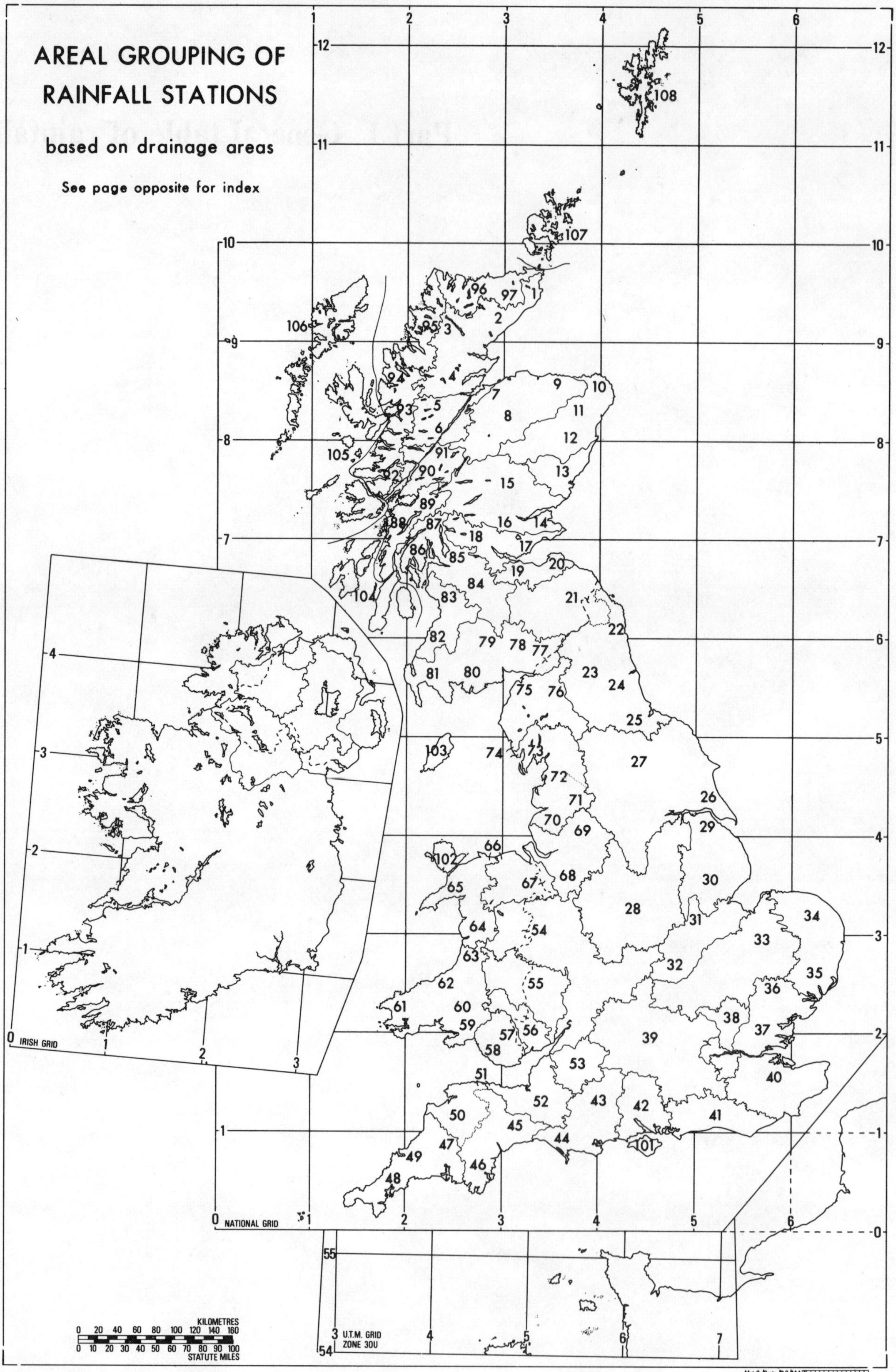

INDEX TO AREAL GROUPING OF RAINFALL STATIONS

Notes. Part of hydrometric area 21 (Tweed) and part of hydrometric area 77 (Esk) are in England but stations within these parts are listed with appropriate stations in Scotland so as to be grouped with the natural drainage areas to which they belong. These stations are Linhope, Lilburn Tower, Middleton Hall, Mindrum, Goldscleugh, Hethpool, Pallinsburn, Etal R.F.M. station and Berwick-on-Tweed under area 21 (Tweed) and Ashycroft and Longtown under area 77 (Esk).

The National Grid reference is used for locating rainfall stations in Great Britain. The Irish Grid reference is used for stations in Northern Ireland and the Universal Transverse Mercator Grid reference (Zone 30U) is used for stations in the Channel Islands.

Station	Grid reference	Altitude m	Av. 1916–1950 mm	Rainfall in millimetres													Highest recorded fall in a day	
				Jan	Feb	Mar	Apr	May	Jun	Jul	Aug	Sep	Oct	Nov	Dec	Year	mm	date

England and Wales

NORTHUMBRIAN RIVER AUTHORITY AREA, 9211 SQUARE KILOMETRES

HYDROMETRIC AREAS 22(including Till) TO 25

Station	Grid reference	Altitude m	Av. 1916–1950 mm	Jan	Feb	Mar	Apr	May	Jun	Jul	Aug	Sep	Oct	Nov	Dec	Year	mm	date
Coastal																		
Newlands	NU 46 116324	58	657	33	63	15	77	34	50	34	186	28	81	42	89	733	93	13 Aug
Seahouses	NU 46 197341	11	...	36	67	12	74	39	59	39	137	46	77	47	86	718	56	13 Aug
Newton House	NU 46 235253	23	690	34	61	13	85	38	68	41	141	35	85	55	83	739	65	13 Aug
Stamford	NU 46 223193	53	706	39	82	17	88	40	58	44	150	30	83	54	84	768	78	13 Aug
Howick Hall	NU 46 246177	34	751	34	82	19	89	47	66	49	116	32	76	52	92	753	51	13 Aug
Aln																		
Callaly Castle	NU 46 051099	104	812	42	104	22	69	56	70	55	139	41	104	73	81	858	55	13 Aug
Whittingham	NU 46 069117	76	762	42	95	18	74	50	65	47	117	36	100	65	76	784	48	13 Aug
Coquet																		
Bygate R.F.M. Sta. Area 59.6 km²	NT 36 870083	776	1053	...	...	...	...	...	...	...	...	...	...	...	...	...	...	...
Usway (Coquet)																		
Shillmoor R.F.M. Sta. Area 21.4 km²	NT 36 886077	207	1103	...	...	...	...	...	...	...	...	...	...	...	...	...	...	...
Coquet																		
Rothbury	NU 46 055013	91	877	47	138	16	93	53	76	37	118	34	98	67	98	874	35	13 Aug
Amble	NU 46 267043	8	647	33	78	13	79	49	63	39	114	41	69	57	70	705	49	13 Aug
Coastal																		
Acklington	NU 46 225007	42	667	32	70	9	85	53	64	38	168	34	69	52	70	743	92	13 Aug
Cockle Park	NZ 45 200912	99	730	43	127	7	107	59	74	50	110	41	85	62	77	843	37	9 Apr
Wansbeck																		
Hart (Wansbeck)																		
Harwood	NZ 45 000903	235	936	56	(125)	27	103	65	103	51	144	63	95	79	104	1015	43	13 Aug
Font (Wansbeck)																		
Font Res.	NZ 45 047932	189	836	48	118	26	97	53	105	56	138	59	88	82	101	973	37	17 Jun
Maiden Hall	NZ 45 138886	84	774	40	109	15	100	63	88	44	153	34	95	58	84	883	...	...
Blyth																		
Pont (Blyth)																		
Great Whittington	NZ 45 007703	161	726	44	67	17	81	57	70	40	119	61	95	68	81	800	...	...
Burnside	NZ 45 047714	128	710	42	67	14	76	51	66	43	126	59	99	59	76	779	...	...
Medburn	NZ 45 129707	101	740	45	85	15	83	53	85	59	122	47	103	55	73	825	41	13 Aug
Blyth																		
Coastal																		
Tynemouth	NZ 45 374694	28	675	37	94	11	84	48	65	52	103	34	83	59	66	737	32	9 Apr
North Tyne																		
Kielder Castle	NY 35 632935	201	1220	84	142	66	106	97	145	52	137	118	109	139	153	1349	38	3 Sep
Whickhope	NY 35 682861	168	1210	71	151	70	110	94	117	55	131	102	105	115	139	1260	34	3 Aug
Tarset R.F.M.Sta. Area 285 km²	NY 35 776861	117	1255	...	...	...	...	...	...	...	...	...	...	...	...	...	...	...
Bower	NY 35 757832	152	1012	55	113	55	92	74	93	46	120	87	84	99	100	1019	44	3 Aug
Rede (North Tyne)																		
Black Cleugh	NT 36 706052	293	1216	79	122	58	96	81	116	61	159	95	127	(152)	(133)	1279	...	...
Game Park	NT 36 715044	268	1253	81	152	55	112	88	123	58	156	99	132	149	130	1336	...	...
Lumsdon	NT 36 734061	317	1174	70	161	39	95	69	130	53	164	82	115	(152)	(133)	1262	...	...
Countess Crag	NT 36 715029	488	1405	91	149	70	134	102	147	69	185	129	129	(165)	161	1530	...	...
Chattlehope	NT 36 734025	305	1236	83	163	50	118	79	129	54	155	108	123	(152)	145	1360	...	...
Catcleugh Nursery	NT 36 749031	229	1168	63	130	41	100	75	123	48	155	92	122	144	134	1227	...	...
Low Byrness	NT 36 781015	210	...	66	144	41	100	82	113	44	142	81	127	131	135	1205	44	3 Aug
North Tyne																		
Stone Haugh	NY 35 795762	198	888	48	104	57	81	85	108	68	138	84	91	97	100	1061	33	3 Aug
Parkend	NY 35 872758	84	818	48	101	38	80	75	81	52	119	79	97	90	90	950	28	3 Aug
Reaverhill R.F.M.Sta. Area 1010 km²	NY 35 906732	65	1062	...	...	...	...	...	...	...	...	...	...	...	...	...	...	...
Barrasford...	NY 35 922729	63	744	41	93	29	74	62	77	53	103	69	97	80	66	845	24	3 Aug 3 Oct
Fawcett	NY 35 967766	160	747	44	107	21	89	55	83	51	118	61	100	63	81	876	...	...
Hallington Res.	NY 35 973762	152	741	43	109	21	90	54	82	47	115	61	97	61	80	859	26	3 Oct
Hallington...	NY 35 978761	142	753	43	107	20	87	53	82	48	117	60	98	61	81	857	...	...
Ryal Cottage	NY 35 995740	141	710	43	109	19	85	54	82	55	123	57	94	58	76	853	...	...
South Tyne (North Tyne)																		
Alston	NY 35 720463	314	1221	63	141	87	67	85	101	61	120	107	121	132	170	1256	43	19 Dec
Nenthead	NY 35 777438	439	1324	74	142	87	86	89	102	70	127	112	147	143	179	1357	40	3 Oct
Haltwhistle	NY 35 681640	122	...	45	98	61	71	59	82	54	120	98	105	90	109	993	27	3 Oct
West Allen (South Tyne)																		
Allenheads	NY 35 860452	411	1252	87	154	84	98	86	95	72	183	90	151	133	171	1405	67	13 Aug
South Tyne (North Tyne)																		
Haydon Bridge	NY 35 838648	82	938	33	75	27	73	43	73	48	135	70	95	70	76	819	34	3 Oct
Haydon Bridge R.F.M.Sta. Area 751 km²	NY 35 856647	59	1172	...	...	...	...	...	...	...	...	...	...	...	...	...	...	...
Tyne																		
Hexham Res.	NY 35 932632	99	704	36	87	35	73	47	71	45	120	63	96	61	77	810	37	3 Oct
Widehaugh Nursery	NY 35 964642	30	672	36	67	26	64	40	71	59	116	60	93	77	63	771	22	13 Aug
Devils Water (Tyne)																		
Tyne																		
Bywell R.F.M.Sta. Area 2180 km²	NZ 45 038617	16	1044	...	...	...	...	...	...	...	...	...	...	...	...	...	...	...
Whittle Dean Res.	NZ 45 064679	116	706	38	69	21	68	47	78	54	109	55	96	65	77	776	33	13 Aug
Whittle Dean Filters	NZ 45 076673	108	709	42	67	22	66	47	68	49	115	55	96	65	75	767	36	13 Aug
Throckley Filters	NZ 45 152669	91	729	41	77	17	78	49	93	52	125	50	99	49	68	798	41	3 Oct
Derwent (Tyne)																		
Quickcleugh	NY 35 889479	421	1142	86	216	63	143	76	86	79	170	81	142	105	161	1410	...	...
Heatheryburn	NY 35 904492	381	1093	83	211	61	137	79	79	71	161	76	129	85	133	1306	...	...
Nookton Burn	NY 35 909465	415	1110	89	136	53	129	72	85	79	170	71	145	98	173	1300	...	...
Little Nookton...	NY 35 931465	393	1068	97	135	51	127	69	83	71	178	72	143	99	119	1243	...	...
Crosshill	NY 35 941483	354	955	70	157	44	105	65	77	68	156	67	129	91	116	1146	46	3 Oct

Station	Grid reference	Altitude m	Av. 1916–1950 mm	Rainfall in millimetres Jan	Feb	Mar	Apr	May	Jun	Jul	Aug	Sep	Oct	Nov	Dec	Year	Highest recorded fall in a day mm	date
NORTHUMBRIAN—*(contd)*																		
Boltshope	NY 35 948468	381	1002	73	137	43	113	68	74	69	163	64	137	97	121	1159	53	3 Oct
Presser Villa	NY 35 959478	397	...	66	130	31	124	66	68	67	171	61	143	92	109	1127	56	3 Oct
Birkside	NY 35 949506	332	876	60	130	33	96	50	70	65	153	52	129	75	100	1014	52	3 Oct
Pennypie House	NY 35 950516	335	...	60	142	35	93	54	74	69	155	57	124	83	111	1055	50	3 Oct
Buckshott	NY 35 973497	293	890	67	142	27	100	59	69	69	147	55	136	89	104	1065	50	3 Oct
Cowbyers	NY 35 972515	323	876	61	127	34	89	58	69	76	161	63	133	76	100	1049	...	...
Acton Burn	NY 35 979537	259	832	63	112	28	81	66	73	71	147	69	117	71	97	995	...	...
Burnhope (Derwent)																		
Belmount	NY 35 973476	384	940	84	152	36	118	71	63	71	166	76	145	100	127	1210	...	...
Edmondbyers Common	NY 35 970459	469	1040	94	165	38	155	71	66	69	169	76	165	98	140	1305	...	...
Feldon Burn	NY 35 999461	375	919	85	149	28	121	60	60	73	163	57	157	84	107	1142	...	...
Harehope Burn	NZ 45 011482	296	892	65	137	24	122	57	53	67	160	49	155	83	99	1072	...	...
Edmondbyers Farm	NZ 45 015502	259	846	57	139	25	104	61	57	71	158	54	131	74	108	1038	48	3 Oct
Derwent (Tyne)																		
Eddy's Bridge R.F.M.Sta. Area 118 km^2	NZ 45 041508	181	932	...	...	...	...	...	...	...	...	...	...	...	...	...	...	...
Muggleswick	NZ 45 046500	213	800	57	136	21	100	55	55	61	134	47	117	69	92	944	46	3 Oct
Hisehope Res.	NZ 45 024461	442	900	70	142	26	123	66	59	79	163	56	153	81	105	1123	53	3 Oct
Smiddy Shaw Res.	NZ 45 039462	344	899	73	139	22	119	66	56	74	138	53	129	77	104	1053	47	3 Oct
Westwood S. Wks.	NZ 45 095548	78	766	49	105	11	93	46	67	42	141	43	104	59	78	839	42	3 Oct
Chopwellwood	NZ 45 136580	136	...	49	100	14	100	47	98	60	123	45	106	62	74	877	42	3 Oct
Rowlands Gill R.F.M.Sta. Area 242 km^2	NZ 45 168581	29	853	...	...	...	...	...	...	...	...	...	...	...	...	...	...	...
Derwenthaugh	NZ 45 191614	15	685	40	97	13	90	53	74	52	122	35	106	61	75	819	46	13 Aug
Team (Tyne)																		
Gateshead	NZ 45 268604	155	701	49	91	9	96	49	63	56	117	33	108	54	46	771	48	13 Aug
Tyne																		
Jesmond Dene	NZ 45 253673	48	...	53	87	11	90	55	74	58	120	41	101	48	63	800	39	13 Aug
Felling-on-Tyne S.Wks.	NZ 45 310603	53	667	44	82	9	83	49	67	66	103	30	84	55	59	730	33	13 Aug
Cleadon P. Sta.	NZ 45 386634	66	638	44	71	9	79	46	71	63	116	28	83	52	55	717	40	13 Aug
South Shields, Whiteleas	NZ 45 364630	26	657	38	84	8	80	47	67	61	110	24	88	49	54	711	34	13 Aug
Moorhouses Res.	NZ 45 334697	59	680	43	84	12	89	51	63	59	127	39	96	51	64	778	38	13 Aug
Coastal																		
South Shields, South Pier	NZ 45 373678	5	650	35	83	10	79	42	62	46	107	31	83	53	59	690	32	9 Apr
South Shields, Bents Park	NZ 45 374672	11	675	37	86	11	82	43	65	46	104	29	84	53	60	701	34	9 Apr
Whitburn	NZ 45 398615	28	...	43	74	6	77	41	56	55	131	24	80	54	56	697	58	13 Aug
Wear																		
Lodgegill Mine	NY 35 802375	594	1670	76	279	127	203	114	127	63	178	140	165	152	292	1918	...	...
Grass Meres	NY 35 825372	564	1517	63	241	114	203	102	127	63	165	121	165	140	279	1784	...	...
Plover Hill	NY 35 831401	546	1523	63	241	114	178	95	114	57	152	114	140	127	254	1651	...	...
St.John's Chapel	NY 35 885381	299	...	70	177	59	106	84	81	42	142	79	119	108	149	1216	39	1 Dec
Westgate Railway Sta.	NY 35 910380	269	1001	63	167	68	113	79	66	40	143	67	147	106	145	1206	39	13 Aug
Rookhope (Wear)																		
Rookhope	NY 35 938425	334	1096	64	157	53	136	76	59	57	153	68	139	108	142	1213	44	3 Oct
Eastgate R.F.M.Sta. Area 36.5 km^2	NY 35 952390	241	1156	...	...	...	...	...	...	...	...	...	...	...	...	...	...	...
Wear																		
Stanhope R.F.M.Sta. Area 172 km^2	NY 35 984391	202	1318	...	...	...	...	...	...	...	...	...	...	...	...	...	...	...
Waskerley (Wear)																		
Waskerley No.3	NZ 45 004433	436	930	83	159	31	157	63	59	74	153	(58)	156	106	114	1213	...	...
Waskerley	NZ 45 022444	359	903	92	157	26	153	63	61	77	163	54	153	89	110	1198	50	3 Oct
Waskerley No.1	NZ 45 026451	401	909	71	152	25	106	64	61	81	165	60	155	89	112	1141	...	...
Waskerley No.6	NZ 45 042430	375	900	73	154	22	149	60	58	77	134	40	152	87	114	1119	...	...
Tunstall Res.	NZ 45 063407	221	875	70	161	22	125	59	62	65	135	37	137	86	103	1062	43	3 Oct
Wear																		
Bedburn (Wear)																		
The Grove	NZ 45 066299	178	883	63	127	26	115	75	65	59	117	47	131	84	74	983	...	...
Dryderdale	NZ 45 087331	209	...	53	139	23	(112)	(66)	72	53	120	(41)	123	65	71	937	18	1 Dec
Bedburn R.F.M.Sta. Area 74.9 km^2	NZ 45 118322	109	854	...	...	...	...	...	...	...	...	...	...	...	...	...	...	...
Wear																		
Mayland West	NZ 45 077294	282	...	55	117	23	113	67	72	63	113	41	120	63	66	913	31	3 Oct
Low Wadsworth S.Wks.	NZ 45 177310	78	694	54	117	15	113	66	53	55	120	29	98	51	49	819	43	13 Aug
Gaunless (Wear)																		
Butterknowle	NZ 45 109257	192	811	63	123	30	122	61	69	56	126	48	117	73	73	962	38	3 Oct
West Auckland	NZ 45 181263	109	...	49	109	12	115	86	54	52	109	39	91	52	65	835	25	1 Apr
Bishop Auckland R.F.M.Sta. Area 93.0 km^2	NZ 45 215306	65	752	...	...	...	...	...	...	...	...	...	...	...	...	...	...	...
Wear																		
Willington S.Wks.	NZ 45 208346	55	...	55	104	10	97	63	65	54	116	28	100	49	50	791	29	11 May
Sunderland Bridge R.F.M.Sta. Area 658 km^2	NZ 45 264376	40	945	...	...	...	...	...	...	...	...	...	...	...	...	...	...	...
Browney (Wear)																		
Satley	NZ 45 117437	226	786	67	105	17	107	62	66	59	127	33	118	71	80	914	36	3 Oct
Crookhall S.Wks.	NZ 45 125504	206	...	56	107	8	111	56	79	59	141	31	114	71	75	909	41	3 Oct
Waterhouses	NZ 45 189413	134	766	60	113	10	110	71	74	63	117	27	97	67	62	871	31	11 May
Ushaw College	NZ 45 219436	181	742	58	109	10	105	68	69	61	127	30	100	61	59	856	27	11 May
Burn Hall R.F.M.Sta. Area 178 km^2	NZ 45 259387	44	770	...	...	...	...	...	...	...	...	...	...	...	...	...	...	...
Wear																		
Houghall	NZ 45 283412	37	708	53	104	8	85	56	69	73	108	30	95	53	54	789	27	3 Oct
Durham	NZ 45 267416	102	668	51	98	9	80	57	62	67	113	27	101	49	49	763	28	3 Oct
Washington, Glebe School	NZ 45 313560	49	663	49	89	9	77	55	69	71	100	25	105	54	62	764	26	9 Apr
Coastal																		
Sunderland, Thorneholme	NZ 45 393562	40	642	56	81	10	88	51	75	61	115	26	92	59	58	774	34	13 Aug
Ryhope P.Sta.	NZ 45 404524	69	659	47	71	9	84	43	66	84	108	20	97	58	60	746	36	9 Apr
Dalton P.Sta.	NZ 45 411468	105	708	50	85	9	93	49	64	83	120	23	110	60	57	805	36	9 Apr
Mill Hill Res.	NZ 45 413425	154	704	54	94	13	97	57	66	81	136	26	113	63	61	862	35	9 Apr
New Winning P.Sta.	NZ 45 406384	123	743	62	119	13	112	58	66	75	129	28	103	67	59	892	33	7 Feb
Shotton Hall	NZ 45 417399	124	732	61	117	12	107	56	61	80	137	26	117	67	60	902	33	9 Apr
Hart Res.	NZ 45 484343	52	582	52	87	14	81	47	77	62	115	26	105	80	53	798	33	3 Oct
West Hartlepool W.Wks.	NZ 45 508334	11	609	51	92	13	88	41	70	53	109	25	102	77	51	772	32	3 Oct
Hartlepool	NZ 45 510327	9	623	47	100	10	86	38	67	52	104	25	99	73	49	750	31	3 Oct
West Hartlepool, Catcote Road	NZ 45 494299	21	...	46	98	8	85	38	(71)	50	105	23	92	74	52	742	25	3 Oct

GENERAL TABLE, ENGLAND AND WALES

Station	Grid reference	Altitude m	Av. 1916–1950 mm	Jan	Feb	Mar	Apr	May	Jun	Jul	Aug	Sep	Oct	Nov	Dec	Year	Highest recorded fall in a day mm	date
NORTHUMBRIAN—*(contd)*																		
Trout (Tees)																		
Moor House	NY 35 757328	556	2062	103	242	186	116	171	163	121	171	181	179	205	298	2138	74	19 Dec
Moor House R.F.M.Sta. Area 11.4 km²	NY 35 759336	533	2182	...	...	...	...	...	...	...	...	...	...	...	...	...	...	...
Tees																		
Greenhills Farm	NY 35 838320	444	...	86	205	107	95	110	123	93	133	111	143	152	214	1572	41	19 Dec
Forest-in-Teesdale	NY 35 872295	372	1393	97	177	94	101	104	103	74	145	92	149	127	185	1448	38	3 Oct
Newbiggin	NY 35 915276	271	1143	91	173	73	113	104	97	60	134	81	144	119	157	1346	42	3 Oct
Dent Bank R.F.M.Sta. Area 217 km²	NY 35 932260	227	1717	...	...	...	...	...	...	...	...	...	...	...	...	...	...	...
Lune (Tees)																		
Fishlake	NY 35 851228	458	1578	95	163	161	103	137	143	101	175	145	161	148	266	1797	73	19 Dec
Selset	NY 35 918217	329	1077	76	150	81	102	89	102	58	118	79	121	102	157	1234	...	...
Grassholme Res.	NY 35 946224	285	1041	71	162	66	102	82	94	58	112	69	118	93	149	1177	32	3 Oct
Tees																		
Eggleston	NY 35 999239	253	...	74	156	40	(102)	(76)	84	48	101	53	120	95	115	1064	32	3 Oct
Balder (Tees)																		
Balderhead Embankment	NY 35 928187	343	--	79	152	72	118	78	96	63	132	82	122	(102)	161	1254	17	1 Jan
Hury Res.	NY 35 967192	263	965	60	134	46	93	68	80	49	111	61	110	84	121	1017	33	3 Oct
Tees																		
Lartington Filters	NZ 45 012179	224	834	57	133	35	120	74	66	48	107	56	109	76	98	979	29	3 Oct
Barnard Castle, Bowes Museum	NZ 45 056164	171	800	54	134	21	124	65	69	44	111	43	111	65	79	920	30	3 Oct
Greta (Tees)																		
Old Spittal Farm	NY 35 910122	423	...	61	165	47	123	76	92	74	155	80	115	95	156	1240	...	...
Bar Gap Farm	NY 35 960107	374	1137	62	169	53	115	81	74	61	151	85	125	108	151	1235	39	19 Dec
Bowes, New site	NY 35 986139	331	...	(53)	150	46	(103)	75	69	57	135	65	121	91	113	1080	...	...
Rutherford Bridge R.F.M.Sta. Area 86.2 km²	NZ 45 034122	223	1179	...	...	...	...	...	...	...	...	...	...	...	...	...	...	...
Brignall	NZ 45 071122	201	841	54	154	18	132	72	60	42	146	38	108	72	71	967	44	13 Aug
Tees																		
Raby Castle	NZ 45 128221	140	769	53	103	11	111	49	52	52	109	41	103	47	59	791	30	3 Oct
Selaby Hall	NZ 45 152184	112	709	49	90	12	110	56	53	48	108	35	92	54	52	758	28	13 Aug
Darlington, Broken Scar	NZ 45 256141	48	669	56	101	6	95	66	64	41	106	26	92	50	43	745	28	3 Oct
Darlington W.Wks.	NZ 45 258139	49	641	48	90	6	94	64	62	41	103	26	91	47	37	709	27	3 Oct
Broken Scar R.F.M.Sta. Area 818 km²	NZ 45 259137	37	1207	...	...	...	...	...	...	...	...	...	...	...	...	...	...	...
Skerne (Tees)																		
Hurworth Burn Res.	NZ 45 408333	109	660	59	107	13	102	65	75	63	119	35	107	81	56	882	31	9 Apr
Shildon S.Wks	NZ 45 243253	117	...	52	87	10	96	83	60	53	115	40	100	52	52	798	26	3 Oct
Aycliffe	NZ 45 283232	79	677	52	94	7	92	69	63	55	106	37	98	45	46	766	25	3 Oct
Walworth	NZ 45 231189	85	662	(51)	93	7	93	77	62	43	107	30	91	47	41	742	26	3 Oct
Darlington, Coniscliffe Road	NZ 45 280141	53	654	54	109	7	101	58	60	46	104	26	98	45	40	751	29	3 Oct
South Park R.F.M.Sta. Area 250 km²	NZ 45 284129	34	671	...	...	...	...	...	...	...	...	...	...	...	...	...	...	...
Clow (Tees)																		
Croft R.F.M.Sta. Area 78.2 km²	NZ 45 282101	29	758	...	...	...	...	...	...	...	...	...	...	...	...	...	...	...
Tees																		
Croft Hall	NZ 45 288095	31	...	48	108	7	98	62	63	40	119	24	91	54	37	750	25	7 Feb 3 Oct
Fighting Cocks Res.	NZ 45 342138	52	651	49	97	7	98	58	68	45	105	26	87	57	37	733	26	3 Oct
Worsall Hall	NZ 45 393101	24	638	49	93	7	95	56	60	44	115	25	85	58	36	724	25	3 Aug
Leven (Tees)																		
Kildale Hall	NZ 45 609094	182	930	44	89	18	99	43	100	74	136	25	110	104	59	900	31	21 Jun
Pinchinthorpe House	NZ 45 580148	104	735	53	90	13	88	39	85	76	118	28	122	86	46	844	31	26 Oct
Crathorne Hall	NZ 45 444086	56	649	43	89	11	87	54	76	44	128	23	93	75	41	764	26	3 Oct
Leven Bridge R.F.M.Sta. Area 196 km²	NZ 45 445122	5	790	...	...	...	...	...	...	...	...	...	...	...	...	...	...	...
Tees																		
Long Newton Res	NZ 45 364166	56	651	49	91	5	89	49	88	46	102	29	90	64	39	743	30	17 Jun
Stockton-on-Tees	NZ 45 432186	19	600	55	90	9	93	42	89	48	103	28	92	71	40	759	25	17 Jun
Billingham (Tees)																		
Billingham	NZ 45 461224	11	601	40	73	7	89	40	75	52	106	25	89	75	38	710	25	3 Oct
Tees																		
Middlesborough, Albert Park	NZ 45 494189	9	618	56	83	9	88	34	83	57	103	25	93	67	39	737	22	3 Oct
Nunthorpe	NZ 45 538149	87	706	50	81	12	94	41	84	53	119	27	109	83	38	792	31	1 Apr
South Bank	NZ 45 542198	12	608	44	82	11	81	33	74	63	102	23	100	83	41	738	24	3 Oct
Crookfoot Res.	NZ 45 430309	89	632	54	91	11	94	55	76	56	105	33	111	73	51	809	30	3 Oct
Coastal																		
Redcar	NZ 45 599249	8	608	33	75	9	89	20	62	60	123	17	95	75	47	706	30	21 Aug
Guisborough W.Wks.	NZ 45 638155	183	784	43	104	10	83	35	70	68	116	26	113	72	61	800	34	21 Aug
Lockwood Res.	NZ 45 668141	193	867	44	100	15	97	35	75	81	146	30	129	91	67	912	39	21 Aug
Skinningrove Iron Works	NZ 45 707201	61	716	30	81	7	92	19	64	65	109	14	91	80	51	703	34	21 Aug

YORKSHIRE OUSE AND HULL RIVER AUTHORITY AREA, 13491 SQUARE KILOMETRES

HYDROMETRIC AREAS 26 AND 27

Station	Grid reference	Altitude m	Av. 1916–1950 mm	Jan	Feb	Mar	Apr	May	Jun	Jul	Aug	Sep	Oct	Nov	Dec	Year	Highest recorded fall in a day mm	date
Coastal																		
Scaling Res.	NZ 45 741121	187	...	50	112	15	91	33	75	87	148	29	116	95	74	924	25	3 Oct
Mulgrave Castle	NZ 45 845125	125	735	41	98	20	102	36	75	60	142	23	125	79	75	875	35	3 Oct
Esk																		
Westerdale	NZ 45 664052	232	991	53	156	32	128	68	90	72	145	31	115	105	95	1089	28	21 Aug
Danby Lodge	NZ 45 717084	122	954	54	139	24	104	50	68	76	148	28	115	110	80	996	27	7 Feb
Whitby	NZ 45 895109	41	669	35	95	18	105	34	73	62	111	22	117	61	67	800	28	9 Apr 3 Oct
Coastal																		
Scarborough	TA 54 044884	36	649	34	89	25	100	61	65	79	114	23	68	69	47	774	23	21 Aug
Filey P.Sta.	TA 54 114807	37	...	34	101	19	90	53	58	66	91	21	75	59	56	724	33	21 Aug
Filey	TA 54 113801	37	685	33	107	19	87	57	59	63	96	24	74	60	59	737	27	7 Feb
Gypsey Race																		
High Mowthorpe	SE 44 888685	175	778	48	142	20	84	57	79	88	115	29	74	78	61	875	33	7 Feb
West Lutton	SE 44 933695	92	...	41	131	23	83	66	72	69	102	28	64	85	57	821	31	7 Feb
Sledmere House	SE 44 933648	121	778	45	142	27	84	70	75	86	111	26	58	66	66	857	31	7 Feb

Station	Grid reference	Altitude m	Av. 1916–1950 mm	Jan	Feb	Mar	Apr	May	Jun	Jul	Aug	Sep	Oct	Nov	Dec	Year	Highest recorded fall in a day mm	date
YORKSHIRE OUSE AND HULL—*(contd)*																		
Rudston	TA 54 097676	27	...	33	124	17	77	68	66	93	89	30	62	(76)	58	795	24	7 Feb
Bridlington	TA 54 175689	48	696	42	129	21	104	67	72	82	104	35	70	72	59	857	31	22 Jun
Coastal																		
Burton Agnes P.Sta	TA 54 110634	16	672	30	123	24	94	63	80	78	95	34	65	69	60	816	33	7 Feb
Hornsea Town Hall	TA 54 207482	11	616	29	104	25	73	59	60	71	117	57	73	60	49	776	31	21 Aug
Burton Pidsea	TA 54 246314	7	629	33	111	22	73	53	63	91	77	41	62	71	52	749	38	22 Jun
Withernsea High School	TA 54 337281	9	605	28	101	27	58	47	83	117	62	38	60	59	44	724	50	22 Jun
Kilnsea	TA 54 417161	12	...	38	85	24	62	37	67	73	61	25	63	69	46	652	30	22 Jun
Sunk Island	TA 54 266189	2	617	33	106	22	55	45	78	105	89	35	77	68	45	757	47	22 Jun
Preston	TA 54 190297	5	617	41	107	24	66	40	61	107	71	42	64	66	54	743	33	22 Jun
Holderness Drain																		
Brandesburton Hall	TA 54 112478	8	667	38	119	26	77	63	64	79	123	40	66	65	43	803	33	21 Aug
Hull																		
Wetwang	SE 44 933589	67	758	44	157	29	89	53	73	81	125	35	55	81	63	884	41	7 Feb
Wansford Bridge R.F.M.Sta. Area 192 km^2	TA 54 064560	5	754	...	...	...	...	...	...	...	...	...	...	...	...	...	...	...
Kelk (Hull)																		
Langtoft	TA 54 009670	79	706	44	135	23	84	74	63	71	106	24	70	86	65	844	31	7 Feb
Foston Mill R.F.M.Sta. Area 57.2 km^2	TA 54 093548	6	719	...	...	...	...	...	...	...	...	...	...	...	...	...	...	...
Beeford	TA 54 129541	9	643	42	115	25	78	75	65	73	100	38	62	66	41	781	30	7 Feb
Driffield Canal (Kelk)																		
Driffield, Glaxo Laboratory	TA 54 031575	18	...	42	131	22	90	62	61	71	103	28	63	65	48	787	35	21 Aug
Hull																		
Hutton Cranswick	TA 54 023523	20	664	45	130	25	85	59	67	70	106	29	67	71	51	806	36	7 Feb
Hempholme Lock	TA 54 080499	3	659	36	121	27	83	64	64	78	114	34	67	69	50	806	38	21 Aug
Hempholme Lock R.F.M.Sta. Area 378 km^2	TA 54 080498	3	724	...	...	...	...	...	...	...	...	...	...	...	...	...	...	...
Dalton Holme	SE 44 965452	34	723	43	141	37	93	57	72	75	129	31	72	66	56	872	36	7 Feb
Leconfield	TA 54 026438	7	650	36	124	28	69	57	72	72	117	34	64	55	48	777	31	7 Feb
South Bullock P.Sta.	TA 54 053426	3	655	38	116	27	75	56	67	70	118	34	67	52	51	772	31	21 Aug
Beverley, Lairgate	TA 54 032392	10	651	40	126	33	77	56	78	86	148	49	72	56	47	869	33	7 Feb
Little Weighton	SE 44 982338	87	677	43	127	36	75	56	81	79	137	43	72	53	58	860	33	7 Feb
Dunswell P.Sta.	TA 54 067350	2	630	41	118	25	70	47	65	71	121	51	72	49	47	777	29	7 Feb
Coastal																		
Hull	TA 54 084302	2	645	40	122	26	74	43	68	77	103	52	72	62	49	786	31	7 Feb
Hull, Ringrose Street	TA 54 067281	3	639	41	121	30	73	43	69-	71	108	51	77	69	49	801	32	7 Feb
Springhead	TA 54 040294	4	649	40	116	29	70	48	67	76	102	53	84	61	54	799	31	7 Feb
Foulness																		
Water End	SE 44 791384	7	676	41	117	31	69	41	68	79	129	27	62	57	47	769	34	7 Feb
Bursea House	SE 44 812337	5	...	36	108	33	61	44	65	60	121	24	59	53	48	710	34	7 Feb
Ure																		
Moorland Cottage	SD 34 807923	343	1831	95	264	156	106	180	158	104	164	177	133	215	298	2052	69	22 May
Hawes S.Wks.	SD 34 874902	227	1451	77	249	(102)	(102)	(127)	107	83	137	112	100	177	189	1561	32	4 Feb
Bainbridge	SD 34 934902	216	--	74	219	73	104	95	96	68	111	76	104	127	138	1285	39	22 May
West Burton	SE 44 016866	180	...	58	190	63	133	107	79	65	108	57	(102)	(102)	(114)	1177	36	22 May
Bolton Hall	SE 44 075897	128	886	52	169	27	114	104	56	58	108	43	123	81	86	1020	33	1 Dec
Harmby	SE 44 127896	136	...	57	164	24	119	74	59	54	111	38	109	69	63	939	32	11 May
Horse House	SE 44 047813	251	...	75	269	85	141	150	85	95	135	81	125	145	143	1529	39	22 May
Burn (Ure)																		
Low Houses	SE 44 137813	211	1026	63	210	45	147	98	69	66	131	55	131	105	94	1213	44	3 Oct
Healey Pasture	SE 44 157808	189	970	62	205	45	147	98	64	65	126	57	127	90	86	1172	41	3 Oct
Breary Bank	SE 44 157804	183	996	60	209	45	144	98	66	66	127	57	129	89	86	1176	44	3 Oct
Towler Hill	SE 44 153798	250	983	61	209	45	143	100	66	67	130	63	130	93	88	1194	45	3 Oct
Agill (Burn)																		
Agill Head	SE 44 114766	421	1198	81	239	56	189	121	81	89	156	70	164	86	127	1459	...	...
Grey Ridge	SE 44 120756	405	1186	76	234	51	189	(114)	77	75	132	62	(151)	79	137	1378	...	...
Low Ash Head	SE 44 140763	258	1119	74	229	56	184	112	67	76	127	62	135	75	150	1346	...	...
Arnagill Beck West	SE 44 145746	369	1099	83	221	52	177	111	70	76	141	66	142	81	142	1361	...	...
Arnagill Beck East	SE 44 162756	351	1065	81	229	48	171	107	76	78	152	65	165	83	145	1400	...	...
Somerside	SE 44 143773	283	1054	61	211	54	149	95	75	66	132	56	129	83	107	1217	44	3 Oct
High Sour Mire	SE 44 143790	269	1013	59	214	49	140	100	65	69	137	63	130	92	99	1218	...	...
Leighton Res.	SE 44 161790	193	953	59	202	45	142	97	59	72	129	58	125	92	87	1168	43	3 Oct
Laver (Ure)																		
Hambleton Hill	SE 44 160730	337	1037	69	202	49	141	97	67	63	135	60	145	89	103	1219	...	...
Hawsett	SE 44 184729	250	982	58	201	48	146	96	73	71	142	56	139	77	99	1208	45	3 Oct
Stone and Ears Wham	SE 44 168751	341	1038	68	215	47	168	102	72	58	137	60	146	91	107	1272	...	...
Bagwith Brae	SE 44 195746	214	980	54	197	44	138	92	69	60	129	55	132	77	94	1140	45	3 Oct
Drover's Inn	SE 44 213722	176	991	56	198	51	138	98	70	63	135	54	135	103	90	1192	44	3 Oct
Lumley Moor Res.	SE 44 224706	172	947	58	189	51	143	96	62	56	144	48	127	91	97	1162	41	3 Oct
Harper Hill	SE 44 197702	250	981	62	207	47	145	97	71	74	139	60	141	83	92	1218	50	3 Oct
Ure																		
Ripon S.Wks.	SE 44 324707	21	...	41	133	28	85	61	60	54	111	29	92	56	44	793	28	21 Aug
Westwick Lock R.F.M.Sta. Area 914 km^2	SE 44 356667	15	1163	...	...	...	...	...	...	...	...	...	...	...	...	...	...	...
Swale (Ure)																		
Keld	NY 35 893011	320	...	85	209	85	106	119	105	66	139	90	127	141	184	1454	47	1 Dec
Rowleth End	SD 34 971975	213	...	72	189	53	122	94	93	69	137	76	112	123	127	1266	36	1 Dec
Arkengarthdale	NZ 45 001029	282	1128	70	190	65	118	96	82	75	131	(70)	123	125	135	1281	33	1 Dec
Grinton Lodge	SE 44 049976	259	1045	67	189	49	131	89	83	77	119	56	115	135	93	1203	...	...
Richmond R.F.M.Sta. Area 381 km^2	NZ 45 146006	106	1300	...	...	...	...	...	...	...	...	...	...	...	...	...	...	...
Richmond, Green Howard Road	NZ 45 172016	189	...	61	157	20	129	94	61	42	131	26	123	77	60	978	39	3 Oct
Scruton	SE 44 299923	35	658	54	126	19	102	61	68	49	109	28	95	56	36	802	30	3 Oct
Bedale (Swale)																		
Craggs Lane Farm	SE 44 199945	168	752	57	136	19	109	66	71	44	124	32	112	61	61	891	33	3 Oct
Leeming	SE 44 305890	32	637	37	107	14	93	54	56	43	101	27	91	44	30	697	29	3 Oct
Swale (Ure)																		
Maunby Hall	SE 44 355862	30	669	40	106	14	93	74	56	42	115	30	85	52	35	743	28	3 Oct
Wiske (Swale)																		
Northallerton	SE 44 371947	43	...	58	96	17	71	56	68	54	106	31	104	61	36	759	27	3 Oct
Cod (Swale)																		
Whorlton Moor	SE 44 480988	235	815	50	129	18	80	67	92	64	132	38	103	86	43	904	...	...
Cod Beck Res.	SE 44 463983	191	786	47	116	13	79	64	82	64	117	36	104	86	37	845	...	...
Oakdale W.Wks.	SE 44 466966	183	807	47	116	19	77	68	87	63	129	40	109	81	38	871	...	...
Oakdale W.Wks, New site	SE 44 466966	183	816	47	125	18	77	70	87	63	128	40	106	80	37	877	...	...

GENERAL TABLE, ENGLAND AND WALES

Station	Grid reference	Altitude m	Av. 1916–1950 mm	Jan	Feb	Mar	Apr	May	Jun	Jul	Aug	Sep	Oct	Nov	Dec	Year	Highest recorded fall in a day mm	date
YORKSHIRE OUSE AND HULL—*(contd)*																		
Osmotherley Filters	SE 44 458967	147	772	45	113	13	79	64	84	58	119	32	108	80	39	835	28	21 Aug
Thirsk	SE 44 426810	35	...	36	105	14	80	59	77	47	128	29	65	52	41	732	27	21 Aug
Thirsk, Ashton House	SE 44 432809	34	675	35	103	14	81	63	77	53	131	31	73	60	39	759	27	21 Aug
Boltby	SE 44 491866	140	797	42	123	13	85	74	87	(63)	138	39	78	64	51	857	...	...
Mount St.John	SE 44 474848	165	733	42	120	12	83	70	(76)	(63)	132	33	68	74	41	816	...	...
Topcliffe	SE 44 405790	25	644	35	110	12	88	55	59	48	112	25	72	46	36	697	31	1 Apr
Swale (Ure)																		
Leckby Grange R.F.M.Sta. Area 1350 km²	SE 44 415748	18	879	...	...	...	...	...	...	...	...	...	...	...	...	...	...	...
Coxwold	SE 44 533773	79	734	43	134	22	81	68	95	62	137	37	73	52	51	853	28	17 Jun
Ure becomes Ouse																		
Easingwold	SE 44 528692	29	674	33	134	15	79	70	63	57	126	25	66	47	43	756	28	7 Feb
Linton upon Ouse	SE 44 492613	14	...	35	121	17	85	61	56	50	135	22	65	39	42	728	36	21 Aug
Nidd (Ouse)																		
West Houses	SE 44 043766	355	1467	76	263	95	156	150	124	81	169	110	134	171	215	1744	50	17 Dec
Scar House Res.	SE 44 065766	331	1230	73	257	92	152	147	118	76	159	97	134	151	164	1619	42	21 Aug
New Houses	SE 44 094768	230	1175	67	241	69	137	133	(89)	66	137	69	151	107	114	1380	...	...
How Stean (Nidd)																		
Cross Gill	SE 44 039709	533	...	79	244	105	157	154	131	95	157	129	140	155	211	1757	...	...
Great Blowing Gill	SE 44 032721	555	1650	97	279	142	170	151	155	122	151	160	127	169	241	1965	...	...
Bain Grain Beck	SE 44 032728	533	1590	95	283	122	159	163	142	108	157	135	137	178	212	1891	...	...
Howstean	SE 44 046726	427	1617	86	272	122	146	183	133	104	135	135	138	202	229	1885	...	...
Meugher Dike	SE 44 043699	533	1551	89	231	105	178	151	128	107	150	154	142	177	227	1839	...	...
Calside Dyke	SE 44 051699	518	1526	90	244	119	170	145	132	108	152	121	137	170	205	1793	...	...
Little Stangate	SE 44 059716	465	1569	100	292	131	170	169	126	89	171	128	145	178	220	1919	...	...
Backstean Gill	SE 44 058723	373	...	88	302	122	159	168	116	83	140	140	138	194	216	1865	...	...
Aygill Beck	SE 44 050744	472	1474	95	265	123	145	155	113	80	147	105	146	173	201	1749	...	...
Low West Moor	SE 44 057739	396	1501	90	274	131	131	157	112	84	138	135	137	187	203	1779	...	...
Busker Beck	SE 44 068741	343	1437	86	261	127	135	159	98	75	137	127	134	179	214	1733	61	17 Dec
Rain Stang	SE 44 079758	451	1290	95	248	90	133	137	99	89	137	91	127	127	156	1530	...	...
Nidd (Ouse)																		
Raygill House	SE 44 110704	283	1173	77	235	90	141	140	101	72	133	88	127	162	156	1521	41	17 Dec
Ramsgill	SE 44 120709	137	1125	69	242	83	139	124	91	56	109	70	137	147	139	1407	39	17 Dec
Holme House	SE 44 128699	141	1049	63	229	83	129	104	86	50	94	50	129	129	145	1291	...	...
Gouthwaite Res. R.F.M. Sta. Area 114 km²	SE 44 142683	123	1367	...	...	...	...	...	...	...	...	...	...	...	...	...	...	...
Pateley Bridge	SE 44 166646	120	...	55	194	44	121	98	66	56	130	51	108	98	99	1120	33	7 Feb
Swarcliffe Hall	SE 44 235593	158	922	55	187	40	126	93	77	59	148	45	116	87	99	1129	32	3 Oct
Birstwith Hall	SE 44 241583	99	939	53	184	38	130	93	77	51	145	45	118	86	94	1114	32	7 Feb
Knabbs Moor	SE 44 237557	213	844	54	159	42	124	104	74	(53)	143	44	107	81	98	1081	...	...
Oak (Nidd)																		
John o' Gaunts	SE 44 215544	183	...	55	189	39	137	100	79	50	141	47	102	86	98	1124	...	...
Long Stoop	SE 44 212552	213	873	54	175	41	135	105	64	55	148	46	106	87	99	1113	...	...
Beaverdike Res.	SE 44 229547	183	877	58	171	42	127	94	75	51	148	46	109	79	96	1096	33	7 Feb
Ling Cottage	SE 44 228521	233	864	53	179	34	151	91	71	53	145	48	109	84	89	1106	...	...
Cobblers Cottage	SE 44 239525	223	862	54	179	33	146	93	77	55	148	48	106	81	89	1107	...	...
Ten Acres Res.	SE 44 249533	189	881	55	167	36	118	87	83	46	155	44	113	75	87	1066	33	20 Aug
Irongatebridge Res.	SE 44 287548	139	819	45	152	30	96	84	90	47	136	38	95	68	75	955	36	20 Aug
Harlow Hill Res.	SE 44 289542	167	...	45	149	30	93	78	90	49	137	38	93	66	75	942	37	20 Aug
Harrogate	SE 44 303579	63	792	41	147	24	106	63	70	49	138	32	98	62	63	893	31	7 Feb
Nidd (Ouse)																		
Knaresborough, Water Tower	SE 44 348567	73	728	41	147	19	104	64	72	42	131	29	82	68	76	875	29	7 Feb
Hunsingore Weir R.F.M. Sta. Area 484 km²	SE 44 428530	18	993	...	...	...	...	...	...	...	...	...	...	...	...	...	...	...
Crimple (Nidd)																		
Nidd (Ouse)																		
Ouse																		
Long Marston	SE 44 498515	21	...	(38)	133	16	95	67	65	78	130	27	67	47	62	825	31	21 Aug
York, Poppleton	SE 44 575529	15	...	40	128	15	81	49	53	72	129	21	64	41	58	750	29	7 Feb
York, Acomb Landing	SE 44 581527	9	629	38	127	15	82	55	50	58	126	19	61	40	51	720	28	7 Feb
Foss (Ouse)																		
York, Heslington	SE 44 630510	20	...	35	118	15	72	48	69	64	120	24	66	47	50	729	31	7 Feb
Ouse																		
York, The Retreat	SE 44 615508	18	675	38	132	16	76	53	66	67	128	27	71	49	56	778	31	7 Feb
Wharfe (Ouse)																		
Kettlewell	SD 34 970723	212	1436	75	261	113	112	153	138	94	112	106	116	172	216	1669	49	17 Dec
Skirfare (Wharfe)																		
Litton	SD 34 908740	251	1530	79	262	121	123	149	120	109	121	124	130	176	206	1721	41	17 Dec
Fountains Fell	SD 34 869709	657	...	95	279	136	140	180	155	130	109	171	137	207	261	2001	...	...
Wharfe (Ouse)																		
Thorpe Fell	SE 44 007597	503	1326	76	205	78	116	130	149	106	171	96	134	136	209	1605	...	...
Grassington	SE 44 005635	186	...	57	184	77	86	123	121	81	125	78	104	143	175	1353	46	17 Dec
Sandbed Beck	SE 44 019614	206	1141	63	170	70	90	123	135	71	120	64	95	114	145	1260	...	...
Priest's Tarn	SE 44 028695	515	...	93	232	98	183	143	123	94	149	111	146	147	212	1731	...	...
Bullfront Moss	SE 44 059678	451	...	77	225	95	122	133	119	97	142	89	143	165	196	1603	...	...
Trunla Hill	SE 44 069657	349	1248	63	193	80	142	127	102	91	140	71	132	132	142	1416	...	...
Jack Hole	SE 44 089657	402	...	72	211	74	145	132	99	77	152	76	138	151	147	1476	...	...
Upper Barden Res.	SE 44 015575	355	1270	69	189	85	110	118	156	91	148	95	125	126	204	1517	...	...
Barden Res.	SE 44 035563	227	1149	65	166	77	105	115	130	85	158	91	122	128	198	1441	35	17 Dec
Bolton Abbey	SE 44 071539	112	951	53	154	45	99	103	94	62	145	64	94	90	119	1121	36	11 May
Ilkley R.F.M. Sta. Area 443 km²	SE 44 112481	71	1382	...	...	...	...	...	...	...	...	...	...	...	...	...	...	...
Ilkley	SE 44 113476	96	957	47	165	34	109	103	79	69	157	58	96	93	111	1120	36	20 Aug
March Ghyll Res.	SE 44 122509	215	...	60	159	38	110	97	94	64	150	52	91	93	107	1113	37	29 Aug
Otley S. Wks.	SE 44 209460	52	...	44	135	30	102	81	63	52	132	47	77	75	73	911	42	20 Aug
Washburn (Wharfe)																		
High House	SE 44 136610	326	1172	61	182	35	130	89	74	52	156	57	107	90	116	1149	31	7 Feb
Lane Head	SE 44 129588	290	1067	66	177	37	131	90	74	56	151	57	101	97	121	1159	30	7 Feb
Spittle Ings	SE 44 135567	295	1129	59	187	39	127	99	71	56	155	57	102	96	123	1173	32	7 Feb
Blubberhouses	SE 44 166553	175	1003	60	185	34	137	93	75	48	145	52	109	91	104	1133	32	3 Oct
Fewston	SE 44 187543	160	908	58	178	35	138	92	72	45	136	49	103	83	95	1084	31	3 Oct
Broad Dubb	SE 44 219530	221	879	53	188	42	136	88	66	51	147	46	107	85	93	1103	...	...
Swinsey	SE 44 197525	140	882	55	174	32	130	88	68	47	133	46	91	77	91	1031	29	7 Feb
Lindley Wood	SE 44 217486	95	809	49	152	28	112	88	68	57	130	46	87	72	85	973	33	20 Aug
Lindley Wood Res. R.F.M. Sta. Area 87.3 km²	SE 44 220487	91	1021	...	...	...	...	...	...	...	...	...	...	...	...	...	...	...

Station	Grid reference	Altitude m	Av. 1916–1950 mm	Jan	Feb	Mar	Apr	May	Jun	Jul	Aug	Sep	Oct	Nov	Dec	Year	Highest recorded fall in a day mm	date
YORKSHIRE OUSE AND HULL—*(contd)*																		
Wharfe (Ouse)																		
Arthington	SE 44 261452	42	...	44	145	24	102	77	68	47	125	45	85	73	72	906	29	20 Aug
Eccup	SE 44 309418	114	818	50	148	20	112	88	65	44	121	43	95	69	72	928	28	7 Feb
Flint Mill Weir R.F.M. Sta. Area 759 km^2	SE 44 422473	14	1161	...	...	...	...	...	...	...	...	...	...	...	...	...	...	...
Bramham	SE 44 441412	54	672	43	123	18	90	58	75	54	113	31	82	50	57	795	27	7 Feb
Church Fenton	SE 44 528380	8	628	41	122	20	88	57	75	65	117	27	67	44	51	777	30	7 Feb
Askham Bryan	SE 44 552475	27	661	41	125	19	81	59	62	58	129	28	59	43	59	764	30	7 Feb
Ouse																		
Riccall	SE 44 608373	5	619	38	107	19	83	41	54	79	133	27	58	46	48	734	32	29 Aug
Selby Dam (Ouse)																		
Brayton Barff	SE 44 585305	53	657	40	117	20	97	45	65	75	113	33	71	57	60	794	35	21 Aug
Derwent (Ouse)																		
Silpho Moor	SE 44 957946	203	854	66	173	33	149	72	84	82	152	31	104	95	91	1132	43	7 Feb
Osgodby P. Sta.	TA 54 055840	51	683	40	128	22	106	64	58	75	102	34	76	85	49	838	35	7 Feb
Irton P. Sta.	TA 54 004840	29	692	42	118	22	104	45	59	85	99	29	75	81	56	814	30	21 Aug
Ganton, Golf Club	SE 44 982781	30	709	36	110	17	93	52	57	85	91	31	60	77	56	763	32	21 Aug
Wykeham Nursery	SE 44 949860	145	736	54	123	23	116	53	83	91	109	34	68	84	67	904	28	7 Feb
Scampston Hall	SE 44 864756	31	722	39	113	20	70	41	74	93	86	26	52	66	51	729	24	26 Jul
Rye (Derwent)																		
Hawnby	SE 44 569925	123	883	56	147	29	91	70	78	77	143	38	85	90	62	966	25	7 Feb
East Ness	SE 44 695787	43	...	42	149	28	81	52	64	66	111	35	56	51	58	793	26	7 Feb
Dove (Rye)																		
Farndale No.10	SE 44 642997	175	1099	91	165	28	89	85	80	81	163	47	110	109	101	1151	...	...
Farndale No.13	SE 44 673995	322	1063	81	165	25	91	81	81	89	167	49	127	104	95	1156	...	...
Hodge (Dove)																		
Bransdale Weir R.F.M. Sta. Area 18.9 km^2	SE 44 627944	156	1057	...	...	...	...	...	...	...	...	...	...	...	...	...	...	...
Lund Court	SE 44 669857	78	745	47	153	19	74	68	64	65	101	29	60	54	54	788	27	7 Feb
Welburn Hall	SE 44 680846	41	742	48	160	22	87	64	63	57	95	30	54	56	54	790	30	7 Feb
Holbeck (Rye)																		
Ampleforth	SE 44 598789	95	763	51	141	26	97	66	79	69	125	42	59	52	49	857	31	7 Feb
Leysthorpe Hall	SE 44 636789	70	716	47	153	26	86	65	72	53	117	38	60	47	55	819	27	7 Feb
Hovingham Hall	SE 44 665757	36	739	43	140	26	87	61	71	75	113	35	61	53	57	821	22	7 Aug
Seven (Rye)																		
Spiers House	SE 44 757918	157	829	57	149	31	71	82	85	75	115	36	74	87	77	939	26	7 Feb
Rye (Derwent)																		
Little Habton R.F.M. Sta. Area 679 km^2	SE 44 743771	23	851	...	...	...	...	...	...	...	...	...	...	...	...	...	...	...
Costa (Rye)																		
Pickering (Costa)																		
Newton on Rawcliffe	SE 44 824938	168	891	58	150	35	79	74	102	74	129	39	105	81	80	1006	27	7 Feb
Pickering	SE 44 795843	44	703	44	127	22	70	42	82	75	88	29	58	65	52	753	31	7 Feb
Derwent (Ouse)																		
Birdsall House	SE 44 819650	94	772	46	147	19	86	56	82	79	112	28	68	76	55	853	33	7 Feb
Welburn	SE 44 737673	34	696	43	124	17	75	57	90	70	92	29	(63)	(63)	(51)	775	27	7 Feb
Stamford Bridge R.F.M. Sta. Area 1630 km^2	SE 44 714557	8	787	...	...	...	...	...	...	...	...	...	...	...	...	...	...	...
Pocklington (Derwent)																		
Warter	SE 44 861502	88	755	50	126	28	87	50	84	104	120	24	61	73	61	869	33	7 Feb
Aire (Ouse)																		
Malham Tarn	SD 34 894672	395	1531	73	245	112	111	154	143	121	138	119	118	138	244	1715	49	17 Dec
Lee Gate House	SD 34 926645	378	1310	62	215	104	87	137	146	97	142	103	119	140	225	1576	59	17 Dec
Salterforth	SD 34 882442	219	1259	60	171	87	87	120	114	88	129	101	132	138	202	1431	42	3 Oct
Earby S.Wks.	SD 34 909472	128	1039	49	179	60	73	94	96	65	136	74	112	96	151	1186	37	3 Oct
Elslack	SD 34 937485	198	1060	56	182	52	84	99	112	71	137	73	118	108	144	1236	46	7 Feb
Skipton, Town Hall	SD 34 991518	112	899	47	147	50	101	89	83	54	119	64	102	79	93	1029	31	7 Feb
Lower Coppy	SD 34 955409	297	1131	60	167	61	90	99	112	93	152	97	137	103	153	1325	40	3 Oct
Cringles Res.	SE 44 042491	218	872	53	146	39	117	90	(76)	63	135	61	112	76	114	1082	...	...
Keighley, Hard Ings Lane	SE 44 068421	88	924	41	163	44	113	89	74	66	119	61	97	85	113	1065	32	5 Apr
Worth (Aire)																		
Watersheddles Res.	SD 34 968379	340	1398	75	199	95	111	130	132	113	160	116	139	137	226	1632	41	17 Dec
Oldfield Filters No.2	SE 44 003379	298	1236	68	190	83	111	124	125	97	173	101	144	131	206	1553	45	3 Oct
Harbour Lodge	SD 34 996353	355	1344	70	206	92	106	124	145	105	158	106	147	118	228	1606	47	3 Oct
Lower Laithe Res.	SE 44 015367	236	1232	68	210	74	121	125	123	85	161	95	140	117	195	1515	47	3 Oct
Nan Scar	SE 44 037334	336	1257	70	218	94	116	121	123	84	166	90	131	123	220	1556	...	...
Leeming Res.	SE 44 040340	259	1178	61	204	77	120	114	128	81	168	87	132	118	189	1478	36	20 Aug
Leeshaw Res.	SE 44 011351	267	1298	73	230	89	116	120	134	103	155	97	137	121	216	1590	44	3 Oct
Stairs Top	SE 44 004339	415	1372	71	208	99	123	123	135	98	155	106	169	120	203	1612	...	...
Keighley Moor	SD 34 981390	411	1336	74	217	85	129	136	132	107	169	107	129	128	207	1618	...	...
Clough Hey	SD 34 999394	328	1272	60	209	84	125	122	117	90	157	107	129	129	196	1526	...	...
Keighley Moor No.2	SD 34 994408	312	1324	63	212	88	129	122	125	102	142	109	125	125	206	1547	...	...
Newsholme Dean	SE 44 020404	206	1118	55	193	70	111	105	107	(83)	152	85	126	121	168	1376	...	...
Aire (Ouse)																		
Marley S.Wks.	SE 44 087412	83	...	46	167	35	102	87	73	59	122	57	89	85	117	1039	33	11 May
Thornton Moor Filters	SE 44 051334	363	1226'	69	231	83	131	120	118	81	179	87	132	115	192	1537	37	7 Feb
Stubden Res.	SE 44 057332	326	1207	68	228	81	123	130	120	84	169	95	135	119	207	1561	44	17 Dec
Doe Park Res.	SE 44 077341	230	1082	73	227	72	153	125	109	86	177	79	139	113	186	1538	39	20 Aug
Hewenden Res.	SE 44 072355	215	1062	63	200	59	130	116	90	79	142	72	127	106	153	1335	35	13 Oct
Gilstead Filters	SE 44 118393	174	880	53	169	33	126	103	81	51	151	55	97	85	121	1125	37	11 May
Graincliffe Res., New Site	SE 44 120422	259	...	55	157	36	115	99	85	72	142	61	93	84	119	1117	36	11 May
Clayton (Aire)																		
Brayshaw Res.	SE 44 122306	299	974	54	173	49	118	108	83	59	159	65	118	94	141	1219	43	20 Aug
Chellow Heights Res.	SE 44 116352	258	935	54	178	48	126	110	79	62	160	58	107	96	129	1207	32	20 Aug
Chellow Dean Res.	SE 44 124345	197	932	60	173	46	129	107	85	65	168	56	117	99	133	1237	...	...
Horton Park	SE 44 146316	182	891	54	174	48	120	102	79	44	157	57	105	92	122	1154	53	20 Aug
Bowling Cemetery	SE 44 174308	183	890	53	172	47	134	97	87	46	139	56	107	80	106	1125	33	1 Apr
Bradford, Peel Park	SE 44 168348	137	838	49	158	31	124	93	75	56	156	41	97	88	98	1067	55	20 Aug
Bradford	SE 44 149352	134	853	47	166	35	121	106	94	54	161	48	89	86	105	1111	54	20 Aug
Heaton Res.	SE 44 145354	162	867	51	167	33	122	101	86	52	155	47	92	80	110	1095	43	20 Aug
Aire (Ouse)																		
Weecher Res.	SE 44 134421	251	873	55	155	34	120	93	97	72	147	63	95	82	123	1135	32	11 May
Reva Res.	SE 44 153425	221	816	57	159	31	116	89	81	68	159	61	92	84	119	1114	33	20 Aug
Guiseley Moor	SE 44 192435	229	770	49	159	29	127	84	87	63	147	54	92	71	73	1033	46	20 Aug
Rawdon, Micklefield House	SE 44 207397	162	769	55	149	23	116	85	63	65	115	48	86	74	71	950	28	29 Aug
Rawdon S.Wks.	SE 44 213380	46	716	55	150	24	117	86	64	65	117	49	87	75	72	962	28	29 Aug

GENERAL TABLE, ENGLAND AND WALES

Station	Grid reference	Altitude m	Av. 1916–1950 mm	Jan	Feb	Mar	Apr	May	Jun	Jul	Aug	Sep	Oct	Nov	Dec	Year	Highest recorded fall in a day mm	Highest recorded fall in a day date
YORKSHIRE OUSE AND HULL—*(contd)*																		
Armley R.F.M. Sta. Area 692 km²	SE 44 281340	31	1059	...	...	...	...	...	...	...	...	...	...	...	...	...	...	...
Pudsey S.Wks.	SE 44 234330	101	716	38	135	28	96	84	58	47	120	38	80	73	79	877	26	29 Aug
Farnley Hall	SE 44 248323	123	718	38	136	26	107	77	49	45	134	37	84	72	69	875	31	29 Aug
Lawnswood Cemetery	SE 44 267391	143	774	52	134	23	113	90	55	48	113	45	95	77	73	919	27	7 Feb
Weetwood Filters	SE 44 271371	100	748	44	138	22	113	77	50	42	113	38	90	75	68	870	29	21 Aug
Woodhouse Moor	SE 44 291350	94	709	42	126	19	101	67	49	42	105	35	91	65	70	811	25	7 Feb 21 Aug
Allerton, Grange Rise	SE 44 305379	137	738	44	143	20	117	77	56	47	116	44	93	71	63	891	27	7 Feb
Leeds, Knostrop	SE 44 324313	23	611	39	124	20	91	61	39	33	97	33	86	67	59	750	25	21 Aug
Leeds, Roundhay Park	SE 44 330383	130	756	47	137	20	113	78	69	51	113	49	90	65	61	895	29	21 Aug
Seacroft, Hospital	SE 44 349346	72	...	45	132	19	101	70	58	58	110	32	76	60	51	813	28	7 Feb
Lemonroyd S.Wks.	SE 44 377280	20	623	39	116	23	102	51	69	48	110	33	73	57	51	772	28	7 Feb
Walsden (Calder)																		
Calf Lee	SD 34 953201	288	1276	71	246	85	180	145	129	123	190	109	143	139	197	1757	53	7 Feb
Warland Res.	SD 34 954215	379	1335	69	250	88	177	157	138	126	183	124	153	155	262	1881	47	3 Oct
Ramsden Res.	SD 34 915214	303	1359	70	253	100	127	178	127	121	184	122	134	143	236	1798	54	7 Feb
Ramsden Hill	SD 34 917208	404	1353	73	242	110	125	187	134	121	185	132	142	145	256	1853	...	...
Reaps Moss	SD 34 897229	378	1511	66	242	116	129	181	150	125	172	136	145	153	246	1860	...	...
Gorpley Res.	SD 34 907229	274	1464	69	254	101	127	182	132	103	170	114	139	136	238	1766	49	7 Feb
Calder (Aire)																		
Todmorden, Lower Laith	SD 34 943244	152	1313	59	234	82	142	132	122	80	154	88	127	126	177	1523	49	20 Aug
Todmorden, Hospital	SD 34 965244	198	1311	68	221	75	127	127	122	82	150	87	141	117	167	1483	45	20 Aug
Eastwood S.Wks.	SD 34 970260	112	1223	63	226	82	108	124	120	87	152	92	141	124	193	1512	44	20 Aug
Hebden (Calder)																		
Walshaw Dean No.3	SD 34 962344	360	1469	81	219	102	139	119	139	117	154	118	137	142	271	1738	...	...
Walshaw Dean Lodge	SD 34 962335	317	...	73	223	95	128	113	127	95	154	105	146	121	220	1600	44	3 Oct
Walshaw Dean No.4	SD 34 956335	434	...	77	233	99	123	113	132	118	145	120	141	134	243	1680	...	...
Graining (Hebden)																		
Widdop Res.No.3	SD 34 923331	323	1352	73	203	79	130	121	137	105	128	109	149	121	190	1545	...	...
Widdop Res.No.4	SD 34 927327	323	1385	73	209	82	141	127	137	111	130	111	144	120	194	1580	...	...
Widdop Res.No.2	SD 34 931331	325	1377	75	204	89	110	122	139	106	141	113	157	126	223	1607	46	13 Oct
Widdop Res.No.5	SD 34 942348	390	1441	79	209	96	139	131	143	125	136	114	147	128	241	1689	...	...
Gorple Res.No.3	SD 34 918311	354	1359	73	235	99	125	143	133	105	146	127	143	146	224	1700	...	...
Gorple Res.No.2	SD 34 930301	389	1463	81	230	89	119	137	131	111	146	121	147	131	203	1646	...	...
Gorple Res.	SD 34 955311	313	1403	76	256	90	138	129	121	82	152	111	159	132	222	1668	44	7 Feb
Coppy	SD 34 957305	346	1357	68	249	81	125	145	114	93	143	122	135	137	207	1617	...	...
Hebden (Calder)																		
High Greenwood R.F.M. Sta. Area 36.0 km²	SD 34 973308	191	1415	...	...	...	...	...	...	...	...	...	...	...	...	...	...	...
Calder (Aire)																		
Mytholmroyd, Redacre	SE 44 008263	96	...	49	209	63	104	92	97	72	124	75	104	106	153	1249	40	7 Feb
Whiteholme	SD 34 975203	377	1263	66	237	103	158	145	129	132	168	105	143	157	231	1773	52	7 Feb
Byron Edge	SD 34 973192	382	1326	73	237	103	161	152	135	116	165	114	131	157	238	1782	51	7 Feb
Turvin No.1	SD 34 984197	366	1282	63	231	68	132	141	123	111	173	112	148	144	182	1629	...	...
Turvin No.2	SD 34 998214	304	...	57	227	68	132	120	113	105	164	102	143	132	177	1541	...	...
Warley Moor	SE 44 032312	404	1213	58	210	66	137	106	151	85	141	92	122	128	177	1472	...	...
Midgley Moor	SE 44 016306	411	1194	57	209	65	137	101	150	87	143	109	115	133	167	1472	...	...
Castle Carr	SE 44 026297	323	1122	61	199	66	139	92	131	83	144	80	133	113	176	1416	36	7 Feb
Ryburn (Calder)																		
Rishworth Drain	SD 34 981176	397	1282	72	234	95	167	149	128	119	184	107	147	159	218	1779	53	7 Feb
Baitings	SE 44 004190	287	1329	64	240	79	137	123	125	101	158	111	152	146	174	1609	...	...
Manshead	SE 44 004198	373	1324	63	239	75	128	132	121	101	156	113	147	147	209	1633	...	...
Greystones	SE 44 002179	363	1361	68	250	75	136	139	118	113	174	105	162	135	183	1658	...	...
Round Ing	SE 44 014178	267	...	63	257	75	122	122	128	116	165	117	150	127	168	1612	...	...
Stansfield Hey	SE 44 024189	213	...	57	209	67	132	105	111	89	(165)	89	135	116	141	1415	48	7 Feb
Ryburn Res. R.F.M. Sta. Area 10.7 km²	SE 44 025187	163	1326	...	...	...	...	...	...	...	...	...	...	...	...	...	...	...
Booth Dean (Ryburn)																		
Linsgreave	SD 34 986143	385	1361	65	228	93	134	127	120	125	167	111	156	140	187	1652	...	...
Waystone Drain	SD 34 998148	366	1412	59	213	77	141	129	122	117	181	106	168	152	203	1668	...	...
Green Withens Res.	SD 34 990166	367	...	72	212	90	135	128	106	122	182	109	164	137	194	1650	55	7 Feb
Great Wolden Edge No.1	SD 34 996156	346	1327	68	215	81	139	127	108	131	177	102	161	134	186	1629	57	7 Feb
Great Wolden Edge No.2	SD 34 996156	346	1342	75	225	84	132	125	108	125	180	107	169	146	198	1673	60	7 Feb
Long Clough	SE 44 017156	308	1391	63	249	88	130	119	111	123	182	105	157	165	214	1705	...	...
Lodge Hole	SE 44 021146	442	...	62	232	86	125	112	115	126	190	100	158	167	210	1683	...	...
Pike End	SE 44 017168	373	...	65	249	93	129	121	109	119	187	103	161	164	201	1700	...	...
Upper Moss	SE 44 027158	305	...	58	240	88	126	114	109	123	174	100	151	156	188	1629	...	...
Rishworth Moor Lodge No.2	SE 44 026165	290	1265	67	228	96	117	118	108	113	180	101	160	141	181	1609	59	7 Feb
Rishworth Moor Lodge No.1	SE 44 026165	290	...	67	238	92	119	116	107	111	173	102	159	138	187	1607	61	7 Feb
Stott Hall	SE 44 032161	285	1251	56	227	94	131	111	108	115	173	100	147	154	189	1603	...	...
Booth Wood Mill Res. R.F.M. Sta. Area 15.9 km²	SE 44 033166	207	1359	...	...	...	...	...	...	...	...	...	...	...	...	...	...	...
Ringstone Res.	SE 44 048178	293	1233	63	249	86	135	126	124	114	183	104	152	157	187	1680	57	7 Feb
Calder (Aire)																		
Ovenden Moor	SE 44 049316	419	1217	62	210	88	120	117	135	85	149	99	135	130	208	1539	...	...
Ramsden Wood	SE 44 059271	248	1095	54	185	65	111	90	114	82	140	69	106	105	161	1281	37	20 Aug
Albert Res.	SE 44 074256	242	979	52	189	62	118	91	104	73	155	72	107	100	142	1264	39	20 Aug
Halifax, Library	SE 44 085250	191	903	43	167	54	112	78	107	65	143	57	103	91	114	1134	44	20 Aug
Deanhead Res.	SE 44 039151	307	1326	57	226	95	134	123	101	118	162	102	148	161	213	1639	51	7 Feb
Elland R.F.M. Sta. Area 342 km²	SE 44 124219	59	1219	...	...	...	...	...	...	...	...	...	...	...	...	...	...	...
Colne (Calder)																		
Redbrook Res.	SE 44 026096	360	1322	70	216	95	140	131	121	133	141	116	156	164	182	1665	52	7 Feb
Marsden Res.	SE 44 041119	213	1375	67	240	99	143	119	113	127	163	108	156	175	190	1700	56	7 Feb
Wessenden (Colne)																		
Wessenden Head	SE 44 068077	366	1273	68	237	99	140	125	113	126	170	113	151	165	189	1696	...	...
Holly Bank Moss	SE 44 064082	325	1376	65	229	83	152	125	116	121	178	107	147	175	180	1678	...	...
Wessenden Old Res.	SE 44 056087	287	1330	63	233	91	150	122	114	116	163	106	144	173	175	1649	...	...
Butterly Moss	SE 44 047096	338	1276	65	217	87	157	132	114	132	167	99	145	167	159	1640	...	...
Bobus	SE 44 035097	366	1282	71	234	93	161	125	111	144	157	106	154	157	154	1666	...	...
Colne (Calder)																		
Deerhill	SE 44 070113	350	1363	59	201	97	127	121	99	96	156	105	150	162	168	1541	45	7 Feb
Blackmoorfoot	SE 44 096130	244	1094	44	214	79	146	109	89	77	152	87	127	147	164	1436	43	7 Feb
Huddersfield, Oakes	SE 44 113177	235	956	42	195	56	140	78	93	75	125	69	123	108	125	1229	48	1 Apr
Holme (Colne)																		
Ramsden	SE 44 116051	262	...	65	207	115	140	121	117	120	148	113	137	196	242	1721	51	19 Dec

Station	Grid reference	Altitude m	Av. 1916–1950 mm	Jan	Feb	Mar	Apr	May	Jun	Jul	Aug	Sep	Oct	Nov	Dec	Year	Highest recorded fall in a day mm	date
YORKSHIRE OUSE AND HULL—*(contd)*																		
Netherly (Holme)																		
Lightens Road	SE 44 102038	449	1596	71	220	132	144	138	142	136	142	127	145	215	262	1874	...	...
Great Heys	SE 44 104042	366	1720	79	221	127	141	134	137	135	158	118	152	213	269	1884	53	19 Dec
Red Cabin	SE 44 115036	460	1631	70	218	129	142	132	139	134	141	125	146	211	251	1840	...	...
Yateholme	SE 44 111043	308	1616	71	207	118	137	127	124	126	155	116	144	202	250	1778	52	19 Dec
Hey Springs	SE 44 117046	335	1619	73	211	121	137	132	132	131	155	113	148	209	257	1819	53	19 Dec
Holme (Colne)																		
Far Naze End	SE 44 098048	457	1666	73	225	135	146	140	144	138	151	132	149	220	265	1917	...	...
Bilberry Res.	SE 44 102068	265	1310	52	211	94	125	114	109	96	144	106	131	164	182	1528	...	...
Snape Res.	SE 44 115087	328	1199	47	197	98	120	110	95	91	146	106	127	169	173	1479	37	7 Feb 21 Aug
Boshaw Whams Res.	SE 44 153057	283	1138	45	197	81	134	95	87	85	(142)	94	120	148	154	1382	...	...
Brow Grains	SE 44 080105	274	1263	59	219	106	126	118	99	106	171	106	134	156	187	1587	47	7 Feb
Meltham Grange	SE 44 087114	259	1228	50	201	81	123	114	82	92	171	100	128	147	162	1451	41	7 Feb
Harden Moss	SE 44 098086	369	1345	69	260	115	133	131	118	105	160	121	147	188	169	1716	...	...
Colne (Calder)																		
Huddersfield	SE 44 163165	99	794	37	161	41	116	59	61	55	116	53	102	88	87	974	37	1 Apr
Colnebridge R.F.M. Sta. Area 245 km^2	SE 44 174199	48	1107	...	...	...	...	...	...	...	...	...	...	...	...	...	...	...
Calder (Aire)																		
Bradford, Harold Park	SE 44 150292	213	838	49	172	42	110	82	85	47	140	52	109	77	96	1059	38	20 Aug
Staincliffe Res.	SE 44 229236	150	694	39	125	25	100	61	63	64	113	49	93	67	65	864	38	21 Aug
Heckmondwike, The Bottoms	SE 44 220224	50	695	38	131	27	106	56	57	58	106	48	89	69	61	845	31	7 Feb
Dewsbury, Crow Nest Park	SE 44 234213	94	692	43	135	33	121	61	61	70	120	52	95	73	68	931	38	1 Apr
Wilton Park	SE 44 235258	130	...	41	119	33	101	69	56	65	106	41	86	58	65	839	23	21 Aug
Mitchell Laithes S.Wks.	SE 44 257202	35	649	38	126	25	117	53	48	52	114	41	86	59	57	817	36	1 Apr
Horbury, Carr Lodge	SE 44 297185	76	654	39	125	27	110	53	55	53	116	35	95	58	58	821	32	1 Apr
Ardsley Res., Upper	SE 44 287248	120	661	37	112	28	107	65	68	58	130	43	80	56	53	839	33	1 Apr
Ardsley Res., Lower	SE 44 287247	119	...	40	122	29	107	68	68	57	132	44	78	55	55	854	31	29 Aug
Jaw Hill Filters	SE 44 292233	96	654	35	115	26	104	60	66	50	132	43	84	57	52	824	30	7 Feb 1 Apr
Wakefield S.Wks.	SE 44 346202	24	636	37	111	23	108	48	65	51	117	35	65	49	44	753	31	1 Apr
Kirkthorpe Weir R.F.M. Sta. Area 899 km^2	SE 44 358212	19	1026	...	...	...	...	...	...	...	...	...	...	...	...	...	...	...
Normanton S.Wks.	SE 44 395241	16	614	34	105	22	96	41	70	48	108	39	78	54	47	743	29	21 Aug
Aire (Ouse)																		
Ings Lane P.Sta.	SE 44 436265	12	615	34	102	21	97	37	86	51	104	32	71	46	47	726	32	15 Jun
Pontefract	SE 44 454213	78	606	33	104	17	90	38	63	61	109	25	63	40	48	691	28	7 Feb
Beal Weir R.F.M. Sta. Area 1930 km^2	SE 44 534255	6	975	...	...	...	...	...	...	...	...	...	...	...	...	...	...	...
Roall W.Wks.	SE 44 567243	13	625	36	100	18	89	49	53	67	109	26	61	49	50	706	31	21 Aug
Ouse																		
Goole S.Wks.	SE 44 748252	3	640	32	107	26	60	43	69	87	139	37	68	49	42	759	37	7 Feb
Don (Ouse)																		
Swiner Clough Moss	SE 44 126028	458	1711	119	220	137	145	140	136	131	157	125	132	205	303	1950	...	...
End Springs	SE 44 130022	442	1722	104	215	118	159	127	132	109	146	133	113	196	302	1854	...	...
Grains Plantation	SE 44 143025	351	...	100	220	104	155	119	117	98	155	113	107	185	241	1714	...	...
Dearden Moss	SE 44 137015	457	1652	132	205	118	218	125	127	108	154	111	146	169	209	1821	...	...
Dunford Bridge Res.	SE 44 151023	329	1442	66	201	107	156	118	115	91	136	104	123	164	228	1608	48	19 Dec
Reaps Slack	SE 44 130037	454	1658	97	223	123	185	125	109	112	159	107	138	171	251	1801	...	...
Brown's Clough No.10	SE 44 145004	428	1607	116	236	116	207	135	129	102	146	133	136	188	265	1909	...	...
Woodland Clough No.11	SE 44 153002	438	...	99	226	104	171	126	116	98	140	123	135	161	211	1709	...	...
Stoney Gate	SE 44 157010	331	1428	72	226	108	164	125	119	96	143	119	125	181	252	1729	51	19 Dec
Long Grain	SE 44 166019	343	1378	68	216	89	165	115	101	80	137	102	125	155	198	1550	...	...
Cat Clough	SE 44 182025	271	1263	62	203	88	173	106	85	72	132	91	129	149	189	1479	44	1 Apr
Town Brook Intake	SE 44 177037	315	1264	61	196	88	159	100	89	75	134	84	128	149	189	1452	...	...
Scout Dike (Don)																		
Spring Head	SE 44 179068	355	1078	56	178	81	143	81	69	75	139	63	112	118	111	1226	...	...
Broadstone Res.	SE 44 195062	314	1124	52	180	74	155	89	68	72	142	74	123	129	137	1296	36	1 Apr
Greenley Carr	SE 44 201050	332	1093	47	169	66	127	83	70	69	141	64	119	121	133	1209	...	...
Spicer House	SE 44 204056	300	...	48	177	70	132	83	65	69	140	63	116	122	122	1207	...	...
Ingbirchworth Res.	SE 44 213056	260	990	45	165	62	151	74	62	72	140	59	118	110	110	1167	33	1 Apr
Royd Moor	SE 44 222045	255	973	49	166	59	154	76	58	70	129	59	115	102	125	1162	38	1 Apr
Scout Dike Res.R.F.M. Sta. Area 10.5 km^2	SE 44 236047	207	1034	...	...	...	...	...	...	...	...	...	...	...	...	...	...	...
Little Don (Don)																		
Fiddlers Green No.2	SK 43 150993	451	1439	87	205	94	163	138	103	91	161	125	127	152	164	1611	...	...
Laund Clough	SK 43 157990	462	1524	91	204	93	162	135	101	92	160	123	129	155	167	1611	...	...
Catclough	SK 43 172984	462	1471	88	205	94	163	132	102	91	159	126	131	153	170	1613	...	...
Mickleden No.1	SK 43 192988	371	1368	71	165	87	163	103	97	86	136	88	153	157	174	1479	...	...
Border Hill	SE 44 177011	350	1267	61	186	77	141	106	93	84	133	94	122	155	200	1452	...	...
Swinden Lodge	SE 44 189009	315	1182	51	167	63	134	96	82	74	125	89	130	142	164	1317	41	14 Sep
Flouch Road	SE 44 205004	293	1119	48	170	61	140	88	75	65	124	78	127	117	131	1223	43	21 Aug
Langsett Res.	SE 44 211003	250	1059	49	177	57	153	88	61	64	125	69	129	113	129	1215	43	21 Aug
Upper Midhope	SK 43 215999	282	1066	53	175	63	134	91	64	69	125	78	127	118	137	1231	42	21 Aug
Sheep Howe	SE 44 252004	288	906	42	148	47	131	69	52	72	123	52	116	94	126	1072	...	...
Midhope	SK 43 219994	253	1169	57	197	77	154	98	69	74	131	79	132	138	176	1384	42	1 Dec
Underbank Res.	SK 43 247995	184	949	42	169	54	152	73	56	71	120	57	109	102	126	1128	41	21 Aug
Brockholes	SE 44 227011	275	973	46	157	46	151	65	52	62	119	53	115	87	127	1080	...	...
Underbank Res.R.F.M. Sta. Area 38.6 km^2	SK 43 253992	166	1234	...	...	...	...	...	...	...	...	...	...	...	...	...	...	...
Ewden (Don)																		
Margery Hill	SK 43 189957	544	1403	77	209	85	162	120	106	111	146	77	143	178	143	1558	...	...
Candle Rush	SK 43 203965	427	1324	67	204	82	145	118	100	107	140	73	139	175	140	1491	...	...
Middle Moss	SK 43 207949	488	1345	69	207	75	158	114	103	109	143	74	140	170	136	1499	...	...
Broomhead Moor	SK 43 224956	418	1246	67	201	69	150	112	104	104	139	68	137	169	134	1454	...	...
Brown Edge	SK 43 220969	404	1264	59	198	78	143	116	102	106	136	72	135	171	132	1447	...	...
Broomhead Hall	SK 43 242963	274	1018	47	192	61	135	85	70	65	129	64	117	118	137	1220	...	...
Wightwizzle	SK 43 247956	262	1094	57	181	74	158	85	72	67	133	68	122	134	166	1316	...	...
Canyard Brook	SK 43 261957	178	1033	58	191	68	169	86	68	(70)	(132)	56	116	120	(137)	1270	36	1 Apr
Waldershaigh	SK 43 263965	267	...	45	176	63	168	81	60	69	128	50	116	107	134	1198	...	...
Broomhead Res.No.1	SK 43 268961	178	965	56	178	65	168	74	63	67	129	57	120	122	134	1234	36	21 Aug
Broomhead Res.No.2	SK 43 272959	148	...	50	174	59	166	77	65	(70)	(127)	54	123	123	127	1213	40	1 Apr
White Lee Moor	SK 43 265946	357	1103	61	189	81	189	89	76	71	132	69	119	128	136	1340	...	...
More Hall Res.	SK 43 289957	124	856	44	163	50	141	63	55	61	121	47	101	97	93	1035	36	21 Aug
More Hall Res.R.F.M.Sta. Area 26.4 km^2	SK 43 289957	123	1161	...	...	...	...	...	...	...	...	...	...	...	...	...	...	...
Loxley (Don)																		
Brogging Moss	SK 43 211917	415	1331	69	184	93	153	96	99	76	154	98	140	178	168	1506	...	...
Broadcar	SK 43 223904	306	1197	61	192	91	157	82	84	72	132	77	112	154	173	1387	...	...

GENERAL TABLE, ENGLAND AND WALES

Station	Grid reference	Altitude m	Av. 1916–1950 mm	Jan	Feb	Mar	Apr	May	Jun	Jul	Aug	Sep	Oct	Nov	Dec	Year	Highest recorded fall in a day mm	Highest recorded fall in a day date
YORKSHIRE OUSE AND HULL—*(contd)*																		
Loftshaw Strines	SK 43 242903	306	1137	61	179	65	150	83	80	68	139	68	117	150	140	1299	...	...
Thornseat	SK 43 230923	404	1161	63	184	82	151	93	84	83	131	88	121	157	167	1406	...	...
Dale Dike Res.	SK 43 242917	218	1101	61	182	85	159	83	79	72	128	73	115	151	163	1350	38	21 Aug
Bradfield Filters	SK 43 261916	168	1008	55	182	75	149	83	72	61	123	66	110	131	135	1242	35	21 Aug
Hobson Moss	SK 43 217946	466	1267	66	182	83	138	102	105	105	159	76	137	161	151	1465	...	...
Emlin Moor	SK 43 247934	304	1141	59	173	84	149	85	76	74	134	81	117	152	172	1357	...	...
Rocher Flat	SK 43 267935	337	1013	58	168	62	167	73	67	67	129	58	109	122	126	1207	...	...
Crawshaw Wood	SK 43 252887	358	1071	54	172	60	149	84	84	64	132	70	115	136	138	1258	...	...
Oakes Farm Damflask	SK 43 272903	198	985	50	175	77	144	75	69	57	128	59	112	126	136	1207	...	...
Damflask Res.R.F.M.Sta. Area 43.5 km^2	SK 43 286906	124	1140	...	...	...	...	...	...	...	...	...	...	...	...	...	...	...
Rivelin (Loxley)																		
Oaking Clough No.2	SK 43 246862	361	1124	56	166	74	143	74	87	73	137	51	117	134	136	1249	...	...
Oaking Clough	SK 43 247864	357	...	54	169	72	150	68	85	77	141	51	118	137	126	1249	...	...
Black Clough	SK 43 235876	358	1179	61	186	79	155	86	96	74	143	58	116	141	155	1351	...	...
Hallam Moor	SK 43 243871	356	1157	60	178	82	162	82	93	76	145	55	120	141	146	1340	...	...
Redmires Res.	SK 43 262857	338	1067	60	177	64	181	77	87	79	128	71	134	119	132	1309	37	21 Aug
Redmires Res. No.3	SK 43 262857	338	1081	60	176	64	181	77	87	79	130	71	134	119	132	1311	38	21 Aug
Rivelin	SK 43 287869	172	947	52	169	59	170	70	67	59	104	58	125	106	103	1142	35	1 Apr
Don (Ouse)																		
Crookes	SK 43 330872	192	809	43	159	51	132	58	61	52	100	57	107	99	86	1006	31	1 Apr 21 Aug
Crookes No.2	SK 43 330872	192	796	43	161	52	133	59	61	52	101	58	108	100	87	1014	32	1 Apr
Sheffield	SK 43 339873	131	779	41	154	47	141	59	61	44	93	56	101	94	77	968	38	1 Apr
Badger House	SK 43 286822	347	943	58	169	49	153	78	90	70	121	75	128	107	106	1204	...	...
Norton Lees	SK 43 348838	90	775	49	148	54	144	73	76	49	93	57	103	77	75	998	35	1 Apr
Grenoside	SK 43 336935	171	736	40	154	41	138	48	56	54	103	45	99	88	67	935	33	1 Apr
Wincobank S.Wks.	SK 43 395917	34	620	32	134	37	120	47	57	41	90	40	84	76	53	811	31	1 Apr
Rother (Don)																		
Hipper (Rother)																		
Hunger Hill	SK 43 329679	202	897	65	159	49	151	74	85	59	123	53	122	70	90	1101	32	21 Aug
Whispering Well	SK 43 331694	175	855	57	151	45	142	71	82	54	109	49	115	65	84	1024	33	21 Aug
Birley Farm	SK 43 317728	210	855	57	155	49	147	64	91	55	107	56	114	85	83	1063	31	1 Apr
Upper Linacre Filters	SK 43 338726	159	847	51	152	43	139	61	87	53	102	52	107	77	73	997	30	1 Apr
Rother (Don)																		
Chesterfield	SK 43 384717	91	742	(51)	148	40	122	51	92	47	93	41	99	67	63	913	25	18 Feb
Crowhole Res.	SK 43 320748	184	793	58	144	40	136	61	95	56	107	61	116	87	80	1041	31	9 Apr
Whittington R.F.M.Sta. Area 165 km^2	SK 43 394744	59	800	...	...	...	...	...	...	...	...	...	...	...	...	...	...	...
Chesterfield S.Wks.	SK 43 392746	69	756	41	132	37	120	48	91	45	81	47	98	67	54	861	26	1 Apr
Bolsover	SK 43 463709	86	744	41	131	30	125	46	63	49	70	25	92	62	50	783	25	9 Apr
Woodhouse	SK 43 410844	74	713	31	129	40	126	46	73	52	91	48	93	65	59	853	32	21 Aug
Woodhouse Mill	SK 43 432856	34	...	30	123	39	109	43	57	41	84	42	73	65	53	759	35	21 Aug
Woodhouse Mill R.F.M.Sta. Area 352 km^2	SK 43 432857	29	762	...	...	...	...	...	...	...	...	...	...	...	...	...	...	...
Woodhouse Mill S.Wks.	SK 43 422858	70	...	33	129	41	116	47	61	44	87	44	76	66	(63)	806	34	21 Aug
Ulley Res.No.2	SK 43 455875	56	624	31	119	32	116	47	52	46	92	35	72	71	49	760	30	21 Aug
Ulley Res.No.4	SK 43 452877	40	...	31	117	33	113	45	51	45	90	35	70	69	50	748	30	21 Aug
Ulley Res.No.3	SK 43 452877	40	631	31	119	32	114	45	51	47	91	35	71	69	50	756	30	21 Aug
Don (Ouse)																		
Rotherham R.F.M.Sta. Area 826 km^2	SK 43 427928	23	856	...	...	...	...	...	...	...	...	...	...	...	...	...	...	...
Rotherham	SK 43 435923	86	628	33	118	25	111	36	54	44	93	36	67	63	47	727	36	21 Aug
Thorpe Hesley	SK 43 385964	67	683	32	135	32	108	39	72	49	97	30	86	67	50	796	30	7 Feb
Firsby Res.	SK 43 495958	58	678	36	124	22	115	40	52	61	107	36	73	57	50	772	25	1 Apr
Thrybergh	SK 43 474961	56	645	34	109	27	105	37	52	50	100	31	66	57	54	724	32	21 Aug
Dearne (Don)																		
Bretton Hall	SE 44 283129	98	...	39	129	27	124	49	56	74	116	30	85	75	65	869	31	1 Apr
Cannon Hall	SE 44 273084	113	738	38	137	35	133	48	50	67	122	38	100	85	66	918	39	1 Apr
Barnsley, Jordan Hill	SE 44 336069	140	682	39	119	28	118	50	55	53	120	34	86	65	46	814	35	21 Aug
Barnsley Weir R.F.M. Sta. Area 119 km^2	SE 44 350073	44	767	...	...	...	...	...	...	...	...	...	...	...	...	...	...	...
Barnsley, Queens Road	SE 44 348063	94	676	42	115	24	122	42	57	47	123	30	85	63	49	800	35	21 Aug
Barnsley S. Wks.	SE 44 378071	40	640	33	117	21	111	41	74	53	109	27	71	61	47	764	30	21 Aug
Wortley Res.	SK 43 308998	213	817	38	139	36	132	57	51	47	122	38	96	71	67	894	44	21 Aug
Worsbrough Dale	SE 44 362035	52	673	36	129	27	122	50	65	49	106	29	75	68	49	806	38	21 Aug
Darfield S. Wks.	SE 44 420038	25	630	31	115	20	109	38	64	63	96	27	72	63	45	745	31	21 Aug
Hoyland	SE 44 363009	180	...	37	124	29	120	51	72	57	124	32	89	69	53	857	38	21 Aug
Wath Wood Res.	SK 43 437993	101	619	27	118	26	104	36	50	57	101	27	72	56	45	720	36	21 Aug
Adwick R.F.M. Sta. Area 311 km^2	SE 44 477020	13	701	...	...	...	...	...	...	...	...	...	...	...	...	...	...	...
Don (Ouse)																		
Doncaster R.F.M.Sta. Area 1260 km^2	SE 44 569040	11	798	...	...	...	...	...	...	...	...	...	...	...	...	...	...	...
Doncaster P. Sta.	SE 44 576040	9	625	31	114	16	95	38	81	70	99	31	75	49	42	741	34	7 Feb
Sandall S. Wks.	SE 44 601060	12	600	31	101	16	86	46	81	62	101	33	70	47	48	723	28	7 Feb
Mill (Don)																		
Bentley Colliery	SE 44 571073	7	619	31	106	18	91	36	72	74	94	35	70	51	49	726	31	31 Jul
Don (Ouse)																		
Kirk Bramwith	SE 44 618114	7	586	29	118	18	88	42	90	75	112	32	67	49	54	774	30	22 Jun
Thorne	SE 44 675145	5	578	30	104	17	77	40	76	72	103	33	69	41	45	707	32	7 Feb
Went (Don)																		
Ackworth Moor Top	SE 44 441160	42	633	31	118	19	105	42	63	63	106	33	74	47	51	752	31	7 Feb
Don (Ouse)																		
Goole, Docks Office	SE 44 745235	5	623	32	109	25	66	50	68	89	138	35	66	54	43	775	36	7 Feb

TRENT RIVER AUTHORITY AREA, 10436 SQUARE KILOMETRES

HYDROMETRIC AREA 28

Trent																		
Hanley S. Wks.	SJ 33 886461	118	785	46	86	47	82	72	90	88	76	61	88	64	122	922	25	1 Apr
Etruria	SJ 33 876476	132	818	46	88	45	72	68	95	100	85	62	86	(63)	121	931	24	1 Dec
Knutton	SJ 33 836468	133	...	49	88	51	77	72	89	91	94	52	92	61	129	946	28	21 Aug
Keele	SJ 33 820446	179	802	49	87	46	75	73	90	97	97	51	100	56	126	946	27	20 Aug
Trentham, Meadow Lane	SJ 33 881411	111	...	52	98	47	95	72	104	89	80	51	91	63	131	974	27	22 Jun
Yarlet Hall	SJ 33 912292	123	761	39	98	44	99	70	83	77	110	50	93	59	151	973	28	1 Apr

Station	Grid reference	Altitude m	Av. 1916–1950 mm	Rainfall in millimetres Jan	Feb	Mar	Apr	May	Jun	Jul	Aug	Sep	Oct	Nov	Dec	Year	Highest recorded fall in a day mm	date
TRENT—*(contd)*																		
Great Haywood R.F.M. Sta. Area 324 km²	SJ 33 994231	70	825	...	...	...	...	...	...	...	...	...	...	...	...	...	...	...
Sow (Trent)																		
Whitmore P. Sta.	SJ 33 799401	121	774	49	88	40	102	72	74	93	90	53	101	63	134	959	27	1 Apr
Hatton W. Wks.	SJ 33 829369	108	...	45	88	41	104	73	85	84	88	47	89	53	144	941	29	1 Apr
Mecce Brook (Sow)																		
Sow (Trent)																		
Stone	SJ 33 878321	107	739	38	88	45	89	68	85	72	86	48	87	59	132	898	29	22 Jun
Penk (Sow)																		
Oaken	SJ 33 857031	125	...	33	103	40	89	87	104	81	98	36	110	66	113	960	26	8 May
Codsall	SJ 33 870028	125	740	34	107	40	84	80	109	78	78	36	104	69	115	934	26	1 Dec
Slade Heath P. Sta.	SJ 33 917071	104	706	31	87	35	75	68	71	90	82	40	93	59	103	834	27	1 Dec
Great Wyrley	SJ 33 991076	128	730	33	101	34	83	71	74	89	152	36	104	64	113	953	63	20 Aug
Pye Green	SJ 33 989143	234	774	36	93	38	101	79	74	105	154	49	100	65	135	1029	...	...
Belvide Res.	SJ 33 861103	112	758	31	97	45	76	70	86	86	91	39	99	67	125	912	23	1 Dec
Hollies P. Sta.	SJ 33 816224	110	...	38	82	38	105	63	78	91	94	33	87	61	129	899	26	22 Jun
Sow (Trent)																		
Penk (Sow)																		
Penkridge	SJ 33 920116	101	703	33	84	38	84	69	59	90	90	45	97	58	108	854	23	22 Jun
Sow (Trent)																		
Milford W. Wks.	SJ 33 975212	75	736	40	86	44	98	71	69	100	148	44	103	55	139	997	68	20 Aug
Milford R.F.M. Sta. Area 591 km²	SJ 33 975215	71	754	...	...	...	...	...	...	...	...	...	...	...	...	...	...	...
Trent																		
Rugeley, Brindley Bank	SK 43 037195	77	692	40	97	38	97	66	70	84	177	56	106	55	134	1019	87	20 Aug
Slitting Mill	SK 43 033173	91	738	37	102	41	102	72	74	84	173	55	118	60	141	1060	81	20 Aug
Blithe (Trent)																		
Sheepwash P. Sta.	SJ 33 951452	209	...	56	107	52	98	88	105	93	92	49	98	67	135	1040	36	22 Jun
Wallmires P. Sta.	SJ 33 949464	232	886	58	112	48	103	89	113	88	100	56	101	72	134	1074	...	...
Hulme	SJ 33 934454	216	841	56	102	46	92	88	117	89	87	47	93	74	122	1012	...	...
Meir	SJ 33 937421	188	837	55	102	49	98	86	111	88	77	47	99	64	124	1000	32	22 Jun
Cresswell P. Sta.	SJ 33 974394	164	877	59	110	55	105	90	91	85	101	54	99	69	145	1062	31	1 Apr
Fulford Manor	SJ 33 950386	216	...	55	104	42	102	80	88	89	99	49	102	64	127	1001	...	...
Middleton Green	SJ 33 995354	182	831	53	109	56	113	81	84	89	117	47	103	70	132	1055	...	...
Leigh	SK 43 014371	152	...	55	107	56	106	86	85	89	152	41	95	66	136	1075	...	...
Church Leigh	SK 43 023360	153	810	64	102	55	105	80	76	100	146	49	90	67	140	1074	39	21 Aug
Fradswell	SK 43 007334	137	793	47	106	60	107	74	85	79	135	48	97	71	140	1049	...	...
Gratwich	SK 43 040315	116	775	43	112	53	102	66	72	84	142	51	92	74	132	1024	...	...
Booth	SK 43 042272	110	741	41	102	44	92	68	70	83	144	47	89	71	138	988	...	...
Kingston	SK 43 063295	128	741	41	106	41	96	69	74	84	152	46	91	68	117	983	...	...
Blithfield Res.	SK 43 071226	83	728	38	88	41	88	61	69	73	167	61	94	62	134	977	79	20 Aug
Port Fields	SK 43 067237	96	725	40	89	39	98	61	72	75	177	58	94	65	121	989	...	...
Hamstall Ridware No.2	SK 43 103193	73	667	37	94	37	83	64	70	74	119	54	91	58	121	902	...	...
Hamstall Ridware R.F.M. Sta. Area 162 km²	SK 43 109191	64	790	...	...	...	...	...	...	...	...	...	...	...	...	...	...	...
Trent																		
Yoxall R.F.M. Sta. Area 1230 km²	SK 43 131177	57	775	...	...	...	...	...	...	...	...	...	...	...	...	...	...	...
Gentleshaw	SK 43 049123	227	795	40	112	49	110	69	72	97	179	42	121	58	139	1089	89	20 Aug
Hanch Res.	SK 43 103136	79	684	36	112	38	95	63	68	77	109	42	100	46	118	904	28	21 Aug
Lichfield, Sandfields	SK 43 110083	92	...	39	109	34	84	56	73	73	83	39	104	47	114	856	21	9 Dec
Alrewas	SK 43 169149	55	663	35	97	33	81	54	74	67	103	41	97	45	106	832	29	29 Aug
Tame, Left (Trent)																		
Willenhall S. Wks. No.2	SO 32 978985	119	...	32	97	32	75	62	94	68	154	36	110	54	103	916	67	20 Aug
Darlaston S. Wks.	SO 32 985978	120	671	39	98	34	69	61	89	67	141	37	103	55	107	900	57	20 Aug
Tame, Right (Tame, Left)																		
Bilston, The Lunt Outfall	SO 32 966960	125	704	31	96	29	71	70	89	68	138	34	103	56	99	883	63	20 Aug
Bradley	SO 32 956950	144	...	37	102	32	82	75	88	74	129	32	106	53	97	908	50	20 Aug
Tipton Branch (Tame, Right)																		
Dudley, Shavers End	SO 32 933912	227	728	33	90	31	78	76	93	70	107	23	89	54	92	837	31	10 Jun
Dudley, Tipton Road	SO 32 950916	149	708	30	81	27	75	61	78	73	103	27	87	57	95	793	28	20 Aug
Oldbury, Barnford Hill Park	SO 32 997876	185	783	39	107	36	82	66	76	80	103	33	98	58	87	863	33	10 Jun
Tividale School	SO 32 974903	152	...	36	97	31	81	59	90	71	119	32	95	56	99	866	44	20 Aug
Tividale S. Wks.	SO 32 972910	132	738	34	95	34	82	62	90	75	113	29	97	63	108	883	31	20 Aug
Toll End S. Wks.	SO 32 978932	122	719	35	89	31	78	66	81	72	128	32	92	57	99	859	44	20 Aug
Tame (Trent)																		
Brownhills, Combe House	SK 43 038058	150	726	36	99	37	76	66	81	91	147	42	110	(57)	112	953	57	20 Aug
Walsall, Aldridge	SK 43 058010	163	...	40	103	33	91	59	76	81	128	38	101	55	109	914	50	20 Aug
Goscote S. Wks.	SK 43 021017	133	722	35	96	31	77	63	79	77	116	39	101	52	102	868	38	20 Aug
Goscote S. Wks.	SK 43 021017	133	...	36	102	32	83	61	77	78	118	40	102	55	109	893	38	20 Aug
Bescot S. Wks.	SP 42 007966	118	719	34	91	32	76	59	78	73	103	31	95	51	96	819	29	20 Aug
West Bromwich, Ray Hall S. Wks.	SP 42 025946	109	...	40	(95)	(32)	78	54	65	79	104	31	93	48	97	817	35	20 Aug
West Bromwich, Central Nurseries	SP 42 018934	134	...	37	104	32	84	59	68	84	99	35	99	57	101	857	31	20 Aug
Handsworth, Howard Road	SP 42 060908	112	741	34	109	33	77	61	73	89	87	33	103	50	97	846	22	29 Aug
Edgbaston	SP 42 046864	163	780	35	113	35	81	80	86	72	95	27	109	58	99	891	30	10 Jun
Warley Park	SP 42 010861	210	...	38	112	43	81	72	88	75	99	35	107	65	107	921	36	10 Jun
Handsworth Park	SP 42 048900	129	754	36	122	36	83	68	72	87	95	31	105	57	91	884	25	29 Aug
Great Barr	SP 42 051941	140	...	36	117	29	82	63	79	90	92	30	99	50	105	873	21	20 Aug
Erdington	SP 42 102917	135	727	37	104	39	77	61	76	89	98	29	105	48	102	865	25	13 Oct 29 Aug
Rea (Tame)																		
Rubery Hill	SO 32 992779	207	...	38	113	38	94	83	81	73	(102)	36	115	60	97	930	31	10 Jun
Birmingham Golf Course	SP 42 039784	168	...	42	103	49	99	89	77	86	109	41	119	62	108	985	...	...
Wast Hill	SP 42 038768	183	...	41	119	50	98	88	99	88	109	33	117	64	111	1016	36	10 Jun
Lifford Res.	SP 42 057800	137	...	41	115	43	93	86	81	73	122	34	117	58	110	974	47	29 Aug
Frankley	SP 42 007801	174	818	36	111	39	90	86	67	73	95	30	112	56	97	893	29	29 Aug
Bartley Res.	SP 42 010812	183	833	39	114	41	98	88	72	68	98	31	113	58	88	908	32	29 Aug
Bromwich Wood	SO 32 999810	186	836	38	114	41	98	87	72	67	98	31	113	57	87	905	32	29 Aug
Birmingham, King Edwards School	SP 42 051835	137	769	35	108	37	80	80	93	67	103	35	122	50	107	918	38	10 Jun
Edgbaston, Botanical Gardens	SP 42 048852	137	...	35	108	37	81	75	96	76	100	26	109	64	100	907	33	10 Jun
Saltley S. Wks.	SP 42 104895	89	716	40	107	41	87	70	86	87	99	29	107	59	102	912	28	29 Aug
Tame (Trent)																		
Erdington, Spring Lane	SP 42 113911	118	724	33	110	42	85	64	82	88	99	26	106	49	96	880	24	1 Dec
Perry Barr Res.	SP 42 084952	157	736	37	115	34	84	63	95	85	89	31	106	52	93	883	27	29 Aug
Sutton Coldfield Park	SP 42 116963	113	717	42	115	39	88	65	86	83	90	33	95	53	105	894	22	13 Oct
Sutton Coldfield, Bedford Road	SP 42 130968	125	...	38	110	34	85	60	76	86	89	31	101	49	102	861	22	31 Jul
Water Orton R.F.M. Sta. Area 407 km²	SP 42 169915	75	742	...	...	...	...	...	...	...	...	...	...	...	...	...	...	...

GENERAL TABLE, ENGLAND AND WALES

Station	Grid reference	Altitude m	Av. 1916–1950 mm	Jan	Feb	Mar	Apr	May	Jun	Jul	Aug	Sep	Oct	Nov	Dec	Year	Highest recorded fall in a day mm	date
TRENT—*(contd)*																		
Minworth	SP 42 165924	79	654	35	84	31	80	51	67	75	89	26	106	46	90	778	31	29 Aug
Coleshill	SP 42 194912	73	648	31	84	34	85	60	70	56	93	32	92	47	88	771	42	29 Aug
Blythe (Tame)																		
Solihull, Station Road	SP 42 148797	129	754	42	116	37	93	70	80	109	120	31	109	43	88	937	50	29 Aug
Cole (Blythe)																		
Highters Heath Res.	SP 42 086792	158	761	43	115	40	98	71	85	78	141	28	129	55	92	975	59	29 Aug
Stechford	SP 42 146885	92	702	39	102	41	92	65	65	71	122	33	110	53	97	892	57	29 Aug
Elmdon	SP 42 171837	97	680	36	100	35	84	66	69	65	113	28	105	43	86	829	60	29 Aug
Tame (Trent)																		
Whitacre W. Wks.	SP 42 225909	82	673	33	89	33	85	59	64	75	85	33	85	49	82	772	31	29 Aug
Lea Marston R.F.M. Sta. Area 795 km²	SP 42 206935	66	734	...	...	...	...	...	...	...	...	...	...	...	...	...	...	...
Langley Mill	SP 42 155967	94	...	38	103	38	81	57	63	83	87	35	96	47	101	830	29	29 Aug
Black (Tame)																		
Bourne Vale	SP 42 074997	128	...	40	105	30	93	57	80	83	89	36	100	52	105	870	24	1 Dec
Sutton Coldfield, Four Oaks	SP 42 112998	160	...	42	120	33	95	66	83	86	85	33	101	46	97	887	23	13 Oct
Cannock Res.	SK 43 041073	145	724	35	106	39	87	66	71	99	166	39	111	(208)	113	1140	55	20 Aug
Sandhills	SK 43 067048	124	722	41	93	39	84	64	68	81	122	38	97	54	115	897	43	20 Aug
Little Hay	SK 43 123030	91	706	38	115	38	81	64	76	74	81	34	97	50	106	854	21	13 Oct
Anker (Tame)																		
Hinckley, Bowling Green Road	SP 42 434941	122	700	31	98	27	79	55	64	77	119	32	116	44	79	821	44	29 Aug
Hinckley S. Wks.	SP 42 422928	99	706	29	99	28	91	57	69	83	100	33	112	44	79	823	40	29 Aug
Nuneaton P. Sta.	SP 42 361923	82	660	30	95	30	87	58	56	83	98	31	96	46	90	800	40	29 Aug
Hartshill S. Wks.	SP 42 330951	88	670	30	92	25	87	59	53	91	101	30	95	41	81	784	39	29 Aug
Sence (Anker)																		
Osbaston Hall	SK 43 425045	119	708	38	93	27	86	68	53	94	121	37	104	47	80	847	41	29 Aug
Anker (Tame)																		
Atherstone S. Wks.	SP 42 318980	71	...	36	96	27	88	56	53	89	109	35	89	44	79	800	39	29 Aug
Tame (Trent)																		
Tamworth S. Wks.	SK 43 197042	55	649	31	103	28	74	55	62	67	84	30	95	47	89	764	22	13 Oct
Elford	SK 43 189104	56	...	30	105	25	81	50	72	77	106	34	91	37	85	792	29	29 Aug
Elford R.F.M. Sta. Area 1480 km²	SK 43 173105	50	714	...	...	...	...	...	...	...	...	...	...	...	...	...	...	...
Mease (Trent)																		
Church Gresley, Maurice Lea Park	SK 43 299187	122	659	31	91	32	80	48	71	74	103	40	96	49	85	799	28	20 Aug
Chilcote	SK 43 284103	73	638	25	84	27	79	45	53	72	106	28	84	36	83	722	55	29 Aug
Trent																		
Stanton S. Wks.	SK 43 269192	64	...	31	97	32	91	53	73	80	104	42	97	57	87	845	43	29 Aug
Outwoods	SK 43 227240	94	693	36	93	29	95	64	74	79	100	53	93	54	88	856	33	29 Aug
Shobnall	SK 43 234231	48	648	37	90	32	93	62	67	78	103	47	97	54	86	845	39	29 Aug
Winshill P. Sta.	SK 43 264229	94	...	38	99	31	100	57	68	77	85	47	99	58	90	850	28	29 Aug
Stretton P. Sta.	SK 43 261257	44	...	39	95	34	96	59	65	79	86	53	99	58	86	849	28	13 Oct 29 Aug
Dove (Trent)																		
Wallnook	SK 43 034692	482	...	97	171	105	147	114	156	138	129	123	119	173	249	1722	...	...
Colshaw	SK 43 041679	431	...	92	168	102	148	119	155	124	132	121	115	164	241	1680	...	...
Brand Top	SK 43 043687	424	...	101	172	101	142	122	162	136	134	125	114	183	271	1762	...	...
Booth	SK 43 056680	373	...	91	164	98	148	119	142	114	134	113	118	161	225	1627	...	...
Hollinsclough	SK 43 066665	291	1205	77	162	101	116	106	140	105	128	111	103	155	244	1547	42	14 Sep
Biggin	SK 43 156593	302	1034	72	168	63	126	91	117	85	(127)	74	121	108	148	1301	33	1 Apr
Manifold (Dove)																		
Hulme End	SK 43 106596	216	...	71	162	81	119	86	129	90	138	81	104	137	191	1390	31	1 Apr
Leek, Mermaid Inn	SK 43 037604	448	–	81	140	76	102	88	145	119	125	91	109	(122)	(174)	1372	...	...
Onecote	SK 43 035576	411	1093	79	141	63	131	84	128	105	129	92	111	116	164	1342	30	22 Jun
Dove (Trent)																		
Ashbourne W.Wks.	SK 43 187459	179	862	67	134	64	94	77	106	64	99	49	100	75	117	1044	29	13 Oct
Rocester Weir R.F.M. Sta. Area 399 km²	SK 43 112397	87	1029	...	...	...	...	...	...	...	...	...	...	...	...	...	...	...
Churnet (Dove)																		
Middle Hills	SK 43 030639	451	...	79	129	86	99	98	142	131	113	91	112	(122)	(174)	1377	...	...
Stonycliff	SK 43 018602	294	1029	71	127	58	99	81	132	104	119	86	102	127	163	1270	...	...
Withens, New site	SK 43 012618	323	1019	63	129	57	127	89	148	128	104	77	107	142	177	1350	...	...
Leek, The Hollies	SK 43 004602	212	...	66	114	61	97	79	127	103	109	84	86	127	178	1231	28	22 Jun
Alder Lee	SJ 33 989605	209	...	67	107	53	114	76	125	97	102	79	90	119	164	1193	...	...
Leek, Lower Haddon	SJ 33 977605	267	...	69	109	62	116	81	129	104	98	(83)	89	112	174	1225	...	...
Tittesworth W.Wks.	SJ 33 994585	183	...	64	111	57	113	80	109	100	100	82	87	114	172	1189	28	22 Jun
Rudyard Res.	SJ 33 951583	160	...	62	105	52	87	76	102	91	86	72	88	96	151	1069	25	1 Apr
Wall Grange P.Sta.	SJ 33 966536	139	...	52	107	49	89	73	95	90	97	77	82	79	149	1039	25	9 Dec 1 Apr
Stanley Res.	SJ 33 929519	177	...	55	103	52	94	78	99	97	88	72	101	76	160	1075	27	1 Dec
Rocester R.F.M. Sta. Area 238 km²	SK 43 103389	82	968	...	...	...	...	...	...	...	...	...	...	...	...	...	...	...
Dove (Trent)																		
Dilhorne	SJ 33 968470	280	892	60	109	42	91	95	123	115	104	64	97	81	152	1133	...	...
Hardiwick Farm	SJ 33 956442	247	825	55	107	46	93	91	101	92	90	49	98	65	122	1008	...	...
Tean (Dove)																		
Dove (Trent)																		
Uttoxeter, Oldfields School	SK 43 085336	101	...	45	105	56	93	60	67	70	125	49	99	74	123	965	40	20 Aug
Foston Hall	SK 43 185316	64	...	45	109	42	103	65	73	61	118	47	101	63	91	916	31	20 Aug
Marston-on-Dove R.F.M. Sta. Area 883 km²	SK 43 234289	48	950	...	...	...	...	...	...	...	...	...	...	...	...	...	...	...
Sutton (Dove)																		
Shirley House	SK 43 215419	137	...	55	125	53	95	60	83	64	93	44	105	57	113	946	32	13 Oct
Rodsley P.Sta.	SK 43 197413	126	794	59	128	57	97	66	94	74	100	48	105	66	120	1013	29	13 Oct
Tutbury	SK 43 219310	54	706	46	109	38	102	58	71	68	129	49	103	63	98	936	33	20 Aug
Trent																		
Repton	SK 43 302267	70	646	47	109	40	98	66	75	79	89	51	114	60	84	912	35	13 Oct
Derby, Brayfield Road	SK 43 344339	71	...	52	106	33	99	51	74	77	84	49	118	53	81	877	42	13 Oct
Stanton-by-Bridge W. Wks.	SK 43 374271	44	675	39	103	31	94	44	56	69	93	39	117	49	74	808	39	13 Oct
Kings Newton	SK 43 390268	37	...	41	98	31	96	44	60	69	89	43	106	52	(83)	811	36	13 Oct
Shardlow R.F.M. Sta. Area 4400 km²	SK 43 448299	29	775	...	...	...	...	...	...	...	...	...	...	...	...	...	...	...
Derwent (Trent)																		
Barrow Clough	SK 43 142976	405	1587	89	218	107	161	117	119	106	147	114	162	196	265	1803	...	...
Broadhead Clough	SK 43 169962	308	1407	94	227	112	149	114	104	95	132	90	127	180	220	1645	...	...

Station	Grid reference	Altitude m	Av. 1916–1950 mm	Jan	Feb	Mar	Apr	May	Jun	Jul	Aug	Sep	Oct	Nov	Dec	Year	Highest recorded fall in a day mm	Highest recorded fall in a day date
TRENT—*(contd)*																		
Sandy Lee ...	SK 43 180953	402	1433	92	212	(114)	(156)	116	108	99	135	105	143	185	230	1695	...	...
Slippery Stones	SK 43 169951	277	1338	93	214	107	146	116	111	106	141	94	140	190	213	1672	...	...
Ronksley Moor	SK 43 147951	506	1579	94	216	119	137	122	125	107	151	132	134	197	314	1848	...	...
Deep Grain	SK 43 123951	415	1570	97	226	117	135	127	127	109	(148)	116	135	187	299	1823	...	...
Blacklow Cote ...	SK 43 145936	296	1403	84	213	97	127	122	122	105	150	113	134	197	257	1720	...	...
Ridge Farm	SK 43 161931	269	1409	88	222	109	141	123	121	106	150	111	130	202	230	1734	...	...
Howden Clough, No.64A	SK 43 177936	308	1268	91	203	108	147	97	104	97	138	86	126	179	212	1589	...	...
Howden Dam ...	SK 43 168924	258	1285	79	213	84	137	108	105	94	142	95	123	185	200	1565	39	7 Feb
Howden Dean ...	SK 43 185926	293	1296	81	193	(95)	(145)	99	103	99	137	94	107	183	196	1533	...	...
Birchinlee ...	SK 43 165918	271	1343	75	204	109	142	112	108	93	141	109	117	214	222	1645	...	...
Ouzelden Clough	SK 43 160907	270	1360	76	208	109	146	117	114	98	146	107	111	223	227	1683	...	...
Derwent Dam ...	SK 43 175899	249	...	71	198	81	143	103	108	97	144	95	106	189	203	1538	...	...
Fairholmes	SK 43 173893	209	1287	68	193	83	137	97	107	100	147	96	108	191	205	1533	41	14 Sep
Mill Brook	SK 43 188894	235	1235	61	183	73	134	97	106	93	142	88	100	170	196	1443	...	...
Crook Hill	SK 43 185870	335	1207	60	181	78	143	89	104	91	138	91	99	177	166	1415	...	...
Ashop (Derwent)																		
Ashop Head	SK 43 068907	509	...	113	203	87	119	124	159	143	144	115	116	182	255	1760	...	...
Sandy Heys	SK 43 075894	622	1496	81	198	75	152	111	149	137	142	103	109	164	241	1663	...	...
Nether Red Clough	SK 43 084903	418	1553	107	200	86	117	129	143	130	174	118	138	191	265	1796	...	...
Urchin Clough ...	SK 43 101908	363	...	93	207	84	119	130	136	119	168	94	130	184	248	1712	...	...
Featherbed Top	SK 43 089928	507	...	102	202	93	162	126	141	121	171	121	139	195	228	1801	...	...
Snake Inn ...	SK 43 112905	314	...	85	217	92	147	127	147	119	151	118	132	207	257	1799	52	14 Sep
Middle Seal Clough	SK 43 111897	402	1635	97	205	95	148	134	145	137	165	138	129	225	316	1933	...	...
Blackdean Brook	SK 43 127890	326	1600	102	209	104	139	132	150	129	155	125	126	221	299	1891	...	...
Ashop Weir	SK 43 140895	242	1464	74	205	83	136	121	138	111	147	109	115	186	220	1644	...	...
Bleaklow Hill ...	SK 43 102960	625	...	106	224	93	138	125	136	144	149	124	136	180	232	1787	...	...
Alport Moor	SK 43 106943	546	1692	104	209	95	156	123	142	138	156	129	144	193	234	1821	...	...
Lady Clough Moor ...	SK 43 112926	524	...	101	210	97	157	119	148	119	151	100	141	189	212	1742	...	...
Alport Castle ...	SK 43 133919	297	...	76	208	(89)	129	107	117	(110)	161	111	119	170	220	1617	...	...
Hagg Farm	SK 43 160885	213	1295	67	183	81	129	104	117	99	151	108	109	185	214	1546	...	...
Grimbocar	SK 43 172872	213	1271	65	180	80	129	90	116	92	139	97	100	173	207	1465	...	...
Derwent (Trent)																		
Ladybower	SK 43 205869	300	1166	54	186	79	154	79	107	81	135	86	106	153	157	1377	...	...
Yorkshire Bridge	SK 43 198850	171	1049	51	169	58	119	70	93	73	127	73	92	124	136	1183	32	21 Aug
Yorkshire Bridge R.F.M. Sta. Area 127 km²	SK 43 198851	159	1410	...	...	...	...	...	...	...	...	...	...	...	...	...	...	...
Bamford Filters	SK 43 212830	201	1020	52	157	59	125	68	93	63	130	69	103	134	134	1186	36	21 Aug
Noe (Derwent)																		
Jacobs Ladder ...	SK 43 090860	340	1520	84	173	79	(108)	94	152	132	122	109	113	155	247	1568	...	...
Grain Clough	SK 43 096854	304	...	83	188	72	84	90	144	127	118	99	121	163	239	1528	...	...
Upper Booth	SK 43 104854	290	1267	80	183	65	81	92	135	125	127	97	107	145	214	1449	...	...
Mam Nick	SK 43 123837	417	1257	69	185	112	80	80	134	111	130	88	127	131	176	1422	...	...
Grindslow ...	SK 43 119867	287	1416	81	192	84	127	95	140	126	152	95	102	168	228	1589	...	...
Edale Mill ...	SK 43 134854	221	1269	63	189	74	105	84	125	100	139	94	106	157	193	1428	42	14 Sep
Nether Moor	SK 43 143875	445	1420	89	184	87	127	97	137	114	148	112	94	165	222	1577	...	...
Edale End ...	SK 43 162863	193	1147	66	173	71	114	88	113	86	137	91	104	155	193	1392	...	...
Killhill	SK 43 171838	171	1055	56	161	70	103	67	108	73	121	86	96	129	142	1212	42	21 Aug
Rowter Farm ...	SK 43 132820	425	...	69	191	113	105	95	132	97	127	95	117	141	188	1469	...	...
Little Hucklow	SK 43 162784	308	...	59	186	69	121	80	116	75	118	71	105	123	183	1304	...	...
Moss Rake Head	SK 43 144799	437	1194	88	201	128	143	96	133	97	128	93	119	137	191	1555	...	...
Bradwell	SK 43 162811	323	1146	70	178	95	117	81	121	85	135	89	109	140	185	1405	...	...
Hope ...	SK 43 168822	177	...	61	175	70	106	76	115	79	137	93	100	136	159	1304	44	14 Sep
Bamford	SK 43 202829	155	1011	50	159	61	118	73	102	68	123	67	101	135	124	1179	31	21 Aug
Derwent (Trent)																		
Redmires No. 2	SK 43 248842	425	...	62	165	63	141	75	92	77	137	71	134	115	137	1270	...	...
Redmires ...	SK 43 248840	425	...	63	168	64	143	76	91	78	138	72	134	118	139	1284	...	...
Burbage (Derwent)																		
Upper Burbage Check	SK 43 263828	416	...	58	155	41	155	67	83	69	123	67	130	113	89	1150	...	...
Higger Lodge ...	SK 43 273816	373	966	55	154	45	131	75	90	63	124	70	115	107	91	1121	...	...
Parsons House ...	SK 43 271810	398	947	62	155	49	136	73	90	65	124	71	135	117	102	1179	...	...
Lower Burbage	SK 43 258802	282	944	48	150	53	134	73	90	65	111	77	105	112	98	1117	...	...
White Edge	SK 43 264786	369	934	62	152	61	142	76	90	69	131	75	122	117	107	1204	...	...
Derwent (Trent)																		
Barbrook Res. ...	SK 43 281770	323	...	69	157	58	156	74	104	73	129	79	130	106	112	1247	33	1 Apr
Chatsworth Gardens	SK 43 262701	133	828	59	126	43	127	61	88	51	116	60	105	93	89	1018	37	29 Aug
Wye (Derwent)																		
Buxton	SK 43 059738	307	1232	72	184	101	119	110	126	132	116	101	120	151	219	1550	40	26 Mar
Ashwood Dale S. Wks. ...	SK 43 067729	280	1229	70	182	88	124	105	119	115	125	97	120	133	198	1476	42	1 Apr
Stanley Moor	SK 43 041712	396	1353	75	167	86	125	99	133	132	131	107	120	139	234	1548	39	14 Sep
Leap Edge	SK 43 043696	453	...	97	174	106	125	118	158	138	135	122	121	179	276	1749	...	...
Green Fairfield	SK 43 087747	341	1152	68	166	86	112	100	111	123	128	91	107	146	206	1441	...	...
Dove Holes	SK 43 073777	334	...	52	154	83	93	85	116	123	119	88	111	128	170	1320	...	...
Monks Dale (Wye)																		
Rushup Edge	SK 43 103824	379	1259	65	155	(83)	107	79	127	136	137	95	115	133	184	1415	...	...
Eldon Hill ...	SK 43 121806	411	1178	84	148	77	159	87	118	112	123	93	109	133	175	1418	...	...
Peak Forest	SK 43 111792	320	1134	53	154	75	126	88	105	123	128	91	106	129	171	1348	...	...
Middle Hill	SK 43 102772	329	...	57	164	76	99	88	104	(121)	119	79	114	114	178	1313	...	...
Wheston No. 112A ...	SK 43 131770	347	...	62	182	89	125	77	103	84	135	104	127	122	159	1368	...	...
Hargate Hall	SK 43 117752	347	1118	56	145	76	113	70	106	93	127	84	95	124	186	1274	...	...
Wye (Derwent)																		
Whiterake ...	SK 43 143781	372	1108	72	183	83	154	86	120	86	128	88	111	119	178	1409	...	...
Tideswell ...	SK 43 149752	305	1024	58	148	68	129	70	108	73	150	95	106	113	155	1272	...	...
Stanley Lodge ...	SK 43 175763	277	...	59	152	61	131	73	94	68	132	66	110	121	135	1202	...	...
Priestcliffe School ...	SK 43 137712	370	...	69	171	84	117	84	116	94	140	98	110	132	172	1385	...	...
Bakewell ...	SK 43 206692	149	...	62	141	67	113	67	92	70	122	81	103	108	126	1151	43	21 Aug
Monyash ...	SK 43 152664	269	1046	75	157	71	115	79	121	106	144	101	110	133	165	1378	34	1 Apr
Derwent (Trent)																		
Matlock, Wolds Res.	SK 43 306609	206	...	56	161	53	150	77	73	58	111	53	118	87	88	1084	34	1 Apr
Matlock Bath R.F.M. Sta. Area 689 km²	SK 43 297586	83	1105	...	...	...	...	...	...	...	...	...	...	...	...	...	...	...
Whatstandwell W. Wks. ...	SK 43 325554	75	858	57	160	51	139	83	84	59	114	64	124	79	97	1111	31	13 Oct
Amber (Derwent)																		
Milltown ...	SK 43 357622	160	...	59	150	53	128	71	77	60	104	49	124	70	81	1027	34	13 Oct
Ogston Res.	SK 43 380598	102	...	54	151	43	119	86	70	57	82	51	112	72	73	971	30	13 Oct
Press Res. ...	SK 43 354657	232	831	63	162	50	130	79	91	64	121	53	127	69	100	1109	32	13 Oct
Lindway Res. ...	SK 43 356582	181	842	62	165	44	136	84	83	59	100	59	129	71	87	1078	31	13 Oct
South Wingfield	SK 43 379535	78	...	53	143	44	125	60	80	65	80	56	107	66	77	956	30	13 Oct
Derwent (Trent)																		
Belper Meadows P.Sta. ...	SK 43 340477	69	...	59	145	51	121	66	83	59	93	56	115	79	98	1026	33	13 Oct

Station	Grid reference	Altitude m	Av. 1916–1950 mm	Rainfall in millimetres Jan	Feb	Mar	Apr	May	Jun	Jul	Aug	Sep	Oct	Nov	Dec	Year	Highest recorded fall in a day mm	date
TRENT—*(contd)*																		
Belper	SK 43 346468	62	731	56	139	36	124	46	82	53	90	52	120	67	81	947	33	13 Oct
Wirksworth, Wash Green, New site	SK 43 297537	244	...	(61)	(165)	56	136	88	80	74	122	63	139	81	108	1173	37	13 Oct
Little Eaton, Lower P.Sta., New site	SK 43 364406	52	...	50	120	32	109	58	76	59	81	49	115	56	89	892	37	13 Oct
Morley	SK 43 385401	107	748	55	122	32	109	56	73	60	81	48	122	51	82	891	38	13 Oct
Little Eaton, Higher P.Sta.	SK 43 372413	128	...	55	130	36	119	61	78	63	93	51	123	60	97	966	38	13 Oct
Derby	SK 43 359367	48	699	51	115	31	96	51	68	57	77	53	109	49	81	839	37	13 Oct
Markeaton Park	SK 43 331377	58	730	54	124	36	104	64	72	66	86	58	116	58	93	929	36	13 Oct
Mackworth	SK 43 319362	89	...	52	117	35	94	71	69	63	89	54	121	54	85	905	40	13 Oct
Longbridge Weir R.F.M.Sta. Area 1119 km²	SK 43 357364	45	996	...	...	...	...	...	...	...	...	...	...	...	...	...	...	...
Derby S.Wks.	SK 43 389351	41	666	46	118	31	100	46	68	69	87	47	113	53	76	853	34	13 Oct
Draycott P.Sta.	SK 43 449328	36	...	46	107	31	95	43	70	57	85	37	104	59	69	802	30	13 Oct
Soar (Trent)																		
Frolesworth	SP 42 506903	113	650	34	99	19	85	57	73	88	111	30	127	43	78	843	59	29 Aug
Sence (Soar)																		
Wistow Hall	SP 42 641957	94	648	30	91	17	82	49	44	65	96	25	102	41	81	722	28	29 Aug 30 Aug
Kilby Bridge	SP 42 609970	78	...	28	91	15	79	55	49	74	91	30	113	47	81	753	32	29 Aug
Soar (Trent)																		
Leicester, De Montfort Square	SK 43 593037	76	613	30	82	21	82	40	56	69	72	35	103	41	78	708	24	13 Oct
Leicester, Ainsdale Road	SK 43 564040	76	641	30	87	22	89	49	65	74	72	37	121	44	83	776	24	13 Oct
Gilroes Cemetery	SK 43 562064	90	683	30	91	22	93	50	55	78	75	36	124	48	81	782	24	13 Oct
Leicester, Stoughton Drive North	SK 43 609031	73	648	33	77	22	89	47	55	75	88	33	104	45	81	747	26	29 Aug
Wreak (Soar)																		
Waltham-on-the-Wolds	SK 43 804251	174	647	40	86	21	94	51	65	85	79	29	86	57	82	776	23	22 Jun
Scalford	SK 43 764241	105	693	41	88	24	92	52	65	79	85	39	95	71	90	821	24	30 Aug
Melton Mowbray, Brook Lane	SK 43 757186	72	634	36	89	24	85	49	63	81	77	32	95	63	81	774	25	30 Aug
Sysonby Lodge	SK 43 745209	104	...	40	88	26	87	45	61	85	69	29	86	59	81	757	24	19 Apr
Brooksby Hall	SK 43 671160	67	...	30	81	21	82	51	64	74	74	29	100	48	77	733	31	30 Aug
Burrough-on-the-Hill	SK 43 750101	163	...	40	92	19	94	49	69	65	71	30	87	61	83	759	25	22 Jun
Lowesby Hall	SK 43 722074	134	671	39	93	20	97	55	64	66	72	29	83	43	77	738	23	30 Aug
Barkby Hall	SK 43 637098	69	668	32	85	18	86	43	53	63	82	34	101	44	76	716	28	30 Aug
Rothley (Soar)																		
Thornton Res.	SK 43 473072	112	722	34	93	26	99	64	64	89	104	41	125	57	89	885	38	29 Aug
Soar (Trent)																		
Ulverscroft	SK 43 497117	195	776	35	107	26	102	58	80	99	88	37	113	49	88	882	31	5 Jul
Newtown Linford	SK 43 530095	119	687	32	99	21	95	54	68	69	81	33	107	49	87	796	27	13 Oct
Cropston Res.	SK 43 549114	75	733	34	99	25	107	63	63	58	75	33	116	53	88	814	24	13 Oct
Swithland Res.	SK 43 555148	52	712	37	117	25	106	66	60	59	82	30	110	52	89	833	24	18 Feb
Charley Knoll	SK 43 491158	171	...	44	113	32	111	64	72	78	114	37	107	53	84	909	...	...
Bawdon Castle	SK 43 496143	221	...	41	107	25	110	67	77	78	101	36	110	45	82	879	...	...
Nanpantan Res.	SK 43 508171	82	715	39	107	29	112	55	77	76	91	32	106	47	78	850	27	21 Aug
Loughborough S.Wks.	SK 43 529209	41	680	44	97	25	93	45	66	74	75	34	95	43	75	766	23	13 Oct
Black Brook (Soar)																		
Timberwood Hill	SK 43 472149	232	...	47	118	33	115	63	65	87	106	38	105	53	90	920	...	...
Charnwood Lodge	SK 43 474156	140	...	44	111	28	118	59	63	83	107	34	104	52	80	884	...	...
Lubcloud Farm	SK 43 478163	148	...	44	106	27	116	57	64	84	103	34	107	50	77	867	...	...
Mount St.Bernard Abbey	SK 43 459158	186	...	46	105	28	125	58	57	93	108	32	107	48	85	893	33	13 Oct
Gun Hill	SK 43 455168	155	...	43	105	32	110	54	58	103	109	33	102	49	79	877	...	...
Blackbrook Res.	SK 43 456178	107	709	42	101	31	111	48	54	89	102	35	102	43	78	837	33	5 Jul
Kingston (Soar)																		
Six Hills	SK 43 645209	134	...	37	92	32	102	55	59	72	68	31	87	52	83	769	...	...
Old Dalby	SK 43 660225	125	...	36	94	32	102	63	61	78	72	34	85	52	81	790	...	...
Wysall	SK 43 606274	76	...	44	87	29	90	44	63	63	58	29	96	49	74	726	...	...
Fox Hill	SK 43 542270	73	...	41	79	23	85	36	62	79	65	30	84	49	63	695	...	...
White Hill	SK 43 542249	66	...	42	87	28	88	39	71	75	61	29	90	46	70	727	...	...
Moulter Hill	SK 43 517269	52	...	42	77	25	90	37	61	64	83	31	87	45	69	710	...	...
Sutton Bonington	SK 43 507261	48	583	44	81	25	100	37	58	62	95	28	94	47	71	743	34	21 Aug
Kingston Hall	SK 43 508277	40	654	45	84	23	99	38	61	62	91	30	93	46	69	740	32	21 Aug
Erewash (Trent)																		
Codnor Park	SK 43 442517	74	776	59	141	43	109	59	81	58	94	55	101	63	76	938	...	...
Eastwood W.Wks.	SK 43 460468	74	735	56	129	41	110	55	63	49	76	54	103	64	88	885	26	13 Oct
Ilkeston S.Wks.	SK 43 482395	41	--	47	112	27	98	38	65	79	79	39	97	49	72	803	25	13 Oct
Morley Hall	SK 43 394410	136	...	57	120	25	104	56	68	57	76	45	102	45	73	828	28	13 Oct
New Stanton	SK 43 468395	43	...	46	110	27	113	45	63	87	75	42	104	51	77	839	26	13 Oct
Breaston, Woodlands Avenue	SK 43 469335	35	640	44	112	31	97	45	101	53	85	38	103	54	72	835	29	20 Jun
Long Eaton, Town Hall	SK 43 490339	32	...	45	107	28	96	40	86	46	86	37	100	(57)	71	799	24	13 Oct
Trent																		
Bramcote Res.	SK 43 507375	100	645	47	111	26	107	47	81	49	85	40	102	50	77	822	27	21 Aug
Lenton	SK 43 538382	49	631	48	110	25	100	46	73	57	75	39	100	48	69	789	33	21 Aug
Fairham (Trent)																		
Broughton Lodge	SK 43 650253	113	...	38	93	32	100	54	66	66	77	31	84	58	73	773	...	...
Nottingham Training College	SK 43 547352	51	...	38	107	23	95	38	66	49	65	25	97	59	71	734	26	21 Aug
Wilford Hill Res.	SK 43 579351	90	...	40	102	22	90	38	69	53	57	29	94	59	70	723	20	18 Feb
Leen (Trent)																		
Papplewick P.Sta.	SK 43 583521	97	699	64	133	30	104	52	59	49	90	46	98	57	74	856	28	21 Aug
Bulwell Hall	SK 43 532468	77	692	56	136	36	104	53	89	56	102	53	94	58	76	913	36	20 Aug
Watnall	SK 43 503456	119	691	53	133	34	101	56	105	68	89	55	96	57	79	926	30	20 Jun
Arnold County High School	SK 43 594448	93	...	47	117	26	112	48	64	61	75	35	96	49	54	784	24	20 Aug
Basford W.Wks.	SK 43 565429	51	646	49	113	23	101	45	60	65	81	41	100	52	74	805	29	21 Aug
Trent																		
West Bridgford	SK 43 588376	24	...	44	110	23	95	39	63	53	61	33	91	52	72	735	23	18 Feb
Sneinton P.Sta.	SK 43 592393	25	609	46	111	23	97	40	63	61	61	36	88	47	74	746	22	7 Feb
Colwick Hill Res.	SK 43 596399	96	614	44	116	22	95	42	64	55	68	31	90	48	74	751	23	18 Feb
Holme Sluice No.2	SK 43 615393	23	...	45	106	21	92	41	58	49	64	27	84	42	69	697	22	18 Feb
Colwick R.F.M.Sta. Area 7490 km²	SK 43 620399	17	785	...	...	...	...	...	...	...	...	...	...	...	...	...	...	...
Colwick	SK 43 617401	21	599	44	108	24	92	37	60	50	60	28	90	36	69	699	22	18 Feb
Cotgrave Res.	SK 43 667349	97	...	39	99	20	89	37	55	39	59	26	92	52	64	671	22	18 Feb
Netherfield, Chandos School	SK 43 622414	24	...	49	106	22	90	42	59	53	59	29	88	44	68	710	22	18 Feb
Stoke Bardolph	SK 43 647416	20	612	52	123	26	103	42	69	57	69	29	95	47	73	785	23	7 Feb
Burton Joyce W.Wks.	SK 43 654443	20	...	52	113	28	98	45	71	62	65	32	91	49	72	777	22	18 Feb
Lambley P.Sta.	SK 43 627451	62	...	51	116	28	101	47	63	63	66	35	95	45	73	783	25	18 Feb
Dover (Trent)																		
Blidworth	SK 43 562553	122	724	63	138	33	113	65	69	63	91	60	92	61	78	926	27	7 Feb
Ramsdale Hill Res.	SK 43 596483	153	699	57	127	28	107	61	59	51	85	47	99	55	69	845	25	18 Feb

Station	Grid reference	Altitude m	Av. 1916–1950 mm	Jan	Feb	Mar	Apr	May	Jun	Jul	Aug	Sep	Oct	Nov	Dec	Year	Highest recorded fall in a day mm	date
TRENT—*(contd)*																		
Trent																		
Oxton Res.	SK 43 649518	126	...	62	120	26	94	56	64	48	65	44	99	51	69	798	22	18 Feb
Halam P.Sta.	SK 43 668535	47	...	57	122	24	94	50	54	48	68	36	99	45	67	765	22	5 Apr 7 Feb
Southwell	SK 43 697542	38	670	52	132	24	93	48	58	54	69	33	95	45	69	771	22	7 Feb
Devon (Trent)																		
Woolsthorpe	SK 43 838351	57	596	37	78	19	79	41	71	55	77	21	72	47	63	661	26	22 Jun
Trent																		
Devon (Trent)																		
Smite (Devon)																		
Barnstone	SK 43 736349	32	...	39	80	22	84	34	58	46	54	25	80	53	67	642	21	22 Jun
Devon (Trent)																		
Syerston	SK 43 729477	63	...	41	110	19	80	36	57	49	55	27	76	45	25	621	21	7 Feb
Balderton	SK 43 823516	17	...	41	101	18	77	33	61	48	54	27	74	47	54	634	21	7 Feb
Balderton S.Wks.	SK 43 813509	12	...	40	97	21	77	35	68	47	63	29	77	49	54	659	23	22 Jun
Newark, The Friary	SK 43 802540	15	610	45	111	21	79	43	68	61	60	29	84	50	56	706	22	18 Feb
Kelham	SK 43 792552	12	567	40	104	21	79	45	68	59	57	32	80	47	51	683	21	22 Jun
Newark S.Wks.	SK 43 802558	10	...	45	106	22	75	41	67	61	54	29	72	46	54	672	22	18 Feb
Trent																		
Torksey P.Sta....	SK 43 836780	5	583	38	98	14	72	57	57	60	74	24	64	58	48	663	21	22 Jun
Gate Burton Hall	SK 43 838831	30	595	35	95	17	80	39	55	77	80	33	62	57	42	673	22	20 Aug
Idle (Trent)																		
Sutton-in-Ashfield S.Wks.	SK 43 510595	139	735	55	135	38	110	53	60	52	83	41	91	70	69	857	25	21 Aug
Mansfield	SK 43 543618	110	698	52	144	32	109	59	71	52	77	41	92	68	64	861	23	21 Aug
Clipstone W.Wks.	SK 43 603643	69	649	49	132	27	103	55	72	51	80	33	81	65	55	804	24	20 Aug
Edwinstowe	SK 43 624671	66	640	50	122	23	107	56	86	53	89	37	81	67	52	824	32	20 Jun
Rainworth (Idle)																		
Norman's Hollow P.Sta.	SK 43 539555	143	...	64	137	31	109	63	75	60	83	56	96	63	77	914	25	7 Feb
Rainworth W.Wks....	SK 43 587586	106	689	57	136	33	112	59	64	56	90	53	84	58	75	877	27	21 Aug
Rufford P.Sta....	SK 43 632611	84	663	53	129	28	107	58	58	59	80	44	89	59	59	825	22	31 Jul
Idle (Trent)																		
Boughton P.Sta.	SK 43 667692	43	630	50	109	19	89	52	50	51	101	37	83	72	48	760	23	20 Aug
Markham Clinton P.Sta.	SK 43 710726	28	...	49	96	16	95	56	47	56	101	28	86	67	43	739	32	20 Aug
Rainworth (Idle)																		
Meden (Idle)																		
Warsop	SK 43 591699	46	587	44	110	25	102	46	67	41	76	35	79	66	46	735	25	20 Aug
Budby P.Sta.	SK 43 604702	49	...	43	112	25	104	49	62	41	81	36	76	68	49	745	28	20 Aug
Bevercotes Colliery	SK 43 696735	24	631	43	91	20	91	49	45	53	86	29	91	70	42	711	31	20 Aug
Poulter (Meden)																		
Whaley Well	SK 43 509718	103	654	45	131	27	111	45	63	50	81	27	91	52	34	757	29	21 Aug
Carburton P.Sta.	SK 43 594721	43	634	44	113	19	108	49	75	50	83	32	79	65	41	760	27	20 Aug
Elkesley P.Sta....	SK 43 664759	37	616	42	95	18	95	61	45	58	80	28	77	61	42	702	28	20 Aug
Idle (Trent)																		
Retford Secondary School	SK 43 710807	15	...	49	100	18	93	48	56	59	84	30	98	59	49	743	22	7 Feb
Babworth Hall	SK 43 686810	30	586	48	98	21	91	55	60	65	84	32	89	64	49	755	23	20 Aug
Wiseton	SK 43 711893	6	...	37	83	21	93	40	75	62	84	31	87	52	41	706	25	20 Aug
Ryton (Idle)																		
Kiveton Park Farm...	SK 43 498834	119	...	29	110	31	100	42	51	50	88	38	73	66	46	725	22	22 Jun
Sunnyside W.Wks.	SK 43 589806	74	629	39	111	20	111	42	68	55	73	31	79	49	40	719	22	20 Aug 7 Feb
Bracebridge S.Wks.	SK 43 597790	30	630	42	112	21	106	42	61	52	68	33	78	56	46	716	21	9 Apr 18 Feb
Carlton-in-Lindrick	SK 43 589838	30	...	39	104	22	101	49	65	59	72	35	70	56	47	720	21	18 Feb
Idle (Trent)																		
Finningley...	SK 43 658995	10	566	31	95	15	93	55	110	63	90	47	72	51	48	769	24	15 Jun
Gringley-on-the-Hill	SK 43 734908	62	642	43	90	20	95	39	79	79	101	33	99	50	46	775	26	21 Aug
Eau (Trent)																		
Kirton-in-Lindsey	SK 43 934990	25	640	43	101	27	69	37	78	81	127	36	64	62	46	771	30	7 Feb
Trent																		
Messingham S.Wks.	SE 44 875058	4	618	34	96	20	66	30	93	68	105	39	74	61	42	729	27	7 Feb
West Butterwick	SE 44 835065	6	628	31	99	22	69	35	89	79	99	46	77	58	45	749	27	7 Feb
Torne (Trent)																		
Tickhill, Dadsley Road	SK 43 590938	18	613	36	111	23	102	44	56	63	93	43	69	58	44	740	26	7 Feb
Rossington Colliery	SK 43 604984	7	612	33	108	20	103	52	76	61	97	39	74	54	47	765	25	7 Feb
Balby S.Wks.	SE 44 569010	9	596	33	105	22	99	44	73	65	97	32	71	51	38	730	29	21 Aug
Low Pasture P.Sta....	SE 44 589022	7	595	33	109	19	97	52	79	65	103	34	76	51	43	760	25	21 Aug
Rossington Bridge P.Sta.	SK 43 630999	12	...	29	100	15	99	52	85	55	92	35	79	49	47	738	24	7 Feb
Candy Farm P.Sta....	SE 44 698030	1	...	35	97	14	87	42	78	69	108	44	72	53	51	749	27	29 Aug
Finningley P.Sta.	SE 44 677002	6	...	28	97	15	94	46	88	67	95	39	78	53	49	749	23	7 Feb
Wroot P.Sta.	SE 44 732017	2	581	28	91	10	82	35	72	66	105	41	73	53	44	701	27	29 Aug
Doncaster, Runnymede Road	SE 44 597035	11	...	31	100	17	92	50	85	63	106	34	74	48	40	739	25	7 Feb
Nutwell P.Sta....	SE 44 633031	7	...	26	99	15	94	51	95	59	103	34	81	44	45	747	28	21 Aug
Lindholme...	SE 44 682063	5	...	31	93	15	88	44	101	63	98	39	65	43	17	695	31	15 Jun
Crowle P.Sta.	SE 44 749098	3	599	30	95	15	74	38	61	73	101	39	68	40	41	675	29	21 Aug
Keadby P.Sta.	SE 44 835112	6	625	34	100	23	64	37	71	68	111	41	72	44	40	705	29	7 Feb
Trent																		

LINCOLNSHIRE RIVER AUTHORITY AREA, 5271 SQUARE KILOMETRES

HYDROMETRIC AREAS 29 AND 30

Station	Grid reference	Altitude m	Av. 1916–1950 mm	Jan	Feb	Mar	Apr	May	Jun	Jul	Aug	Sep	Oct	Nov	Dec	Year	Highest recorded fall in a day mm	date
Ancholme																		
Osgodby	TF 53 077926	21	607	41	93	23	54	36	71	93	66	31	77	66	53	705	28	22 Jun
Caistor	TA 54 082019	21	...	42	100	23	63	39	63	96	72	40	69	77	52	735	26	7 Feb
Scawby Hall	SE 44 968057	24	645	44	102	26	62	43	68	87	88	41	62	67	48	737	29	7 Feb
Brigg Sugar Factory	SE 44 987062	5	613	46	105	26	63	36	59	83	95	(38)	65	68	49	732	33	7 Feb
Brigg, Wrawby Road	TA 54 005074	7	620	44	102	26	63	33	62	83	96	42	64	68	49	732	31	7 Feb
Brigg	TA 54 002083	3	...	43	106	27	68	36	60	89	96	42	65	74	51	757	31	7 Feb
Coastal																		
Immingham	TA 54 208149	4	602	35	95	19	54	31	70	90	90	38	69	61	43	694	34	22 Jun
Freshney																		
Little Coates P. Sta.	TA 54 252086	4	...	33	89	27	51	32	82	86	88	31	78	60	45	701	34	22 Jun

Station	Grid reference	Altitude m	Av. 1916–1950 mm	Jan	Feb	Mar	Apr	May	Jun	Jul	Aug	Sep	Oct	Nov	Dec	Year	Highest recorded fall in a day mm	Highest recorded fall in a day date
LINCOLNSHIRE—*(contd)*																		
Coastal																		
Cleethorpes	TA 54 314079	7	595	33	90	24	55	34	87	87	67	25	75	66	45	688	49	22 Jun
Cleethorpes P. Sta.	TA 54 321071	8	565	33	95	26	55	39	93	106	62	26	66	58	44	703	51	22 Jun
Waithe																		
Binbrook	TF 53 196958	114	...	45	116	30	70	49	68	73	75	29	90	73	66	783	30	7 Feb
Thoresway	TF 53 168967	62	...	35	96	30	62	54	61	86	65	33	118	64	59	763	27	7 Feb
Beelsby	TA 54 207016	64	...	41	104	33	54	39	73	82	74	31	104	60	56	751	29	22 Jun
Brigsley R.F.M. Sta. Area 108 km²	TA 54 253016	15	698	...	...	...	...	...	...	...	...	...	...	...	...	...	...	...
Coastal																		
Marshchapel	TF 53 354997	3	...	33	97	17	66	43	113	73	71	21	78	64	35	711	47	22 Jun
Lud																		
Louth, Technical School	TF 53 331875	21	...	39	111	25	66	45	96	113	56	26	64	66	58	766	51	22 Jun
Great Eau																		
Driby P. Sta.	TF 53 388745	41	...	49	122	29	79	51	79	88	59	31	70	87	65	808	45	22 Jun
Belleau	TF 53 402783	15	...	33	115	26	73	51	83	87	62	32	63	80	60	764	17	31 Jul
Claythorpe Mill R.F.M. Sta. Area 77.4 km²	TF 53 416793	7	635	...	...	...	...	...	...	...	...	...	...	...	...	...	...	...
Withern P. Sta.	TF 53 434824	10	613	35	105	23	63	42	84	83	63	28	54	70	53	703	47	22 Jun
Manby	TF 53 391869	17	579	35	98	20	58	44	93	110	59	28	68	69	54	736	50	22 Jun
Coastal																		
Strubby	TF 53 445806	13	596	36	109	23	58	45	89	77	55	30	57	71	56	707	47	22 Jun
Maltby-le-Marsh P. Sta.	TF 53 472826	5	...	36	109	20	61	45	87	81	51	27	57	75	51	701	43	22 Jun
Trusthorpe P. Sta.	TF 53 513840	2	...	33	102	18	56	43	90	83	50	28	55	74	56	689	48	22 Jun
Mumby P. Sta.	TF 53 514746	9	...	39	102	19	63	48	81	76	61	27	70	72	61	719	41	22 Jun
Anderby	TF 53 514755	3	610	40	108	21	65	49	86	74	59	28	67	71	59	727	41	22 Jun
Willoughby High																		
Chapel St.Leonards P. Sta.	TF 53 559729	3	...	40	90	19	63	51	80	100	60	32	62	80	61	738	34	22 Jun
Coastal																		
Skegness	TF 53 569631	5	568	43	85	16	59	45	102	83	65	37	96	81	66	777	42	22 Jun
Skegness, Laythorpe Avenue	TF 53 566621	5	...	44	85	15	61	45	99	89	63	37	91	76	68	774	39	22 Jun
Steeping																		
Tetford	TF 53 329748	59	...	42	116	29	68	67	87	97	55	27	52	89	64	792	47	22 Jun
Hagworthingham	TF 53 354691	62	649	47	119	29	63	53	87	83	58	31	58	83	70	781	46	22 Jun
Partney Mill R.F.M. Sta. Area 61.6 km²	TF 53 402676	15	673	...	...	...	...	...	...	...	...	...	...	...	...	...	...	...
Fordington W. Wks	TF 53 415714	49	...	55	127	27	82	51	106	82	65	33	75	84	69	854	50	22 Jun
Witham																		
Cottesmore	SK 43 909154	138	...	38	100	18	34	61	78	73	112	36	68	55	91	765	36	30 Aug
Great Ponton W. Wks.	SK 43 924296	72	...	42	83	27	97	49	68	64	110	30	79	55	77	781	32	21 Aug
Saltersford P. Sta.	SK 43 926334	57	627	44	84	23	96	43	83	72	99	29	81	60	68	782	30	21 Aug
Grantham Park	SK 43 916368	53	624	45	80	23	89	45	69	70	89	25	77	56	65	734	28	22 Jun
Marston	SK 43 893434	30	...	39	78	23	73	39	70	56	82	22	74	52	67	673	35	21 Aug
Foston (Witham)																		
Sedgebrook	SK 43 853378	44	626	42	83	25	79	40	77	42	86	21	74	52	64	687	28	22 Jun
Witham																		
Claypole Mill R.F.M. Sta. Area 298 km²	SK 43 842480	17	625	...	...	...	...	...	...	...	...	...	...	...	...	...	...	...
Brant (Witham)																		
Brant Broughton	SK 43 917546	11	609	43	89	25	79	40	62	61	61	29	79	46	59	672	26	22 Jun
Witham																		
Lincoln, Brant Road	SK 43 963666	10	600	45	106	19	77	40	59	78	49	28	72	52	52	676	26	22 Jun
Skellingthorpe Hall	SK 43 930715	12	612	48	101	(15)	69	56	58	79	55	23	87	54	59	705	26	22 Jun
Till (Witham)																		
Lincoln	SK 43 962719	6	603	44	102	13	70	41	75	104	47	27	74	52	53	703	36	22 Jun
Lincoln, Hollands Row	SK 43 968716	6	...	45	101	15	73	44	65	91	49	27	75	47	52	684	26	22 Jun
Witham																		
Barlings Eau (Witham)																		
Hackthorn Hall	SK 43 992825	32	620	44	111	20	71	50	65	81	79	27	87	57	54	747	27	22 Jun
Holton	TF 53 116816	29	...	38	94	17	55	43	76	80	67	24	60	53	52	659	20	29 Aug
Riseholme	SK 43 985756	38	...	44	101	16	73	49	99	78	61	26	77	54	54	731	37	22 Jun
Langworth Bridge R.F.M. Sta. Area 210 km²	TF 53 066766	4	617	...	...	...	...	...	...	...	...	...	...	...	...	...	...	...
Reepham	TF 53 040737	18	599	37	99	15	67	42	73	83	60	27	69	43	53	668	31	22 Jun
Bardney	TF 53 113688	5	599	30	89	12	60	56	69	72	69	23	55	49	50	634	28	22 Jun
Witham																		
Potterhanworth	TF 53 055661	18	621	44	104	14	72	45	85	66	73	32	59	43	49	687	31	22 Jun
Southrey	TF 53 140664	6	...	34	91	15	59	43	70	71	88	20	57	(51)	51	650	31	22 Jun
Baumber	TF 53 222740	64	...	44	94	17	58	47	86	100	70	20	56	61	56	710	31	22 Jun
Woodhall Spa W. Wks.	TF 53 223617	14	592	41	97	19	51	47	76	68	90	29	56	63	57	693	35	22 Jun
Bain (Witham)																		
Wykeham Hall	TF 53 227881	102	730	47	120	27	92	48	95	99	60	28	86	70	65	835	34	22 Jun
Donington-on-Bain	TF 53 236830	70	703	52	108	27	86	53	85	119	67	30	72	63	63	823	32	22 Jun
Goulceby	TF 53 251792	73	...	46	98	26	67	42	88	94	64	26	70	67	57	746	36	22 Jun
Horncastle	TF 53 251695	41	651	40	97	24	52	43	82	80	70	24	51	67	51	681	37	22 Jun
Revesby Res.	TF 53 303634	38	672	44	101	23	65	49	82	68	78	28	68	71	59	735	42	22 Jun
Fulsby Lock R.F.M. Sta. Area 197 km²	TF 53 241611	10	671	...	...	...	...	...	...	...	...	...	...	...	...	...	...	...
Witham																		
Slea (Witham)																		
Witham																		
Boston, Grand Sluice	TF 53 323444	5	588	43	85	15	70	40	85	53	65	24	75	(49)	52	656	50	22 Jun
South Forty Foot (Witham)																		
Guthram Gowt	TF 53 171225	5	582	29	81	11	85	46	79	46	87	28	70	49	59	669	39	22 Jun
Billingborough	TF 53 112342	11	614	35	80	15	86	50	72	70	84	22	73	47	54	688	37	21 Aug
Newton Hall	TF 53 047363	52	618	39	71	24	81	47	93	68	101	28	71	57	65	746	36	21 Aug
Culverthorpe Hall	TF 53 022402	58	604	42	86	22	86	46	72	58	84	28	64	58	67	713	28	21 Aug
Heckington	TF 53 143443	8	...	40	77	17	74	48	77	53	78	25	77	45	52	662	34	22 Jun
Boston, Cooks Lock	TF 53 313432	3	...	39	79	17	63	42	96	58	61	24	75	50	54	659	53	22 Jun
Skirbeck, Black Sluice	TF 53 325428	3	612	40	81	17	64	47	93	54	65	28	77	(51)	57	674	48	22 Jun
Witham																		
Twentylands P. Sta.	TF 53 387646	33	...	49	104	24	70	55	107	88	66	34	64	78	57	795	51	22 Jun
East Kirkby Manor	TF 53 334618	11	649	45	96	23	67	49	95	72	62	29	62	76	61	737	45	22 Jun

Station	Grid reference	Altitude m	Av. 1916–1950 mm	Jan	Feb	Mar	Apr	May	Jun	Jul	Aug	Sep	Oct	Nov	Dec	Year	Highest recorded fall in a day mm	date
LINCOLNSHIRE—*(contd)*																		
Carrington Grange	TF 53 312545	3	...	42	86	23	67	52	102	70	72	29	74	61	66	742	42	22 Jun
Froghall	TF 53 253524	4	...	41	79	20	62	48	101	63	71	30	69	59	60	701	43	22 Jun
Cowbridge	TF 53 328472	3	...	43	88	22	69	50	87	62	70	27	79	53	59	709	47	22 Jun
Boston, Church Road P. Sta.	TF 53 335435	2	605	37	82	18	60	45	96	57	67	26	74	51	56	670	51	22 Jun
Boston, Frieston Road	TF 53 339441	5	617	44	89	20	63	47	96	59	78	28	77	53	59	711	50	22 Jun
Tytton Hall	TF 53 329417	3	610	42	82	20	66	50	93	52	67	26	75	53	57	683	50	22 Jun
Wyberton	TF 53 341415	3	...	44	84	22	66	54	100	50	65	31	77	57	61	710	51	22 Jun
Frampton S. Wks.	TF 53 314397	3	...	33	79	15	58	43	93	48	63	23	74	51	53	634	47	22 Jun
Kirton	TF 53 302380	4	604	37	78	14	63	46	93	51	65	31	75	49	54	657	49	22 Jun
Eastville	TF 53 404602	1	561	42	86	20	72	48	105	76	59	28	69	69	51	726	52	22 Jun
Lade Bank Bridge	TF 53 378544	4	566	43	81	19	57	41	123	78	69	28	70	56	63	730	46	22 Jun
Leverton	TF 53 422484	3	569	42	80	16	56	44	109	47	58	31	69	52	52	657	51	22 Jun
Hobhole P. Sta.	TF 53 365399	16	...	40	81	19	62	48	104	47	72	38	75	56	58	699	46	22 Jun

WELLAND AND NENE RIVER AUTHORITY AREA, 3970 SQUARE KILOMETRES

HYDROMETRIC AREAS 31 AND 32

Station	Grid reference	Altitude m	Av. 1916–1950 mm	Jan	Feb	Mar	Apr	May	Jun	Jul	Aug	Sep	Oct	Nov	Dec	Year	Highest recorded fall in a day mm	date
Welland																		
Sibbertoft	SP 42 679826	171	691	31	98	10	95	63	56	77	110	33	97	53	80	803	47	29 Aug
Husbands Bosworth W. Wks.	SP 42 644843	155	666	30	102	12	90	66	47	77	111	38	95	50	79	797	40	29 Aug
Market Harborough, Northampton Road	SP 42 735869	79	640	25	90	10	81	54	51	56	114	29	82	47	71	709	58	29 Aug
Market Harborough, Burnmill Road	SP 42 732879	96	623	26	94	12	83	53	51	58	119	31	87	47	77	737	61	29 Aug
Great Bowden	SP 42 747888	76	641	27	91	13	86	53	57	63	108	31	79	47	81	737	55	29 Aug
Kibworth Harcourt	SP 42 681943	124	658	26	89	14	89	44	45	67	109	25	83	44	75	710	45	29 Aug
Ashley	SP 42 796908	76	619	30	82	12	85	50	57	61	99	29	74	46	73	697	46	29 Aug
Hallaton	SP 42 789966	110	...	30	81	13	84	50	62	66	81	31	75	53	86	711	26	22 Jun
Eyebrook (Welland)																		
Eye (Welland)																		
Wardley	SK 43 832001	113	656	32	84	14	95	42	65	59	78	31	70	63	79	710	25	22 Jun
Uppingham	SP 42 849988	163	672	35	83	15	101	42	68	67	93	35	69	53	82	743	28	21 Aug
Eye Brook Res. R.F.M. Sta. Area 60.1 km²	SP 42 853942	55	671	...	...	...	...	...	...	...	...	...	...	...	...	...	...	...
Eye Brook Res.	SP 42 855941	57	603	31	78	8	90	38	52	63	92	35	65	47	74	673	35	21 Aug
Caldecott	SP 42 864932	53	576	24	67	10	77	38	53	55	98	35	60	44	68	629	38	21 Aug
Chater (Welland)																		
Welland																		
Seaton Mill	SP 42 908976	41	598	33	66	12	79	37	58	58	93	41	63	50	74	665	34	21 Aug
Tixover P. Sta.	SK 43 974002	38	593	36	83	16	91	40	91	53	109	45	82	59	90	796	47	21 Aug
Stamford S. Wks.	TF 53 041074	23	577	29	77	12	80	38	69	47	92	41	72	50	73	680	33	30 Aug
Gwash (Welland)																		
Gunthorpe Hall	SK 43 868056	119	...	33	97	15	90	41	57	58	79	34	69	49	77	700	26	22 Jun
Oakham, Catmose	SK 43 863085	107	627	32	94	18	92	47	58	52	89	36	67	49	84	718	25	22 Jun
Oakham, Stamford Road	SK 43 863088	106	...	34	97	20	93	54	63	55	89	38	64	48	82	736	27	22 Jun
Tickencote Hall	SK 43 988094	42	–	30	76	13	81	35	78	43	118	32	64	(51)	(89)	709	43	21 Aug
Welland																		
Deeping St. James	TF 53 148096	6	...	31	80	12	87	37	77	59	106	37	77	50	75	729	40	22 Jun
Etton	TF 53 142048	11	525	32	75	11	84	38	70	69	96	38	91	45	68	717	34	22 Jun
Marholm	TF 53 145020	21	...	36	78	11	84	39	71	71	95	34	80	45	73	718	36	22 Jun
Deeping St. Nicholas	TF 53 213157	3	564	28	79	9	78	44	81	44	110	37	72	43	60	685	40	30 Aug
Pode Hole No.3	TF 53 214219	3	600	30	76	10	80	45	83	50	79	47	68	50	56	673	33	22 Jun
Pode Hole No.2	TF 53 215220	3	584	31	77	11	78	45	83	49	76	48	67	50	57	672	34	30 Aug 22 Jun
Spalding Grammar School	TF 53 246222	5	562	27	79	10	76	47	106	47	90	43	70	51	57	704	39	22 Jun
Spalding Depot	TF 53 244225	9	...	28	78	10	72	45	112	47	90	38	67	51	58	698	39	22 Jun
Spalding, Marsh Road	TF 53 258238	4	...	28	80	12	74	47	110	51	74	41	68	49	58	693	41	22 Jun
Glen (Welland)																		
Eden (Glen)																		
Welby	SK 43 974381	101	...	36	80	20	86	45	81	64	91	31	85	49	68	736	30	22 Jun
Glen (Welland)																		
Ropsley	SK 43 993342	77	...	38	83	24	93	49	85	68	103	28	73	63	70	777	32	29 Aug
Eden (Glen)																		
Irnham Hall	TF 53 022265	67	621	36	91	18	89	45	90	56	115	23	77	57	66	763	35	30 Aug
Wilsthorpe	TF 53 081148	15	565	32	82	17	91	36	84	44	116	37	67	51	69	725	39	30 Aug
Kate's Bridge R.F.M. Sta. Area 342 km²	TF 53 106149	6	612	...	...	...	...	...	...	...	...	...	...	...	...	...	...	...
Glen (Welland)																		
Tongue End	TF 53 151184	3	...	31	79	11	88	38	77	47	109	37	71	51	65	704	50	30 Aug
Bourne P. Sta	TF 53 101201	7	593	35	90	15	93	40	77	45	128	26	68	58	64	739	47	30 Aug
Pinchbeck Marsh	TF 53 261261	3	...	31	77	13	72	49	93	53	65	47	67	47	69	682	42	22 Jun
Surfleet	TF 53 280292	4	...	31	80	12	75	51	91	53	67	44	69	51	63	689	44	22 Jun
Welland																		
Irby Hall Farm	TF 53 323224	3	...	26	82	13	70	61	117	71	76	44	90	75	71	796	55	22 Jun
Majors Farm	TF 53 345309	4	...	37	85	15	71	50	92	57	74	48	74	63	71	737	42	22 Jun
Hovenden House	TF 53 397261	5	583	36	84	17	59	53	105	61	93	47	77	66	64	760	43	22 Jun
Bingham Lodge Farm	TF 53 391322	3	...	38	75	14	59	57	111	51	67	44	70	58	61	705	43	22 Jun
Nene																		
Kislingbury (Nene)																		
Norton Junction	SP 42 602657	109	674	29	82	20	79	54	72	61	94	33	108	47	97	776	38	13 Oct
Dodford R.F.M. Sta. Area 107 km²	SP 42 627607	79	673	...	...	...	...	...	...	...	...	...	...	...	...	...	...	...
Nene																		
Litchborough	SP 42 633542	148	660	27	79	20	93	55	75	61	78	41	123	51	95	796	33	13 Oct
Bugbrooke Mill	SP 42 681586	69	599	28	75	10	85	41	82	55	82	37	105	40	82	721	34	29 Aug
Upton Lodge	SP 42 713606	91	...	28	81	14	89	50	87	61	108	43	109	45	95	809	42	29 Aug
Upton R.F.M. Sta. Area 223 km²	SP 42 721592	62	668	...	...	...	...	...	...	...	...	...	...	...	...	...	...	...
Blisworth	SP 42 720550	90	648	26	81	13	80	41	81	59	70	46	112	41	92	742	27	22 Jun
Brampton (Nene)																		
Pitsford Res.	SP 42 758687	101	618	30	76	7	82	47	70	62	109	33	102	48	79	746	34	30 Aug
Lamport Hall	SP 42 759745	146	690	29	79	10	83	44	59	58	110	37	90	48	84	732	37	29 Aug
Hollowell	SP 42 687722	122	648	31	73	11	86	49	67	71	122	44	93	44	95	787	58	29 Aug
Ravensthorpe	SP 42 682704	98	647	30	74	9	83	49	63	64	119	41	94	45	91	760	49	29 Aug
Althorp Park	SP 42 682652	122	673	29	80	13	87	49	78	56	118	35	99	45	95	784	39	29 Aug
St. Andrew's Mill R.F.M. Sta. Area 233 km²	SP 42 747617	59	655	...	...	...	...	...	...	...	...	...	...	...	...	...	...	...

Station	Grid reference	Alti-tude m	Av. 1916–1950 mm	Rainfall in millimetres Jan	Feb	Mar	Apr	May	Jun	Jul	Aug	Sep	Oct	Nov	Dec	Year	Highest recorded fall in a day mm	date
WELLAND AND NENE—*(contd)*																		
Duston	SP 42 717612	104	664	29	76	13	85	49	93	61	105	38	101	43	87	780	44	29 Aug
Nene																		
Northampton, Hardingstone	SP 42 762598	58	583	28	77	10	85	46	79	58	97	39	93	46	76	734	46	29 Aug
Northampton, Cliftonville Road	SP 42 767602	72	612	28	74	11	81	47	73	56	100	40	93	47	83	733	50	29 Aug
Northampton, Stimpson Avenue	SP 42 768615	100	607	29	79	11	89	45	81	59	117	38	88	49	86	772	48	29 Aug
Sywell Res.	SP 42 830653	83	617	28	73	10	79	44	72	65	125	46	75	48	72	736	39	29 Aug
Yardley Hastings	SP 42 852573	101	637	37	84	9	84	55	71	58	105	53	87	54	84	782	44	29 Aug
Wollaston P. Sta.	SP 42 889641	43	582	28	71	6	74	45	63	55	81	54	70	53	68	668	28	22 Jun
Wollaston Hall	SP 42 909631	84	...	30	81	7	75	47	63	53	77	58	73	52	63	678	30	22 Jun
Wellingborough, Nene Wharf	SP 42 899664	41	586	30	70	7	81	45	62	62	79	54	68	52	70	679	29	22 Jun
Ise (Nene)																		
Kelmarsh Hall	SP 42 735795	126	719	36	84	15	89	51	59	76	135	47	110	56	93	851	64	29 Aug
Clover Hill W. Wks.	SP 42 869799	105	638	29	77	13	86	40	53	58	84	41	84	53	74	690	33	30 Aug
Thorpe Malsor	SP 42 830795	107	646	28	77	10	91	48	55	51	99	41	85	44	74	704	37	30 Aug
Cransley	SP 42 833782	84	641	29	77	12	89	46	49	53	89	39	86	47	76	692	36	30 Aug
Orlingbury	SP 42 844715	118	646	25	81	6	97	52	57	71	99	42	89	53	81	755	31	30 Aug
Harrowden Hall	SP 42 882709	88	629	25	73	6	96	41	63	59	83	46	75	49	80	698	28	22 Jun
Harrowden Old Mill R.F.M. Sta. Area 194 km²	SP 42 898715	46	648	...	...	...	...	...	...	...	...	...	...	...	...	...	...	...
Wellingborough, Swanspool	SP 42 894675	57	623	32	77	7	94	50	64	63	87	53	76	54	78	735	30	22 Jun
Nene																		
Rushden Cemetery	SP 42 965664	79	584	33	80	5	78	47	73	56	103	49	90	52	67	734	38	22 Jun
Raunds	SP 42 991721	59	579	32	81	9	83	53	67	55	94	39	89	56	64	722	34	22 Jun
Woodford P. Sta.	SP 42 971759	32	566	25	74	13	87	36	78	56	87	46	76	50	63	691	35	22 Jun
Harpers (Nene)																		
Brigstock	SP 42 944852	52	...	31	96	11	93	44	83	61	102	34	75	54	75	759	33	21 Aug
Sudborough	SP 42 968821	43	633	29	88	14	99	45	90	63	98	41	81	57	75	782	36	22 Jun
Old Mill Bridge R.F.M. Sta. Area 74.3 km²	SP 42 983799	31	617	...	...	...	...	...	...	...	...	...	...	...	...	...	...	...
Nene																		
Barnwell P. Sta.	TL 52 043872	21	...	24	79	8	81	39	80	66	101	45	74	45	61	703	40	21 Aug
Oundle, Head Office	TL 52 042883	32	...	27	84	9	86	42	78	65	102	47	76	49	71	736	38	21 Aug
Oundle, Nene Wharf	TL 52 043888	20	586	28	82	7	84	41	73	62	100	45	75	49	68	713	37	22 Jun 21 Aug
Willow (Nene)																		
Corby	SP 42 901885	97	607	26	88	10	84	44	66	64	96	38	88	49	81	733	33	29 Aug
Bulwick	SP 42 957940	68	...	28	83	8	88	40	70	62	113	37	85	60	67	741	49	21 Aug
Apethorpe	TL 52 022961	39	594	25	72	11	84	42	81	54	112	39	85	54	74	733	41	21 Aug
Fotheringhay R.F.M. Sta. Area 89.6 km²	TL 52 067933	15	602	...	...	...	...	...	...	...	...	...	...	...	...	...	...	...
Nene																		
Wittering	TF 53 048032	80	583	28	75	11	85	41	69	61	110	34	80	50	76	719	36	22 Jun
Orton R.F.M. Sta. Area 1630 km²	TL 52 166972	4	627	...	...	...	...	...	...	...	...	...	...	...	...	...	...	...
Orton Longueville	TL 52 174971	15	...	30	72	10	82	37	80	61	83	30	58	54	64	662	44	22 Jun
Woodston	TL 52 179975	11	550	29	69	8	77	36	83	58	89	30	55	55	66	654	38	22 Jun
Peterborough S. Wks.	TL 52 201984	3	543	27	68	9	74	35	73	61	98	35	51	51	62	643	39	22 Jun
Thorney, North House	TF 53 307034	1	536	27	68	11	65	37	76	60	111	41	51	47	64	657	43	22 Jun
Thorney Toll	TF 53 355028	1	555	27	74	13	68	42	106	69	114	51	53	55	76	748	50	30 Aug
Thorney, Nutsgrove Farm	TF 53 328058	4	529	33	74	12	75	50	75	68	122	44	55	49	65	722	44	22 Jun
Clough's Cross	TF 53 368093	2	542	27	73	12	60	42	97	70	118	46	47	51	66	710	53	30 Aug
Tydd Manor	TF 53 429166	3	526	33	84	12	65	44	98	61	93	35	54	55	64	700	49	22 Jun
Tydd	TF 53 467181	3	604	33	86	13	58	49	112	58	106	31	47	49	65	707	45	22 Jun
Clifton's Bridge	TF 53 378189	2	...	38	91	15	60	58	104	69	111	43	70	(63)	(63)	785	44	22 Jun
Sutton Bridge	TF 53 476200	6	581	33	94	17	61	63	122	67	102	35	49	56	77	776	56	22 Jun
Sutton Bridge Experimental Sta.	TF 53 481203	4	...	32	93	16	57	49	119	63	95	35	42	54	63	717	53	22 Jun
Wingland Grange	TF 53 517220	3	524	30	85	17	52	49	105	57	107	35	43	53	66	699	55	29 Aug

GREAT OUSE RIVER AUTHORITY AREA, 8582 SQUARE KILOMETRES

HYDROMETRIC AREA 33

Coastal																		
Terrington St. Clement	TF 53 547187	4	519	27	81	13	54	44	110	77	86	31	48	52	65	688	54	22 Jun
Terrington St. Clement, Manor Farm	TF 53 551203	4	550	31	87	21	56	48	107	74	95	32	50	55	69	724	50	22 Jun
Ouse																		
Hazelborough Woods	SP 42 654427	140	685	28	100	18	90	63	71	70	68	43	116	60	71	799	25	22 Jun
Buckingham, Sandpit Farm	SP 42 701343	82	653	25	86	23	92	63	82	80	73	49	133	47	73	824	32	10 Jun
Twyford (Ouse)																		
Middle Claydon, Catherine Farm	SP 42 714250	93	660	31	75	13	81	68	58	62	97	45	111	39	73	754	47	29 Aug
Calvert	SP 42 688247	90	640	28	72	14	83	62	59	60	97	39	117	36	71	736	...	...
Botolph Claydon, Bernwood Farm	SP 42 732243	122	701	33	82	15	90	61	61	63	97	53	111	47	86	799	47	29 Aug
Claydon (Twyford)																		
Swanbourne	SP 42 801270	123	...	32	83	13	85	55	76	70	97	44	106	54	81	796	46	29 Aug
Shipton Lee, Dry Leys Farm	SP 42 731218	96	663	29	77	16	90	71	67	63	96	51	110	43	78	791	49	29 Aug
North Marston, Hogshaw Farm	SP 42 739228	107	683	31	84	15	85	70	71	59	85	55	106	45	87	793	37	29 Aug
Quainton Hill Res.	SP 42 751213	184	696	33	78	17	92	67	66	63	87	52	107	46	83	791	...	...
Ouse																		
Thornborough Mill R.F.M. Sta. Area 389 km²	SP 42 736353	73	673	...	...	...	...	...	...	...	...	...	...	...	...	...	...	...
Foxcote P. Sta.	SP 42 713357	93	645	26	91	24	83	64	73	77	73	47	111	45	67	780	26	22 Jun
Calverton	SP 42 790389	76	...	27	85	16	75	56	75	61	63	53	88	47	70	716	30	22 Jun
Tove (Ouse)																		
Bradden House	SP 42 645484	128	699	27	94	21	88	61	78	65	62	41	111	55	76	779	29	13 Oct
Cappenham R.F.M. Sta. Area 138 km²	SP 42 714488	76	696	...	...	...	...	...	...	...	...	...	...	...	...	...	...	...
Ouzel (Ouse)																		
Battlesden P. Sta.	SP 42 960281	109	665	30	62	14	84	48	93	86	73	60	117	63	77	809	36	22 Jun
Leighton Buzzard, Heath Road	SP 42 924269	76	670	33	70	15	88	54	101	90	83	56	114	65	74	842	39	29 Aug
Bletchley	SP 42 867336	86	...	32	78	15	72	47	101	58	92	49	89	44	74	753	35	29 Aug
Bletchley S. Wks.	SP 42 886347	67	576	27	79	15	72	43	107	66	88	55	92	46	64	754	38	22 Jun
Willen Weir R.F.M. Sta. Area 277 km²	SP 42 883409	57	655	...	...	...	...	...	...	...	...	...	...	...	...	...	...	...
Woburn	SP 42 964358	89	629	32	76	15	84	48	109	61	97	49	111	63	84	828	44	29 Aug
Newport Pagnell, Wolverton Road	SP 42 870436	55	617	26	85	11	78	52	91	58	69	57	91	53	74	745	36	22 Jun

Station	Grid reference	Altitude m	Av. 1916–1950 mm	Jan	Feb	Mar	Apr	May	Jun	Jul	Aug	Sep	Oct	Nov	Dec	Year	Highest recorded fall in a day mm	date
GREAT OUSE—(contd)																		
Ouse																		
Weston Underwood P. Sta.	SP 42 873510	52	630	30	74	11	70	48	98	74	85	65	100	49	75	779	34	22 Jun
Olney...	SP 42 890514	52	...	33	76	11	78	48	97	71	79	63	88	53	73	769	34	22 Jun
Brayfield House	SP 42 928520	55	628	32	77	11	85	47	93	63	86	56	97	56	77	780	37	22 Jun
Harrold Mill R.F.M. Sta. Area 1320 km^2	SP 42 951565	43	656	...	...	...	...	...	...	...	...	...	...	...	...	...	...	...
Souldrop	SP 42 984615	91	646	29	86	9	79	49	84	66	97	50	106	55	74	787	41	22 Jun
Bedford W. Wks.	TL 52 038515	34	...	24	63	7	77	38	79	52	73	46	93	48	70	670	39	22 Jun
Bromham	TL 52 010506	34	589	26	68	7	75	36	75	52	81	38	91	45	70	667	39	22 Jun
Kempston	TL 52 033477	38	571	28	66	10	78	37	100	44	87	47	97	55	77	726	37	22 Jun
Bedford R.F.M. Sta. Area 1460 km^2	TL 52 055495	24	650	...	...	...	...	...	...	...	...	...	...	...	...	...	...	...
Bedford Park	TL 52 051510	36	529	22	63	5	80	36	82	52	77	50	91	49	66	672	41	22 Jun
Bedford Royal Aircraft Estab.	TL 52 049597	85	566	22	69	9	73	38	78	63	81	38	73	43	65	653	42	22 Jun
Renhold	TL 52 087529	51	573	27	60	6	72	33	72	61	78	53	95	49	67	673	39	22 Jun
Elstow (Ouse)																		
Cardington	TL 52 081463	23	527	25	63	11	76	38	74	52	83	47	91	50	69	678	42	22 Jun
Ouse																		
Mogerhanger Park	TL 52 133485	46	579	31	65	9	79	37	85	65	88	37	77	53	71	697	47	22 Jun
Ivel (Ouse)																		
Baldock, William Road P. Sta.	TL 52 241325	82	594	30	67	9	56	52	85	64	103	41	75	49	86	716	42	22 Jun
Hiz (Ivel)																		
Hitchin, Knightswood	TL 52 191290	85	671	32	72	9	71	48	99	71	84	39	79	52	89	743	41	22 Jun
Hitchin W. Wks.	TL 52 185289	64	636	29	67	9	70	43	93	66	90	35	75	50	89	716	37	22 Jun
Arlesey S. Wks.	TL 52 190363	43	578	24	50	6	61	38	86	(63)	(89)	33	65	42	71	630	45	22 Jun
Letchworth	TL 52 208346	64	574	27	57	9	60	39	89	61	100	30	70	45	82	669	43	22 Jun
Ivel (Ouse)																		
Meppershall	TL 52 137362	76	584	31	58	7	74	50	84	58	74	42	89	54	91	713	39	22 Jun
Clifton S. Wks.	TL 52 177403	34	...	(25)	49	8	62	37	83	(63)	(89)	29	65	46	75	630	46	22 Jun
Old Canal (Ivel)																		
Ivel (Ouse)																		
New Spring P. Sta....	TL 52 210416	53	555	30	51	9	67	42	96	68	93	28	73	45	81	683	49	22 Jun
Potton	TL 52 230494	47	572	32	54	6	69	37	73	66	120	42	73	52	70	695	53	22 Jun
Potton S. Wks....	TL 52 221485	34	571	29	52	6	66	37	71	(63)	(102)	(38)	68	50	67	649	51	22 Jun
Sandy S. Wks....	TL 52 162501	22	...	25	58	6	68	36	74	61	95	37	71	53	65	649	45	22 Jun
Blunham R.F.M. Sta. Area 541 km^2	TL 52 153509	19	597	..	...	...	...	...	...	...	...	...	...	...	...	...	...	...
Ouse																		
Tempsford	TL 52 161535	17	555	28	60	5	70	(36)	87	60	106	35	73	48	68	675	48	22 Jun
St. Neots P. Sta.	TL 52 184581	17	...	25	58	7	72	31	88	67	83	48	66	42	68	654	60	22 Jun
Great Gransden	TL 52 266557	46	566	30	59	7	72	38	63	64	102	49	79	57	71	690	49	21 Aug
St. Neots	TL 52 189605	21	545	26	63	7	82	32	90	61	91	50	70	41	68	681	59	22 Jun
Kym (Ouse)																		
Meagre Farm R.F.M. Sta. Area 138 km^2	TL 52 155631	17	594	...	...	...	...	...	...	...	...	...	...	...	...	...	...	...
Alconbury (Ouse)																		
Ouse																		
Brampton Treatment Wks.	TL 52 223709	9	...	26	66	10	76	37	70	56	98	57	66	47	62	672	41	22 Jun
Houghton	TL 52 288717	7	...	27	69	7	87	40	81	66	93	50	70	46	71	706	45	22 Jun
Wyton	TL 52 284745	43	544	25	63	8	76	35	79	65	96	44	67	45	64	666	50	22 Jun
Boxworth Manor House	TL 52 352645	23	605	29	61	8	76	38	70	60	106	46	66	50	66	677	50	21 Aug
Boxworth	TL 52 347639	48	543	28	60	7	77	36	64	59	98	45	67	49	65	655	42	21 Aug
Brownshill Staunch...	TL 52 371726	3	539	27	64	10	82	45	68	77	98	40	73	54	85	722	32	22 Jun
Stretham	TL 52 516729	4	506	23	50	10	48	31	81	56	52	24	55	41	67	538	36	22 Jun
Cam (Ouse)																		
Wendens Ambo	TL 52 491365	73	...	25	73	9	57	48	56	85	77	29	81	35	82	658	28	22 Jun
Saffron Walden, County High School	TL 52 532378	72	592	23	63	10	48	45	54	69	79	23	75	38	79	606	31	21 Aug
Chesterford Park	TL 52 536422	110	593	24	62	7	56	44	60	66	70	23	73	37	78	599	29	22 Jun
Sewards End	TL 52 570384	120	616	27	67	10	59	46	65	76	59	21	87	41	83	642	31	22 Jun
Dernford R.F.M. Sta. Area 198 km^2	TL 52 466506	15	603	...	...	...	...	...	...	...	...	...	...	...	...	...	...	...
Rhee (Cam)																		
Odsey	TL 52 292380	78	591	28	64	9	59	43	81	70	105	33	78	48	82	701	38	21 Aug
Abington Pigotts Hall	TL 52 304444	30	589	29	51	11	57	28	58	69	97	37	69	47	71	625	36	22 Jun
Royston, East Brooklands	TL 52 355411	60	...	31	67	12	58	39	61	73	109	26	71	44	74	665	47	21 Aug
Bassingbourn	TL 52 338458	24	587	34	48	11	58	31	61	71	139	26	74	39	66	659	73	21 Aug
Melbourn	TL 52 380446	26	557	31	56	10	51	38	63	71	107	27	66	44	71	635	47	21 Aug
Fowlmere P. Sta.	TL 52 427445	30	...	26	(57)	9	57	40	76	81	70	21	66	37	70	611	33	22 Jun
Burnt Mill Weir R.F.M. Sta. Area 303 km^2	TL 52 413522	9	587	...	...	...	...	...	...	...	...	...	...	...	...	...	...	...
Cam (Ouse)																		
Cambridge, Botanic Garden...	TL 52 453572	12	552	28	45	11	54	31	54	89	79	24	66	40	72	594	30	21 Aug
Cambridge P. Sta.	TL 52 465594	7	523	31	50	9	56	33	64	72	71	31	74	40	69	599	30	22 Jun
Cambridge N.I.A.B.	TL 52 434604	24	547	29	57	8	65	36	57	70	75	35	65	40	69	606	31	22 Jun
Milton	TL 52 477629	6	...	28	55	8	51	36	73	70	69	35	71	39	64	599	28	22 Jun
Bottisham R.F.M. Sta. Area 811 km^2	TL 52 508657	3	587	...	...	...	...	...	...	...	...	...	...	...	...	...	...	...
Bottisham Lock	TL 52 507657	3	512	30	54	12	56	39	85	71	52	31	67	46	72	614	33	22 Jun
Fulbourn	TL 52 539549	30	544	26	40	9	53	30	75	100	71	20	52	42	73	591	32	22 Jun
Quy Hall	TL 52 515611	5	...	24	50	7	55	30	89	78	53	29	70	37	70	593	35	22 Jun
Soham (Ouse)																		
Stetchworth Park	TL 52 642591	96	592	26	47	8	52	39	65	85	72	22	74	47	85	622	29	22 Jun
Newmarket S. Wks.	TL 52 635648	25	598	24	42	8	46	34	58	79	61	18	74	42	74	559	25	22 Jun
Ouse																		
Ely, Roswell Pits	TL 52 558805	4	559	22	55	12	49	35	85	75	69	27	62	54	74	619	41	22 Jun
Lark (Ouse)																		
Bury St.Edmunds, Corporation Depot	TL 52 850643	55	633	26	39	15	36	39	37	85	58	20	70	65	88	577	18	9 Dec
Troston Hall	TL 52 902717	49	605	22	47	17	52	44	42	89	54	18	79	65	83	613	20	30 Aug
Broom's Barn	TL 52 753657	75	...	28	45	11	42	38	58	75	64	20	57	45	82	564	21	30 Aug
Temple Weir R.F.M.Sta. Area 272 km^2	TL 52 758730	9	622	...	...	...	...	...	...	...	...	...	...	...	...	...	...	...
Kennet (Lark)																		
Lidgate	TL 52 728576	89	632	26	46	9	49	35	60	70	75	15	65	47	87	582	27	29 Aug
Beck Bridge R.F.M.Sta. Area 102 km^2	TL 52 622733	34	551	...	...	...	...	...	...	...	...	...	...	...	...	...	...	...

Station	Grid reference	Altitude m	Av. 1916–1950 mm	Jan	Feb	Mar	Apr	May	Jun	Jul	Aug	Sep	Oct	Nov	Dec	Year	Highest recorded fall in a day mm	date
GREAT OUSE—*(contd)*																		
Lark (Ouse)																		
Isleham R.F.M. Sta. Area 466 km^2	TL 52 648760	2	620	...	...	...	...	...	...	...	...	...	...	...	...	...	...	...
Isleham Lock	TL 52 648760	3	532	20	45	10	41	27	77	74	72	19	62	51	65	564	33	29 Aug
Mildenhall	TL 52 683778	5	549	18	50	10	42	28	67	77	75	25	86	51	67	597	32	29 Aug
Lark P. Sta.	TL 52 610827	0	...	22	(56)	13	49	39	91	72	74	24	57	53	74	625	38	22 Jun
Prickwillow	TL 52 597825	3	583	25	56	19	47	38	88	77	71	27	57	56	73	636	41	22 Jun
Ouse																		
Littleport, Victoria Street	TL 52 574866	3	...	20	63	13	58	37	89	70	63	31	58	49	71	621	44	22 Jun
Little Ouse (Ouse)																		
County Bridge Euston R.F.M. Sta. Area 129 km^2	TL 52 892801	13	616	...	...	...	...	...	...	...	...	...	...	...	...	...	...	...
Sapiston (Little Ouse)																		
Thurston	TL 52 914653	47	...	24	44	18	46	39	38	121	52	18	72	63	83	618	37	5 Jul
Barton Lodge	TL 52 884664	61	569	25	49	17	43	40	41	95	59	20	79	59	88	617	20	5 Jul
Rectory Bridge Euston R.F.M. Sta. Area 206 km^2	TL 52 896791	15	615	...	...	...	...	...	...	...	...	...	...	...	...	...	...	...
Little Ouse (Ouse)																		
Elvedon Hall	TL 52 822795	40	609	27	49	17	49	37	65	84	67	22	79	82	86	663	23	30 Aug
Thet (Little Ouse)																		
Watton	TF 53 942003	55	678	22	85	22	48	35	65	72	75	20	69	89	74	678	24	29 Aug
Kilverstone Hall	TL 52 891841	17	603	25	52	18	49	37	58	87	54	19	84	88	79	649	19	30 Aug
Melford Bridge R.F.M. Sta. Area 316 km^2	TL 52 880830	11	637	...	...	...	...	...	...	...	...	...	...	...	...	...	...	...
Little Ouse (Ouse)																		
Thetford R.F.M. Sta. Area 702 km^2	TL 52 860832	9	--	...	...	...	...	...	...	...	...	...	...	...	...	...	...	...
Thetford W. Wks.	TL 52 863845	50	624	26	53	19	51	31	71	72	60	21	78	90	72	642	26	30 Aug
Brandon, Santon Downham	TL 52 815879	6	...	22	64	19	50	29	66	78	60	21	86	89	75	659	18	16 Nov
Santon Downham	TL 52 813901	24	666	22	65	20	61	32	65	83	59	22	85	89	79	681	19	16 Nov
Feltwell P. Sta.	TL 52 651869	2	...	22	63	12	53	29	88	69	63	27	54	58	70	607	32	22 Jun
Ouse																		
Southery P.Sta.	TL 52 612932	0	560	21	71	9	51	34	87	74	55	22	55	65	70	613	41	22 Jun
Wissey (Ouse)																		
Catsholm P. Sta.	TL 52 683970	2	592	25	81	19	52	(41)	85	89	69	22	65	83	79	710	30	30 Aug
Hilborough Hall	TF 53 827003	24	628	28	79	22	59	42	81	73	61	20	81	82	88	716	26	17 Jun
Northwold R.F.M. Sta. Area 275 km^2	TL 52 771965	6	660	...	...	...	...	...	...	...	...	...	...	...	...	...	...	...
Swaffham, Mulsey	TF 53 821098	67	695	26	95	31	51	40	92	84	78	19	94	109	87	805	32	17 Jun
Swaffham Cemetery	TF 53 821081	64	708	26	92	25	50	41	87	85	73	18	91	103	81	771	32	17 Jun
Denton Lodge P. Sta.	TL 52 752919	37	630	23	89	14	43	47	83	86	77	17	62	90	(76)	708	34	30 Aug
Denton Lodge P. Sta.	TL 52 752919	37	630	21	66	20	51	36	69	72	63	20	73	80	79	649	20	17 Jun
Ouse																		
Denver Sluice	TF 53 588010	3	578	23	70	19	48	44	79	73	79	24	58	68	75	659	43	22 Jun
Earith	TL 52 387756	5	...	27	60	10	77	33	72	58	74	31	58	53	62	615	39	22 Jun
Puddock Bridge	TL 52 352880	2	530	30	70	13	67	45	90	55	85	60	65	49	70	699	49	30 Sep
Chatteris, The Vicarage	TL 52 395860	9	549	28	69	11	67	44	105	64	68	32	51	46	63	648	43	22 Jun
Littleport	TL 52 563863	20	529	25	64	13	59	37	86	67	64	28	56	53	71	623	43	22 Jun
Abbot's Ripton Hall	TL 52 240778	24	554	25	66	11	77	41	69	57	100	68	70	49	65	698	51	30 Sep
Benwick	TL 52 324899	2	548	30	72	10	66	42	67	50	81	58	61	44	74	655	49	30 Sep
Ransonmoor P. Sta.	TL 52 358922	0	...	33	76	13	63	45	88	69	88	37	53	49	68	682	45	22 Jun
Conington Castle	TL 52 181859	8	581	28	88	10	84	45	70	53	90	40	63	49	71	691	40	22 Jun
Monk's Wood	TL 52 202797	38	...	29	71	10	86	44	70	78	99	48	58	44	60	695	45	22 Jun
Wood Walton Fen	TL 52 232848	2	...	29	73	10	84	41	82	58	118	49	55	47	67	712	49	22 Jun
Whittlesea Mere	TL 52 237904	3	537	33	80	11	97	43	74	63	100	46	66	50	74	736	41	22 Jun
Stanground	TL 52 209962	11	...	27	77	10	81	37	69	64	98	36	59	52	67	677	40	22 Jun
Stanground Sluice	TL 52 208973	5	552	27	70	9	76	36	71	68	98	36	50	53	68	661	42	22 Jun
Whittlesey, Ashlime Sluice	TL 52 272962	5	545	27	66	8	68	38	72	59	91	51	52	43	61	637	41	22 Jun
March, Store Yard	TL 52 420967	2	586	28	75	12	62	44	88	69	86	37	52	49	66	668	45	22 Jun
Downham Market	TF 53 612027	27	...	23	81	16	48	43	79	66	93	22	49	75	70	665	39	22 Jun
Stow Fen P. Sta.	TF 53 598058	2	...	25	86	14	54	44	81	64	93	25	46	67	67	668	42	22 Jun
Crabb's Abbey	TF 53 600078	2	566	23	84	13	53	44	78	63	84	20	44	70	68	643	40	22 Jun
Watlington	TF 53 620104	9	...	23	85	16	52	49	78	66	86	27	45	78	78	682	39	30 Aug
Outwell	TF 53 529029	5	542	24	81	12	48	40	88	63	76	25	49	62	65	633	50	22 Jun
Wiggenhall St. Germans Sluice	TF 53 588143	3	553	26	77	15	53	43	93	60	77	29	44	66	70	653	42	22 Jun
King's Lynn, Freebridge	TF 53 610185	5	574	30	89	22	61	44	93	65	93	28	52	74	72	723	38	22 Jun
Nar (Ouse)																		
Narford Hall	TF 53 766139	15	658	25	77	19	49	39	73	78	83	24	78	98	78	721	...	...
Narborough Hall	TF 53 749130	14	646	27	85	21	45	49	85	75	93	24	75	98	81	758	37	30 Aug
Marham R.F.M. Sta. Area 153 km^2	TF 53 723119	5	691	...	...	...	...	...	...	...	...	...	...	...	...	...	...	...
Marham P.Sta.	TF 53 717107	6	630	23	82	17	43	42	83	75	83	20	62	93	79	702	39	30 Aug
Runcton Bottom	TF 53 647086	21	585	27	90	12	50	49	82	70	74	15	51	73	74	669	35	22 Jun
King's Lynn, Manure Wks.	TF 53 620189	0	567	29	82	18	53	40	85	66	92	26	56	70	65	682	37	22 Jun
Gayton	TF 53 729193	20	646	35	81	26	55	52	89	79	102	25	62	(76)	83	765	38	30 Aug
North Runcton Lodge	TF 53 644163	20	590	30	86	16	55	49	88	63	89	27	54	74	75	705	31	30 Aug
Gaywood (Ouse)																		
West Newton Mill R.F.M. Sta. Area 39.6 km^2	TF 53 696256	40	696	...	...	...	...	...	...	...	...	...	...	...	...	...	...	...
Sandringham House	TF 53 697289	37	691	41	92	29	55	60	85	77	118	33	63	110	85	848	49	30 Aug
Coastal																		

EAST SUFFOLK AND NORFOLK RIVER AUTHORITY AREA, 5294 SQUARE KILOMETRES

HYDROMETRIC AREAS 34 AND 35

Station	Grid reference	Altitude m	Av. 1916–1950 mm	Jan	Feb	Mar	Apr	May	Jun	Jul	Aug	Sep	Oct	Nov	Dec	Year	Highest recorded fall in a day mm	date
Coastal																		
Holme-next-Sea	TF 53 703435	6	563	(38)	(76)	17	47	51	74	81	69	25	62	93	61	695	21	30 Aug
Brancaster	TF 53 775438	12	596	43	78	17	46	35	74	98	69	21	58	96	67	703	21	15 Jul
North Creake	TF 53 854371	21	...	38	77	28	49	55	76	73	89	22	70	120	89	785	24	30 Aug
Wells S. Wks.	TF 53 912440	2	596	32	73	15	47	38	65	78	72	21	52	104	75	673	19	15 Jul
Stiffkey																		
Barney	TF 53 996324	73	...	36	72	32	57	53	57	109	85	15	70	108	69	762	31	20 Jul
Walsingham, Hill House	TF 53 950359	58	...	37	64	18	46	46	61	83	87	12	68	114	73	710	21	30 Aug

Station	Grid reference	Altitude m	Av. 1916–1950 mm	Jan	Feb	Mar	Apr	May	Jun	Jul	Aug	Sep	Oct	Nov	Dec	Year	Highest recorded fall in a day mm	Highest recorded fall in a day date
EAST SUFFOLK AND NORFOLK—*(contd)*																		
Glaven																		
Baconsthorpe Castle	TG 63 121381	62	...	(38)	66	23	65	36	74	101	71	19	83	85	67	729	22	15 Jul
Brinton Grange	TG 63 039358	40	677	32	71	21	53	50	57	109	85	14	60	90	65	707	28	20 Jul
Cley S.Wks.	TG 63 047424	9	...	34	69	15	52	53	64	96	63	16	59	78	69	668	24	15 Jul
Coastal																		
Sheringham Pine Grove	TG 63 163427	24	670	33	69	22	61	36	71	98	65	20	82	84	66	707	20	29 Jul
Cromer	TG 63 208422	54	678	32	70	20	61	30	68	78	61	21	67	88	71	668	(18	16 Nov)
Cromer, Newhaven Court	TG 63 220416	24	686	33	72	19	59	36	65	81	64	24	71	98	77	699	19	16 Nov
Great Yarmouth, Churchill Road	TG 63 527086	3	...	29	66	18	48	26	59	77	45	13	115	106	50	652	26	10 Oct
Yare																		
Mattishall	TG 63 061100	52	...	33	84	18	47	40	55	68	93	13	74	94	68	687	12	8 Feb
Browick Hall	TG 63 130015	46	677	27	79	27	52	32	63	79	96	21	78	96	77	727	35	29 Aug
Marlingford Hall	TG 63 126081	25	669	31	76	21	51	(33)	56	70	86	22	86	94	68	694	35	29 Aug
Colney R.F.M.Sta. Area 232 km²	TG 63 182082	8	688	...	...	...	...	...	...	...	...	...	...	...	...	...	...	...
Swardeston	TG 63 203026	15	...	28	75	23	54	26	67	72	74	24	75	105	73	696	22	16 Nov
Old Lakenham	TG 63 236063	12	...	37	79	26	53	31	66	84	71	22	76	108	67	719	21	16 Nov
Tas (Yare)																		
Wacton Hall	TM 62 180903	55	...	22	59	20	42	42	60	104	77	17	62	97	81	683	37	29 Aug
Shotesham R.F.M.Sta. Area 147 km²	TM 62 226994	10	668	...	...	...	...	...	...	...	...	...	...	...	...	...	...	...
Wensum (Yare)																		
West Raynham	TF 53 847245	73	718	37	84	30	57	44	76	99	114	19	89	136	89	871	32	30 Aug
Fakenham	TF 53 922296	44	674	36	74	34	54	39	74	87	89	19	75	108	69	757	26	30 Aug
Colkirk Hall	TF 53 924266	64	716	31	76	43	69	56	84	101	114	20	95	109	69	866	31	29 Aug
East Dereham Laboratory	TF 53 994132	53	667	32	93	26	60	43	63	81	82	15	81	89	72	738	25	29 Aug
East Dereham, Swanton Drive	TF 53 994138	53	...	34	88	25	57	38	62	74	79	13	75	82	72	698	27	29 Aug
Mileham	TF 53 921195	64	...	34	77	22	57	43	80	76	84	23	79	98	73	746	27	29 Aug
Cawston	TG 63 133238	40	691	32	68	26	47	34	63	83	54	29	93	95	62	688	20	16 Nov
Swanington	TG 63 131193	18	...	31	74	24	45	41	65	77	65	28	82	93	65	690	17	16 Nov
Taverham	TG 63 149156	30	653	29	81	26	49	36	69	77	72	22	104	103	73	744	19	16 Nov
Costessey Mill R.F.M.Sta. Area 536 km²	TG 63 177128	5	688	...	...	...	...	...	...	...	...	...	...	...	...	...	...	...
Tud (Wensum)																		
Costessey Park R.F.M.Sta. Area 73.3 km²	TG 63 170113	9	680	...	...	...	...	...	...	...	...	...	...	...	...	...	...	...
Wensum (Yare)																		
Norwich W. Wks.	TG 63 212096	10	...	30	71	24	46	29	58	68	65	23	68	108	62	651	28	16 Nov
Norwich Cemetery	TG 63 212086	28	650	35	79	23	53	29	67	75	70	26	75	111	71	715	23	16 Nov
Yare																		
Surlingham	TG 63 332060	2	...	34	66	27	48	39	77	84	59	22	61	99	70	688	18	17 Jun
Chet (Yare)																		
Raveningham Hall	TM 62 399965	23	644	37	68	(19)	59	28	62	77	49	21	96	104	72	691	20	16 Jun
Waveney (Yare)																		
Bressingham Hall	TM 62 078806	30	651	26	50	21	49	38	44	88	59	19	74	82	78	630	25	29 Aug
Gissing	TM 62 145844	40	...	(25)	50	21	44	37	45	104	58	24	72	95	79	655	24	29 Aug
Brome	TM 62 155767	40	...	30	42	23	42	29	48	72	58	25	83	98	76	626	29	18 Oct
Gislingham	TM 62 073708	56	...	(25)	45	14	36	32	46	59	62	16	79	79	79	573	29	29 Aug
Brockford, Buses Farm	TM 62 116659	58	...	(25)	(44)	(19)	49	30	49	69	52	20	87	83	81	609	20	18 Oct
Occold	TM 62 156708	49	...	26	41	19	43	31	45	66	58	19	87	86	74	596	24	29 Aug
Worlingworth	TM 62 235685	54	...	(25)	52	20	45	37	49	69	67	17	81	95	86	644	27	29 Aug
Needham Mill R.F.M.Sta. Area 373 km²	TM 62 228811	17	643	...	...	...	...	...	...	...	...	...	...	...	...	...	...	...
Bush Wood House	TM 62 280779	52	...	30	47	20	48	37	65	80	69	19	99	105	83	703	24	29 Aug
Wingfield	TM 62 235782	49	...	27	44	23	45	30	54	77	74	18	80	96	83	650	33	29 Aug
Starston	TM 62 234842	30	...	24	47	17	53	23	57	73	63	18	74	104	73	625	26	29 Aug
Ditchingham	TM 62 340906	5	653	31	54	20	54	33	63	94	54	21	93	94	75	687	27	16 Jun
Woodton	TM 62 295954	37	–	33	69	24	61	41	71	85	55	17	61	100	78	696	27	16 Jun
Barsham	TM 62 406896	15	...	32	57	16	55	33	57	78	64	22	85	107	68	675	20	3 Oct
Hopton	TM 62 530999	7	...	36	65	17	58	33	53	95	51	22	140	109	72	751	28	3 Oct
Fritton Warren	TM 62 463998	13	641	37	66	19	59	29	54	84	62	18	88	120	71	706	19	3 Oct
Bure (Yare)																		
Hope House	TG 63 038295	64	717	36	73	30	58	36	60	86	80	15	68	105	73	720	20	20 Jul
Cropton Hall	TG 63 107284	43	...	36	67	24	50	29	73	73	69	18	75	99	61	675	19	16 Nov
Wolterton Park	TG 63 165318	35	714	37	74	25	54	29	79	96	68	27	93	109	70	761	25	15 Jul
Ingworth R.F.M.Sta. Area 165 km²	TG 63 192296	12	699	...	...	...	...	...	...	...	...	...	...	...	...	...	...	...
Aylsham, Banningham Road	TG 63 200279	15	706	34	67	21	47	38	74	105	57	20	83	99	63	708	24	15 Jul
Woodgate House	TG 63 181259	30	692	32	65	21	42	41	73	91	56	21	79	99	66	687	19	16 Nov
Brampton	TG 63 223240	6	...	31	65	20	44	32	81	85	57	17	79	104	61	676	22	16 Nov
Hevingham	TG 63 172204	37	...	31	72	21	43	37	68	88	61	20	83	99	61	684	21	16 Nov
Coltishall	TG 63 262229	17	...	28	62	18	42	34	66	82	69	25	72	94	59	651	19	15 Jul
Sprowston	TG 63 251123	28	645	31	72	24	48	28	71	87	62	15	68	97	61	664	19	16 Nov
Hoveton Secondary School	TG 63 307187	9	...	32	74	28	39	27	71	88	71	18	68	94	56	666	17	15 Jul
Wroxham	TG 63 306174	3	...	34	77	24	41	30	68	90	63	19	72	95	60	671	15	15 Jul
Horning P.Sta.	TG 63 358164	3	629	32	74	30	46	26	79	93	57	14	76	102	58	688	18	15 Jul
Ant (Bure)																		
North Walsham, The Paston School	TG 63 280299	38	...	30	69	17	46	42	68	70	49	27	87	95	61	661	19	15 Jul
North Walsham	TG 63 285305	34	...	32	71	20	47	41	71	72	48	31	77	92	65	667	20	15 Jul
Edingthorpe	TG 63 304311	30	639	31	70	19	46	39	69	66	52	27	80	81	66	645	19	15 Jul
Barton Turf	TG 63 353217	8	...	27	68	16	40	30	60	93	72	13	76	88	63	646	24	15 Jul
Bure (Yare)																		
Burlingham	TG 63 373101	27	630	31	66	17	49	34	68	90	49	19	65	95	55	638	19	15 Jul
Ormesby St. Michael	TG 63 468152	7	638	34	68	23	52	47	67	90	52	15	87	128	73	736	19	15 Jul
Coastal																		
Gorleston	TG 63 534037	2	607	35	58	18	51	22	60	83	47	17	123	110	66	689	20	10 Oct
Lowestoft	TM 62 543947	25	600	34	55	14	49	28	47	91	59	17	129	116	72	710	25	3 Oct
Lowestoft, St. Margarets Road	TM 62 548942	26	589	22	54	15	47	28	49	89	56	15	126	106	59	665	25	3 Oct
Hundred																		
Ilketshall St.Lawrence	TM 62 382830	43	...	40	45	19	54	37	63	70	63	21	85	101	76	673	27	16 Jun
Sotterley Hall	TM 62 460851	15	...	34	51	19	56	(37)	56	75	72	24	112	109	83	728	35	3 Oct
Benacre P.Sta.	TM 62 536845	3	573	36	49	20	47	37	62	79	85	20	109	126	81	751	24	16 Jun
Blyth																		
Threadbare Hall	TM 62 362722	46	...	30	40	17	49	37	72	79	70	15	84	96	78	666	28	3 Oct
Wenhaston	TM 62 427755	15	...	38	51	17	49	42	73	63	61	17	105	99	78	693	38	3 Oct

Station	Grid reference	Altitude m	Av. 1916–1950 mm	Jan	Feb	Mar	Apr	May	Jun	Jul	Aug	Sep	Oct	Nov	Dec	Year	Highest recorded fall in a day mm	date
EAST SUFFOLK AND NORFOLK—*(contd)*																		
Minsmere																		
Yoxford	TM 62 397688	15	...	31	52	22	51	38	65	108	75	20	98	97	78	736	36	3 Oct
Coastal																		
Upper Abbey	TM 62 453645	18	575	27	47	22	49	39	59	89	84	18	90	96	79	697	31	16 Jun
Aldeburgh	TM 62 461583	6	...	32	51	26	52	42	54	95	75	15	101	117	83	742	37	22 Nov
Alde																		
Badingham	TM 62 309683	33	...	(25)	50	20	48	32	56	85	83	15	80	94	73	662	31	29 Aug
Great Glemham House	TM 62 345618	30	...	26	51	20	54	40	61	115	82	18	97	94	86	744	37	29 Aug
Farnham R.F.M. Sta. Area 64.0 km²	TM 62 360601	5	–	...	...	...	...	...	...	...	...	...	...	...	...	...	...	...
Ore (Alde)																		
Framlingham S.Wks.	TM 62 283622	21	645	28	50	24	59	39	43	121	83	15	88	101	92	742	41	5 Jul
Campsea Ashe	TM 62 339553	29	606	33	51	23	61	43	60	94	76	16	82	89	84	713	27	29 Aug
Alde																		
Aldeburgh, Linden Road	TM 62 452575	12	...	36	54	26	54	37	64	88	75	18	96	120	85	753	38	22 Nov
Aldeburgh, Clover House	TM 62 462564	9	...	37	54	31	62	39	62	95	81	17	111	120	79	788	32	22 Nov
Sudbourne Park	TM 62 407511	8	616	37	54	21	69	37	67	121	85	25	101	110	91	817	38	19 Jul
Butley Abbey Farm	TM 62 375491	8	600	32	49	15	69	35	68	107	77	27	97	93	81	751	38	19 Jul
Hollesley	TM 62 353451	17	...	30	54	11	65	40	64	115	65	22	110	87	86	748	36	19 Jul
Deben																		
Debenham S.Wks.	TM 62 184625	34	...	27	44	19	59	35	44	68	54	18	70	89	83	609	19	16 Jun
Kettleburgh	TM 62 258602	18	...	29	58	20	56	40	47	88	68	18	92	104	91	711	33	3 Oct
Charsfield	TM 62 244567	46	624	25	49	22	59	37	52	71	57	12	83	93	82	642	29	3 Oct
Melton, St.Audry's Hospital	TM 62 283516	27	...	29	50	19	55	37	59	93	67	15	81	93	84	681	25	19 Jul
Melton, Greylands	TM 62 285507	12	626	33	58	14	64	40	63	101	77	19	86	105	92	753	31	19 Jul
Melton, New Road	TM 62 281506	9	...	31	52	19	66	39	56	101	74	17	82	97	84	719	34	19 Jul
Coastal																		
Woodbridge, Warren Hill Road	TM 62 264487	32	616	35	54	19	71	41	61	102	80	15	84	98	86	745	28	29 Aug
Swilland	TM 62 184523	54	...	32	48	20	66	40	57	81	61	18	95	95	79	690	33	3 Oct
Grundisburgh	TM 62 230502	15	617	31	54	17	71	43	55	87	74	22	105	107	92	761	36	3 Oct
Martlesham	TM 62 245472	15	609	34	49	19	70	40	56	102	78	21	83	101	81	734	31	19 Jul
Ipswich, Heathfields Hospital	TM 62 191449	42	...	(31)	49	9	72	43	49	107	69	18	87	75	77	686	33	19 Jul
Trimley St.Martin	TM 62 273380	23	...	40	55	13	68	31	(63)	115	66	23	88	86	(79)	729	31	19 Jul
Gipping																		
Old Newton	TM 62 051630	62	618	28	43	20	50	32	55	67	57	21	100	75	86	634	18	18 Oct
New Bells Farm	TM 62 037640	61	...	22	44	19	42	31	51	61	49	19	82	69	84	571	(17	9 Dec)
Rattlesden School	TL 52 976588	61	...	24	44	14	51	32	47	84	50	23	66	63	79	576	16	3 Oct- 9 Dec
Wattisham	TM 62 026514	89	597	26	47	15	53	39	47	80	52	22	80	69	77	606	21	18 Oct
Cousins Hill	TM 62 117555	55	656	42	54	22	59	40	57	88	62	18	85	86	77	686	24	29 Aug
Westerfield P.Sta.	TM 62 165482	37	611	28	49	14	67	41	51	93	63	16	95	87	75	678	35	3 Oct
Whitton P.Sta.	TM 62 153483	24	634	31	50	15	69	43	51	91	64	17	104	86	77	699	40	3 Oct
Ipswich, Limerick Close	TM 62 140469	35	...	32	46	14	66	42	49	82	63	15	104	81	79	676	44	3 Oct
Ipswich, West End Depot	TM 62 151443	4	...	31	52	12	69	45	53	92	67	15	86	82	75	680	35	19 Jul
Ipswich, Park Road Res.	TM 62 163457	57	619	29	47	11	67	42	47	90	61	17	88	79	73	652	33	3 Oct
Belstead																		
Elmsett	TM 62 060469	61	...	25	43	11	49	33	63	105	51	23	81	70	66	621	43	19 Jul
Belstead Hall	TM 62 126411	38	597	31	49	15	63	49	56	91	57	23	94	77	75	679	25	19 Jul

ESSEX RIVER AUTHORITY AREA, 4192 SQUARE KILOMETRES

HYDROMETRIC AREAS 36 AND 37

Station	Grid reference	Altitude m	Av. 1916–1950 mm	Jan	Feb	Mar	Apr	May	Jun	Jul	Aug	Sep	Oct	Nov	Dec	Year	Highest recorded fall in a day mm	date
Stour																		
Burrough Green	TL 52 634554	116	...	28	49	10	56	37	62	87	69	17	71	49	79	617	28	22 Jun
Little Wratting	TL 52 700475	71	594	25	41	10	50	36	79	87	50	19	59	45	81	581	28	23 Jun
Haverhill W. Wks.	TL 52 659452	85	627	29	63	11	61	40	73	61	53	15	85	43	(80)	614	24	22 Jun
Haverhill S. Wks.	TL 52 680448	67	...	26	57	13	55	36	75	69	51	17	64	41	78	582	26	22 Jun
Poslingford	TL 52 768468	55	...	(25)	(44)	13	55	38	57	72	64	19	71	47	75	582	27	29 Aug
West Mill R.F.M. Sta. Area 224 km²	TL 52 827463	33	617	...	...	...	...	...	...	...	...	...	...	...	...	...	...	...
Glem (Stour)																		
Stradishall	TL 52 724518	113	634	25	44	13	52	39	83	103	62	21	58	49	81	630	33	20 Jun
Clopton Hall	TL 52 766547	110	655	27	47	12	52	44	56	97	68	20	79	58	88	649	18	31 Jul
Hawkedon	TL 52 797528	67	...	(25)	(44)	14	50	44	51	85	61	24	74	55	86	614	15	9 Dec
Boxted Hall	TL 52 828506	49	607	31	46	17	50	38	49	78	55	19	82	58	81	605	16	18 Oct
Glemsford Mill R.F.M. Sta. Area 87.3 km²	TL 52 846472	34	625	...	...	...	...	...	...	...	...	...	...	...	...	...	...	...
Stour																		
Long Melford	TL 52 847456	33	...	30	43	14	42	35	43	81	49	16	72	52	67	545	17	26 Jul
Chad (Stour)																		
Chadacre	TL 52 853521	74	...	28	47	17	49	44	49	83	57	20	70	58	91	613	20	6 Aug
Belchamp (Stour)																		
Halstead, Wickham Hall	TL 52 826371	76	570	31	54	12	57	36	46	79	51	20	80	48	77	593	31	19 Jul
Belchamp Hall	TL 52 826407	49	581	27	54	8	52	29	48	83	49	20	75	45	70	561	21	19 Jul
Bardfield Bridge R.F.M. Sta. Area 58.5 km²	TL 52 848422	27	587	...	...	...	...	...	...	...	...	...	...	...	...	...	...	...
Belchamp St.Paul	TL 52 797435	70	...	30	42	12	45	27	48	87	57	20	77	49	74	567	21	30 Aug
Stour																		
Sudbury, Shrublands	TL 52 868424	31	616	28	52	15	51	38	41	86	55	21	79	56	72	594	20	19 Jul
Sudbury W. Wks.	TL 52 872422	40	588	27	53	11	51	39	39	81	53	19	75	53	68	570	23	19 Jul
Bures, Dunstead House	TL 52 899361	34	564	31	51	11	59	46	41	93	58	17	88	54	73	625	42	19 Jul
Cock Street	TL 52 964364	55	...	31	52	11	70	54	55	91	61	15	88	66	74	667	30	19 Jul
Boxted Lodge	TL 52 998321	49	600	31	54	10	65	43	47	109	53	17	90	67	69	654	35	19 Jul
Boxted	TM 62 004325	47	604	31	54	10	65	43	44	107	48	(15)	(89)	62	66	634	37	5 Jul
Langham W. Wks.	TM 62 017340	13	578	30	47	14	57	44	45	91	53	16	104	63	68	631	28	19 Jul
Langham Mill R.F.M. Sta. Area 578 km²	TM 62 017340	6	610	...	...	...	...	...	...	...	...	...	...	...	...	...	...	...
Box (Stour)																		
Edwardstone	TL 52 944422	56	622	32	55	15	53	44	38	96	56	17	90	60	69	625	35	19 Jul
Polstead Ford R.F.M. Sta. Area 53.9 km²	TL 52 985378	16	617	...	...	...	...	...	...	...	...	...	...	...	...	...	...	...
Brett (Stour)																		
Thorpe Morieux	TL 52 928525	69	...	23	41	16	52	35	40	66	51	25	80	63	75	565	14	6 Aug
Cockfield	TL 52 899556	82	615	26	47	12	53	49	39	69	53	23	73	63	81	588	18	6 Aug

Station	Grid reference	Altitude m	Av. 1916–1950 mm	Jan	Feb	Mar	Apr	May	Jun	Jul	Aug	Sep	Oct	Nov	Dec	Year	Highest recorded fall in a day mm	date
ESSEX—*(contd)*																		
Washmere Green	TL 52 914477	76	...	27	47	14	49	38	35	82	57	17	72	57	74	570	20	26 Jul
Chelsworth	TL 52 983481	38	604	31	49	13	47	45	49	100	57	23	87	64	80	645	22	19 Jul
Hadleigh Mill R.F.M. Sta. Area 156 km^2	TM 62 025429	18	612	...	...	...	...	...	...	...	...	...	...	...	...	...	...	...
Holbecks	TM 62 020417	46	622	29	53	15	68	46	52	103	58	21	86	72	72	674	47	19 Jul
Stour																		
Stratford St. Mary R.F.M Sta. Area 844 km^2	TM 62 042340	6	615	...	...	...	...	...	...	...	...	...	...	...	...	...	...	...
East Bergholt	TM 62 077333	5	600	26	53	12	57	44	51	97	52	21	88	69	72	642	29	19 Jul
Dedham	TM 62 073317	35	584	28	56	11	57	45	60	100	56	28	85	70	75	673	36	19 Jul
Lawford Place	TM 62 093312	34	582	31	58	15	61	36	53	91	55	26	87	71	72	655	30	19 Jul
Coastal																		
Mistley	TM 62 108318	9	561	30	58	13	61	37	53	96	58	25	88	79	74	672	31	19 Jul
Harwich High School	TM 62 254315	15	...	34	58	12	66	28	59	109	68	17	100	89	74	715	32	19 Jul
Dovercourt	TM 62 242301	3	...	28	58	11	69	27	70	120	69	19	89	83	72	715	42	19 Jul
Walton-on-Naze	TM 62 259225	5	...	29	52	15	64	24	71	92	63	14	94	94	76	688	31	16 Jun
Walton-on-Naze, Columbine Gardens	TM 62 249213	15	...	33	55	13	66	26	69	90	63	15	95	88	71	685	24	29 Aug
Holland																		
Horsley Cross	TM 62 123274	37	...	28	58	12	61	31	58	106	57	21	76	74	67	650	37	19 Jul
Clacton Secondary School	TM 62 168159	16	...	39	50	11	68	22	67	89	55	16	81	88	69	653	22	16 Jun
Little Holland	TM 62 216172	4	555	38	53	11	69	26	65	76	81	15	96	89	73	692	37	29 Aug
Coastal																		
Clacton on Sea	TM 62 172143	21	547	28	49	10	63	24	62	89	55	16	87	85	76	643	26	5 Jul
Leewick Farm, St. Osyth	TM 62 107142	6	...	28	48	13	52	24	53	65	57	15	83	79	67	583	17	29 Aug
Ardleigh	TM 62 061281	37	563	35	63	12	63	37	57	98	53	18	91	74	76	678	32	19 Jul
Alresford S. Wks.	TM 62 079205	7	...	32	49	10	58	30	54	86	54	11	78	68	74	604	20	19 Jul
Colne																		
Great Yeldham	TL 52 757386	50	594	27	54	9	53	37	69	89	56	22	63	41	77	597	26	19 Jul
Pool Street R.F.M. Sta. Area 65.0 km^2	TL 52 771364	43	–	...	...	...	...	...	...	...	...	...	...	...	...	...	...	...
Halstead, Fenn Road	TL 52 823304	51	577	28	51	8	57	42	59	85	53	30	85	46	74	618	28	19 Jul
Halstead, Martlet's Hall	TL 52 800298	78	...	33	57	11	71	42	71	92	55	19	98	44	72	665	34	19 Jul
Earls Colne	TL 52 858288	51	608	31	50	9	69	52	56	97	49	19	96	49	75	653	33	19 Jul
Lexden R.F.M. Sta. Area 238 km^2	TL 52 962261	8	599	...	...	...	...	...	...	...	...	...	...	...	...	...	...	...
Colchester W. Wks.	TL 52 991254	12	580	31	60	10	66	41	57	92	59	21	86	64	71	658	30	19 Jul
Colchester, Shrub End	TL 52 981235	37	...	29	65	12	71	37	59	98	67	20	79	59	70	667	35	19 Jul
Colchester, Castle Park	TL 52 999255	27	574	30	60	10	67	38	56	98	53	17	85	61	75	651	31	19 Jul
Ardleigh, Hull Farm	TM 62 043276	37	559	31	61	8	62	41	57	99	58	20	91	73	75	675	33	19 Jul
Roman (Colne)																		
Stanway	TL 52 961237	37	575	35	63	10	63	42	59	91	52	19	95	60	81	672	31	19 Jul
Layer-de-la-Haye	TL 52 965196	44	548	32	67	12	75	39	66	106	55	13	88	63	81	696	35	19 Jul
Coastal																		
Tolleshunt Major	TL 52 897110	21	...	24	51	9	63	38	(57)	88	(51)	14	72	54	73	593	2	27 Mar
Great Totham	TL 52 853121	58	585	31	57	11	65	41	56	86	52	14	81	(51)	(76)	622	25	19 Jul
Pant																		
Copford Hall R.F.M. Sta. Area 62.4 km^2	TL 52 668313	58	614	...	...	...	...	...	...	...	...	...	...	...	...	...	...	...
Blackwater																		
Bocking, Churchstreet	TL 52 754256	46	...	26	52	11	63	41	85	91	57	23	84	39	74	645	33	19 Jul
Braintree, Coggeshall Road	TL 52 772235	68	614	35	48	10	71	47	61	103	67	30	107	44	74	697	30	19 Jul
Harboro Hall	TL 52 900186	55	579	31	56	7	67	47	47	93	63	13	88	54	79	643	34	19 Jul
Tiptree W. Wks.	TL 52 885167	73	572	32	60	9	72	43	54	87	55	15	99	61	79	667	28	19 Jul
Rivenhall	TL 52 823191	44	...	29	54	8	74	43	55	83	59	21	81	42	79	628	23	3 Oct
Appleford Bridge R.F.M. Sta. Area 247 km^2	TL 52 845158	15	605	...	...	...	...	...	...	...	...	...	...	...	...	...	...	...
Brain (Blackwater)																		
Great Lodge	TL 52 694290	79	570	29	56	11	63	39	73	87	71	22	61	38	75	624	24	29 Aug
Naylinghurst	TL 52 733221	70	...	30	52	10	65	44	78	90	54	28	89	36	74	651	37	20 Jun
Faulkbourne Hall	TL 52 803164	24	574	27	47	8	69	41	48	93	51	25	81	42	80	612	23	19 Jul
Guithavon Valley R.F.M. Sta. Area 60.6 km^2	TL 52 818147	16	597	...	...	...	...	...	...	...	...	...	...	...	...	...	...	...
Blackwater																		
Wickham Hall	TL 52 830105	21	544	28	52	10	65	44	51	84	52	14	77	44	74	594	24	19 Jul
Langford R.F.M. Sta. Area 337 km^2	TL 52 836092	6	597	...	...	...	...	...	...	...	...	...	...	...	...	...	...	...
Langford	TL 52 826095	12	553	27	54	10	65	42	49	77	53	16	73	48	74	587	22	19 Jul
Chelmer (Blackwater)																		
Thaxted P. Sta.	TL 52 606307	76	...	26	63	10	54	43	59	90	65	25	75	41	74	625	29	22 Jun
Chickney Hall	TL 52 576280	91	...	26	65	10	52	40	61	87	67	26	76	40	86	635	25	22 Jun
Churchend R.F.M. Sta. Area 72.5 km^2	TL 52 629233	52	–	...	...	...	...	...	...	...	...	...	...	...	...	...	...	...
Great Dunmow	TL 52 636206	49	...	31	59	11	65	45	64	76	63	31	59	42	83	629	27	19 Jul
Felsted School	TL 52 676205	73	599	29	61	10	70	46	58	78	60	37	68	46	90	653	22	19 Jul
Felsted, Sugar Factory	TL 52 664211	61	568	34	58	10	60	43	56	77	52	32	58	43	88	612	29	19 Jul
Howe Street	TL 52 695149	43	...	25	54	11	66	52	70	78	58	25	66	38	80	623	19	22 Jun 19 Jul
Can (Chelmer)																		
Writtle	TL 52 677069	32	586	27	56	8	66	41	51	78	51	32	80	38	73	601	19	22 Jun
Wid (Can)																		
Millfield	TQ 51 571968	101	...	32	64	13	73	38	52	63	56	29	97	44	75	637	22	22 Oct
Hutton. S. Wks.	TQ 51 651957	47	...	32	51	12	85	38	62	74	67	33	93	47	86	680	25	29 Aug
Writtle Mill R.F.M. Sta. Area 136 km^2	TL 52 686060	27	620	...	...	...	...	...	...	...	...	...	...	...	...	...	...	...
Chelmer (Blackwater)																		
Beach's Mill R.F.M. Sta. Area 228 km^2	TL 52 690072	23	617	...	...	...	...	...	...	...	...	...	...	...	...	...	...	...
Longstomps Res.	TL 52 703045	69	601	32	57	10	71	49	62	85	56	27	91	45	79	665	25	3 Oct
Sandford Mill W. Wks.	TL 52 738060	21	554	28	55	8	64	40	76	89	49	19	91	45	74	639	20	19 Jul
Galleywood	TL 52 712042	58	601	33	58	8	71	45	62	81	55	23	91	46	76	649	19	19 Jul
Sandon (Chelmer)																		
Downham P. Sta.	TQ 51 723970	56	594	33	54	9	86	43	68	70	61	18	79	51	81	654	19	19 Apr
South Hanningfield Res.	TQ 51 745991	48	582	32	55	9	78	38	69	69	57	17	95	49	77	646	20	19 Jul
Sandon Bridge R.F.M. Sta. Area 60.6 km^2	TL 52 755055	20	577	...	...	...	...	...	...	...	...	...	...	...	...	...	...	...

Station	Grid reference	Altitude m	Av. 1916–1950 mm	Jan	Feb	Mar	Apr	May	Jun	Jul	Aug	Sep	Oct	Nov	Dec	Year	Highest recorded fall in a day mm	date
ESSEX—*(contd)*																		
Chelmer (Blackwater)																		
Hatfield Peverel, Rickstones	TL 52 784101	30	...	29	55	7	67	40	62	83	55	16	88	47	79	628	22	19 Jul
Sandon (Chelmer)																		
Rushes Lock R.F.M. Sta. Area 534 km^2	TL 52 794090	11	602	...	...	...	...	...	...	...	...	...	...	...	...	...	...	...
Ter (Chelmer)																		
Whitelands	TL 52 783135	40	575	33	58	11	69	43	61	86	63	26	98	47	86	681	25	29 Aug
Crabb's Bridge R.F.M. Sta. Area 77.7 km^2	TL 52 786107	15	587	...	...	...	...	...	...	...	...	...	...	...	...	...	...	...
Coastal																		
Maldon	TL 52 863065	4	525	28	50	9	69	37	46	71	52	15	88	47	76	587	20	19 Jul
Purleigh	TL 52 827023	52	...	31	58	8	79	43	55	75	60	15	94	51	82	653	18	19 Jul
Bradwell on Sea	TL 52 979054	24	559	27	51	9	76	35	58	80	58	9	77	62	67	609	19	29 Aug
Southminster Hall	TQ 51 965994	12	...	28	55	8	73	33	58	77	70	12	79	64	71	629	22	22 Oct
Burnham Farm	TQ 51 962958	2	542	28	51	7	75	28	64	83	78	10	78	66	74	642	20	19 Apr 22 Oct
Burnham, Laboratory	TQ 51 947957	0	523	26	53	6	75	24	62	77	69	10	75	64	65	607	20	22 Oct
Crouch																		
Basildon S. Wks.	TQ 51 737908	12	...	31	53	11	92	32	87	79	54	15	87	45	72	658	25	20 Jun
Thundersley	TQ 51 787890	76	...	33	55	9	91	33	55	97	68	16	88	51	71	666	22	6 Aug
Wickford	TQ 51 772921	14	...	28	51	9	87	31	69	93	60	15	90	49	72	652	22	19 Apr
Battlesbridge	TQ 51 777956	23	566	29	54	8	84	38	71	92	55	18	95	53	84	680	28	5 Jul
Coastal																		
Prittlewell S. Wks.	TQ 51 880875	15	...	31	48	9	79	27	52	96	58	17	76	65	74	630	38	19 Jul
Rayleigh, Alanbank	TQ 51 800902	80	587	33	57	9	95	33	57	109	71	17	90	45	77	695	23	19 Apr
Eastwood, Progress Road	TQ 51 832888	46	...	33	51	9	86	31	57	93	70	17	95	54	69	664	16	19 Jul
Potton Island	TQ 51 949903	2	562	26	50	5	71	19	61	108	65	12	90	71	70	648	32	19 Jul
Shoeburyness	TQ 51 948857	3	515	33	48	8	66	20	53	103	67	11	93	64	70	636	39	19 Jul
Thorpe Bay	TQ 51 916861	8	...	35	51	9	72	19	64	104	63	15	80	67	73	653	40	19 Jul
Southend W. Wks.	TQ 51 874853	37	537	29	50	10	73	28	54	100	58	15	77	65	71	630	38	19 Jul
Canvey Island P. Sta.	TQ 51 800828	0	...	27	52	13	66	28	47	99	63	14	85	51	69	612	31	19 Jul
Vange P. Sta.	TQ 51 709860	13	531	33	56	7	90	31	58	93	64	13	82	45	62	635	27	19 Jul
Stanford-le-Hope S.Wks.	TQ 51 686816	6	...	30	56	13	90	33	51	102	81	22	84	41	71	675	34	6 Aug
Marsh Farm S. Wks.	TQ 51 654756	3	518	30	47	10	79	31	53	95	60	19	77	39	74	615	25	19 Jul
Tilbury, Beaconsfield Road	TQ 51 640766	2	532	36	51	6	83	26	60	95	60	20	63	35	61	597	25	19 Jul
Mar																		
Upminster, Howard Road	TQ 51 562868	24	...	38	56	9	95	31	69	71	79	32	84	36	71	671	30	29 Aug
Bury Farm S. Wks.	TQ 51 606867	7	518	28	50	10	101	29	46	71	75	31	77	41	77	636	28	19 Apr
Grays	TQ 51 620796	24	520	29	47	7	76	27	54	90	69	25	58	36	64	581	25	19 Jul
Stifford P. Sta.	TQ 51 590802	5	...	29	59	11	95	26	58	83	74	31	61	38	71	636	24	22 Jun
Coastal																		
Rainham, Recreation Ground	TQ 51 540824	7	560	39	61	8	88	(32)	61	74	(76)	27	80	35	72	654	28	22 Jun
Inglebourne																		
Harold Wood	TQ 51 549901	28	557	(31)	52	9	94	33	(63)	59	(76)	29	98	44	79	667	(22	3 Oct)
Upminster, Acacia Drive	TQ 51 552860	17	...	38	58	9	97	30	61	68	79	29	82	33	72	655	30	29 Aug
Upminster, Roxburgh Avenue	TQ 51 567858	34	...	41	63	12	96	31	51	75	87	32	91	47	75	700	33	29 Aug
Beam																		
Romford, Raphael Park	TQ 51 519901	33	573	32	50	13	86	34	79	65	72	26	99	39	68	663	31	22 Jun
Romford, Danes Road	TQ 51 502879	17	...	32	55	10	93	36	69	64	72	24	92	39	70	657	33	22 Jun
Romford, Castellan Avenue	TQ 51 531901	37	...	33	54	11	91	31	77	63	72	23	93	(38)	72	659	30	22 Jun
Hornchurch, Langtons Gardens	TQ 51 537875	27	558	33	59	11	(89)	31	66	68	76	25	86	35	73	652	29	22 Jun
Bretons Farm S. Wks.	TQ 51 514850	6	...	33	65	9	98	33	57	69	70	28	95	36	76	669	31	22 Jun
Dagenham, Salvage Depot	TQ 51 499863	16	...	35	51	8	94	32	63	65	61	27	93	33	74	636	32	22 Jun
Dagenham, Marley School	TQ 51 496839	3	585	31	59	9	99	33	62	66	70	27	102	37	73	668	31	22 Jun
Roding																		
Great Canfield	TL 52 581185	82	613	31	51	8	76	47	59	77	66	30	65	37	79	627	26	29 Aug
Fyfield	TL 52 573078	51	...	31	59	13	75	49	68	78	73	30	66	42	81	665	26	29 Aug
High Ongar Bridge R.F.M. Sta. Area 95.1 km^2	TL 52 561041	41	—	...	...	...	...	...	...	...	...	...	...	...	...	...	...	...
Hastingwood	TL 52 479071	98	...	30	72	15	77	44	75	81	72	36	81	39	79	701	31	22 Jun
Little End	TQ 51 546999	38	636	33	57	15	69	38	58	74	73	27	79	46	78	647	27	22 Jun
Theydon Bois	TQ 51 457987	75	...	29	60	12	81	42	70	67	80	32	76	41	68	657	34	22 Jun
Birch Hall	TQ 51 439991	85	621	33	66	11	79	47	75	74	81	37	87	43	73	705	33	22 Jun
High Beech Res.	TQ 51 410980	115	674	35	72	12	82	56	90	77	86	33	86	44	76	750	41	22 Jun
Woodford, Eastern S. Wks.	TQ 51 418907	13	...	34	59	9	85	39	73	58	77	30	81	40	70	654	37	22 Jun
Redbridge R.F.M. Sta. Area 303 km^2	TQ 51 415884	6	635	...	...	...	...	...	...	...	...	...	...	...	...	...	...	...
Wanstead, Redbridge	TQ 51 415882	8	577	32	53	9	88	36	58	66	78	31	82	37	75	644	33	22 Jun
Clayhall Park	TQ 51 425893	33	610	35	58	10	88	39	72	66	73	32	93	42	74	682	38	22 Jun
City of London Cemetery	TQ 51 418863	18	577	38	53	12	95	40	62	73	86	35	96	39	78	708	36	22 Jun
Valentines Park	TQ 51 434872	13	576	35	59	9	88	33	62	77	73	31	88	40	75	669	35	22 Jun
Loxford Park	TQ 51 442852	9	600	36	61	11	94	38	61	63	71	29	89	38	77	671	33	22 Jun
East Ham Well	TQ 51 438842	5	563	35	59	11	88	32	56	57	72	30	93	37	76	647	31	22 Jun
East Ham S. Wks.	TQ 51 434831	6	...	33	56	9	91	33	54	60	67	32	96	36	72	639	33	22 Jun
Barking P. Sta.	TQ 51 447830	2	568	37	61	9	92	35	58	60	68	28	93	(38)	74	652	33	22 Jun
Goodmayes Park	TQ 51 468868	12	585	33	58	13	93	33	64	71	70	(29)	81	43	75	664	36	22 Jun
Coastal																		
Barking Outfall	TQ 51 452822	7	556	37	57	6	94	31	54	61	69	27	95	35	67	633	31	22 Jun

LEE CONSERVANCY AREA, 1419 SQUARE KILOMETRES

HYDROMETRIC AREA 38

Station	Grid reference	Altitude m	Av. 1916–1950 mm	Jan	Feb	Mar	Apr	May	Jun	Jul	Aug	Sep	Oct	Nov	Dec	Year	Highest recorded fall in a day mm	date
Lea																		
Sundon Park	TL 52 052254	122	649	36	71	14	74	53	97	78	99	40	105	52	77	795	41	29 Aug
Luton, Runley Wood	TL 52 064217	137	673	40	67	13	88	52	95	98	93	38	90	54	87	813	44	29 Aug
Luton, Runley Wood	TL 52 064217	137	673	31	72	15	77	54	100	107	77	43	104	55	78	815	38	29 Aug
Chaul End	TL 52 056215	187	...	36	79	13	77	58	101	96	88	47	109	56	86	847	43	29 Aug
Luton P.Sta.	TL 52 121179	101	669	34	87	20	76	52	90	91	81	37	98	52	85	802	37	10 Jun
Hatfield, Clarke's Road	TL 52 229090	82	669	29	86	16	81	63	85	61	89	34	(89)	(44)	(76)	752	38	29 Aug
Bedwell Park	TL 52 277073	110	679	34	87	14	84	58	95	60	90	30	113	41	71	778	39	22 Jun
Mimram (Lea)																		
Breachwood Green	TL 52 155222	146	673	35	81	15	80	51	102	93	105	33	83	55	86	818	38	10 Jun
Luton	TL 52 118216	158	...	40	67	13	88	52	95	98	93	38	90	54	87	813	...	...
Kimpton	TL 52 175182	88	662	34	100	14	83	59	93	89	86	35	77	49	86	804	33	10 Jun
Welwyn, Mill Lane	TL 52 232160	67	...	30	91	10	81	56	88	74	168	31	87	47	79	842	84	21 Aug
Welwyn S.Wks.	TL 52 233157	79	...	29	87	12	76	50	(89)	65	154	33	79	40	76	790	75	21 Aug

Station	Grid reference	Altitude m	Av. 1916–1950 mm	Jan	Feb	Mar	Apr	May	Jun	Jul	Aug	Sep	Oct	Nov	Dec	Year	Highest recorded fall in a day mm	Highest recorded fall in a day date
LEE—*(contd)*																		
Digswell P.Sta.	TL 52 239153	65	645	28	84	11	74	55	94	69	143	32	83	39	78	791	63	21 Aug
Panshanger Park R.F.M.Sta. Area 134 km²	TL 52 282133	47	665	...	...	...	...	...	...	...	...	...	...	...	...	...	...	...
Beane (Lea)																		
Ardeley Bury	TL 52 300270	109	678	32	88	7	65	54	84	86	102	31	81	47	98	776	33	22 Jun
Walkern Bury	TL 52 305262	122	...	28	90	8	62	56	90	82	112	26	85	46	94	779	34	22 Jun
Whitehall P.Sta.	TL 52 284220	70	646	26	85	13	70	53	96	70	102	26	84	44	87	756	35	22 Jun
Broomin Green	TL 52 229241	94	648	33	86	9	67	49	83	74	107	35	75	42	82	741	41	21 Aug
Pin Green	TL 52 250257	138	656	34	90	9	64	49	(89)	(81)	113	34	82	43	84	771	37	21 Aug
Knebworth	TL 52 254199	92	652	32	91	(13)	72	57	82	(70)	(140)	26	79	48	83	793	(21	5 Nov)
Rib (Lea)																		
Chipping	TL 52 355323	104	637	27	73	9	58	44	75	88	98	27	74	39	81	695	40	22 Jun
Meesden	TL 52 426327	137	654	33	76	8	63	48	63	95	78	24	92	48	84	711	33	22 Jun
Anstey	TL 52 405329	125	...	26	79	8	50	52	65	99	79	24	89	44	88	703	33	22 Jun
Stonebury	TL 52 387281	117	679	25	92	9	63	55	76	91	94	21	93	46	90	754	43	29 Aug
Braughing, Friars	TL 52 420245	116	667	23	72	10	65	46	78	81	98	26	97	46	94	735	43	29 Aug
St.Edmund's College	TL 52 371221	117	669	23	76	10	(66)	49	79	69	84	27	99	40	84	705	38	22 Jun
Standon P.Sta.	TL 52 382226	81	...	23	75	13	65	51	91	77	79	26	102	43	85	728	39	22 Jun
Barwick	TL 52 385194	58	632	27	69	12	71	46	88	69	77	26	98	45	79	707	38	22 Jun
Herts Training School R.F.M.Sta. Area 148 km²	TL 52 335158	41	650	...	...	...	...	...	...	...	...	...	...	...	...	...	...	...
Lea																		
Hertford S.Wks.	TL 52 338134	37	623	27	79	9	75	52	77	69	76	33	93	42	74	706	31	22 Jun
Ware, Chadwell Spring	TL 52 350137	43	605	24	77	7	69	51	77	68	75	33	91	41	71	683	33	22 Jun
Amwell End P.Sta.	TL 52 360139	32	612	25	81	9	70	48	76	68	77	32	100	48	79	713	33	22 Jun
Ash (Lea)																		
Hadham Hall	TL 52 453228	103	643	28	69	15	59	42	72	76	102	29	89	(46)	(83)	710	50	29 Aug
Moor Place	TL 52 422187	82	658	28	77	13	66	41	83	82	79	28	82	47	86	713	36	22 Jun
Hadham P.Sta.	TL 52 420166	52	...	24	76	12	63	40	98	75	70	29	82	44	79	693	37	22 Jun
Easneye R.F.M.Sta. Area 85.2 km²	TL 52 380138	32	648	...	...	...	...	...	...	...	...	...	...	...	...	...	...	...
Lea																		
Hoddesdon	TL 52 371105	47	...	36	73	10	72	47	89	70	71	33	87	42	73	700	36	22 Jun
Stort (Lea)																		
Clavering Hall	TL 52 477307	91	614	29	71	7	64	44	69	77	77	26	82	40	85	672	36	22 Jun
Burton Bower	TL 52 526241	102	...	29	(65)	10	52	42	54	84	84	27	80	42	93	662	35	29 Aug
Stansted Nurseries	TL 52 510241	79	655	27	65	11	54	39	53	82	80	31	83	42	87	654	33	29 Aug
Maze Green P.Sta.	TL 52 474214	102	637	26	72	12	58	42	73	81	87	31	84	40	95	699	29	22 Jun
Bishop's Stortford College	TL 52 481213	76	635	25	71	13	68	35	70	71	77	38	83	42	91	683	30	22 Jun
Bishop's Stortford, Castle Mound ...	TL 52 490214	68	637	21	73	11	56	39	63	72	88	31	81	41	88	663	40	29 Aug
Stansted	TL 52 531226	101	632	29	78	11	62	40	57	81	80	26	77	43	89	672	37	29 Aug
Sheering Mill Lock	TL 52 488144	47	633	31	66	8	63	39	77	74	74	38	82	39	76	666	28	29 Aug
Hatfield Broad Oak	TL 52 553177	91	640	29	61	10	64	40	69	81	73	33	89	44	83	678	29	22 Jun
Sawbridgeworth P.Sta.	TL 52 464130	55	...	26	62	8	62	37	81	90	72	33	75	39	82	666	33	22 Jun
Eastwick Lodge	TL 52 445109	40	...	24	63	12	75	43	77	87	72	38	82	43	69	685	36	22 Jun
Harlow	TL 52 470101	79	617	24	57	15	66	40	71	84	69	37	82	40	73	657	31	22 Jun
Roydon P.Sta.	TL 52 412102	37	591	25	63	10	74	42	73	82	78	39	77	41	68	671	34	22 Jun
Stanstead Abbots	TL 52 393105	30	...	26	68	10	73	43	81	68	73	31	73	40	72	657	33	22 Jun
Lea																		
Feilde's Weir R.F.M.Sta. Area 1040 km²	TL 52 391092	28	649	...	...	...	...	...	...	...	...	...	...	...	...	...	...	...
Feilde's Weir	TL 52 392092	29	614	31	67	11	76	44	86	71	72	31	76	40	(76)	682	31	22 Jun
Hoddesdon, High Street	TL 52 372085	40	...	32	72	15	79	41	85	76	62	31	71	40	69	673	33	22 Jun
Broxbournebury	TL 52 364073	53	...	34	71	13	81	45	81	72	70	35	84	43	79	706	32	22 Jun
Cheshunt Yard	TL 52 353035	35	629	31	71	12	75	45	81	81	82	36	79	38	72	702	29	22 Jun 29 Aug
Cheshunt Cemetery	TL 52 346021	38	...	33	74	12	85	45	81	84	84	35	88	38	75	734	30	22 Jun
Waltham Abbey, Gunpowder Factory	TL 52 375013	20	617	30	65	12	73	45	79	82	72	32	77	38	68	674	30	22 Jun
Waltham Abbey P.Sta.	TL 52 375003	19	607	29	65	10	74	42	83	89	75	31	84	36	67	684	29	22 Jun 29 Aug
Epping Upland	TL 52 437049	94	652	27	66	11	74	47	83	85	67	33	77	42	69	681	36	22 Jun
Darnicle Hill P.Sta.	TL 52 309048	73	674	32	84	12	94	53	104	78	89	36	98	38	77	794	34	22 Jun
Northaw	TL 52 283024	91	687	32	79	13	90	57	112	85	93	41	105	37	73	817	38	22 Jun
Whitewebbs Park	TQ 51 326989	37	638	34	77	13	89	51	83	(76)	87	31	88	37	78	743	32	22 Jun
Chingford P.Sta.	TQ 51 374949	13	...	31	65	12	76	41	68	75	85	31	87	37	67	675	29	22 Jun
Enfield, Southbury Road	TQ 51 332965	32	641	32	72	13	83	45	74	108	75	32	84	34	71	723	46	5 Jul
Chingford Mill	TQ 51 363925	12	599	33	67	11	80	40	62	57	69	33	87	37	72	649	29	22 Jun
Chingford, Crescent Road	TQ 51 394945	55	623	36	70	12	84	47	77	67	80	31	82	40	73	699	35	22 Jun
Buckhurst Hill Res.	TQ 51 407938	84	639	37	69	10	84	43	70	54	74	33	85	39	75	673	36	22 Jun
Pymmes Brook (Lea)																		
East Barnet, Sports Ground	TQ 51 262968	70	660	30	74	13	95	56	71	95	102	29	100	37	81	783	38	29 Aug
Muswell Hill	TQ 51 278909	64	681	34	84	10	101	52	73	92	110	35	93	43	80	805	39	29 Aug
Broomfield Park	TQ 51 304925	41	652	35	81	12	97	53	75	95	91	33	87	44	81	785	34	22 Jun
Hadley Road P.Sta.	TQ 51 303980	49	...	32	72	13	91	46	87	89	81	30	95	39	77	751	33	22 Jun
Southgate	TQ 51 299952	67	636	34	75	13	89	50	74	94	77	30	80	38	74	729	34	22 Jun
Crouch End, Priory Park	TQ 51 300891	36	...	39	83	11	98	52	85	77	107	32	93	42	83	801	42	22 Jun
Hornsey, High Street	TQ 51 307894	33	645	37	77	8	95	49	84	77	99	31	83	38	79	757	38	29 Aug
Lea																		
Woodford, Haggar Lane	TQ 51 392901	54	622	38	70	7	86	46	58	64	93	30	84	45	77	698	37	29 Aug
Walthamstow, Ferry Lane	TQ 51 350892	7	590	35	77	8	88	39	65	61	72	33	93	39	79	689	29	22 Jun
Stoke Newington Filters	TQ 51 319868	31	607	38	74	8	87	44	71	65	89	33	92	39	81	723	36	29 Aug
Stoke Newington Res.	TQ 51 325874	32	608	35	72	9	82	43	74	61	82	33	85	37	78	692	31	29 Aug
Lea Bridge W.Wks.	TQ 51 358867	7	584	37	64	9	100	36	61	52	71	31	95	34	69	658	29	22 Jun
Leyton S.Wks.	TQ 51 376862	8	611	41	68	12	101	41	55	55	73	29	96	42	83	697	27	29 Aug
Hackney, Clapton Pond	TQ 51 349860	21	608	37	69	10	96	42	63	58	82	33	100	37	79	707	29	29 Aug
Old Ford, Penstock House	TQ 51 369839	12	...	35	64	12	94	41	65	54	88	31	92	39	81	697	31	29 Aug
West Ham, Abbey Mills	TQ 51 387832	5	615	37	71	9	97	41	57	59	90	29	92	41	85	707	30	29 Aug

LONDON, LEFT BANK AREA, 432 SQUARE KILOMETRES

HYDROMETRIC AREA 39 (Part)

Station	Grid reference	Altitude m	Av. 1916–1950 mm	Jan	Feb	Mar	Apr	May	Jun	Jul	Aug	Sep	Oct	Nov	Dec	Year	Highest recorded fall in a day mm	Highest recorded fall in a day date
Coastal																		
Bethnal Green, Victoria Park	TQ 51 355835	13	641	42	71	13	107	44	70	60	88	31	95	39	84	743	33	29 Aug
Camden Square	TQ 51 296845	37	639	36	72	9	96	40	73	65	90	34	90	37	80	723	37	29 Aug
Finsbury, New River Head	TQ 51 314828	26	...	38	63	10	100	46	75	58	83	31	89	35	76	704	35	22 Jun
Regent's Park	TQ 51 283826	39	654	41	72	14	110	43	69	68	97	32	100	41	81	770	37	29 Aug
Buckingham Palace	TQ 51 287793	11	619	40	68	10	95	42	75	75	100	29	90	38	79	741	42	29 Aug
St.James's Park	TQ 51 298800	8	595	42	65	14	97	40	77	81	90	33	90	36	80	746	34	29 Aug

Station	Grid reference	Altitude m	Av. 1916–1950 mm	Jan	Feb	Mar	Apr	May	Jun	Jul	Aug	Sep	Oct	Nov	Dec	Year	Highest recorded fall in a day mm	date
LONDON, LEFT BANK—*(contd)*																		
Westminster P.Sta.	TQ 51 286781	5	570	35	62	8	84	35	71	82	107	29	80	32	68	693	47	29 Aug
Kensington Palace	TQ 51 259801	25	638	39	68	8	94	41	62	70	94	33	89	37	78	713	42	29 Aug
Chelsea Physic Garden	TQ 51 277778	6	592	40	63	9	94	40	69	80	95	33	92	35	76	726	41	29 Aug
Holland Park	TQ 51 246795	21	626	38	66	10	93	42	63	67	97	32	83	34	73	699	46	29 Aug
Campden Hill Res.	TQ 51 250801	42	610	34	63	9	88	42	62	65	93	33	82	34	74	679	42	29 Aug
Chelsea College	TQ 51 259773	4	...	46	60	11	79	33	60	68	82	(33)	(91)	(36)	75	674	33	29 Aug
Fulham, Bishop's Park	TQ 51 241764	5	602	39	67	9	93	39	61	68	90	35	84	33	78	695	40	29 Aug
Hammersmith P.Sta.	TQ 51 232782	6	...	40	70	11	100	46	68	70	91	36	87	39	84	741	41	29 Aug
Acton P.Sta.	TQ 51 212797	7	...	33	72	13	100	47	68	72	88	30	95	35	87	740	35	29 Aug
Acton Cemetery	TQ 51 205818	29	...	33	73	7	86	47	68	68	84	29	87	31	72	683	33	22 Jun
Brent																		
Totteridge...	TQ 51 259942	70	703	43	82	14	112	56	78	102	104	33	105	49	95	873	38	29 Aug
Mill Hill Cemetery	TQ 51 231917	78	...	41	82	10	108	55	65	89	115	33	102	45	83	828	54	29 Aug
Hampstead Garden Suburb	TQ 51 257883	79	...	37	81	9	100	49	73	98	109	32	99	45	82	814	42	29 Aug
Finchley, Avenue House	TQ 51 252902	85	682	42	89	12	106	59	74	88	109	36	107	49	89	859	40	29 Aug
Hampstead	TQ 51 262863	137	685	40	78	10	97	47	65	91	111	31	98	40	82	793	43	29 Aug
Golders Hill Park	TQ 51 256870	94	679	42	84	10	104	54	63	89	102	31	97	42	81	799	41	29 Aug
Hendon Depot...	TQ 51 239875	48	665	36	79	8	94	47	55	84	102	31	99	40	75	751	42	29 Aug
Stanmore, Uxbridge Road	TQ 51 159920	86	703	38	76	11	106	58	79	96	91	30	133	42	78	838	42	29 Aug
Edgware, Chandos Park	TQ 51 189911	55	...	35	74	8	101	52	69	86	90	26	117	40	80	777	43	29 Aug
Edgwarebury Park	TQ 51 189932	67	...	33	72	11	100	54	76	116	99	29	135	42	85	851	49	29 Aug
Brent Res....	TQ 51 208870	42	639	36	75	9	93	55	70	72	94	29	105	43	78	758	39	29 Aug
Harrow, Courtfield Avenue	TQ 51 158885	57	...	36	74	8	104	58	64	81	84	27	118	38	76	769	36	29 Aug
Brent Town Hall	TQ 51 197866	57	614	39	74	9	99	54	73	65	95	31	109	38	75	761	41	29 Aug
Willesden Res....	TQ 51 226842	59	653	35	83	11	94	48	58	76	96	27	102	39	74	744	41	29 Aug
Stonebridge Park	TQ 51 204848	30	...	35	81	8	101	48	58	70	92	30	113	36	68	740	36	29 Aug
Ealing, Castlebar	TQ 51 170819	37	...	38	88	12	97	56	72	77	85	33	117	38	82	795	32	22 Jun
Hanwell, Brentside School	TQ 51 151821	16	...	37	78	9	99	52	65	69	90	28	102	38	77	743	35	29 Aug
Harrow, Oakhurst	TQ 51 148863	90	...	39	82	9	105	57	67	77	87	33	124	40	75	795	34	22 Jun
South Harrow	TQ 51 144860	65	...	39	80	9	108	52	65	77	91	33	126	39	76	793	31	22 Jun
Coastal																		
Lampton Park Nursery	TQ 51 134763	23	...	34	68	0	94	50	61	83	80	28	97	36	68	698	28	29 Aug
Crane																		
Pinner Cemetery	TQ 51 130893	57	...	35	70	9	97	55	81	95	88	27	130	42	74	803	42	29 Aug
Eastcote	TQ 51 110881	53	...	35	69	9	96	57	74	95	84	29	118	40	78	784	37	29 Aug
Northolt	TQ 51 097846	40	635	41	75	9	89	55	71	78	98	32	123	40	75	784	45	29 Aug
Harrow, Roxeth	TQ 51 140869	37	...	40	76	11	99	54	67	76	86	31	126	43	77	786	34	29 Aug
Wood End Nurseries	TQ 51 094813	40	...	35	76	12	98	48	69	87	87	31	127	41	81	791	35	29 Aug
Heathrow	TQ 51 076768	24	603	33	73	12	99	51	67	78	86	37	122	42	63	761	37	29 Aug
Harmondsworth	TQ 51 055773	24	640	31	66	12	85	53	71	84	86	36	125	32	62	743	35	29 Aug
Longford	TQ 51 060759	23	615	31	74	12	90	54	69	72	85	35	122	39	60	741	33	29 Aug
Isleworth, Mogden Wks.	TQ 51 154753	10	607	30	71	11	97	51	66	73	79	30	101	37	70	715	33	29 Aug
Coastal																		
Teddington, Stanley Road	TQ 51 150718	13	...	35	69	8	97	52	62	72	100	28	103	37	67	729	33	29 Aug

THAMES CONSERVANCY AREA, 9950 SQUARE KILOMETRES

HYDROMETRIC AREA 39

Station	Grid reference	Altitude m	Av. 1916–1950 mm	Jan	Feb	Mar	Apr	May	Jun	Jul	Aug	Sep	Oct	Nov	Dec	Year	Highest recorded fall in a day mm	date
Thames																		
Shorncote S. Wks	SU 41 034971	94	...	49	100	17	102	56	37	54	93	41	130	47	83	808	39	29 Aug
Churn (Thames)																		
Baunton W. Wks.	SP 42 019047	121	831	55	107	28	97	63	33	69	111	35	110	58	101	865	55	29 Aug
Cirencester, Chesterton Grove	SP 42 022009	123	817	55	109	23	100	64	31	61	103	40	116	56	(83)	842	44	29 Aug
South Cerney	SU 41 050993	111	...	45	90	16	83	58	31	52	44	41	128	45	32	664	21	18 Oct
Latton W. Wks.	SU 41 080968	84	726	35	94	13	91	47	37	48	97	35	130	41	78	746	48	29 Aug
Thames																		
Purton	SU 41 080872	134	691	31	96	19	84	49	47	57	121	35	125	43	77	785	72	29 Aug
Wroughton	SU 41 151802	137	752	37	96	16	90	55	42	59	104	42	123	44	68	776	31	21 Aug
Swindon, Town Gardens	SU 41 151834	145	737	31	103	15	78	52	41	62	75	34	119	41	64	715	37	29 Aug
Kempsford	SU 41 148972	79	720	39	101	15	84	58	42	57	86	40	121	39	75	756	38	29 Aug
Coln (Thames)																		
Brockhampton Park	SP 42 034225	186	...	40	107	22	82	61	48	65	85	36	119	60	127	853	29	18 Oct
Bibury W. Wks.	SP 42 111072	119	...	46	106	26	93	62	36	67	89	35	124	53	99	834	36	29 Aug
Bibury R.F.M. Sta. Area 107 km²	SP 42 122062	101	945	...	...	...	...	...	...	...	...	...	...	...	...	...	...	...
Cole (Thames)																		
Chisledon	SU 41 190800	177	...	48	104	20	94	69	45	77	87	36	144	48	76	847	22	18 Oct
Great Coxwell	SU 41 269939	116	653	33	99	15	99	51	53	61	68	44	111	43	76	753	22	22 Jun
Thames																		
Lechlade	SU 41 222990	72	665	39	106	12	100	51	42	69	83	43	148	42	76	811	29	4 Oct
Leach (Thames)																		
Kelmscott	SU 41 245993	70	655	36	91	11	89	45	40	59	72	42	133	43	74	737	25	6 Aug
Thames																		
Buckland	SU 41 342981	92	...	29	95	14	105	44	51	71	72	45	129	40	68	762	25	29 Aug
Windrush (Thames)																		
Dickler (Windrush)																		
Upper Swell W. Wks.	SP 42 175270	146	...	38	112	31	91	57	45	68	76	24	104	49	113	807	23	9 Dec
Windrush (Thames)																		
Little Rissington	SP 42 209191	226	772	41	113	19	80	62	47	68	76	26	110	43	101	785	23	29 Aug
Milton Under Wychwood	SP 42 237157	162	734	40	111	19	82	64	54	63	80	24	127	44	93	801	24	29 Aug
Newbridge R.F.M. Sta. Area 363 km²	SP 42 402019	63	785	...	...	...	...	...	...	...	...	...	...	...	...	...	...	...
Thames																		
Farmoor	SP 42 452064	63	...	25	79	10	88	52	58	75	81	58	108	31	79	743	31	29 Aug
Swinford P. Sta.	SP 42 446085	62	650	29	79	12	93	53	51	78	77	54	107	31	83	746	31	30 Sep
Eynsham R.F.M. Sta. Area 1620 km²	SP 42 445087	60	770	...	...	...	...	...	...	...	...	...	...	...	...	...	...	...
Evenlode (Thames)																		
Moreton in Marsh S. Wks.	SP 42 216320	128	...	30	97	21	81	48	42	56	73	22	100	42	101	713	32	29 Aug
Adlestrop	SP 42 237278	143	--	34	99	25	86	51	55	63	73	29	106	51	102	773	26	29 Aug
Chipping Norton, Stonecroft	SP 42 312264	210	740	35	113	17	85	60	47	70	78	25	118	42	93	782	27	29 Aug
Chadlington	SP 42 333219	125	...	36	109	16	85	69	52	54	66	24	120	40	92	762	24	29 Aug

Station	Grid reference	Altitude m	Av. 1916–1950 mm	Jan	Feb	Mar	Apr	May	Jun	Jul	Aug	Sep	Oct	Nov	Dec	Year	Highest recorded fall in a day mm	date
THAMES—*(contd)*																		
Glyme (Evenlode)																		
Heythrop College	SP 42 363265	183	763	32	111	18	99	65	50	59	81	30	124	39	91	799	31	29 Aug
Woodstock	SP 42 455172	95	683	29	97	15	87	66	52	68	80	53	101	29	85	763	31	29 Aug
Thames																		
Oxford	SP 42 509072	63	652	31	75	10	81	54	57	81	79	52	122	28	80	750	35	29 Aug
Osney Lock	SP 42 503059	57	636	31	74	11	90	56	63	74	77	56	122	32	83	770	28	29 Aug
Cherwell (Thames)																		
Chipping Warden	SP 42 493492	137	...	(25)	78	17	76	47	61	64	68	40	105	41	87	709	28	13 Oct
Grimsbury P. Sta.	SP 42 458417	91	...	19	74	15	67	47	52	64	83	28	98	42	71	661	47	29 Aug
Sor (Cherwell)																		
Bodicote P. Sta.	SP 42 460371	91	677	26	83	20	81	50	49	63	86	35	117	39	78	727	42	29 Aug
West Adderbury	SP 42 466355	103	666	24	92	20	81	54	54	67	85	39	118	40	83	757	45	29 Aug
Cherwell (Thames)																		
Enslow Mill R.F.M.Sta. Area 552 km²	SP 42 482183	65	715	...	...	...	...	...	...	...	...	...	...	...	...	...	...	...
Hampton Poyle	SP 42 499154	64	644	32	79	16	100	57	57	80	84	71	105	33	85	797	38	29 Aug
Ray (Cherwell)																		
Woodham (Ray)																		
Doddershall, Lower South Farm	SP 42 725194	82	658	28	78	16	86	62	60	71	91	48	111	39	80	771	44	29 Aug
Shipton Lee (Woodham)																		
Shipton Lee, Middle Farm	SP 42 728211	93	--	28	67	15	80	64	63	68	98	50	107	42	71	754	50	29 Aug
Shipton Lee, Woodlands Farm	SP 42 714213	84	...	31	77	15	89	61	72	68	107	49	123	43	77	813	58	29 Aug
Middle Claydon, Knowlhill Farm	SP 42 707238	91	...	29	74	14	81	69	58	69	100	47	109	38	72	760	49	29 Aug
Grendon Underwood, Lower Greatmoor Farm	SP 42 705222	76	--	29	84	15	85	69	65	60	105	45	117	41	75	790	63	29 Aug
Finmerhill House	SP 42 715224	128	...	31	76	16	80	59	65	63	115	47	113	43	74	782	65	29 Aug
Edgcott, Prune Farm	SP 42 692223	76	664	29	83	16	85	69	67	64	102	43	123	43	76	798	56	29 Aug
Grendon Underwood R.F.M.Sta. Area 18.6 km²	SP 42 680211	66	660	...	...	...	...	...	...	...	...	...	...	...	...	...	...	...
Kingswood	SP 42 689191	78	650	30	70	15	77	63	73	65	99	39	137	37	82	787	49	29 Aug
Edgcott, Lawn House	SP 42 681222	76	...	26	68	15	80	69	62	60	96	40	130	38	73	756	47	29 Aug
Grendon Underwood	SP 42 677215	67	632	26	80	15	83	67	63	65	91	37	132	39	74	770	45	29 Aug
Ray (Cherwell)																		
Middleton Park	SP 42 526234	99	692	24	96	17	85	67	54	60	88	59	112	39	89	789	40	29 Aug
Cherwell (Thames)																		
Oxford, Norham Road	SP 42 514078	61	649	35	74	12	92	59	57	81	80	55	127	31	85	786	31	29 Aug
Headington	SP 42 557071	101	--	40	82	13	102	63	56	94	91	43	119	36	91	831	45	29 Aug
Thames																		
Brasenose Res.	SP 42 557061	112	633	35	71	9	89	59	56	93	81	38	108	31	85	755	35	29 Aug
Littlemore P. Sta.	SP 42 526029	59	...	31	69	12	93	61	68	79	69	48	113	33	81	757	27	29 Aug
Kennington Nursery	SP 42 518029	69	652	35	75	11	94	64	66	73	71	51	113	33	85	772	31	29 Aug
Sandford S. Wks.	SP 42 542019	64	...	28	63	13	82	53	63	70	59	40	105	31	72	678	25	29 Aug
Sandford Lane S. Wks.	SP 42 528008	67	628	30	69	9	90	53	63	74	60	45	104	31	81	710	25	29 Aug
Wootton W. Wks.	SP 42 481007	81	638	32	65	9	88	58	61	76	81	47	114	35	74	741	43	29 Aug
Ock (Thames)																		
Kingstone Lisle	SU 41 328882	128	...	42	101	15	105	52	54	65	59	67	(140)	(51)	(76)	827	28	30 Sep
Letcombe Regis	SU 41 381865	107	741	41	87	16	90	66	66	71	69	55	145	41	82	827	36	29 Aug
Frilford Heath	SU 41 447983	73	624	33	70	11	93	50	54	67	77	56	120	36	75	742	37	29 Aug
Abingdon R.F.M.Sta. Area 234 km²	SU 41 486969	51	681	...	...	...	...	...	...	...	...	...	...	...	...	...	...	...
Abingdon, The Firs	SU 41 486970	55	612	34	69	10	91	55	54	65	81	49	116	36	73	734	47	29 Aug
Abingdon	SU 41 482990	64	611	33	66	9	90	55	61	72	77	47	119	35	73	737	41	29 Aug
Abingdon, Albert Park	SU 41 491973	58	...	35	70	10	91	53	54	64	82	48	119	37	76	738	45	29 Aug
Thames																		
Abingdon S. Wks.	SU 41 493952	50	...	29	68	14	83	54	52	58	79	43	109	35	71	697	45	29 Aug
Wantage, Kitford Gardens	SU 41 428867	122	737	38	86	17	98	63	61	81	83	49	131	39	74	820	52	29 Aug
Steventon	SU 41 466921	64	645	26	67	12	84	54	50	53	73	42	106	(38)	(76)	681	39	29 Aug
Drayton S. Wks.	SU 41 482933	58	587	27	64	11	76	54	45	53	73	42	113	33	72	663	36	29 Aug
Didcot Power Sta.	SU 41 514916	53	...	26	65	10	78	54	50	83	77	47	111	33	71	705	45	29 Aug
Didcot S. Wks.	SU 41 527908	52	657	28	79	17	89	67	54	86	68	37	130	37	86	780	33	29 Aug
Day's Weir R.F.M. Sta. Area 3440 km²	SU 41 568935	46	734	...	...	...	...	...	...	...	...	...	...	...	...	...	...	...
Thame (Thames)																		
Dancers End P. Sta.	SP 42 906089	198	...	33	93	19	93	70	86	88	93	49	126	53	104	907	44	29 Aug
Bulbourne	SP 42 931136	122	737	29	76	18	88	61	97	87	76	49	107	52	72	811	40	29 Aug
Mentmore Gardens	SP 42 907196	125	671	26	62	14	88	58	90	70	72	60	114	58	75	787	34	22 Jun
Aylesbury S. Wks.	SP 42 791145	76	649	29	63	18	78	49	66	58	61	66	131	47	91	757	40	30 Sep
Quinton	SP 42 748202	102	678	32	76	16	89	68	68	67	88	50	104	46	79	784	39	29 Aug
Green Hailey	SP 42 828031	245	...	45	97	19	93	70	96	117	71	58	138	56	91	952	40	22 Jun
Princes Risborough	SP 42 798029	101	703	31	92	12	78	63	90	71	69	56	123	50	87	821	36	22 Jun
Monks Risborough W. Wks.	SP 42 814046	80	713	31	86	13	77	62	84	81	70	56	113	52	81	807	34	22 Jun
Wheatfield	SU 41 687996	79	662	26	71	10	89	57	77	95	78	61	152	46	83	847	35	1 Oct
Shirburn Castle	SU 41 696959	107	...	30	74	11	87	55	71	84	67	72	135	55	74	815	37	30 Sep
Britwell Salome	SU 41 673934	113	680	32	74	15	91	54	90	83	67	61	156	54	77	855	34	30 Sep
Pyrton Manor	SU 41 686956	99	685	28	74	13	88	60	76	87	65	67	141	56	79	833	34	22 Jun
Cuxham	SU 41 664953	84	642	26	68	13	87	56	72	91	68	56	153	48	79	818	28	29 Aug
Thames																		
Benson	SU 41 631909	67	617	27	69	12	73	47	69	75	63	48	159	45	64	751	29	22 Jun
Cholsey, School House	SU 41 587868	46	683	30	81	11	71	55	65	86	59	50	158	37	75	778	26	22 Jun
Cholsey S. Wks.	SU 41 590872	46	...	31	78	11	75	52	67	84	57	53	149	39	73	767	28	22 Jun
Goring	SU 41 599806	48	...	35	91	15	85	53	61	70	66	62	143	43	70	795	27	22 Jun
Pang (Thames)																		
Compton	SU 41 519802	103	...	34	83	16	89	65	74	108	112	46	140	39	68	874	67	29 Aug
Cold Ash	SU 41 518696	146	...	41	107	15	93	65	68	71	108	47	130	49	69	863	41	29 Aug
Yattendon Court	SU 41 558743	104	699	40	95	15	105	65	64	91	88	56	127	46	76	868	30	29 Aug
Bradfield P. Sta.	SU 41 599731	67	659	32	88	15	105	47	68	67	83	55	119	41	65	786	29	22 Jun
Pangbourne	SU 41 635763	43	...	31	90	14	116	50	65	75	77	62	128	38	67	813	30	19 Apr
Thames																		
Englefield	SU 41 625720	53	689	36	88	17	113	52	81	76	78	58	121	44	69	833	29	22 Jun
Reading, Highways Depot	SU 41 708739	40	...	33	85	13	96	51	60	78	85	44	127	41	65	777	28	22 Jun
Caversham Lock	SU 41 720740	37	636	33	81	12	91	50	60	80	81	41	127	40	63	757	31	22 Jun
Kennet (Thames)																		
Overton Down	SU 41 130706	232	...	47	100	13	124	73	75	83	150	42	163	62	95	1026	...	
Marlborough	SU 41 185686	129	853	51	106	15	106	85	59	75	140	44	165	51	83	980	48	21 Aug
Aldbourne	SU 41 260756	140	807	51	110	21	108	71	65	95	101	53	155	47	80	956	40	29 Aug
Kintbury, Holt Road	SU 41 390665	111	...	53	114	17	101	65	58	55	104	52	160	46	74	900	29	2 Oct

GENERAL TABLE, ENGLAND AND WALES

Station	Grid reference	Altitude m	Av. 1916–1950 mm	Jan	Feb	Mar	Apr	May	Jun	Jul	Aug	Sep	Oct	Nov	Dec	Year	Highest recorded fall in a day mm	date
THAMES—*(contd)*																		
Inkpen	SU 41 380646	149	857	54	113	16	108	69	58	57	115	48	162	49	78	928	35	29 Aug
Newbury, Gloucester Road	SU 41 465669	76	768	47	118	18	115	67	77	66	119	52	139	46	76	939	43	29 Aug
Newbury S.Wks.	SU 41 477674	70	740	42	115	17	113	65	78	59	117	52	127	47	73	905	42	29 Aug
Lambourn (Kennet)																		
Fognam Down P.Sta.No.1	SU 41 296802	145	...	45	104	15	95	70	61	77	65	48	151	48	78	858	25	18 Oct
East Shefford	SU 41 385737	119	...	43	87	18	96	60	76	70	119	61	141	46	78	895	32	29 Aug
Peasemore House	SU 41 461770	145	753	48	100	20	97	62	74	108	129	56	162	49	77	983	65	29 Aug
Shaw R.F.M. Sta. Area 234 km²	SU 41 470682	76	805	...	...	...	...	...	...	...	...	...	...	...	...	...	...	...
Priors Court	SU 41 487738	116	706	41	99	17	91	66	80	78	93	47	135	46	70	864	39	29 Aug
Embourn (Kennet)																		
Kingsclere	SU 41 504580	162	865	61	135	23	121	80	92	70	138	61	139	50	83	1053	37	29 Aug
Kennet (Thames)																		
Theale R.F.M. Sta. Area 1030 km²	SU 41 649708	44	808	...	...	...	...	...	...	...	...	...	...	...	...	...	...	...
Tilehurst Res.	SU 41 662729	91	...	35	87	15	94	55	66	73	84	40	122	40	68	779	25	22 Jun
Reading W. Wks.	SU 41 702727	66	643	36	89	15	97	55	61	84	84	47	133	37	59	799	25	22 Jun
Thames																		
Reading, Seed Trial Grounds	SU 41 738738	43	616	33	83	12	91	49	62	84	85	40	123	42	63	767	31	22 Jun
Reading	SU 41 738748	45	653	33	88	13	94	53	65	82	83	39	131	39	61	782	33	22 Jun
Sonning	SU 41 762753	51	...	33	77	9	91	51	62	78	65	38	121	40	63	727	31	22 Jun
Loddon (Thames)																		
West Ham P. Sta.	SU 41 625523	98	738	42	125	15	110	74	75	93	75	37	106	47	73	873	26	22 Jun
Basingstoke Cemetery	SU 41 625518	93	767	46	126	16	117	75	75	91	76	33	114	46	72	886	25	22 Jun
Nately Hatch	SU 41 708533	99	740	48	119	17	108	68	56	77	63	45	126	49	72	848	26	22 Jun
Heckfield	SU 41 714626	78	668	48	102	21	117	54	68	79	81	42	137	42	63	852	29	22 Jun
Blackwater (Loddon)																		
Aldershot Res.	SU 41 859502	120	745	41	118	9	96	64	65	95	71	43	105	53	77	839	...	...
Aldershot W. Wks.	SU 41 863493	77	751	38	116	9	90	68	67	94	69	44	107	49	66	818	35	5 Jul
Frimley P. Sta.	SU 41 887562	65	708	40	108	15	111	59	75	89	78	37	124	54	73	863	35	22 Jun
South Farnborough	SU 41 867544	69	685	35	104	13	96	53	62	94	69	34	111	50	62	783	33	22 Jun
Camberley S. Wks.	SU 41 862597	60	686	32	87	14	102	63	69	65	70	28	97	47	65	740	38	22 Jun
Camberley, Woodlands Road	SU 41 867600	66	...	39	94	16	109	68	72	70	75	29	102	50	72	795	36	22 Jun
Broadmoor	SU 41 852640	91	692	38	81	12	117	65	66	78	78	27	107	43	66	779	34	22 Jun
Finchampstead	SU 41 793633	65	...	38	94	13	112	62	72	79	65	31	116	45	67	793	33	22 Jun
Whitewater (Blackwater)																		
Upton Grey House Farm	SU 41 696484	117	...	49	126	13	105	66	62	78	75	44	130	46	85	878	25	19 Jul
Swainshill	SU 41 720441	183	836	52	144	19	109	74	73	92	87	47	122	57	96	973	...	...
Long Sutton	SU 41 749467	156	796	51	122	13	100	75	81	98	70	33	133	54	81	912	30	22 Jun
Greywell P. Sta.	SU 41 722513	80	717	45	118	14	102	61	58	79	55	37	120	49	69	806	28	22 Jun
Lodge Farm R.F.M. Sta. Area 44.5 km²	SU 41 735524	72	790	...	...	...	...	...	...	...	...	...	...	...	...	...	...	...
Odiham S. Wks.	SU 41 739521	73	697	44	113	13	104	66	63	101	67	39	135	47	73	866	41	19 Jul
Odiham	SU 41 737494	116	...	37	114	12	97	70	72	87	63	34	125	49	68	827	31	22 Jun
Hartley Wintney S. Wks.	SU 41 768580	55	693	29	95	14	98	67	65	76	58	32	129	46	61	770	30	22 Jun
Blackwater (Loddon)																		
Swallowfield R.F.M.Sta. Area 355 km²	SU 41 731648	42	719	...	...	...	...	...	...	...	...	...	...	...	...	...	...	...
Loddon (Thames)																		
Shinfield	SU 41 734686	61	655	30	81	13	89	44	67	72	71	39	125	38	60	728	31	22 Jun
Shinfield Grange	SU 41 740688	45	...	33	84	14	90	49	64	70	67	38	116	38	61	723	31	22 Jun
Winnersh	SU 41 778709	43	...	32	78	13	88	53	61	80	76	42	105	42	63	733	30	22 Jun
Reading, Earley	SU 41 737708	78	...	33	91	17	92	49	64	82	79	41	138	40	63	789	31	22 Jun
Wokingham, Schiehallion	SU 41 826685	64	...	34	86	19	95	57	80	77	90	36	116	43	65	800	34	22 Jun
Wellington College	SU 41 833635	88	691	38	87	13	105	65	62	70	68	27	100	43	61	739	31	22 Jun
Wokingham, Bucks Copse	SU 41 804681	53	642	33	87	17	88	59	69	86	84	39	107	43	64	776	33	5 Jul
Wokingham, Reading Road	SU 41 794697	55	646	37	83	15	98	58	66	77	79	48	105	44	65	775	31	22 Jun
Beenham's Heath	SU 41 850750	42	663	35	77	11	99	55	69	67	71	38	115	46	73	757	34	22 Jun
Hare Hatch	SU 41 798779	55	681	34	83	12	105	55	77	89	76	40	126	51	78	826	35	22 Jun
Wargrave, Piggott School	SU 41 785774	38	...	32	81	7	88	51	70	82	71	35	118	43	61	740	33	22 Jun
Thames																		
Stoke Row	SU 41 677840	176	755	35	102	14	96	70	61	80	81	70	175	52	81	917	37	4 Oct
Henley on Thames, Greys Lane	SU 41 757822	37	671	33	97	10	108	57	67	79	80	44	118	52	73	819	36	22 Jun
Henley on Thames, Longlands	SU 41 762827	46	666	33	102	13	110	57	64	77	78	44	120	53	73	824	33	22 Jun
Christmas Common	SU 41 715931	241	...	42	95	13	88	73	79	79	87	61	134	58	92	902	...	...
Greenlands	SU 41 774855	37	...	28	99	12	103	56	70	87	87	43	127	51	75	837	35	22 Jun
Hambleden Lock	SU 41 782851	32	715	29	99	11	105	52	70	77	83	39	130	52	73	819	34	22 Jun
Hamble Brook (Thames)																		
Wellground	SU 41 746951	151	...	38	102	13	101	71	87	85	83	64	135	56	101	935	...	...
Turville Wood	SU 41 750919	107	...	40	109	14	95	(63)	81	83	(86)	63	141	54	98	927	...	...
Northend Farm	SU 41 735928	213	...	42	111	15	93	66	85	86	87	65	141	56	97	946	...	...
Fingest	SU 41 772908	78	...	34	104	14	88	62	83	79	84	50	139	51	90	878	...	...
Cholsey Grange	SU 41 754938	213	...	41	113	14	96	70	86	81	86	64	141	57	105	953	...	...
Frieth	SU 41 794904	177	...	36	101	15	90	(63)	(81)	91	95	47	142	52	92	907	...	...
Colstrope	SU 41 780882	58	...	34	99	12	91	58	75	87	98	44	153	52	75	878	...	...
Thames																		
Medmenham	SU 41 805839	30	...	32	88	12	101	50	72	77	98	40	120	49	70	809	35	22 Jun
Marlow, Bovingdon Place	SU 41 833872	91	...	37	95	15	100	56	83	83	90	47	134	47	86	872	41	22 Jun
Hurley	SU 41 823829	43	665	33	84	10	102	48	77	78	83	38	119	44	64	781	37	22 Jun
Marlow Bottom	SU 41 843893	79	...	36	98	17	95	51	80	93	96	47	132	47	89	880	37	22 Jun
Bourne End P. Sta.	SU 41 892881	38	694	29	79	14	93	51	79	98	90	42	109	42	77	802	34	22 Jun
Wye (Thames)																		
Gurdons Farm	SU 41 749974	236	792	40	100	12	91	76	86	93	87	71	149	52	102	959	...	...
Bolter End	SU 41 796925	191	782	37	106	17	89	63	92	89	86	53	131	51	111	926	...	...
Hughenden P. Sta.	SU 41 864964	92	...	35	109	21	92	74	92	95	91	54	120	52	105	940	36	22 Jun
Hedsor Mill R.F.M. Sta. Area 137 km²	SU 41 896867	27	772	...	...	...	...	...	...	...	...	...	...	...	...	...	...	...
Thames																		
Cookham	SU 41 902850	27	662	33	75	12	96	48	85	88	82	39	109	42	69	777	33	22 Jun
Boulter's Lock	SU 41 903824	25	650	32	75	13	91	44	81	85	79	35	125	43	65	767	35	22 Jun
Taplow Court	SU 41 907821	65	647	34	79	13	93	42	80	85	72	37	119	45	67	767	33	22 Jun
The Cut (Thames)																		
Bracknell	SU 41 874693	74	...	35	83	15	102	55	72	73	79	35	119	43	66	777	31	22 Jun
Bracknell S. Wks.	SU 41 858718	49	...	33	76	10	95	53	64	67	82	32	110	42	61	724	31	22 Jun
Maidenhead W. Wks.	SU 41 882813	38	...	32	81	13	90	46	82	77	79	33	115	48	71	767	...	...
Maidenhead S. Wks.	SU 41 891805	25	629	32	75	13	96	45	81	76	68	31	121	47	68	756	33	22 Jun

Station	Grid reference	Altitude m	Av. 1916–1950 mm	Jan	Feb	Mar	Apr	May	Jun	Jul	Aug	Sep	Oct	Nov	Dec	Year	Highest recorded fall in a day mm	date
THAMES—*(contd)*																		
Thames																		
Bray Weir R.F.M. Sta. Area 6920 km²	SU 41 909796	21	731	...	...	...	...	...	...	...	...	...	...	...	...	...	...	...
Bray	SU 41 908795	23	...	33	71	13	101	45	81	78	67	37	120	48	67	761	33	22 Jun
Burnham W. Wks.	SU 41 929829	55	653	32	79	13	99	41	80	89	81	37	125	42	70	790	33	22 Jun
Eton, Tangier Island	SU 41 968777	20	658	33	71	11	103	49	89	65	73	45	132	41	62	776	37	22 Jun
Cippenham S. Wks.	SU 41 948794	22	...	29	71	13	96	49	82	66	74	38	122	39	60	739	28	29 Aug
Slough, Stoke Road Cemetery	SU 41 976812	29	...	28	72	10	90	49	95	69	77	41	119	40	68	758	33	22 Jun
Slough, Ragstone Road	SU 41 969793	20	...	33	72	14	99	49	85	67	71	39	117	41	64	751	37	22 Jun
Upton	SU 41 978791	24	645	34	68	13	96	49	88	54	75	43	118	42	65	748	36	22 Jun
Windsor Royal Gardens	SU 41 979754	21	655	36	80	12	96	57	82	62	79	59	109	37	58	767	35	30 Sep
Slough, St. Bernard's Convent	SU 41 990796	24	...	28	74	13	96	50	98	61	75	39	122	42	63	761	34	22 Jun
Ditton Park	TQ 51 001778	21	654	34	72	15	99	56	79	65	78	42	109	37	62	747	30	22 Jun
Old Windsor	SU 41 987741	18	638	36	77	13	102	54	68	65	75	43	97	41	63	734	31	22 Jun
Colne Brook (Thames)																		
Beaconsfield, Twin Oaks	SU 41 939912	103	...	32	87	18	106	55	81	90	102	45	109	47	85	859	37	22 Jun
Hedgerley Hill	SU 41 967864	91	700	38	77	12	97	54	88	100	89	43	124	42	83	846	31	22 Jun
Fulmer	TQ 51 030865	73	...	39	84	11	99	51	92	93	100	33	138	42	85	867	35	29 Aug
Thames																		
Egham W. Wks.	TQ 51 022717	15	658	31	95	11	106	53	59	77	71	44	118	38	70	773	27	22 Jun 29 Aug
Colne (Thames)																		
Barnet, Old Fold View	TQ 51 233966	110	722	35	81	13	103	62	69	101	125	(33)	115	33	81	853	...	...
Potters Bar, Playing Fields	TL 52 243024	79	662	39	78	15	96	63	97	83	102	39	120	38	78	847	33	29 Aug
North Mimms P. Sta.	TL 52 232036	77	653	33	76	13	91	59	81	68	86	35	113	43	75	774	33	29 Aug
Tyttenhanger P. Sta.	TL 52 198057	73	647	27	79	11	85	58	79	60	79	34	87	36	70	705	26	29 Aug
Sandridge (Colne)																		
Hatfield P. Sta.	TL 52 220077	104	671	30	79	16	84	57	77	64	83	33	92	40	70	725	33	29 Aug
Roe Green	TL 52 214073	81	...	27	74	14	80	53	79	58	79	33	86	35	69	686	29	29 Aug
Rothamsted	TL 52 132134	128	701	32	100	17	85	65	79	76	80	37	87	47	84	788	31	29 Aug
Rothamsted No.2	TL 52 132134	128	722	33	100	18	83	67	79	79	81	40	90	50	88	808	31	29 Aug
St.Albans	TL 52 182076	83	639	28	82	15	85	55	91	65	81	35	82	37	70	726	27	10 Jun
St.Albans, Hospital	TL 52 176068	90	...	32	84	15	90	59	94	66	83	38	90	41	71	764	27	10 Jun
Colne (Thames)																		
Bushey Heath Res.	TQ 51 152941	158	682	34	80	10	97	60	86	91	111	34	124	42	77	846	54	29 Aug
Boreham Wood, Furzehill Road	TQ 51 193964	92	726	35	81	15	106	67	78	117	96	37	134	45	88	899	45	5 Jul
Boreham Wood, Shenley Road	TQ 51 194965	91	...	36	76	8	100	63	74	108	92	35	127	42	72	834	39	5 Jul
Ver (Colne)																		
Markyate W. Wks.	TL 52 051158	164	744	29	88	18	72	54	88	72	90	39	99	53	96	797	35	29 Aug
Flamstead	TL 52 090149	107	...	32	88	17	76	50	86	83	86	39	102	52	82	792	33	29 Aug
Studham Hall	TL 52 012153	194	...	31	82	20	84	56	77	86	91	44	120	55	85	831	41	29 Aug
Stonecross P. Sta.	TL 52 151081	128	670	29	86	15	84	61	110	71	73	38	89	37	74	769	31	22 Jun
St.Albans, Holywell Hill	TL 52 146066	76	...	35	97	17	98	60	97	79	83	42	99	42	80	828	31	22 Jun
Hansteads R.F.M. Sta. Area 132 km²	TL 52 151016	61	719	...	...	...	...	...	...	...	...	...	...	...	...	...	...	...
Colne (Thames)																		
Garston	TL 52 123016	78	686	34	81	15	87	51	90	79	98	38	103	38	75	791	45	29 Aug
Watford, Callowland	TQ 51 112993	76	...	31	71	13	84	51	83	85	94	42	104	36	74	770	39	29 Aug
Berrygrove R.F.M. Sta. Area 352 km²	TQ 51 123982	55	699	...	...	...	...	...	...	...	...	...	...	...	...	...	...	...
Hilfield Park Res.	TQ 51 153960	85	...	33	76	12	91	59	81	121	105	34	113	41	77	843	63	5 Jul
Aldenham School	TQ 51 157972	94	...	36	75	15	95	57	90	85	82	34	109	40	78	797	29	22 Jun
Watford, Aldenham Road	TQ 51 120958	62	668	33	77	12	84	54	75	85	106	38	106	39	74	782	55	29 Aug
Watford W. Wks.	TQ 51 114958	52	...	32	77	11	96	56	81	80	108	40	105	40	73	797	55	29 Aug
Northwood, St.Mary's Avenue	TQ 51 091923	70	685	35	78	9	91	52	68	92	95	39	116	41	77	793	...	...
Gade (Colne)																		
Little Gaddesden P. Sta.	SP 42 997132	186	...	36	91	21	84	60	84	97	113	45	111	59	91	893	61	29 Aug
Ashridge Park	SP 42 996113	168	...	35	95	21	74	61	86	101	105	45	109	53	97	884	53	29 Aug
Piccotts End	TL 52 048092	91	731	31	105	20	76	62	76	77	82	38	119	46	89	823	33	22 Jun
New Ground P. Sta.	SP 42 950107	132	750	29	96	20	86	68	97	90	84	52	116	55	93	885	37	29 Aug
Tring, Cowroast Lock	SP 42 957104	119	751	31	86	22	72	63	92	96	89	45	111	54	94	857	40	29 Aug
Berkhamsted P. Sta.	SP 42 992078	149	...	29	98	16	71	56	77	107	91	37	120	48	95	847	43	29 Aug
Kings Langley	TL 52 073026	85	760	36	90	17	97	65	77	84	99	51	117	40	88	861	50	29 Aug
Hunton Bridge P. Sta.	TL 52 083007	66	...	40	89	16	91	55	79	84	89	43	116	42	91	834	39	29 Aug
The Grove W. Wks.	TQ 51 087990	58	...	33	81	14	92	57	88	96	95	44	117	40	83	840	38	29 Aug
Bovingdon	TL 52 007038	157	734	32	104	20	91	74	73	85	91	39	112	47	97	865	41	29 Aug
Chess (Colne)																		
Hawridge P. Sta.	SP 42 956055	127	791	33	109	19	87	60	81	95	102	46	109	51	107	900	51	29 Aug
The Lee	SP 42 899042	191	791	37	118	15	90	69	73	93	92	43	127	56	111	923	44	29 Aug
Chesham S. Wks.	SP 42 975000	91	...	35	99	16	91	74	74	97	86	41	118	44	86	863	34	29 Aug
Amersham	SU 41 969986	152	816	36	104	15	101	69	82	114	87	45	123	47	94	917	37	22 Jun
Chorley Wood P. Sta.	TQ 51 037977	67	...	36	85	12	86	61	67	72	111	43	123	47	83	827	47	29 Aug
Colne (Thames)																		
Batchworth P. Sta.	TQ 51 063940	48	695	35	76	12	95	53	63	93	110	45	123	44	81	829	52	29 Aug
Batchworth Heath	TQ 51 076925	111	...	32	63	8	91	47	70	94	95	41	119	43	64	766	45	29 Aug
Maple Lodge S. Wks.	TQ 51 037922	42	...	31	76	11	89	50	56	93	91	40	119	44	76	777	32	5 Jul
Denham R.F.M. Sta. Area 743 km²	TQ 51 052864	34	726	...	...	...	...	...	...	...	...	...	...	...	...	...	...	...
Misbourne (Colne)																		
Wendover P. Sta.	SP 42 879039	148	791	38	109	19	98	70	75	103	93	42	122	56	106	932	42	29 Aug
Amersham P. Sta.	SU 41 961970	89	...	32	94	15	95	62	83	102	90	46	121	47	85	873	40	29 Aug
Little Chalfont	SU 41 988968	130	759	36	92	17	98	64	69	99	102	45	122	44	89	879	41	29 Aug
Chalfont St. Peter	TQ 51 004925	104	741	38	90	14	97	54	72	93	108	43	116	46	88	859	39	29 Aug
Denham	TQ 51 033879	64	660	37	76	8	106	52	78	98	100	41	(121)	(41)	79	837	40	29 Aug
Fray (Colne)																		
Cowley	TQ 51 053818	29	...	33	78	10	92	58	71	73	88	36	120	41	70	771	39	29 Aug
Ruislip, Manor Farm	TQ 51 090877	45	666	35	77	10	101	56	68	103	92	36	126	44	78	825	42	29 Aug
Ickenham P. Sta.	TQ 51 074875	46	...	34	70	9	95	49	67	95	84	36	116	44	77	775	38	29 Aug
Hillingdon, Bishopshalt School	TQ 51 069827	55	...	33	70	10	82	53	69	74	86	35	113	43	69	737	25	22 Jun
West Drayton P. Sta.	TQ 51 062802	30	...	34	76	12	90	50	68	74	81	37	113	39	73	747	34	29 Aug
Thames																		
Abbey Mead	TQ 51 046677	16	599	30	74	11	101	46	67	79	87	25	96	37	65	717	33	29 Aug
Bourne (Thames)																		
Sunninghill Cedars	SU 41 939686	84	...	37	91	14	124	54	68	89	77	53	109	46	76	839	31	22 Jun
Blacknest P. Sta.	SU 41 959687	40	...	39	87	13	125	56	62	79	73	45	104	45	71	800	29	22 Jun
Chertsey, Council Depot	TQ 51 049662	13	...	31	77	14	102	47	68	81	91	25	100	45	65	747	35	29 Aug

Station	Grid reference	Altitude m	Av. 1916–1950 mm	Jan	Feb	Mar	Apr	May	Jun	Jul	Aug	Sep	Oct	Nov	Dec	Year	Highest recorded fall in a day mm	date
THAMES—*(contd)*																		
Thames																		
Shepperton Lock	TQ 51 074659	12	638	33	71	16	94	50	64	79	76	26	104	46	71	731	32	29 Aug
Wey (Thames)																		
Lasham P. Sta.	SU 41 681412	124	864	50	150	15	110	72	65	90	76	45	146	56	96	971	27	22 Jun
Alton W. Wks.	SU 41 722389	134	...	53	171	15	109	78	63	67	81	42	135	65	92	971	32	9 Feb
Rotherfield Park	SU 41 693322	169	955	61	178	20	123	89	75	79	90	45	151	74	89	1074	33	29 Aug
Four Marks	SU 41 679356	220	...	63	181	19	122	86	74	71	95	47	145	64	97	1064	34	29 Aug
Alton, Treloar Hospital	SU 41 703382	137	860	53	161	17	106	77	64	72	82	43	131	62	95	963	29	9 Feb
Alton S. Wks.	SU 41 729394	102	...	54	155	16	106	72	69	62	77	41	131	73	97	951	30	22 Jun
Farnham, Broomleaf Road	SU 41 852470	76	743	43	118	11	90	65	62	87	57	43	111	52	75	815	34	22 Jun
Alice Holt Lodge	SU 41 804429	115	794	51	137	14	99	70	62	67	66	33	112	54	91	855	32	22 Jun
Rowledge	SU 41 820434	108	...	54	136	16	104	74	65	69	65	34	111	54	88	869	35	22 Jun
Boundstone	SU 41 835441	113	...	49	124	11	93	70	60	76	67	37	113	48	81	830	35	22 Jun
Farnham, Little Park	SU 41 851455	122	...	43	121	11	92	70	64	95	56	41	120	51	79	845	36	22 Jun
South Wey (Wey)																		
Shottermill S. Wks.	SU 41 879325	117	894	60	158	22	116	75	74	76	89	38	168	62	77	1017	37	22 Jun
Selborne	SU 41 740336	116	969	62	164	20	117	94	71	71	90	55	130	63	90	1028	32	6 Aug
Selborne, Temple Manor	SU 41 761331	140	...	59	176	19	118	80	73	71	90	40	147	61	92	1026	(33	22 Oct)
Churt	SU 41 858376	140	...	55	161	14	100	71	65	63	76	37	154	61	80	938	35	22 Jun
Wey (Thames)																		
Tilford R.F.M. Sta. Area 396 km²	SU 41 874434	48	869	...	...	...	...	...	...	...	...	...	...	...	...	...	...	...
Hindhead W. Wks.	SU 41 889359	256	926	56	165	20	114	77	67	79	76	35	173	62	88	1012	35	22 Oct
Seale, Greensand	SU 41 900479	91	...	49	120	(11)	103	78	59	98	68	37	138	60	78	899	29	22 Jun
Milford	SU 41 935422	55	766	46	131	15	87	74	58	60	66	30	144	52	73	837	30	18 Oct
Godalming P. Sta.	SU 41 967444	37	748	40	122	11	80	60	61	107	65	34	113	54	74	823	31	22 Jun
Puttenham	SU 41 943471	104	...	32	115	10	86	68	57	86	63	26	130	55	78	806	31	22 Jun
Unstead S. Wks.	SU 41 993454	37	721	35	113	14	74	56	51	91	61	42	109	48	74	768	29	22 Jun
Cranleigh S. Wks.	TQ 51 041392	47	761	42	131	19	86	64	52	94	81	39	132	47	71	857	31	22 Oct
Godalming, Munstead Rough	SU 41 988426	125	...	49	137	15	88	68	62	81	77	40	122	55	83	878	32	22 Jun
Tilling (Wey)																		
Abinger	TQ 51 115463	155	...	42	124	17	99	79	58	97	75	35	122	50	83	881	27	6 Aug
Holmbury St. Mary	TQ 51 113441	155	892	51	143	25	109	73	59	115	89	37	142	55	104	1000	31	22 Oct
Gomshall P. Sta.	TQ 51 079478	76	...	41	125	15	86	73	51	72	80	36	126	51	89	844	33	29 Aug
Shere	TQ 51 069479	90	778	40	115	12	85	75	52	108	72	34	135	56	83	867	27	29 Aug
Peaslake	TQ 51 089448	137	879	51	141	19	112	78	57	109	88	43	150	53	98	996	32	6 Aug
Wey (Thames)																		
Guildford, Wodeland Avenue	SU 41 989492	68	...	41	110	14	86	58	52	78	57	40	115	50	79	780	28	22 Jun
Guildford, Joseph's Road	SU 41 994509	30	...	33	95	13	81	51	53	77	57	40	109	45	64	719	29	22 Jun
Guildford, Woking Road	TQ 51 000515	30	...	31	89	12	83	53	56	82	55	43	113	44	63	724	31	22 Jun
Merrow	TQ 51 017507	67	...	38	94	12	81	64	55	86	63	41	113	47	64	758	28	22 Jun
Woking, The Grange	TQ 51 021570	25	706	31	91	13	(80)	62	56	92	68	24	103	39	62	723	27	22 Jun
Pirbright, Bakersgate	SU 41 951544	37	690	36	93	13	93	60	65	90	56	22	106	46	67	748	33	22 Jun
Pirbright Institute	SU 41 952542	37	...	35	93	11	89	63	63	82	53	21	103	42	65	719	33	22 Jun
Normandy	SU 41 923520	61	...	(36)	(97)	11	86	64	70	89	55	27	110	48	69	761	31	22 Jun
Woking, Council Offices	TQ 51 003581	40	690	31	86	14	90	66	57	89	64	29	102	37	61	728	28	22 Jun
Old Woking S. Wks.	TQ 51 026573	23	645	32	81	11	77	59	55	94	71	26	106	40	69	722	28	22 Jun
West Clandon	TQ 51 047506	97	747	36	(102)	11	83	71	54	93	77	35	122	54	78	816	32	29 Aug
East Clandon	TQ 51 059517	88	...	36	104	15	87	67	52	91	78	34	121	52	75	812	33	29 Aug
Horsley P. Sta.	TQ 51 080523	87	...	34	(102)	10	78	70	49	89	76	32	119	50	76	785	34	29 Aug
East Horsley	TQ 51 092532	85	757	35	104	11	84	66	50	93	76	34	111	49	73	786	31	29 Aug
Wisley R.H.S. Gardens	TQ 51 063583	32	661	34	78	12	78	59	47	94	73	30	103	41	65	715	29	5 Jul
Wisley S. Wks.	TQ 51 059597	16	...	33	77	10	84	51	46	90	60	28	98	42	64	683	25	5 Jul 29 Aug
West Byfleet, Scotland Bridge	TQ 51 047616	20	661	29	69	12	93	61	56	(86)	82	26	101	42	64	723	28	29 Aug
Addlestone	TQ 51 061643	14	...	30	70	13	95	49	63	86	83	25	102	43	68	727	35	29 Aug
Bourne (Wey)																		
Bagshot	SU 41 907626	82	721	38	99	17	114	63	75	73	73	43	111	50	73	830	38	22 Jun
West Byfleet, Golf Club	TQ 51 029612	30	661	34	85	13	99	54	56	88	77	27	105	39	65	741	26	29 Aug
Thames																		
Weybridge, Thames Street	TQ 51 075655	12	...	33	69	15	94	52	65	86	76	26	110	46	71	743	32	29 Aug
Walton on Thames W. Wks.	TQ 51 085663	10	604	33	67	13	89	52	58	80	72	23	100	43	68	698	28	29 Aug
Staines P. Sta.	TQ 51 042719	14	615	33	82	10	101	50	69	73	80	40	106	38	67	749	29	29 Aug
Staines Res.	TQ 51 046730	16	633	35	80	11	104	51	71	75	82	39	109	38	67	760	31	29 Aug
Queen Mary Res.	TQ 51 059696	13	619	29	69	10	105	45	63	72	101	34	97	38	68	732	35	29 Aug
Walton W. Wks.	TQ 51 117684	11	601	32	68	11	86	47	53	70	74	23	97	35	64	661	27	29 Aug
Kempton W. Wks.	TQ 51 111710	12	578	33	76	11	110	46	58	74	88	27	93	37	69	723	29	29 Aug
Hampton	TQ 51 131695	12	587	34	65	11	91	49	64	73	81	28	90	36	68	690	28	29 Aug
Hampton, Compressor Sta.	TQ 51 131715	18	...	33	74	12	101	49	67	73	100	26	97	37	72	743	42	29 Aug
Mole (Thames)																		
Broadfield	TQ 51 263345	91	857	35	150	26	105	61	77	93	81	32	157	57	89	963	31	22 Jun
Gatwick	TQ 51 265407	59	769	34	123	19	103	57	67	86	65	44	139	46	77	859	37	22 Oct
Crawley, The Beeches	TQ 51 271361	76	816	36	149	23	100	68	74	92	70	31	153	53	82	931	36	18 Oct 22 Oct
Stanford Brook (Mole)																		
Tinsley S. Wks.	TQ 51 289402	64	744	38	135	19	99	63	63	81	62	41	145	53	81	881	37	22 Oct
Mole (Thames)																		
Burstow S. Wks.	TQ 51 305437	58	...	29	123	13	85	56	59	77	59	45	137	56	76	817	34	22 Oct
Redhill (Mole)																		
Brewer Street	TQ 51 325526	125	...	36	106	10	120	53	58	120	70	37	128	57	107	901	...	...
Pendell Court	TQ 51 316519	101	...	32	103	10	120	51	56	107	81	37	123	54	88	863	28	29 Aug
Merstham Lodge	TQ 51 283546	152	855	39	113	14	109	57	62	124	72	44	139	57	108	937	28	6 Aug
Merstham S. Wks.	TQ 51 295526	81	...	33	116	12	104	56	57	109	69	39	127	58	101	880	27	22 Oct
Redhill, London Road	TQ 51 281512	82	767	32	111	12	109	54	51	100	69	31	130	52	90	842	27	22 Oct
Earlswood S. Wks.	TQ 51 275483	62	727	26	112	11	90	55	53	87	72	34	128	49	82	799	29	22 Oct
Mole (Thames)																		
Reigate, Wray Park Road	TQ 51 262511	101	794	36	120	13	106	58	53	100	74	36	140	54	91	881	29	22 Jun
Reigate, Castle Grounds	TQ 51 253504	98	...	34	113	13	92	57	49	90	56	33	114	47	81	778	28	22 Jun
Reigate, Clifton's Lane	TQ 51 239512	85	...	36	119	13	91	52	51	81	53	39	121	55	94	804	...	...
Headley Res.	TQ 51 207527	192	826	39	127	13	89	69	58	98	82	40	139	51	99	904	32	29 Aug
Pixham Lane S. Wks.	TQ 51 177506	42	794	34	120	12	83	66	55	99	67	35	116	47	90	825	28	22 Jun
Great Bookham Res.	TQ 51 145526	141	...	38	104	10	94	76	51	86	79	36	103	47	81	805	...	...
Denbies	TQ 51 149504	189	874	41	139	17	107	83	58	101	78	41	134	49	100	949	28	29 Aug
Mickleham	TQ 51 173527	55	751	37	118	12	86	64	51	91	55	34	119	52	83	801	27	22 Oct
Leatherhead, Yarm Way	TQ 51 175556	91	...	37	107	11	81	65	49	113	65	38	112	47	77	802	43	5 Jul
Leatherhead, Elmer Wks.	TQ 51 160557	51	733	31	100	8	72	67	51	89	69	39	112	43	73	755	25	22 Jun
Leatherhead S. Wks.	TQ 51 149578	32	689	33	87	10	77	52	47	81	65	34	99	41	65	691	27	29 Aug
Effingham	TQ 51 125531	102	...	35	114	10	89	71	55	95	77	36	109	45	78	813	29	29 Aug
Cobham	TQ 51 124599	37	649	36	85	13	82	60	58	92	68	34	99	45	75	747	27	29 Aug

Station	Grid reference	Altitude m	Av. 1916–1950 mm	Jan	Feb	Mar	Apr	May	Jun	Jul	Aug	Sep	Oct	Nov	Dec	Year	Highest recorded fall in a day mm	date
THAMES—*(contd)*																		
Cobham S. Wks.	TQ 51 100607	22	633	32	75	13	82	68	50	83	72	31	99	44	67	717	30	29 Aug
Weyland S. Wks.	TQ 51 125656	15	591	32	65	10	87	55	49	80	68	26	99	38	66	675	25	29 Aug
Esher S. Wks.	TQ 51 135661	13	597	37	72	11	93	57	54	79	68	26	99	39	69	705	24	22 Jun
Thames																		
Surbiton, Seething Wells	TQ 51 172671	10	626	33	63	8	80	51	74	83	77	31	90	39	64	694	35	29 Aug
Hogsmill (Thames)																		
Tadworth Depot	TQ 51 232561	177	...	46	109	11	101	72	61	102	74	43	149	52	94	915	30	29 Aug
Tadworth P. Sta.	TQ 51 223568	123	773	33	103	9	87	71	62	96	71	38	135	43	78	827	31	29 Aug
Epsom, Downs Road	TQ 51 217588	122	705	33	106	8	79	68	57	79	81	35	119	43	76	784	32	29 Aug
Epsom College	TQ 51 223601	82	656	34	93	11	79	63	54	77	78	31	117	43	74	754	30	29 Aug
Epsom W. Wks.	TQ 51 211612	46	670	31	90	8	72	55	59	76	72	29	108	41	74	714	33	29 Aug
Banstead, Green Curve	TQ 51 245599	137	...	37	100	9	85	67	58	84	75	38	125	43	79	801	28	22 Jun
Ewell Court	TQ 51 213639	29	...	37	81	12	81	52	59	83	75	34	106	46	76	741	32	29 Aug
Chessington, Bolton Road	TQ 51 180638	45	...	35	81	11	86	53	82	84	73	36	98	40	73	750	33	20 Jun
Malden S. Wks.	TQ 51 197685	16	...	31	62	8	79	47	57	81	90	31	104	38	66	695	44	29 Aug
Hogsmill Valley S. Wks.	TQ 51 194682	12	...	33	68	8	78	49	62	79	86	33	97	37	68	696	44	29 Aug
Kingston-upon-Thames R.F.M. Sta. Area 69.2 km²	TQ 51 182688	6	691	...	...	...	...	...	...	...	...	...	...	...	...	...	...	...
Thames																		
Kingston, Canbury Gardens	TQ 51 179700	7	598	37	64	9	85	45	60	67	93	32	97	34	66	690	39	29 Aug
Teddington Weir R.F.M. Sta. Area 9870 km²	TQ 51 170713	5	735	...	...	...	...	...	...	...	...	...	...	...	...	...	...	...

LONDON, RIGHT BANK AREA, 561 SQUARE KILOMETRES

HYDROMETRIC AREA 39 (Part)

Station	Grid reference	Altitude m	Av. 1916–1950 mm	Jan	Feb	Mar	Apr	May	Jun	Jul	Aug	Sep	Oct	Nov	Dec	Year	Highest recorded fall in a day mm	date
Coastal																		
Park Gate House	TQ 51 187716	9	...	38	68	10	95	49	65	71	93	32	105	38	70	735	36	29 Aug
Richmond, Terrace Gardens	TQ 51 181740	9	622	29	58	8	98	47	69	70	85	31	104	34	76	709	32	29 Aug
Kew	TQ 51 171757	5	608	32	69	11	92	48	66	69	77	28	100	33	74	699	30	29 Aug
Kew Gardens	TQ 51 183773	5	614	34	71	12	100	51	58	78	77	30	95	36	81	722	31	29 Aug
Barn Elms	TQ 51 230775	5	619	36	61	9	94	39	64	77	88	31	81	33	79	693	36	29 Aug
Beverley																		
Worcester Park S.Wks.	TQ 51 228667	21	627	33	69	13	83	43	57	83	74	33	99	39	69	694	32	29 Aug
Banstead, Sutton Lane	TQ 51 258601	149	...	47	109	12	89	73	58	92	77	43	133	48	90	871	28	29 Aug
Sutton W.Wks.	TQ 51 263642	49	...	30	79	10	77	53	58	77	70	33	100	40	72	698	27	29 Aug
Sutton S.Wks.	TQ 51 249655	26	625	28	70	10	74	47	52	76	74	32	101	42	65	670	34	29 Aug
Merton, Recreation Ground	TQ 51 240688	15	615	35	66	10	79	40	49	77	93	30	93	41	67	681	49	29 Aug
Raynes Park P.Sta.	TQ 51 237695	15	620	35	65	8	83	46	50	78	97	32	98	43	68	705	50	29 Aug
Wimbledon Common R.F.M. Sta. Area 43.5 km²	TQ 51 216717	11	640	...	...	...	...	...	...	...	...	...	...	...	...	...	...	...
Putney Heath Res.	TQ 51 234737	56	...	34	68	8	99	42	58	74	93	37	90	34	81	719	39	29 Aug
Wandle																		
Margery Hall Res.	TQ 51 252526	211	852	40	118	13	100	65	60	113	81	43	155	54	102	944	31	22 Jun
Burgh Heath Res.	TQ 51 237581	177	773	37	117	13	91	69	64	98	79	47	146	50	90	902	30	22 Jun
Kingswood, The Chase	TQ 51 255563	165	802	40	118	12	106	67	63	116	69	44	144	53	94	927	28	22 Jun
Banstead, The Council House	TQ 51 248588	168	...	39	109	12	95	69	62	98	75	46	141	49	90	886	30	22 Jun
How Green Res.	TQ 51 283581	161	796	39	117	10	103	69	54	103	68	46	136	51	98	895	25	22 Jun
Alderstead Heath	TQ 51 302550	176	...	41	108	13	113	58	65	121	81	50	130	57	98	934	28	29 Aug
Coulsdon Allotments	TQ 51 282593	104	792	43	107	12	93	74	53	104	72	43	130	49	97	879	26	22 Jun
Woodcote Park	TQ 51 289603	137	738	30	106	12	88	65	53	97	75	42	126	(51)	(89)	834	27	29 Aug 22 Jun 6 Aug
Purley P. Sta.	TQ 51 312614	65	766	38	105	10	80	65	51	93	71	42	131	45	86	816	30	29 Aug
Woldingham (Wandle)																		
Woldingham W.Wks.	TQ 51 375572	146	839	39	103	12	112	53	68	121	73	43	138	61	97	922	36	22 Jun
Woldingham Res.	TQ 51 378547	256	883	38	112	12	116	51	67	116	86	48	153	63	99	962	...	...
Caterham, Dome Hill Peak	TQ 51 331543	198	...	38	113	12	117	57	63	137	81	42	127	61	107	953	30	29 Aug
Caterham Res.	TQ 51 331545	215	845	40	106	14	115	54	63	129	73	41	139	63	103	940	31	22 Jun
Caterham, Whiteknobs Way	TQ 51 345547	152	871	39	120	15	121	59	64	135	85	45	150	68	108	1009	30	29 Aug
Kenley, Refuse Wks.	TQ 51 335592	87	782	35	100	8	85	55	61	115	65	42	128	49	86	831	29	22 Jun
Kenley P. Sta.	TQ 51 325601	84	751	38	85	9	81	47	50	106	58	37	131	51	89	783	24	22 Jun 29 Aug
Wandle																		
Purley, Oaks Depot	TQ 51 321623	58	735	37	94	9	85	61	60	84	74	38	129	45	81	796	31	29 Aug
Sanderstead, Beechwood Road	TQ 51 331628	98	770	39	103	9	86	64	63	91	77	36	136	51	89	843	32	29 Aug
Waddon	TQ 51 314639	59	693	31	91	9	76	52	59	76	82	35	114	43	79	749	36	29 Aug
Beddington R.F.M. Sta. Area 122 km²	TQ 51 296655	33	800	...	...	...	...	...	...	...	...	...	...	...	...	...	...	...
Banstead Hospital	TQ 51 261611	125	...	31	83	(11)	76	57	54	82	72	37	113	44	83	745	26	22 Jun
Beddington S. Wks.	TQ 51 303661	39	625	32	72	15	80	58	55	75	86	32	112	49	72	738	38	29 Aug
Carshalton P. Sta.	TQ 51 278663	26	641	32	76	11	75	52	55	78	76	31	97	43	70	697	36	29 Aug
Morden Hall	TQ 51 261685	16	638	31	70	10	80	43	54	79	87	31	91	45	67	687	42	29 Aug
Mitcham, London Road	TQ 51 278701	19	601	32	71	11	84	39	55	83	86	28	90	40	72	691	39	29 Aug
Connolly's Mill R.F.M. Sta. Area 176 km²	TQ 51 265705	10	754	...	...	...	...	...	...	...	...	...	...	...	...	...	...	...
Streatham, Recreation Ground	TQ 51 286712	40	633	41	61	10	81	40	52	89	91	30	90	42	83	711	41	29 Aug
Wimbledon P.Sta.	TQ 51 258717	18	588	36	58	8	79	46	52	79	86	29	94	38	79	683	37	29 Aug
Southfields, Coronation Gardens	TQ 51 251734	11	608	36	73	9	97	43	58	77	91	37	95	36	81	733	41	29 Aug
Coastal																		
Streatham Hill Res.	TQ 51 306732	62	599	35	65	8	87	41	57	85	103	29	92	33	81	716	49	29 Aug
Battersea Park	TQ 51 275773	4	603	41	68	7	103	37	55	73	154	34	84	39	80	775	82	30 Aug
Rock Hill Res.	TQ 51 339718	111	...	42	71	10	88	41	58	95	118	26	102	36	90	777	50	29 Aug
Dulwich, Belair	TQ 51 329732	34	632	36	69	14	95	47	55	95	115	31	97	35	85	772	49	29 Aug
Brockwell Park	TQ 51 315740	44	600	35	65	9	94	40	65	90	97	29	96	38	87	746	42	29 Aug
Brixton Hill Res.	TQ 51 305743	37	595	34	60	10	91	41	60	83	90	28	87	33	81	696	39	29 Aug
Nunhead Res.	TQ 51 351754	41	579	33	65	10	89	41	64	96	86	29	98	33	74	720	31	29 Aug
Telegraph Hill	TQ 51 357761	30	...	35	66	11	96	42	69	103	89	31	103	38	76	758	30	29 Aug
Southwark Park	TQ 51 351791	3	607	(34)	63	9	102	42	64	80	79	36	93	38	79	721	31	29 Aug
Ravensbourne																		
Keston	TQ 51 419646	119	...	40	91	9	90	51	67	109	71	41	105	55	95	822	37	22 Jun
Biggin Hill (Ravensbourne)																		
Keston Court Farm	TQ 51 415630	106	752	41	103	10	102	60	70	126	76	40	106	54	85	874	36	22 Jun
Warlingham Park Hospital	TQ 51 373595	179	803	38	113	11	94	63	72	133	80	40	140	63	102	949	34	22 Jun
Hares Bank P.Sta.	TQ 51 370627	101	757	38	94	9	88	55	60	102	84	35	122	55	89	832	30	22 Jun
Addington, Abbotswood	TQ 51 373645	119	727	40	93	10	86	56	61	102	80	35	119	53	89	821	28	22 Jun
West Wickham P.Sta.	TQ 51 379645	73	689	35	82	9	85	51	61	103	73	32	121	57	96	805	29	22 Jun

Station	Grid reference	Altitude m	Av. 1916–1950 mm	Jan	Feb	Mar	Apr	May	Jun	Jul	Aug	Sep	Oct	Nov	Dec	Year	Highest recorded fall in a day mm	date
LONDON, RIGHT BANK—*(contd)*																		
Ravensbourne																		
Bromley, Henville Road	TQ 51 409696	70	618	38	72	16	75	54	81	80	95	29	113	40	88	781	38	29 Aug
Bromley	TQ 51 399693	65	638	38	84	20	81	49	71	85	86	32	122	47	91	807	35	29 Aug
Pool (Ravensbourne)																		
Beckenham, Kelsey Park	TQ 51 374692	33	...	39	77	15	85	45	63	79	91	29	114	47	(91)	775	33	29 Aug
Beckenham, Foxgrove Road	TQ 51 377700	43	...	39	74	10	70	50	73	93	97	30	118	47	93	794	33	29 Aug
Addington	TQ 51 352643	144	735	41	102	12	90	53	68	95	74	36	132	52	90	846	32	22 Jun
Croydon, Northampton Road	TQ 51 342658	64	663	37	85	13	81	53	65	79	92	33	124	53	82	797	32	22 Jun
Anerley, Oakfield Road	TQ 51 348700	47	632	39	76	11	85	48	52	84	91	28	103	(46)	(86)	749	36	29 Aug
Penge, Recreation Ground	TQ 51 352703	40	652	39	76	13	86	45	55	88	102	30	104	43	(81)	764	36	29 Aug
Ravensbourne																		
Eltham, High Street	TQ 51 433745	75	611	38	64	11	86	36	70	109	87	29	125	40	79	775	29	22 Jun
Deptford W.Wks.	TQ 51 375766	6	586	34	62	10	89	40	69	92	79	36	102	37	84	734	26	22 Jun
Deptford P.Sta.	TQ 51 377771	5	574	33	64	10	81	35	67	91	80	31	99	36	76	701	26	22 Jun
Coastal																		
Greenwich Maritime Museum	TQ 51 387776	7	575	35	62	10	89	38	73	96	83	33	104	36	75	735	28	22 Jun

KENT RIVER AUTHORITY AREA, 4784 SQUARE KILOMETRES

HYDROMETRIC AREA 40

Station	Grid reference	Altitude m	Av. 1916–1950 mm	Jan	Feb	Mar	Apr	May	Jun	Jul	Aug	Sep	Oct	Nov	Dec	Year	Highest recorded fall in a day mm	date
Coastal																		
Cross Ness	TQ 51 485809	8	534	26	54	8	72	32	55	72	60	23	89	33	61	584	31	22 Jun
Erith S. Wks.	TQ 51 523777	4	562	31	50	9	87	27	52	87	75	23	65	32	67	605	30	22 Jun
Darent																		
Westwood P. Sta.	TQ 51 425541	127	816	35	104	11	100	50	71	114	58	39	113	65	100	860	40	22 Jun
Westerham	TQ 51 446536	128	823	41	110	12	105	50	74	105	63	45	117	66	103	892	35	22 Jun
Westerham Hill P. Sta.	TQ 51 429558	174	863	48	110	14	108	57	81	135	73	46	117	65	99	952	37	22 Jun
Knockholt W. Wks.	TQ 51 466583	239	891	53	119	15	116	56	81	126	85	53	130	70	113	1016	15	4 Sep
Sundridge P. Sta.	TQ 51 489556	82	760	35	92	10	97	44	64	106	69	42	110	59	95	823	30	22 Jun
Chevening Gardens	TQ 51 484577	107	828	40	104	12	109	50	74	109	71	39	113	52	93	866	33	22 Jun
Sevenoaks, Oak Lane	TQ 51 528541	130	821	56	115	15	109	44	78	113	81	40	134	64	101	950	(30	22 Oct)
Cramptons Road P. Sta.	TQ 51 530570	68	743	35	94	16	87	41	72	110	89	39	114	59	91	847	35	21 Aug
Kemsing P. Sta.	TQ 51 569576	89	752	36	99	14	91	40	79	112	84	37	131	59	92	874	31	21 Aug
Oak Bank Res.	TQ 51 563561	167	769	38	109	14	105	40	78	107	76	40	135	62	96	900	28	22 Oct
Stone Street	TQ 51 572549	162	832	42	117	15	107	43	79	112	77	47	144	66	104	953	31	22 Oct
Highfield	TQ 51 541604	198	771	41	98	13	97	43	73	138	84	41	108	57	102	896	32	21 Aug
Eynsford P. Sta.	TQ 51 535655	40	671	35	79	8	75	39	67	125	83	42	95	47	86	780	36	19 Jul
Horton Kirby	TQ 51 561685	15	586	33	71	12	73	36	63	116	73	35	79	37	73	701	38	19 Jul
Sutton-at-Hone	TQ 51 559703	18	...	35	66	17	78	35	69	128	71	37	74	34	73	718	41	19 Jul
Crockenhill	TQ 51 518668	101	669	35	86	12	77	46	68	125	79	42	95	43	83	793	38	19 Jul
Swanley	TQ 51 512701	54	609	32	60	14	68	35	65	97	67	48	77	41	71	676	30	19 Jul
Hawley R.F.M. Sta. Area 191 km²	TQ 51 551718	11	754	...	...	...	...	...	...	...	...	...	...	...	...	...	...	...
Fairseat Manor	TQ 51 622614	213	798	40	107	14	92	42	63	131	92	39	135	56	104	914	29	21 Aug
Hartley	TQ 51 617664	82	673	33	76	13	81	36	61	160	69	34	116	48	83	812	30	19 Jul
Wilmington W. Wks.	TQ 51 543728	8	584	32	56	10	65	30	73	100	63	35	72	31	72	639	28	22 Jun
Dartford	TQ 51 543737	5	558	35	56	12	76	27	73	93	66	33	69	32	69	640	28	22 Jun
Orpington W. Wks.	TQ 51 459652	67	705	36	89	9	84	51	66	119	69	38	93	56	95	805	33	22 Jun
Danson Park	TQ 51 472755	50	614	33	56	14	76	33	63	102	72	27	85	40	79	680	29	22 Jun
Dartford, Christchurch Road	TQ 51 531740	30	...	37	63	12	87	33	73	95	69	35	72	35	74	685	29	22 Jun
Coastal																		
Dartford S. Wks.	TQ 51 552765	4	571	31	51	8	79	24	57	89	70	29	66	32	65	601	28	19 Jul
Northfleet P. Sta.	TQ 51 635691	49	642	30	68	14	73	39	54	138	58	26	98	42	73	712	36	19 Jul
Southfleet	TQ 51 611705	30	...	29	61	9	68	33	56	133	53	28	82	38	67	657	32	19 Jul
Southfleet W. Wks.	TQ 51 610724	25	593	30	62	11	74	28	53	114	59	26	69	29	63	618	28	19 Jul
Jeskyns Court	TQ 51 660690	80	652	28	68	13	70	41	57	125	60	27	102	38	75	704	33	19 Jul
Gravesend S. Wks.	TQ 51 666739	4	...	28	56	10	76	31	65	101	52	20	79	37	65	621	24	19 Jul
Isle of Grain	TQ 51 871749	2	507	37	54	11	67	23	43	97	61	13	111	57	62	634	24	19 Jul
Medway																		
Turner's Hill	TQ 51 335361	137	...	34	154	20	105	68	79	104	77	44	161	69	95	1011	...	...
Duckyls	TQ 51 352338	183	898	36	159	25	112	64	86	108	87	46	162	66	103	1054	36	22 Oct
Whillett's Bridge	TQ 51 376344	76	...	35	156	27	108	59	95	117	86	45	164	69	107	1068	...	...
Weir Wood Res. R.F.M. Sta. Area 26.9 km²	TQ 51 406355	62	871	...	...	...	...	...	...	...	...	...	...	...	...	...	...	...
Weir Wood Res.	TQ 51 406354	69	843	33	153	26	104	58	86	105	71	38	144	69	84	970	37	22 Jun
Ashurstwood S. Wks.	TQ 51 432382	76	...	30	159	17	99	65	99	99	67	41	142	67	95	977	43	22 Jun
Jarvis Brook	TQ 51 540297	104	939	42	159	31	108	62	99	116	112	39	170	73	122	1133	27	22 Jun
Tunbridge Wells, Madeira Park	TQ 51 584385	85	807	37	139	14	89	54	76	115	68	36	164	67	86	945	27	9 Feb
Tunbridge Wells S. Wks.	TQ 51 548378	63	...	35	149	17	86	60	81	115	71	34	157	66	94	964	27	9 Feb 22 Oct
Falconhurst	TQ 51 470426	110	817	33	133	12	112	50	80	92	67	42	150	69	93	931	40	22 Jun
Chafford R.F.M. Sta. Area 255 km²	TQ 51 517405	31	874	...	...	...	...	...	...	...	...	...	...	...	...	...	...	...
Saints Hill	TQ 51 523414	38	731	29	(133)	9	92	47	77	86	62	37	145	56	81	855	31	22 Oct
Tunbridge Wells, Ridgemereton	TQ 51 560400	99	...	42	136	13	93	56	71	115	77	36	160	71	94	963	32	22 Oct
Great Bounds Res.	TQ 51 573436	146	748	36	(125)	11	91	45	71	97	70	31	164	58	90	889	33	22 Oct
Eden (Medway)																		
Godstone P. Sta.	TQ 51 352522	108	...	36	108	11	119	52	63	114	69	38	125	61	96	892	29	22 Oct
Godstone S. Wks.	TQ 51 365505	79	...	28	105	11	98	49	68	92	63	50	115	59	91	829	32	22 Jun
Limpsfield and Oxted S. Wks.	TQ 51 398501	70	...	32	102	14	95	50	79	92	59	41	114	57	93	829	35	22 Jun
Felbridge S. Wks.	TQ 51 362412	65	...	27	133	11	87	57	69	79	65	36	131	58	69	822	36	22 Oct
Lingfield S. Wks.	TQ 51 384450	47	758	30	115	13	87	53	75	94	56	44	131	60	79	838	35	22 Oct
South Park Farm	TQ 51 342488	104	751	32	114	17	103	58	67	92	69	44	122	58	88	866	28	22 Oct
South Godstone S. Wks.	TQ 51 363478	67	...	22	109	10	90	51	72	79	54	43	121	58	95	805	31	22 Jun
Paines Hill	TQ 51 412515	94	795	35	103	11	90	51	76	95	61	41	114	55	95	828	38	22 Jun
Edenbridge P. Sta.	TQ 51 452468	40	735	30	111	14	96	52	68	92	54	44	125	59	83	829	38	22 Jun
Kent Hatch Res.	TQ 51 436515	202	...	36	97	13	91	48	74	97	63	43	103	58	102	825	...	...
Penshurst R.F.M. Sta. Area 224 km²	TQ 51 520437	28	775	...	...	...	...	...	...	...	...	...	...	...	...	...	...	...
Medway																		
Penshurst Place	TQ 51 528440	40	734	31	121	10	99	45	77	84	69	41	136	56	83	853	35	22 Oct
Bayleys Hill Res.	TQ 51 519521	212	804	35	103	12	102	43	77	106	81	46	120	57	97	880	31	22 Oct
Hildenborough	TQ 51 571488	48	718	34	96	10	86	31	77	111	83	37	138	57	83	842	31	22 Oct
Northern S. Wks.	TQ 51 601423	59	...	37	117	10	89	50	81	102	85	29	159	71	88	919	31	22 Oct
Bourne (Medway)																		
Borough Green P. Sta.	TQ 51 611573	85	...	37	99	15	92	38	68	103	84	42	132	67	100	877	27	22 Oct
Hadlow R.F.M. Sta. Area 49.7 km²	TQ 51 632497	23	724	...	...	...	...	...	...	...	...	...	...	...	...	...	...	...

Station	Grid reference	Altitude m	Av. 1916–1950 mm	Jan	Feb	Mar	Apr	May	Jun	Jul	Aug	Sep	Oct	Nov	Dec	Year	Highest recorded fall in a day mm	date
KENT—*(contd)*																		
Hadlow S. Wks.	TQ 51 631491	22	695	39	93	12	71	(33)	71	(97)	62	33	154	61	77	803	31	22 Oct
Medway																		
Pembury W. Wks.	TQ 51 627425	60	743	41	(117)	9	89	43	67	109	79	30	163	68	86	901	31	22 Oct
Teise (Medway)																		
Bayham Abbey	TQ 51 640371	104	816	42	129	14	86	51	79	96	85	29	182	72	88	953	29	9 Feb
Wadhurst	TQ 51 643320	134	869	50	159	24	124	55	80	95	98	30	189	85	118	1107	25	3 Aug
Scotney Castle	TQ 51 685354	64	742	51	145	17	107	42	78	91	82	27	173	90	98	1000	25	22 Oct
Goudhurst	TQ 51 722333	88	834	54	148	19	105	43	92	82	97	24	187	81	99	1029	25	25 Feb
Bedgebury Park	TQ 51 719341	79	839	55	144	20	105	43	87	85	97	22	180	83	100	1021	25	22 Oct
Goudhurst P. Sta.	TQ 51 711367	37	755	45	141	15	87	40	66	86	99	20	160	70	83	911	32	21 Aug
Stone Bridge R.F.M. Sta. Area 136 km^2	TQ 51 718399	25	797	...	...	...	...	...	...	...	...	...	...	...	...	...	...	...
Paddock Wood S. Wks.	TQ 51 677453	16	...	31	87	7	73	35	66	90	51	24	180	64	74	782	30	22 Oct
Medway																		
Yalding	TQ 51 690498	12	...	39	89	11	71	35	72	126	53	26	163	66	79	829	33	22 Oct
Beult (Medway)																		
High Halden	TQ 51 890370	52	694	39	113	15	79	40	77	103	71	19	200	83	93	930	25	9 Feb
Cranbrook	TQ 51 771360	91	...	(51)	141	16	87	45	67	80	99	20	196	79	88	970	27	9 Feb
Cranbrook S. Wks.	TQ 51 784361	69	...	50	137	16	83	44	74	76	96	21	193	81	91	962	...	...
Headcorn S. Wks.	TQ 51 817443	20	...	48	97	10	69	31	61	121	55	19	166	88	83	848	23	28 Jul
Great Tong	TQ 51 831465	34	654	44	93	11	64	32	63	119	56	16	169	92	79	836	23	5 Nov
Staplehurst S. Wks.	TQ 51 788445	19	...	39	99	6	60	38	85	134	56	20	173	90	(84)	883	24	28 Jul
Stile Bridge R.F.M. Sta. Area 277 km^2	TQ 51 758478	11	696	...	...	...	...	...	...	...	...	...	...	...	...	...	...	...
Medway																		
Teston R.F.M. Sta. Area 1260 km^2	TQ 51 708530	7	759	...	...	...	...	...	...	...	...	...	...	...	...	...	...	...
East Farleigh Lock	TQ 51 735535	9	...	44	92	15	78	37	74	116	49	24	171	70	86	855	33	22 Oct
Barming W. Wks.	TQ 51 735548	75	667	43	91	13	83	35	67	127	49	27	163	73	96	867	33	22 Oct
Langley	TQ 51 799523	108	712	58	97	13	78	31	81	143	57	20	170	84	97	929	34	22 Oct
Maidstone, College Road	TQ 51 757549	9	665	45	89	15	72	35	68	127	49	23	167	78	90	858	31	22 Oct
Len (Medway)																		
Lenham Court	TQ 51 884519	116	770	60	105	23	83	41	68	153	57	21	180	101	105	998	25	19 Jul
Thurnham P. Sta.	TQ 51 816559	69	741	57	93	14	80	33	73	149	52	17	168	99	88	923	32	19 Jul
Hockers Lane P. Sta.	TQ 51 788571	65	709	49	89	14	71	41	63	141	44	22	163	88	88	873	31	22 Oct
Maidstone, Mill Street	TQ 51 761555	11	679	44	82	12	69	33	59	120	47	24	165	80	83	819	33	22 Oct
Medway																		
Maidstone, Brenchley Gardens	TQ 51 759561	15	677	37	86	12	73	33	61	123	47	25	153	81	89	821	32	22 Oct
Boxley W. Wks.	TQ 51 774594	97	691	44	87	13	79	38	53	142	56	21	159	79	81	851	34	22 Oct
Allington Lock	TQ 51 746581	6	...	37	86	12	75	33	52	126	55	22	149	74	85	806	33	22 Oct
Forstal P. Sta.	TQ 51 739587	6	653	41	79	13	76	32	49	131	49	21	145	66	75	777	33	22 Oct
East Malling	TQ 51 711568	37	682	37	88	12	82	35	61	121	53	32	155	64	95	834	33	22 Oct
Trottiscliffe P. Sta.	TQ 51 640594	76	725	36	97	15	87	41	59	124	58	38	129	65	100	849	27	22 Oct
West Malling	TQ 51 677553	93	...	38	93	13	88	37	73	118	58	36	153	67	91	862	34	22 Oct
Halling P. Sta.	TQ 51 697643	23	652	38	87	13	88	37	46	118	49	31	134	58	90	791	29	22 Oct
Luddesdown Res.	TQ 51 662661	70	...	29	85	13	82	40	54	129	57	28	118	49	83	767	25	22 Oct
Cuxton P. Sta.	TQ 51 689669	27	661	35	80	13	81	39	59	126	59	23	128	49	79	771	26	22 Oct
Woolmans Wood	TQ 51 746635	143	685	43	87	12	86	42	51	145	61	25	149	61	84	848	31	22 Oct
Nashenden P. Sta.	TQ 51 733655	38	677	41	76	15	85	35	50	119	49	29	132	54	76	760	30	22 Oct
Strood W. Wks.	TQ 51 729693	34	616	33	67	13	78	32	50	110	54	25	111	45	73	692	24	22 Oct
Luton W. Wks.	TQ 51 775662	27	...	39	79	14	86	35	53	116	53	20	130	56	72	753	30	22 Oct
Shorne Res.	TQ 51 703706	109	...	31	72	12	73	36	51	130	56	23	113	39	71	708	36	19 Jul
Lodge Hill	TQ 51 757734	49	579	37	71	14	84	33	52	125	55	27	95	49	71	713	34	19 Jul
Wainscott	TQ 51 743720	18	...	36	75	15	81	31	49	118	60	27	105	49	72	719	31	19 Jul
Frindsbury	TQ 51 745711	15	617	33	69	13	76	29	49	113	61	24	103	47	70	688	25	19 Jul
Coastal																		
Gillingham	TQ 51 785692	3	615	31	69	12	73	29	49	111	51	25	120	51	65	687	26	22 Oct
Motney Hill	TQ 51 829684	9	577	28	65	12	71	36	38	101	57	14	122	56	60	659	32	22 Oct
Matts Hill P. Sta.	TQ 51 820625	69	727	50	95	15	93	38	59	152	74	22	162	89	87	936	33	22 Oct
Barnland	TR 61 011720	55	521	49	51	12	62	20	62	98	71	19	126	78	65	713	28	22 Oct
Church Marshes S. Wks.	TQ 51 912648	2	573	47	72	10	68	24	43	128	53	14	137	67	64	727	30	22 Oct
Highsted W. Wks.	TQ 51 910607	42	689	58	101	17	83	36	52	141	52	23	164	83	85	893	30	22 Oct
Oare Creek																		
Newnham P. Sta.	TQ 51 975597	33	668	61	86	15	71	33	51	132	53	16	177	86	79	860	33	22 Oct
Faversham	TR 61 007593	48	...	46	84	15	65	30	59	135	49	17	175	87	79	840	32	22 Oct
Faversham S. Wks.	TR 61 023617	7	634	51	75	16	64	28	55	135	51	17	168	84	77	822	30	22 Oct
Coastal																		
Whitstable	TR 61 112665	3	549	41	55	10	62	21	69	104	66	11	149	90	76	753	28	21 Aug
Herne Bay	TR 61 171681	6	578	56	52	7	65	18	65	93	56	8	171	97	75	764	28	22 Oct
Manston	TR 61 335666	44	560	49	49	6	60	23	84	90	68	11	250	112	75	878	43	29 Oct
Margate	TR 61 368714	16	540	53	45	6	60	14	119	95	63	7	240	98	71	872	52	16 Jun
Stone House	TR 61 396690	40	569	41	54	7	70	22	103	127	61	12	261	120	82	959	51	29 Oct
Broadstairs	TR 61 391681	40	586	52	53	8	68	17	100	120	65	10	250	109	79	931	52	29 Oct
Ramsgate	TR 61 385648	24	590	53	53	6	65	17	99	125	82	8	265	114	86	974	53	29 Oct
Stour																		
Lenham Quarry	TQ 51 915524	169	...	43	97	18	78	32	71	137	81	17	198	106	99	979	31	19 Jul
Pett Place	TQ 51 961491	91	785	63	87	16	72	33	64	145	72	16	198	106	106	978	29	3 Oct
Ashford, Albert Road	TR 61 009430	56	742	48	98	19	79	34	81	107	75	18	242	105	113	1020	37	3 Oct
Lympne	TR 61 112355	104	...	56	92	22	89	33	70	72	93	23	258	104	110	1023	43	3 Oct
Mersham	TR 61 055399	70	756	52	99	21	86	33	75	118	76	22	241	116	118	1057	44	3 Oct
Ashford, Hythe Road	TR 61 018425	43	745	44	97	17	77	34	69	100	73	18	231	107	111	979	35	3 Oct
Ashford	TR 61 013430	58	760	48	95	18	76	36	84	100	74	19	241	102	114	1006	37	3 Oct
Ashford, Belmore	TR 61 007433	61	730	45	95	18	75	35	82	108	76	17	243	104	117	1016	37	3 Oct
Wye R.F.M. Sta. Area 230 km^2	TR 61 049470	29	749	...	...	...	...	...	...	...	...	...	...	...	...	...	...	...
Wye	TR 61 057469	56	...	52	86	18	72	33	63	122	91	23	237	113	103	1014	36	3 Oct
Godmersham P. Sta.	TR 61 073516	25	746	62	94	19	85	33	59	120	85	21	224	118	100	1019	32	3 Oct
Elmsted	TR 61 113439	175	853	53	104	27	83	39	77	154	85	33	231	121	126	1131	...	...
Chartham	TR 61 097554	24	711	45	87	18	82	29	59	128	73	20	213	114	96	965	28	3 Oct
Horton R.F.M. Sta. Area 345 km^2	TR 61 116554	13	766	...	...	...	...	...	...	...	...	...	...	...	...	...	...	...
Canterbury S. Wks.	TR 61 166596	5	688	44	68	12	70	24	69	96	69	15	192	103	81	843	31	28 Oct
Stelling Minnis	TR 61 143474	145	...	53	113	22	89	49	75	171	79	23	259	123	115	1171	43	3 Oct
Little Stour (Stour)																		
Skeete P. Sta.	TR 61 143409	136	830	53	101	23	85	36	86	139	93	20	266	128	134	1165	42	3 Oct
Barham P. Sta.	TR 61 198509	53	769	51	97	15	82	38	57	127	93	18	247	112	101	1039	40	3 Oct
Littlebourne S. Wks.	TR 61 213576	7	...	57	73	11	72	31	78	110	65	14	205	107	80	902	18	21 Aug
Nonington	TR 61 260522	43	718	47	81	11	75	35	91	123	70	13	265	113	82	1007	44	3 Oct 29 Oct

Station	Grid reference	Altitude	Av. 1916–1950	Rainfall in millimetres													Highest recorded fall in a day	
		m	*mm*	Jan	Feb	Mar	Apr	May	Jun	Jul	Aug	Sep	Oct	Nov	Dec	Year	*mm*	*date*
KENT—*(contd)*																		
Woodnesborough P. Sta.	TR 61 300566	17	...	46	69	10	70	26	87	122	69	12	259	113	77	960	45	29 Oct
Wingham W. Wks.	TR 61 244553	42	714	41	87	13	77	33	81	111	78	17	242	105	96	980	47	3 Oct
Stour																		
West Stourmouth P. Sta.	TR 61 250629	2	...	61	65	11	72	24	79	93	65	12	223	114	80	898	35	3 Oct
Nash Farm	TR 61 268590	23	...	40	61	9	58	27	77	105	60	11	241	109	73	869	42	3 Oct
Elmstone	TR 61 256604	9	...	52	67	9	66	25	77	97	68	12	222	104	83	883	41	3 Oct
Ash, Corking Farm...	TR 61 277599	15	577	46	64	7	58	24	77	94	61	11	247	102	79	871	43	3 Oct
Minster	TR 61 309650	23	...	55	46	8	65	22	73	97	62	9	252	117	72	877	41	3 Oct
Sandwich, Jacobs Churchyard	TR 61 328583	5	625	(44)	52	9	64	19	74	101	62	11	242	93	69	839	55	29 Oct
Martin Mill P. Sta.	TR 61 337471	87	793	63	111	21	79	33	86	141	83	23	258	149	112	1161	48	29 Oct
Waldershare Park	TR 61 291482	76	833	63	110	20	82	38	100	132	70	19	294	128	95	1151	49	29 Oct
Coastal																		
Deal	TR 61 377536	2	...	48	89	12	62	23	86	117	61	15	222	131	85	949	37	29 Oct
Walmer W. Wks.	TR 61 364509	34	718	47	84	13	59	26	91	117	65	17	226	135	70	949	36	29 Oct
Dover, Royal Military School	TR 61 327439	123	848	70	108	21	92	31	70	159	95	23	263	166	125	1223	39	29 Oct
Lower Standen	TR 61 240404	76	834	58	111	30	82	45	78	156	97	23	289	139	135	1244	47	19 Oct
Dover W. Wks.	TR 61 322420	61	776	68	93	21	74	32	67	148	81	12	249	155	103	1104	38	19 Oct
Dover	TR 61 320410	6	732	67	98	22	73	41	73	146	90	13	238	161	111	1132	35	29 Oct
Folkestone	TR 61 227358	39	670	58	90	20	66	36	75	107	79	10	230	124	103	998	35	29 Oct
Folkestone, Cherry Garden	TR 61 210374	70	780	57	94	22	71	33	76	113	95	18	264	129	120	1092	37	29 Oct
Hythe, Oaklands	TR 61 162345	3	715	55	97	21	63	32	79	91	93	13	232	113	108	995	45	3 Oct
Sandling Park	TR 61 142368	79	778	56	105	20	86	35	81	125	89	19	255	117	126	1113	53	3 Oct
Dymchurch	TR 61 102296	2	647	51	97	16	70	33	71	64	90	19	202	96	89	898	42	3 Oct
Denge P. Sta.	TR 61 068201	4	...	52	95	17	83	26	84	74	69	14	175	103	83	875	33	3 Oct
Dungeness	TR 61 094169	6	636	45	91	19	67	23	76	74	62	14	174	102	81	829	32	3 Oct
Rother																		
Cranesden...	TQ 51 586262	91	846	55	149	21	95	54	93	85	85	25	192	74	108	1036	29	9 Feb
Stonegate	TQ 51 652288	70	863	48	158	25	95	53	106	91	81	24	205	79	96	1060	29	9 Feb
Crowhurst Bridge	TQ 51 682265	30	785	49	145	22	90	45	74	66	83	19	181	79	93	945	29	9 Feb
Bellhurst Farm...	TQ 51 724285	73	802	48	133	19	103	50	79	78	109	22	191	83	93	1009	27	8 Aug
Woodman's Farm	TQ 51 684227	98	877	52	165	22	102	51	73	76	78	33	227	81	113	1073	32	9 Feb
Brightling	TQ 51 683211	168	921	60	158	24	112	53	73	88	89	29	220	79	111	1096	31	9 Feb
Robertsbridge	TQ 51 737242	18	...	48	149	25	103	45	55	76	67	27	201	91	91	977	28	22 Oct
Darwell (Rother)																		
Darwell Hill House	TQ 51 704189	143	...	55	148	23	103	45	78	85	84	22	207	91	109	1049	28	6 Aug
Darwell Res. R.F.M. Sta. Area 9.63 km²	TQ 51 723217	21	904	...	...	...	...	...	...	...	...	...	...	...	...	...	...	...
Darwell Res.	TQ 51 723212	24	830	48	143	24	108	45	61	77	75	24	200	95	103	1003	27	22 Oct
Darvell Hall	TQ 51 729229	24	802	48	139	22	93	44	56	77	69	25	198	90	92	953	28	9 Feb
Rother																		
Udiam R.F.M. Sta. Area 196 km²	TQ 51 773245	2	851	...	...	...	...	...	...	...	...	...	...	...	...	...	...	...
Bodiam	TQ 51 785249	22	804	47	143	19	91	41	74	69	64	23	196	92	89	949	26	9 Feb
Field Green	TQ 51 786295	61	751	42	128	15	89	36	68	84	106	23	188	(89)	(89)	957	27	9 Feb
Great Dixter	TQ 51 820251	64	770	44	127	20	89	40	80	71	68	23	178	90	91	920	28	19 Oct
Tenterden, Tophill	TQ 51 891344	64	...	42	105	17	81	36	79	81	71	19	200	81	90	903	32	3 Oct
Biddenden...	TQ 51 850362	59	712	40	108	18	88	35	87	83	80	16	172	(84)	94	904	...	...
Iden	TQ 51 917241	46	...	47	126	16	80	37	86	61	78	20	219	95	97	962	47	3 Oct
Playden, Scots Float	TQ 51 932226	4	742	43	113	17	79	34	78	65	63	22	204	93	97	909	48	3 Oct
Playden, Oakfield	TQ 51 925232	44	...	44	116	15	78	34	79	61	63	20	213	86	91	899	47	3 Oct
Playden, Houghton Cottage	TQ 51 927227	15	...	43	116	15	77	34	80	63	63	21	214	89	96	912	48	3 Oct
Midrips	TQ 51 999185	5	...	42	101	17	85	28	72	58	59	17	180	99	94	852	35	3 Oct
Dean Court	TQ 51 970254	4	...	39	100	16	76	36	66	59	68	18	204	93	93	868	47	3 Oct
Brede (Rother)																		
Staplecross	TQ 51 781213	108	...	49	147	23	106	50	76	83	72	21	190	99	100	1016	26	9 Feb 3 Oct
Great Sanders	TQ 51 783208	76	...	51	146	(23)	102	51	77	71	78	23	194	103	102	1020	...	...
Powder Mill Res.	TQ 51 799192	18	...	51	146	23	94	47	72	91	83	27	200	103	105	1041	28	19 Oct
Brede P. Sta.	TQ 51 814178	8	781	46	129	23	93	44	66	81	77	28	199	95	94	976	33	3 Oct
Fairlight Res.	TQ 51 842121	157	...	52	122	19	104	33	55	66	69	29	224	88	109	971	57	3 Oct
Icklesham	TQ 51 879165	38	762	47	127	22	88	37	58	69	83	31	182	87	100	930	36	3 Oct
Pickham Farm	TQ 51 868148	27	...	49	133	23	93	32	51	68	83	39	200	91	119	981	49	3 Oct
Brede Sluice	TQ 51 919199	5	...	46	128	19	89	34	76	66	63	29	199	97	104	951	51	3 Oct
Udimore	TQ 51 886194	28	769	45	133	19	83	42	66	82	69	38	211	93	107	989	45	3 Oct
Coastal																		
Fairlight	TQ 51 878121	30	741	51	107	19	89	32	50	63	73	33	199	90	104	910	53	3 Oct

SUSSEX RIVER AUTHORITY AREA, 3085 SQUARE KILOMETRES

HYDROMETRIC AREAS 41E AND 41W

Station	Grid reference	Altitude *m*	Av. 1916–1950 *mm*	Jan	Feb	Mar	Apr	May	Jun	Jul	Aug	Sep	Oct	Nov	Dec	Year	*mm*	*date*
Coastal																		
Halton Res.	TQ 51 826104	90	761	61	110	19	87	32	55	64	63	30	213	90	118	942	60	3 Oct
Hastings, Wykeham Road	TQ 51 813098	40	...	52	103	16	88	30	51	67	63	22	209	84	90	874	53	3 Oct
Hastings	TQ 51 809094	45	725	52	102	17	86	31	50	69	59	24	204	85	92	872	52	3 Oct
Hastings, Newgate	TQ 51 807102	74	739	57	114	19	87	38	49	70	71	21	223	85	104	938	53	3 Oct
Combe																		
Catsfield Manor	TQ 51 729133	70	...	48	144	21	93	45	88	71	80	31	188	79	93	980	34	6 Aug
Buckholt Farm...	TQ 51 747110	34	...	46	125	19	89	36	68	63	77	27	195	78	96	918	35	6 Aug
Battle W. Wks.	TQ 51 742144	34	...	51	134	20	95	46	76	75	81	25	204	84	101	993	32	6 Aug
Filsham W. Wks.	TQ 51 778093	11	729	59	120	19	94	34	59	64	73	20	206	83	97	925	52	3 Oct
Baldslow Res.	TQ 51 796132	129	...	55	124	20	96	39	63	75	86	32	204	94	119	1007	41	3 Oct
St. Leonards, Ashbrooke Lodge	TQ 51 796120	70	794	52	115	19	98	37	56	69	79	28	201	84	99	937	43	3 Oct
Coastal																		
Bexhill	TQ 51 737072	4	716	51	113	21	86	32	47	60	67	24	173	74	82	829	38	6 Aug
Ash Bourne																		
Ashburnham Forge...	TQ 51 684161	27	...	47	131	18	89	40	96	83	86	22	207	81	99	997	...	...
Hammer Wood Bridge R.F.M. Sta. Area 18.7 km²	TQ 51 684142	6	859	...	...	...	...	...	...	...	...	...	...	...	...	...	...	...
Hugletts Stream (Ash Bourne)																		
Peartree Farm	TQ 51 658156	60	...	48	132	19	91	42	93	83	82	23	191	77	96	977	30	6 Aug
Nunningham (Ash Bourne)																		
Summertree Farm	TQ 51 641162	74	...	53	135	22	89	51	93	91	90	23	207	79	93	1026	28	6 Aug
Hazards Green P. Sta.	TQ 51 681123	23	...	54	120	20	87	39	76	69	75	24	204	75	93	937	30	19 Oct
Pevensey																		
Hellingly Hospital	TQ 51 597124	43	799	44	116	17	81	42	98	61	68	19	187	66	80	878	33	18 Jun
Amberstone	TQ 51 598113	9	771	50	119	19	82	48	87	63	76	26	192	61	85	907	28	22 Oct

Station	Grid reference	Altitude m	Av. 1916–1950 mm	Jan	Feb	Mar	Apr	May	Jun	Jul	Aug	Sep	Oct	Nov	Dec	Year	Highest recorded fall in a day mm	date
SUSSEX—*(contd)*																		
New Stream (Pevensey)																		
Polegate S. Wks.	TQ 51 593046	8	860	70	126	27	102	42	66	67	82	35	200	69	100	987	35	6 Aug
Eastbourne	TV 50 611980	7	783	59	106	21	99	38	53	73	70	31	199	73	103	925	33	22 Oct
Eastbourne, Churchdale Road	TQ 51 620005	8	...	56	102	21	91	34	55	77	63	28	203	65	104	899	34	19 Oct
Eastbourne Coast Guard	TV 50 629006	6	781	(58)	104	19	89	31	56	80	61	34	212	68	105	917	34	22 Oct
Coastal																		
Pevensey Bay	TQ 51 661042	5	...	53	116	23	84	37	51	65	56	28	191	68	89	860	34	19 Oct
Cuckmere																		
Birchford Farm	TQ 51 622198	109	851	56	142	22	91	56	88	88	87	29	201	79	107	1045	36	9 Feb
Cowbeech P. Sta.	TQ 51 610149	30	...	49	131	19	80	46	93	81	76	22	190	70	90	946	27	22 Oct
Heathfield	TQ 51 564202	93	887	55	144	19	87	52	93	91	89	31	192	71	102	1025	37	9 Feb
East Hoathly, Hesmonds	TQ 51 526162	61	...	49	133	16	78	44	94	75	85	28	170	63	88	923	...	...
Swansbrook Nursery	TQ 51 573150	44	...	49	133	17	77	44	98	71	86	31	201	69	89	966	...	...
Michelham Priory	TQ 51 559093	6	...	60	137	22	96	54	92	70	83	33	217	69	95	1028	29	22 Oct
Sherman Bridge R.F.M. Sta. Area 131 km²	TQ 51 053051	4	825	...	...	...	...	...	...	...	...	...	...	...	...	...	...	...
Milton Street	TQ 51 537048	30	...	67	122	28	97	45	67	69	68	25	209	72	101	966	35	22 Oct
Litlington	TQ 51 523018	6	812	68	107	21	101	41	59	66	73	22	211	68	94	931	33	22 Oct
Jevington	TQ 51 563013	84	839	71	120	25	105	45	69	77	78	42	199	74	118	1022	35	22 Oct
Friston, Water Tower	TV 50 553984	119	816	77	125	23	111	46	74	90	61	39	207	88	125	1066	31	22 Oct
Coastal																		
Seaford, Newlands	TV 50 495997	46	743	60	102	18	83	38	70	61	61	26	193	63	86	860	33	22 Jun
Poverty Bottom W. Wks.	TQ 51 467022	18	777	56	105	20	79	43	77	75	65	21	181	57	91	869	37	22 Jun
Ouse																		
Holmsted Manor	TQ 51 280267	69	...	39	127	19	115	51	80	94	90	36	140	56	92	939	33	22 Oct
Balcombe W. Wks.	TQ 51 290312	142	885	39	136	20	122	66	87	128	88	42	155	59	95	1037	39	22 Oct
Wakehurst Place	TQ 51 338314	125	...	(25)	(140)	(25)	111	67	85	115	83	39	149	72	98	1010	34	22 Oct
Lullings	TQ 51 328304	106	946	35	131	23	100	66	103	111	90	41	162	71	97	1030	35	22 Oct
Shell Brook W. Wks.	TQ 51 334284	39	849	40	137	20	120	61	92	120	91	35	153	65	98	1032	31	22 Oct
Cinder Hill	TQ 51 379297	88	...	37	137	21	101	54	95	97	91	33	165	59	94	983	29	6 Aug
Holy Well P. Sta.	TQ 51 369276	38	818	36	135	24	104	59	93	102	96	28	172	63	97	1009	33	6 Aug
Furners Green	TQ 51 406263	64	...	32	120	20	94	55	85	98	97	26	163	65	86	942	36	22 Jun
North Chailey	TQ 51 372212	76	794	38	139	15	82	60	105	68	85	27	151	59	77	907	34	22 Jun
Newick	TQ 51 413211	46	808	39	147	17	81	57	109	70	86	25	172	66	81	950	36	22 Jun
Gold Bridge R.F.M. Sta. Area 182 km²	TQ 51 429214	12	859	...	...	...	...	...	...	...	...	...	...	...	...	...	...	...
Horney Common	TQ 51 449252	52	...	34	123	18	80	52	100	82	97	31	163	67	86	933	36	22 Jun
Uck (Ouse)																		
Burnt Oak	TQ 51 510269	58	...	38	137	22	92	55	105	83	105	36	173	72	103	1022	...	...
Hadlow Down	TQ 51 526241	130	...	39	132	16	88	52	99	79	90	26	169	67	90	947	...	...
East Hoathly	TQ 51 512147	40	...	50	137	15	82	49	97	62	73	27	167	59	84	904	39	22 Jun
Isfield R.F.M. Sta. Area 88.7 km²	TQ 51 459190	11	850	...	...	...	...	...	...	...	...	...	...	...	...	...	...	...
Ouse																		
Plumpton	TQ 51 356134	76	...	63	197	29	105	74	102	82	111	36	193	72	114	1179	39	6 Aug
Barcombe Mills	TQ 51 432148	6	842	56	156	20	91	56	95	64	83	29	174	68	88	981	34	22 Jun
Offham P. Sta.	TQ 51 402120	6	868	60	157	23	86	56	89	74	82	28	175	71	98	999	34	22 Jun
Lewes, Brookside	TQ 51 407095	21	...	63	135	26	93	42	121	73	81	28	167	71	92	993	40	18 Jun
Kingston W. Wks.	TQ 51 408092	4	923	61	143	27	94	48	123	70	83	29	170	70	96	1015	43	18 Jun
Coastal																		
Balsdean P. Sta.	TQ 51 378045	57	800	62	127	23	84	45	85	77	69	25	187	65	103	955	35	6 Aug
Ovingdean	TQ 51 359034	49	748	63	111	24	84	38	79	63	69	23	154	59	84	851	32	22 Oct
Moulsecoomb																		
Falmer	TQ 51 341083	55	919	67	143	31	98	55	112	85	103	36	186	72	113	1099	47	6 Aug
Race Hill Res.	TQ 51 335057	139	867	68	126	20	102	51	139	79	93	31	152	65	91	1017	56	18 Jun
Brighton, Lewes Road	TQ 51 319060	42	...	62	120	21	97	45	134	72	86	30	143	58	84	953	55	18 Jun
Brighton	TQ 51 313054	10	688	54	116	20	82	41	79	62	71	22	134	50	77	809	36	22 Oct
Coastal																		
Hove, Goldstone Bottom	TQ 51 285065	53	...	61	129	25	94	47	120	79	81	43	179	58	93	1011	44	18 Jun
Mile Oak	TQ 51 243079	51	801	55	140	26	96	53	101	86	90	29	152	60	88	977	43	6 Aug
Adur																		
Knepp Castle	TQ 51 155218	20	798	47	120	19	117	68	75	75	79	34	141	60	77	912	33	22 Oct
Hatterell Bridge R.F.M. Sta. Area 109 km²	TQ 51 178197	4	785	...	...	...	...	...	...	...	...	...	...	...	...	...	...	...
Washington	TQ 51 118135	23	871	54	152	30	129	79	63	83	105	45	168	65	84	1057	46	6 Aug
Adur East (Adur)																		
Clayton P. Sta.	TQ 51 304137	79	...	56	186	29	104	73	92	67	115	53	193	69	105	1143	38	6 Aug
Hurstpierpoint College, New site	TQ 51 289179	29	...	47	159	20	95	76	84	62	77	34	148	59	78	939	31	22 Jun
Burgess Hill S. Wks.	TQ 51 311204	27	...	38	140	17	90	63	65	71	85	33	141	59	76	878	31	6 Aug
South Lodge, New site	TQ 51 219254	69	...	39	118	20	125	60	76	97	74	35	140	50	86	921	30	22 Jun
Shermanbury Grange, Orchard Site	TQ 51 203192	12	...	42	139	22	96	67	67	67	83	33	150	53	74	893	29	22 Jun
Newtimber Place	TQ 51 270138	48	850	52	163	27	105	74	81	68	97	35	155	61	87	1005	33	6 Aug
Adur																		
Kings Barn Farm	TQ 51 161158	23	...	54	163	26	111	86	59	67	93	40	164	62	82	1007	34	6 Aug
Upper Beeding, Sele Gardens	TQ 51 195105	5	...	56	158	28	119	59	77	85	103	33	168	60	94	1041	44	6 Aug
Beeding, Henfield Road	TQ 51 196102	6	...	48	147	27	115	61	77	84	108	28	164	58	82	1000	44	6 Aug
Applesham Farm	TQ 51 194071	35	...	55	139	27	104	44	76	83	95	37	150	60	90	960	43	6 Aug
Shoreham W. Wks.	TQ 51 210068	3	798	54	128	25	93	52	102	75	89	29	140	55	79	920	41	6 Aug
Coastal																		
North End	TQ 51 119097	64	918	64	158	37	142	69	69	93	119	30	176	71	101	1129	52	6 Aug
Worthing	TQ 51 153029	7	698	50	116	19	97	35	72	55	66	23	144	48	68	793	35	6 Aug
Highdown Tower	TQ 51 096040	30	...	51	113	17	104	52	77	61	86	26	143	50	68	849	40	6 Aug
Ferring	TQ 51 091019	8	732	58	116	19	108	48	81	53	81	24	143	56	68	854	38	6 Aug
Littlehampton	TQ 51 033019	5	717	50	111	20	108	47	79	59	74	30	144	50	62	835	36	6 Aug
Arun																		
Horsham, Pondtail	TQ 51 167304	46	764	40	123	19	110	64	85	97	66	39	148	55	72	919	37	22 Oct
Rusper	TQ 51 190363	79	824	41	139	21	116	65	77	101	73	45	143	52	90	964	38	22 Oct
Horsham, Pondtail Road	TQ 51 174321	50	...	41	118	17	104	61	81	90	63	35	141	50	67	866	38	22 Oct
Jayes Park	TQ 51 141408	88	...	42	143	23	106	72	55	97	74	41	139	48	88	929	31	22 Oct
Dunsfold (Arun)																		
Chiddingfold	SU 41 964355	70	...	49	129	23	102	63	51	80	76	30	152	53	68	875	26	6 Aug
Arun																		
Blackdown Beacon	SU 41 923292	184	...	67	151	23	118	70	65	79	86	42	164	64	87	1016	41	22 Oct
Kirdford	SU 41 999278	46	742	(44)	115	16	101	66	58	79	86	31	150	61	62	870	...	...
Kirdford, Pound Common	TQ 51 015270	24	...	46	125	20	111	62	62	77	79	27	140	54	73	875	37	22 Oct
Brinsbury Farm	TQ 51 067222	18	...	46	123	15	98	55	65	83	61	30	179	65	69	888	38	22 Oct

Station	Grid reference	Altitude m	Av. 1916–1950 mm	Jan	Feb	Mar	Apr	May	Jun	Jul	Aug	Sep	Oct	Nov	Dec	Year	Highest recorded fall in a day mm	date
SUSSEX—*(contd)*																		
Rother (Arun)																		
Greatham	SU 41 773301	79	846	57	162	17	115	71	63	74	76	34	171	54	81	975	...	...
Liss, Kippences	SU 41 777285	76	923	59	164	25	115	69	72	82	79	32	177	67	87	1027	34	22 Oct
Petersfield Res.	SU 41 749245	91	962	67	174	23	132	78	71	77	92	44	185	72	89	1106	35	6 Aug
Petersfield, Town Hall	SU 41 750234	61	...	65	166	25	118	69	63	66	93	40	194	69	86	1055	39	22 Oct
Fyning	SU 41 813240	70	856	62	155	24	122	74	73	76	78	31	172	64	81	1009	45	22 Oct
Midhurst	SU 41 885216	41	867	53	152	25	117	81	66	74	78	32	155	61	82	975	35	22 Oct
Fernhurst	SU 41 908267	57	885	60	144	21	110	70	70	84	84	28	164	58	76	967	45	22 Oct
Lavington House	SU 41 946165	58	...	56	166	31	149	96	89	104	102	41	164	63	93	1155	45	6 Aug
Gorehill	SU 41 985201	38	901	59	150	25	120	81	78	85	71	37	152	63	75	996	40	22 Oct
Petworth Park	SU 41 974216	76	880	59	149	27	123	86	68	81	79	33	147	61	77	992	37	22 Oct
Petworth, Barlavington	SU 41 971161	53	978	62	165	34	145	85	94	95	100	44	151	68	96	1138	47	6 Aug
Hardham P. Sta.	TQ 51 033176	9	--	51	148	27	123	73	84	79	84	45	155	57	79	1007	35	6 Aug
Arun																		
West Chiltington P. Sta.	TQ 51 093171	30	876	54	146	27	121	76	70	74	102	45	143	58	77	992	34	6 Aug
Thakeham	TQ 51 100168	38	842	51	138	26	116	74	69	68	99	42	(152)	(57)	(76)	969	35	6 Aug
Pulborough	TQ 51 059185	19	...	48	135	25	112	66	90	81	79	48	150	55	71	961	34	22 Oct
Parham House	TQ 51 060144	30	871	57	141	31	135	72	67	77	105	54	150	64	83	1035	40	6 Aug
Sutton, The Croft	SU 41 981151	47	...	56	149	33	139	79	86	96	102	43	145	65	91	1085	49	6 Aug
Bignor Park	SU 41 990155	43	971	64	153	31	155	73	88	99	103	45	141	73	93	1119	32	6 Aug
Arundel, River Road	TQ 51 017069	4	...	49	118	23	120	52	70	73	88	37	133	49	71	883	44	6 Aug
Arundel Castle	TQ 51 015072	32	799	45	120	21	113	51	65	72	89	32	137	48	70	866	44	6 Aug
Patching P. Sta.	TQ 51 092074	36	837	51	127	22	120	59	56	83	109	33	175	58	93	984	47	6 Aug
Rustington	TQ 51 046033	3	...	45	106	20	107	53	82	62	83	27	143	50	69	847	38	6 Aug
Court Wick	TQ 51 025035	5	...	44	104	22	105	42	89	60	85	49	141	46	63	849	35	6 Aug
Ryebank																		
Westergate Wood	SU 41 941070	28	...	58	122	27	116	61	80	87	103	28	141	61	77	961	44	6 Aug
Lidsey S. Wks.	SU 41 945033	4	...	45	128	(25)	(114)	53	71	70	92	25	153	57	78	911	44	6 Aug
West Walberton	SU 41 960064	20	785	54	126	28	123	57	79	87	95	28	143	56	81	958	45	6 Aug
Barnham, Lake Lane	SU 41 968046	9	...	49	122	26	116	56	71	77	90	25	147	51	73	903	44	6 Aug
Barnham	SU 41 954048	11	...	52	117	24	120	57	73	81	91	21	140	51	74	900	44	6 Aug
Eartham	SU 41 938093	73	...	63	123	26	122	73	84	93	125	27	135	65	88	1023	49	6 Aug
Coastal																		
Bognor Regis	SZ 40 934989	7	730	65	134	22	136	63	77	62	86	16	130	54	63	907	43	6 Aug
Portfield Depot	SU 41 878051	14	...	57	111	22	118	58	80	87	101	23	125	57	72	911	41	6 Aug
North Mundham	SU 41 875021	6	...	49	110	17	107	56	75	73	90	19	136	50	64	846	43	6 Aug
Selsey S. Wks.	SZ 40 848945	3	...	46	108	16	93	50	73	44	77	23	127	51	58	765	36	22 Jun
Birdham	SU 41 824004	5	723	55	102	15	111	44	81	61	86	17	123	51	63	808	...	...
Lavant																		
West Dean Park	SU 41 864127	58	955	61	148	33	131	85	83	94	119	32	165	73	98	1120	58	6 Aug
Chilgrove House	SU 41 834144	81	926	53	145	32	138	79	89	87	104	31	168	62	95	1083	53	6 Aug
Kingsham School	SU 41 865038	9	...	58	113	18	114	59	77	73	102	22	137	56	75	902	35	22 Jun
Coastal																		
Funtington P. Sta.	SU 41 811081	34	797	54	117	22	121	54	77	72	91	22	135	60	79	905	49	6 Aug
Chidham	SU 41 791036	4	714	52	116	19	99	50	78	78	86	25	143	56	77	878	37	6 Aug
Cobnor House	SU 41 792024	4	760	50	113	16	102	48	75	65	90	22	133	53	65	834	36	6 Aug
Thorney Island	SU 41 758026	3	681	50	103	14	105	50	75	62	83	21	130	54	62	809	33	22 Jun
Ems																		
Compton	SU 41 776148	77	...	63	155	27	131	81	97	90	107	38	189	70	92	1139	49	6 Aug
Stansted Park	SU 41 760101	49	821	53	115	23	120	59	86	72	100	26	184	64	81	985	49	6 Aug

HAMPSHIRE RIVER AUTHORITY AREA, 2735 SQUARE KILOMETRES

HYDROMETRIC AREA 42

Station	Grid reference	Altitude m	Av. 1916–1950 mm	Jan	Feb	Mar	Apr	May	Jun	Jul	Aug	Sep	Oct	Nov	Dec	Year	Highest recorded fall in a day mm	date
Coastal																		
Emsworth	SU 41 748074	11	...	50	112	22	110	56	75	79	104	28	161	56	79	932	45	6 Aug
Ditcham Park	SU 41 748177	169	1003	58	159	27	119	79	90	103	98	42	206	77	97	1155	43	22 Oct
Idsworth Water Tower	SU 41 729114	62	846	55	135	22	118	65	89	79	112	29	187	64	79	1034	...	...
Hayling Island	SZ 40 716988	4	...	60	104	13	97	47	68	58	71	23	133	52	57	784	29	22 Oct
Havant W. Wks.	SU 41 711063	8	720	44	105	21	93	45	79	67	86	21	144	57	68	830	38	6 Aug
Leigh Park Farm	SU 41 720086	32	...	49	106	22	109	54	76	76	93	23	165	62	71	906	46	6 Aug
Farlington Filters	SU 41 679061	46	712	46	108	21	97	51	86	72	83	20	152	60	73	869	42	6 Aug
Farlington Res.	SU 41 679060	36	747	47	109	21	102	53	86	72	83	21	154	60	73	881	42	6 Aug
Farlington P. Sta.	SU 41 678058	24	739	45	107	21	100	51	88	72	82	21	151	60	75	872	42	6 Aug
Eastney	SZ 40 684993	4	...	46	104	12	97	41	75	60	70	22	139	56	57	779	28	22 Oct
Southsea	SZ 40 640990	2	706	54	104	12	90	45	75	48	84	17	128	58	56	771	27	6 Aug
Wallington																		
Catherington Res.	SU 41 696146	130	903	56	149	26	111	79	97	81	95	34	178	70	88	1065	...	...
Southwick	SU 41 637076	46	794	59	122	21	109	58	95	71	95	28	168	63	71	961	...	...
Leydene	SU 41 680190	206	1014	61	182	26	138	85	93	95	104	55	194	78	103	1214	44	6 Aug
Hambledon	SU 41 652155	68	...	49	143	22	106	65	90	71	97	34	171	65	79	995	...	...
Worlds End	SU 41 629118	35	...	49	134	25	97	65	79	54	90	26	161	58	73	911	43	22 Oct
Upper Beckford Farm	SU 41 632110	34	812	52	136	26	100	64	81	61	92	28	168	61	72	941	...	...
North Fareham R.F.M. Sta. Area 111 km²	SU 41 587075	5	843	...	...	...	...	...	...	...	...	...	...	...	...	...	...	...
Fareham W. Wks.	SU 41 589071	12	815	57	126	25	99	51	91	59	97	39	177	53	68	942	47	22 Oct
Coastal																		
Stubbington	SU 41 556034	9	...	64	116	15	99	42	81	65	97	28	151	53	69	878	42	22 Oct
Gosport, Westfield Road	SZ 40 599000	9	...	65	125	13	95	42	83	57	77	35	148	57	61	857	29	6 Aug
Meon																		
West Meon P. Sta.	SU 41 648243	101	900	59	168	25	124	76	72	72	83	46	138	67	99	1029	...	...
Soberton	SU 41 596142	41	832	57	148	22	102	67	87	53	93	45	150	58	77	958	40	22 Oct
Mislingford R.F.M. Sta. Area 72.8 km²	SU 41 589141	30	912	...	...	...	...	...	...	...	...	...	...	...	...	...	...	...
Coastal																		
Hamble																		
Wintershill Hall	SU 41 527184	76	808	43	141	22	101	68	83	61	85	37	143	64	67	915	37	6 Aug
Upper Swanmore	SU 41 581174	96	865	54	144	21	100	68	82	68	85	39	136	58	80	937	39	22 Oct
Hoe P. Sta.	SU 41 567167	34	...	49	142	20	93	63	77	59	80	35	134	65	74	893	39	22 Oct
Curdridge, Sunnyside	SU 41 539137	27	818	54	142	24	90	67	81	65	79	38	149	60	77	924	38	22 Oct
Coastal																		
Itchen																		
Chilland	SU 41 526328	62	800	41	128	21	101	75	84	68	58	37	139	48	68	867	33	22 Jun
Martyr Worthy	SU 41 517338	84	765	45	133	18	105	76	82	68	71	39	126	47	74	884	31	22 Jun
Harestock	SU 41 469318	67	...	51	145	25	120	90	68	72	90	54	132	47	87	982	35	29 Aug

Station	Grid reference	Altitude *m*	Av. 1916–1950 *mm*	Rainfall in millimetres Jan	Feb	Mar	Apr	May	Jun	Jul	Aug	Sep	Oct	Nov	Dec	Year	Highest recorded fall in a day *mm*	*date*
HAMPSHIRE—*(contd)*																		
Winchester	SU 41 484300	38	812	46	141	21	118	75	65	77	85	41	136	(51)	(89)	944	35	29 Aug
Winchester, Romsey Road	SU 41 475294	82	793	41	131	19	117	74	64	72	86	43	124	49	84	906	34	29 Aug
Twyford P. Sta.	SU 41 493249	42	829	39	146	17	105	70	60	62	92	36	136	55	80	897	33	29 Aug
Otterbourne	SU 41 468225	24	802	44	136	22	100	70	79	77	86	34	145	65	73	933	33	29 Aug
Otterbourne W. Wks.	SU 41 469233	34	813	45	138	20	103	72	76	74	83	40	143	59	76	929	31	22 Oct
Allbrook R.F.M. Sta. Area 360 km²	SU 41 461211	18	851	...	...	...	...	...	...	...	...	...	...	...	...	...	...	...
Bishopstoke, Church Road	SU 41 466196	32	...	44	135	25	100	66	88	71	82	32	147	52	62	903	32	22 Oct
Portswood S. Wks....	SU 41 434146	5	756	46	142	17	82	63	67	58	83	41	156	47	68	869	36	22 Oct
Bitterne Park	SU 41 441130	11	...	58	143	22	91	61	74	58	90	42	165	49	68	921	37	22 Oct
Coastal																		
Southampton	SU 41 419119	20	804	53	142	26	97	60	69	56	75	49	150	49	67	894	38	22 Oct
Millbrook S. Wks.	SU 41 387125	5	...	52	148	23	93	53	68	60	77	47	156	46	67	890	40	22 Oct
Rownhams	SU 41 385175	70	...	49	131	19	93	52	63	68	69	44	146	52	69	854	31	22 Oct
Test																		
Wootton House	SU 41 592530	123	795	46	120	20	110	67	80	101	83	33	115	51	82	909	28	22 Jun
Church Oakley	SU 41 564506	108	...	57	138	22	133	87	89	93	105	44	129	51	85	1033	35	29 Aug
Fosbury Manor	SU 41 306585	177	930	63	134	25	116	79	70	84	196	55	183	58	98	1161	55	29 Aug
Linkenholt Manor	SU 41 362581	204	894	52	134	22	108	86	60	73	168	55	167	56	83	1063	53	29 Aug
Conholt Park	SU 41 326550	219	890	63	143	23	129	84	72	96	157	58	181	58	105	1170	57	29 Aug
Bullington (Test)																		
West Stoke Farm	SU 41 477372	84	...	45	140	18	113	84	84	81	87	40	140	45	84	961	33	29 Aug
Wonston Manor	SU 41 471381	71	752	47	141	18	120	89	81	78	83	44	138	47	82	968	35	29 Aug
Test																		
Andover W. Wks.	SU 41 368467	63	733	48	127	15	111	72	62	65	119	65	152	45	76	958	55	29 Aug
Andover S. Wks.	SU 41 359447	61	750	51	124	16	113	71	68	61	115	(63)	159	(46)	75	962	59	29 Aug
Andover, Clarendon Avenue	SU 41 354448	58	...	60	136	16	118	76	68	59	118	64	164	47	81	1007	63	29 Aug
Grateley	SU 41 291436	91	803	60	145	15	116	67	93	62	108	48	193	53	78	1038	35	2 Oct
Leckford	SU 41 393365	117	802	47	125	14	110	71	68	81	148	61	142	57	72	997	72	29 Aug
Houghton Down Farm	SU 41 338352	64	...	52	124	13	127	73	74	69	99	47	155	49	76	957	30	18 Oct
Houghton Lodge	SU 41 344332	34	797	53	127	15	117	68	74	77	78	67	157	48	73	953	27	18 Oct
Sparsholt	SU 41 427319	113	852	49	131	19	113	77	68	76	108	53	136	54	78	961	54	29 Aug
Wallop (Test)																		
Isle of Wight Hill	SU 41 250375	152	...	53	152	15	132	63	66	56	81	38	180	60	79	976	...	...
Garlogs	SU 41 305355	52	...	55	134	17	125	61	68	60	85	67	154	48	77	950	37	29 Sep
Broughton P. Sta.	SU 41 307338	43	...	49	133	17	116	67	60	65	90	71	153	47	78	946	42	29 Sep
Broughton R.F.M. Sta. Area 53.6 km²	SU 41 311330	37	803	...	...	...	...	...	...	...	...	...	...	...	...	...	...	...
Test																		
Timsbury P. Sta.	SU 41 346255	30	792	54	136	18	108	63	66	64	63	54	138	52	77	892	27	29 Sep
Broadlands R.F.M. Sta. Area 1040 km²	SU 41 354188	10	808	...	...	...	...	...	...	...	...	...	...	...	...	...	...	...
Blackwater (Test)																		
Redlynch House	SU 41 199212	102	...	55	141	25	111	71	71	75	74	56	188	61	86	1014	31	18 Oct
Coastal																		
Bartley Nurseries	SU 41 310131	29	927	66	143	22	109	73	67	58	120	77	147	57	79	1017	37	29 Sep
Exbury	SU 41 424000	18	785	65	130	22	111	65	89	53	99	43	157	60	74	968	44	22 Oct
Lyndhurst...	SU 41 296086	42	...	74	153	22	102	68	76	57	152	72	151	63	84	1074	44	29 Aug
Lymington																		
Brockenhurst Park R.F.M. Sta. Area 98.9 km²	SU 41 318019	8	871	...	...	...	...	...	...	...	...	...	...	...	...	...	...	...
Coastal																		
Lymington, Waterford	SZ 40 829948	2	770	65	112	17	96	42	58	50	149	53	145	58	71	915	36	6 Aug
Avon																		
Holmsley	SZ 40 227998	65	895	86	134	17	110	53	59	54	99	68	182	59	79	1001	55	22 Oct
Everton	SZ 40 303937	16	732	66	115	17	111	38	57	47	134	53	149	56	71	915	35	6 Aug
Coastal																		
New Milton	SZ 40 254951	40	...	79	123	20	106	44	62	56	132	54	157	54	73	961	40	22 Oct
Milford-on-Sea...	SZ 40 288924	17	...	65	117	18	103	38	59	59	128	45	149	60	72	913	33	22 Oct

ISLE OF WIGHT RIVER AUTHORITY AREA, 381 SQUARE KILOMETRES

HYDROMETRIC AREA 101

Medina																		
Chillerton W. Wks....	SZ 40 490844	41	862	93	126	20	134	52	95	98	111	33	186	80	88	1116	46	6 Aug
Upper Shide R.F.M. Sta. Area 29.8 km²	SZ 40 503874	10	861	...	...	...	...	...	...	...	...	...	...	...	...	...	...	...
Carisbrooke County Junior School ...	SZ 40 483882	46	...	85	134	20	116	55	82	70	103	36	168	78	82	1030	40	6 Aug
Carisbrooke W. Wks.	SZ 40 487881	17	842	83	136	19	122	56	81	70	106	32	151	67	76	998	43	6 Aug
Osborne	SZ 40 513947	55	...	71	137	17	101	53	79	65	97	33	159	65	69	947	32	6 Aug
Coastal																		
Ryde	SZ 40 595929	5	760	69	121	17	100	54	66	61	105	27	153	63	65	903	35	6 Aug
Yar, East																		
Knighton W. Wks.	SZ 40 567868	18	846	91	126	19	122	63	73	84	89	26	172	77	76	1016	40	6 Aug
Alverstone Mill R.F.M. Sta. Area 57.5 km²	SZ 40 577857	4	874	...	...	...	...	...	...	...	...	...	...	...	...	...	...	...
Shanklin, Victoria Avenue	SZ 40 581812	52	926	86	141	24	131	58	68	70	117	31	182	71	90	1070	53	6 Aug
Sandown P. Sta.	SZ 40 589849	6	833	74	134	19	110	69	68	77	93	20	154	67	75	959	43	6 Aug
Coastal																		
Sandown	SZ 40 602844	4	771	65	129	20	109	58	67	69	89	19	154	64	82	928	40	6 Aug
Shanklin	SZ 40 581808	55	903	88	143	22	134	59	68	76	120	31	173	79	90	1083	53	6 Aug
Ventnor Park	SZ 40 560774	135	824	94	137	19	126	50	79	77	100	24	160	77	90	1032	37	6 Aug
St. Catherine's Point	SZ 40 498753	16	751	87	106	11	122	41	72	79	95	23	139	63	72	910	29	6 Aug
Totland Bay	SZ 40 323867	30	...	66	115	17	103	47	57	79	97	55	125	55	65	883	36	6 Aug
Freshwater W. Wks.	SZ 40 347858	3	762	72	113	17	105	45	64	83	109	58	124	55	68	914	36	6 Aug
Freshwater P. Sta.	SZ 40 341871	3	752	72	112	15	101	45	61	71	103	62	123	56	60	882	37	6 Aug
Calbourne...	SZ 40 425859	70	...	89	124	17	119	53	90	86	106	51	155	70	80	1040	43	6 Aug
Cowes W. Wks.	SZ 40 490952	49	819	69	130	19	102	54	77	58	105	26	142	62	68	912	33	6 Aug

AVON AND DORSET RIVER AUTHORITY AREA, 4297 SQUARE KILOMETRES

HYDROMETRIC AREAS 43 AND 44

Station	Grid reference	Altitude m	Av. 1916–1950 mm	Jan	Feb	Mar	Apr	May	Jun	Jul	Aug	Sep	Oct	Nov	Dec	Year	Highest recorded fall in a day mm	Highest recorded fall in a day date
Avon																		
Bishops Cannings	SU 41 035643	136	803	52	97	15	103	63	58	56	115	28	141	57	75	860	35	21 Aug
Oare	SU 41 154625	152	...	40	93	18	91	64	63	58	116	38	162	57	92	893	42	29 Aug
Alton Priors	SU 41 109624	134	810	52	104	19	101	65	62	58	113	31	170	55	96	925	44	29 Aug
Woodborough	SU 41 115599	125	815	54	104	20	92	69	66	54	101	33	175	57	90	914	41	29 Aug
Everleigh	SU 41 201539	171	793	54	113	18	94	59	79	67	104	34	150	65	79	917	27	29 Aug
Queens Falls R.F.M. Sta. Area 324 km²	SU 41 151414	67	795	...	...	...	...	...	...	...	...	...	...	...	...	...	...	...
Larkhill	SU 41 137447	132	719	53	113	19	101	53	53	55	106	29	165	53	87	886	29	18 Oct
Nadder (Avon)																		
Fonthill House	ST 31 927322	110	...	63	137	38	144	66	71	69	109	34	186	97	120	1133	52	4 Nov
Swallowcliffe	ST 31 973267	125	898	52	150	30	117	62	47	105	82	28	167	85	101	1027	55	4 Nov
Wylye (Nadder)																		
Maiden Bradley House	ST 31 803387	192	994	55	151	40	131	91	61	82	187	40	168	99	124	1229	78	21 Aug
Lower Pertwood	ST 31 883366	183	947	57	145	35	130	67	51	80	95	35	165	91	119	1070	48	4 Nov
Crockerton P. Sta.	ST 31 855423	137	939	49	134	35	97	89	67	72	123	39	157	(89)	(114)	1065	31	1 Oct
Warminster, Kingdown School	ST 31 887447	113	...	53	127	24	109	78	65	66	88	33	153	(83)	93	971	28	1 Oct
Upton Lovell	ST 31 951409	104	823	56	127	27	111	68	40	71	72	27	153	75	98	924	42	4 Nov
Bapton Manor	ST 31 994381	79	...	(51)	112	21	101	57	42	71	93	31	151	71	79	878	43	4 Nov
Fisherton de la Mere	ST 31 998387	104	831	47	118	24	97	60	39	77	91	30	160	70	88	901	37	4 Nov
Steeple Langford	SU 41 039374	73	826	62	131	17	114	58	41	49	97	32	154	65	79	898	36	4 Nov
Till (Wylye)																		
Maddington	SU 41 068443	87	752	57	109	23	98	52	41	55	91	35	164	57	87	869	29	4 Nov
Wylye (Nadder)																		
Nadder (Avon)																		
Salisbury, Folkestone Road	SU 41 139288	58	830	54	139	18	103	64	59	60	86	24	176	50	73	906	33	18 Oct
Bourne (Avon)																		
South Tidworth S. Wks.	SU 41 231468	104	...	55	132	13	100	57	59	64	99	38	179	61	79	935	29	22 Oct
Porton	SU 41 210366	111	747	43	129	14	93	64	58	55	67	35	182	52	65	858	29	18 Oct
Salisbury, Atherton House	SU 41 150307	76	788	45	139	14	95	59	57	55	84	30	168	52	72	871	32	18 Oct
Laverstock Mill R.F.M. Sta. Area 164 km²	SU 41 157303	46	790	...	...	...	...	...	...	...	...	...	...	...	...	...	...	...
Peters Finger S. Wks.	SU 41 159288	43	...	46	140	15	93	60	54	57	75	28	174	47	63	853	31	18 Oct
Ebble (Avon)																		
Broad Chalke	SU 41 039253	82	916	54	154	25	114	60	73	85	114	31	189	75	85	1059	42	4 Nov
Avon																		
East Mills R.F.M. Sta. Area 1450 km²	SU 41 158144	25	851	...	...	...	...	...	...	...	...	...	...	...	...	...	...	...
Martin	SU 41 062200	67	...	61	143	27	118	58	62	73	97	37	184	(76)	(89)	1023	31	6 Aug 18 Oct
Linwood	SU 41 179095	40	892	68	136	19	89	54	60	53	90	46	193	59	69	935	36	22 Oct
Ringwood, Broadshard Lane	SU 41 154058	21	...	83	148	20	97	53	45	51	81	42	205	63	81	969	42	22 Oct
Ringwood R.F.M. Sta. Area 1650 km²	SU 41 142054	14	851	...	...	...	...	...	...	...	...	...	...	...	...	...	...	...
Ringwood, Hightown Road	SU 41 156048	20	...	78	140	18	95	52	44	48	77	41	205	61	77	937	43	22 Oct
Hightown	SU 41 182056	76	848	77	135	18	89	52	53	48	80	47	186	59	79	922	42	22 Oct
Bisterne	SU 41 151006	13	845	90	145	19	99	55	39	58	95	48	208	61	77	995	48	22 Oct
Christchurch	SZ 40 155936	3	771	70	107	19	94	41	38	68	92	49	172	52	65	868	47	22 Oct
Stour (Avon)																		
Stourhead	ST 31 778341	183	1000	61	147	29	137	64	49	85	128	25	159	101	134	1118	56	4 Nov
Penselwood	ST 31 771314	119	885	56	141	31	127	71	53	76	120	29	159	103	119	1085	58	4 Nov
Gillingham	ST 31 812270	82	791	52	121	26	112	61	53	68	125	24	152	90	98	982	54	4 Nov
Stour Provost	ST 31 792215	76	764	53	106	22	128	56	56	78	118	20	174	97	93	1001	60	4 Nov
Wincanton	ST 31 704286	113	878	55	136	26	117	69	63	84	182	24	167	110	97	1130	66	4 Nov
Horsington	ST 31 686233	140	805	56	130	21	128	55	59	57	143	24	162	104	78	1017	66	4 Nov
Lydden (Stour)																		
Little Westrow	ST 31 697107	77	868	61	135	20	99	56	50	59	90	27	173	97	93	960	59	4 Nov
Stock House	ST 31 722130	84	840	60	128	19	90	59	45	50	97	25	170	94	95	932	57	4 Nov
Stour (Avon)																		
Boyne Hollow W. Wks.	ST 31 875217	162	...	60	123	27	134	53	55	85	85	27	169	100	100	1018	57	4 Nov
Shaftesbury	ST 31 861228	207	834	52	107	25	138	59	54	83	97	22	185	98	103	1022	56	4 Nov
Okeford Manor	ST 31 838127	79	885	65	155	20	101	62	46	49	85	19	202	92	95	990	62	4 Nov
Okeford, Fitzpaine P. Sta.	ST 31 809109	70	884	65	160	24	101	70	46	53	97	20	176	90	82	983	59	4 Nov
Shillingstone	ST 31 825113	58	869	71	155	19	106	66	47	45	91	19	193	79	80	973	51	4 Nov
Shillingstone, Green Hills	ST 31 829104	58	...	(71)	(152)	(20)	101	65	39	47	88	21	197	84	85	970	55	4 Nov
Bryanston	ST 31 872067	82	...	72	165	19	113	63	43	42	83	22	200	91	87	1000	57	4 Nov
Bryanston School	ST 31 868074	84	884	75	169	20	111	66	40	41	79	25	203	85	91	1006	52	4 Nov
Blandford, Kings Road	ST 31 891068	59	869	67	149	18	106	58	39	38	89	26	191	85	91	956	49	4 Nov
Winterborne Whitchurch	SY 30 834998	98	...	90	166	26	113	68	53	63	97	31	223	100	99	1130	57	4 Nov
Corfe Mullen W. Wks.	SY 30 973981	25	859	60	119	20	103	52	48	59	99	31	217	68	71	948	49	22 Oct
Corfe Mullen Plantation	SY 30 976963	64	878	83	150	21	115	54	49	61	100	32	212	72	80	1030	53	22 Oct
Allen (Stour)																		
Tollard Royal	ST 31 938177	131	...	64	147	27	121	65	63	68	87	33	198	91	113	1077	46	4 Nov
Stanbridge Mill P. Sta.	SU 41 011089	40	...	65	136	24	114	58	49	58	80	23	218	72	87	985	40	4 Nov
Walsford Bridge P. Sta.	SU 41 006009	21	849	63	138	20	104	48	44	54	90	23	211	75	69	939	47	22 Oct
Wimborne, Clevedon Lodge	SU 41 014001	34	...	73	151	27	118	56	51	56	97	29	220	77	81	1037	47	22 Oct
Stour (Avon)																		
Little Canford	SZ 40 048994	15	...	81	146	23	117	51	46	46	102	31	230	72	75	1020	47	22 Oct
Stapehill	SU 41 057000	34	833	78	135	17	114	43	44	43	107	33	237	73	70	994	47	22 Oct
Moors (Stour)																		
Potterne House	SU 41 095074	24	...	65	134	20	95	49	43	62	69	35	207	63	72	915	37	18 Oct 22 Oct
Hurn	SZ 40 117978	10	774	79	141	18	102	45	47	70	107	42	216	58	69	993	50	22 Oct
Hurnbridge	SZ 40 126967	6	762	79	134	18	100	43	44	71	97	47	219	67	73	993	58	22 Oct
Stour (Avon)																		
Bourne																		
Alderney Res.	SZ 40 045950	69	800	86	131	20	124	42	36	61	106	34	227	70	70	1007	51	22 Oct
Branksome, Bourne Valley	SZ 40 060924	24	799	83	136	24	117	46	42	74	117	41	223	64	77	1044	57	22 Oct
Bournemouth, Rothesay Road	SZ 40 074923	41	787	79	132	24	112	47	41	68	113	40	213	61	75	1004	54	22 Oct
Bournemouth	SZ 40 071917	40	...	74	125	21	116	41	41	70	110	38	216	55	69	976	52	22 Oct
Bournemouth, East Avenue	SZ 40 077926	44	790	75	126	23	111	43	40	70	115	42	213	59	72	989	58	22 Oct

Station	Grid reference	Altitude m	Av. 1916–1950 mm	Jan	Feb	Mar	Apr	May	Jun	Jul	Aug	Sep	Oct	Nov	Dec	Year	Highest recorded fall in a day mm	date
AVON AND DORSET—*(contd)*																		
Coastal																		
Parkstone, Balmoral Road	SZ 40 041914	43	...	79	136	19	127	44	47	68	96	34	209	(63)	(70)	993	56	22 Oct
Poole Gasworks	SZ 40 016907	2	796	81	122	15	115	39	39	65	97	36	200	65	70	945	54	22 Oct
Poole, Henry Harbin School	SZ 40 012916	9	794	78	128	19	117	42	48	57	104	32	211	69	63	965	52	22 Oct
Bloxworth Heath	SY 30 876928	32	902	86	143	22	120	47	57	59	120	32	230	86	87	1089	57	22 Oct
Piddle																		
Piddlehinton S. Wks.	SY 30 724962	76	...	97	167	31	132	71	49	68	119	21	193	101	101	1149	58	4 Nov
Puddletown	SY 30 759944	58	1002	96	156	28	121	70	42	68	111	21	198	95	101	1107	56	4 Nov
Milborne (Piddle)																		
Milton Abbas	ST 31 807030	216	1012	86	164	25	113	75	56	71	98	31	203	97	109	1129	56	4 Nov
Piddle																		
Trigon	SY 30 886889	18	869	84	133	23	128	46	41	68	119	30	198	87	77	1034	54	22 Oct
Baggs Mill R.F.M. Sta. Area 183 km²	SY 30 913876	2	973	...	...	...	...	...	...	...	...	...	...	...	...	...	...	...
Frome																		
Cattistock P. Sta.	ST 31 610001	162	...	97	150	35	130	67	58	67	105	19	202	109	105	1144	67	4 Nov
Hooke P. Sta.	ST 31 534006	150	...	79	164	36	140	91	51	55	107	35	184	101	111	1154	...	...
Minterne House	ST 31 659040	174	1021	87	177	27	131	67	67	68	101	26	190	97	105	1143	52	4 Nov
Dorchester W. Wks.	SY 30 684905	95	981	101	149	34	145	70	34	61	120	17	205	91	111	1137	55	22 Oct
Dorchester, Little Cheam	SY 30 716928	73	...	85	144	30	118	55	39	61	115	17	191	85	100	1041	47	4 Nov
Birkin House	SY 30 712917	70	...	88	147	31	125	59	37	61	110	17	178	86	100	1040	50	4 Nov
Kingston Maurward	SY 30 718912	82	...	96	136	28	127	52	37	57	107	19	200	88	94	1040	53	22 Oct
South Winterbourne (Frome)																		
Martinstown	SY 30 648888	82	978	110	143	37	145	59	41	57	118	22	192	99	113	1138	50	22 Oct
Dorchester, Queens Avenue	SY 30 688898	79	...	96	136	35	143	59	37	59	112	23	207	90	111	1108	57	22 Oct
Frome																		
Warmwell	SY 30 749873	41	911	108	139	34	147	53	39	52	120	29	230	93	95	1140	33	10 Jan
Winfrith	SY 30 820870	26	881	90	129	28	133	48	38	46	108	22	238	91	79	1050	56	4 Nov
East Stoke Mill R.F.M. Sta. Area 414 km²	SY 30 866867	8	996	...	...	...	...	...	...	...	...	...	...	...	...	...	...	...
Creech Grange	SY 30 911823	69	954	109	121	24	136	44	43	44	122	44	183	88	78	1037	49	4 Nov
Coastal																		
Corfe Castle W. Wks.	SY 30 960825	11	...	107	143	26	149	47	52	52	116	38	181	85	93	1086	49	6 Aug
Swanage	SZ 40 030794	10	776	71	119	19	118	39	44	47	109	38	170	58	76	908	40	6 Aug
Friar Waddon	SY 30 652857	65	891	113	115	34	127	51	33	43	113	27	199	91	90	1037	63	22 Oct
Weymouth	SY 30 676791	5	763	94	115	31	111	40	37	33	96	23	190	89	72	931	57	22 Oct
Portland Bill	SY 30 677683	11	675	74	95	14	94	35	38	31	69	21	168	92	62	792	51	4 Nov
Portland	SY 30 687717	94	735	85	111	17	127	39	42	35	84	28	178	88	71	905	54	4 Nov
Wyke Regis	SY 30 667770	12	731	96	119	22	117	37	33	31	91	23	176	92	70	908	44	4 Nov
Abbotsbury, East Cottage	SY 30 578852	43	814	98	107	26	106	39	31	39	89	16	169	108	83	911	58	4 Nov
Abbotsbury, West End	SY 30 565851	49	804	95	109	29	98	43	29	43	91	18	173	111	81	919	64	4 Nov
Litton Cheney	SY 30 550902	58	863	102	131	35	124	45	28	43	111	16	185	105	90	1015	63	4 Nov
Burton Bradstock	SY 30 487898	24	...	72	112	25	100	44	31	44	90	15	179	79	71	859	46	4 Nov
Beaminster	ST 31 482011	56	1018	81	154	38	131	81	54	57	115	24	176	95	99	1105	56	4 Nov
Netherbury	SY 30 470994	47	970	79	143	38	121	73	42	50	128	25	177	91	97	1064	57	4 Nov
Bridport, South Street	SY 30 467927	12	...	81	116	27	115	52	39	39	107	22	171	81	80	929	45	4 Nov
Charmouth S. Wks.	SY 30 371937	10	...	76	121	25	114	55	36	35	100	25	154	68	85	895	40	4 Nov
Pinhay	SY 30 315913	137	909	90	140	28	134	63	50	49	113	25	162	66	104	1023	36	22 Oct

DEVON RIVER AUTHORITY AREA, 6115 SQUARE KILOMETRES

HYDROMETRIC AREAS 45, 46, 50 AND 51D

Devon, south area, 3758 square kilometres

Hydrometric areas 45 and 46

Station	Grid reference	Altitude m	Av. 1916–1950 mm	Jan	Feb	Mar	Apr	May	Jun	Jul	Aug	Sep	Oct	Nov	Dec	Year	Highest recorded fall in a day mm	date
Coastal																		
Rousdon	SY 30 301909	162	901	85	127	26	127	58	48	51	102	22	147	58	90	941	34	22 Oct
Axe																		
Drimpton	ST 31 419049	111	...	79	186	44	127	90	(63)	(51)	(114)	33	204	96	88	1176	59	4 Nov
Cricket St. Thomas	ST 31 371084	134	1040	69	151	35	114	74	73	47	120	26	189	80	79	1058	45	4 Nov
Forde Abbey	ST 31 359051	70	995	77	159	32	129	74	63	39	77	27	182	83	80	1021	44	4 Nov
Hawkchurch	ST 31 358006	81	...	92	158	43	148	90	63	42	111	26	197	78	93	1141	41	4 Nov
Wambrook	ST 31 287067	171	--	93	181	36	137	77	47	47	110	24	168	66	114	1102	41	24 Feb
Axminster, Lyme Road	SY 30 307977	107	...	93	177	44	144	81	58	39	119	36	198	70	99	1157	39	4 Nov
Yarty (Axe)																		
Yarcombe	ST 31 241079	157	...	86	152	37	135	65	51	46	101	25	171	58	116	1041	30	22 Oct
Kilmington	SY 30 267986	67	990	95	166	45	141	83	58	47	110	34	169	54	106	1107	31	14 Oct
Axe																		
Whitford Bridge R.F.M.Sta. Area 298 km²	SY 30 262953	7	1009	...	...	...	...	...	...	...	...	...	...	...	...	...	...	...
Farway	SY 30 186950	76	...	138	186	41	157	90	64	49	135	27	177	61	118	1244	32	24 Feb
Wilmington	ST 31 210000	131	...	110	195	38	161	82	57	53	123	36	182	60	124	1221	36	24 Feb
Seaton Junction	SY 30 248968	35	964	99	147	36	144	70	43	46	111	25	159	51	90	1021	29	10 Jan
Colyford	SY 30 244923	34	931	107	147	32	138	80	68	42	99	24	149	55	94	1033	32	6 Aug
Coastal																		
Beer, Windyridge	SY 30 230894	67	865	95	129	28	126	59	70	41	90	20	136	57	94	945	32	22 Oct
Beer, Set Fair	SY 30 226892	52	...	92	130	27	124	59	63	41	91	19	129	57	99	930	31	22 Oct
Sidbury	SY 30 142932	94	...	141	164	30	148	81	69	49	139	21	165	51	106	1164	41	10 Jan
Chelson	SY 30 169917	194	...	141	164	35	148	78	73	51	131	23	169	48	100	1161	33	10 Jan
Sidmouth	SY 30 124873	10	785	(114)	120	26	120	54	49	43	93	20	120	37	64	862	29	22 Oct
Otter																		
Otterhead	ST 31 224141	247	1128	114	147	41	150	67	104	63	119	31	175	70	131	1212	...	...
Up-Ottery	ST 31 203088	226	1078	109	154	46	157	81	63	54	119	31	175	67	131	1190	29	22 Oct
Honiton, Copper Castle	ST 31 171004	160	1061	129	194	35	151	79	63	48	118	18	187	53	120	1194	35	24 Feb
Feniton Court	SY 30 108994	73	958	106	145	35	132	62	69	42	93	21	165	53	107	1030	33	10 Jan
Broadhembury	ST 31 101048	134	...	96	132	36	129	70	93	55	123	21	167	72	110	1106	28	3 Aug
Greatwell, Ottery St. Mary	SY 30 110954	85	...	132	159	28	134	68	71	46	112	18	159	48	94	1069	40	10 Jan
Ottery St. Mary	SY 30 088952	52	979	127	156	31	137	67	71	46	117	16	168	50	97	1083	41	10 Jan
Venn Ottery	SY 30 084914	46	913	133	164	42	136	78	62	49	113	19	159	53	97	1105	38	10 Jan
Dotton P.Sta.	SY 30 082882	24	...	121	143	33	124	66	51	46	105	21	145	49	81	987	42	10 Jan
Coastal																		
Exmouth Filters	SY 30 026819	73	813	115	144	22	135	56	33	36	74	21	149	45	65	895	43	10 Jan
Exmouth	SY 30 015813	59	846	101	146	23	125	55	37	34	69	23	148	46	61	869	35	10 Jan

Station	Grid reference	Altitude m	Av. 1916–1950 mm	Jan	Feb	Mar	Apr	May	Jun	Jul	Aug	Sep	Oct	Nov	Dec	Year	Highest recorded fall in a day mm	date
DEVON SOUTH—*(contd)*																		
Clyst																		
Rockbeare Manor	SY 30 031940	55	...	117	135	27	132	65	54	39	78	22	158	65	77	967	40	10 Jan
Pinhoe	SX 20 964940	30	...	107	124	29	128	55	46	35	67	23	164	51	86	914	33	10 Jan
Pinhoe, Cotmead	SX 20 969942	23	...	103	128	29	122	67	48	36	67	23	164	50	69	904	29	4 Oct
Exeter	SX 20 995937	32	850	129	124	23	131	55	49	37	67	21	159	51	68	915	57	10 Jan
Redhayes	SX 20 970934	43	820	109	113	28	128	56	49	36	61	21	162	50	76	890	36	10 Jan
Stallcombe	SY 30 038891	110	...	155	170	27	150	67	55	40	95	20	172	55	71	1076	49	10 Jan
Topsham, Red Rock	SX 20 970880	15	...	120	156	35	143	57	47	41	63	15	161	50	68	957	42	10 Jan
Exe																		
Lower Thorne	SS 21 843383	290	1599	112	199	89	172	141	113	87	150	79	177	130	376	1827	85	12 Dec
Exford Rectory	SS 21 857384	293	1511	102	179	64	157	111	101	77	136	64	168	117	319	1595	60	12 Dec
Winsford	SS 21 906348	189	1351	121	170	69	183	94	101	64	117	65	152	107	241	1482	...	...
Enterwell R.F.M.Sta. Area 20.4 km^2	SS 21 919356	188	1584	...	...	...	...	...	...	...	...	...	...	...	...	...	...	...
Goosemoor	SS 21 958354	369	1405	107	161	71	168	97	119	72	111	65	159	105	243	1479	42	10 Jun
Barle (Exe)																		
Great Vintcombe, Lower	SS 21 732395	379	1960	124	211	93	200	180	167	126	161	116	164	160	431	2131	...	...
Kinsford Gate	SS 21 744365	457	2030	119	222	102	212	161	158	110	168	128	177	157	426	2140	...	...
Simonsbath, Thornmead	SS 21 806382	387	...	114	227	92	178	165	119	97	149	102	203	140	391	1976	61	12 Dec
Hawkridge	SS 21 877327	314	1496	122	188	76	196	106	120	90	132	78	141	127	294	1669	47	28 Dec
Litton	SS 21 803337	390	...	115	187	93	193	144	124	87	159	93	167	127	360	1850	...	...
Dulverton, The Cottage	SS 21 911278	137	1342	111	143	65	175	93	103	60	129	59	164	111	258	1471	36	28 Dec
Exe																		
Bittescombe Manor	ST 31 010298	282	...	103	135	49	169	95	123	60	119	61	155	99	200	1367	44	10 Jun
Clayhanger	ST 31 022230	165	...	77	150	55	156	97	104	45	105	44	145	73	163	1213	31	10 Jun
Iron Bridge R.F.M.Sta. Area 422 km^2	ST 21 943178	74	1407	...	...	...	...	...	...	...	...	...	...	...	...	...	...	...
Little Timewell, Morebath	SS 21 958257	168	1167	89	123	55	158	81	102	46	112	42	163	81	155	1208	29	10 Jun
Stoodleigh Court	SS 21 916185	247	...	109	155	61	162	82	78	68	113	59	214	83	216	1401	51	4 Oct
Huntsham	SS 21 990180	241	...	79	135	47	156	90	100	55	121	45	179	70	162	1238	30	4 Oct
Templeton	SS 21 887139	238	...	112	135	49	170	81	74	59	109	38	198	79	179	1284	60	4 Oct
Cadeleigh	SS 21 914062	149	...	111	117	34	161	52	66	40	115	41	139	61	147	1084	31	24 Jan
Thorverton R.F.M.Sta. Area 601 km^2	SS 21 936016	25	1280	...	...	...	...	...	...	...	...	...	...	...	...	...	...	...
Culm (Exe)																		
Westcombe	ST 31 202162	219	1088	93	152	40	140	67	109	56	118	28	169	63	118	1152	49	10 Jun
Dunkeswell	ST 31 136076	253	...	109	143	43	150	72	99	56	125	24	178	61	120	1179	39	3 Aug
Hemyock	ST 31 139126	185	1035	100	152	53	133	79	81	56	128	31	161	66	120	1160	40	3 Aug
Hemyock	ST 31 138128	165	...	97	144	49	129	74	81	55	126	30	158	65	118	1126	39	3 Aug
Culm Davy House	ST 31 124146	156	910	88	155	49	123	69	95	60	106	28	139	(63)	121	1096	29	22 Oct
Woodgate Farm	ST 31 102152	165	895	85	131	40	121	69	83	57	96	26	141	67	115	1029	31	22 Oct
Kentisbeare	ST 31 065075	119	...	89	119	34	125	70	76	52	125	26	149	65	119	1050	29	3 Aug
Hockworthy	ST 31 039194	191	...	75	152	49	156	91	115	53	113	41	156	71	150	1222	43	10 Jun
Tiverton Junction	ST 31 033113	75	919	84	132	38	118	73	72	47	104	29	152	57	131	1039	29	22 Oct
Woodmill R.F.M.Sta. Area 226 km^2	ST 31 021058	54	965	...	...	...	...	...	...	...	...	...	...	...	...	...	...	...
Bradninch	SS 21 994033	69	920	90	123	34	130	54	70	45	100	24	157	57	122	1005	32	3 Aug
Creedy (Exe)																		
Kennerleigh	SS 21 819074	117	...	107	129	36	151	68	71	51	117	34	142	78	140	1123	28	10 Jan
Shobrooke Park	SS 21 854016	90	...	127	117	30	157	67	61	41	114	29	135	67	113	1057	41	10 Jan
Yeo (Creedy)																		
Hittisleigh	SX 20 737953	183	...	131	146	34	193	63	63	52	128	29	134	72	117	1161	45	16 Apr
Horwell Barton	SS 21 760003	91	...	96	119	32	158	64	58	42	124	29	122	68	120	1032	39	10 Jan
Crediton, Okefield	SS 21 832006	91	805	111	106	31	156	58	52	42	107	26	133	68	103	994	38	10 Jan
Creedy (Exe)																		
Newton St.Cyres	SX 20 879980	40	...	115	129	26	153	52	53	38	100	20	152	59	100	997	39	10 Jan
Exe																		
Exeter Institution	SX 20 921926	47	810	145	140	31	143	58	55	31	72	16	139	52	73	955	48	10 Jan
Heavitree	SX 20 943920	34	819	127	123	31	130	53	48	30	63	19	137	48	71	879	43	10 Jan
Ide	SX 20 897903	30	...	158	162	35	155	59	68	32	87	18	135	57	86	1051	61	10 Jan
Shillingford St.George	SX 20 905880	52	923	145	153	39	158	65	56	33	77	24	140	52	77	1020	52	10 Jan
Coastal																		
Kenton	SX 20 959834	18	959	127	172	31	144	58	61	35	81	22	164	49	75	1021	41	10 Jan
Starcross	SX 20 972821	9	862	111	163	28	135	58	44	29	67	20	160	51	66	931	35	10 Jan
Mamhead	SX 20 944800	46	...	135	194	36	167	70	52	34	88	27	185	61	82	1130	41	10 Jan
Dawlish, Beach House	SX 20 969774	24	868	106	165	28	134	60	48	34	68	23	162	53	71	953	37	22 Oct
Luscombe Castle	SX 20 942767	53	1021	124	190	29	159	72	40	34	87	27	187	61	80	1090	37	10 Jan
Holcombe	SX 20 953751	61	954	113	192	33	147	77	51	39	84	25	175	66	82	1083	39	22 Oct
Teign																		
Throwleigh	SX 20 665902	285	...	141	217	57	180	110	72	60	180	49	160	115	240	1580	58	3 Aug
South Teign (Teign)																		
White Ridge	SX 20 650824	488	2235	223	347	101	288	(140)	127	87	211	86	253	163	382	2407	...	...
Fernworthy Res. R.F.M. Sta. Area 9.95 km^2	SX 20 671844	353	2022	...	...	...	...	...	...	...	...	...	...	...	...	...	...	...
Metherall	SX 20 672843	352	...	225	369	97	285	134	109	81	220	77	225	135	337	2294	70	24 Feb
Murchington	SX 20 692881	191	1302	145	233	55	193	95	68	54	148	41	168	104	212	1515	46	24 Feb
Teign																		
Yellam Cottage	SX 20 714870	207	...	176	227	47	218	78	64	44	131	39	169	87	174	1453	54	10 Jan
Newton Barton	SX 20 711921	213	947	120	139	31	155	66	49	51	143	29	132	80	147	1142	38	3 Aug
Clifford Bridge Farm	SX 20 780897	88	1059	176	204	33	225	79	65	42	103	31	165	78	101	1301	54	10 Jan
Smithacott	SX 20 798848	259	1162	190	235	43	236	74	57	43	126	33	191	72	105	1404	...	...
Kennick	SX 20 808838	256	1157	148	218	37	210	69	57	39	121	27	165	73	118	1283	41	18 Feb
Tottiford	SX 20 808824	241	1173	148	197	39	207	65	51	30	107	16	170	68	102	1198	42	18 Feb
Trusham	SX 20 854820	94	1013	121	190	37	175	71	45	37	98	31	158	57	86	1106	35	24 Feb
Chudleigh	SX 20 866793	70	1032	136	201	38	189	70	46	34	86	25	166	62	91	1145	34	10 Jan
Bovey (Teign)																		
Lower Hookney	SX 20 715825	280	...	225	326	70	285	101	85	59	181	58	207	111	245	1953	60	10 Jan
Fursdon, Moretonhampstead	SX 20 752841	226	...	187	203	51	235	61	61	40	131	33	194	78	131	1407	68	10 Jan
Bovey Tracey	SX 20 803772	64	...	150	277	49	203	95	49	32	113	33	179	71	125	1376	44	24 Feb
Teign																		
Preston R.F.M. Sta. Area 381 km^2	SX 20 856746	4	1267	...	...	...	...	...	...	...	...	...	...	...	...	...	...	...
Kingsteignton	SX 20 871749	24	...	122	166	33	180	63	42	36	81	28	170	57	79	1058	35	22 Oct
Newton Abbot	SX 20 828729	83	1005	128	212	37	178	72	43	35	97	31	174	53	97	1156	36	1 Oct
Owlacombe Cross	SX 20 765731	172	...	178	293	60	261	95	39	40	128	36	209	90	171	1601	51	24 Feb
Compton	SX 20 863652	64	...	143	203	30	203	62	49	48	97	31	170	65	87	1188	48	22 Oct

Station	Grid reference	Altitude m	Av. 1916–1950 mm	Jan	Feb	Mar	Apr	May	Jun	Jul	Aug	Sep	Oct	Nov	Dec	Year	Highest recorded fall in a day mm	date
DEVON SOUTH—*(contd)*																		
Coastal																		
Torquay, Lydwell Court	SX 20 927646	125	925	115	175	23	171	61	42	49	85	24	157	65	72	1039	49	22 Oct
Torquay	SX 20 909637	8	888	123	179	25	180	63	42	43	79	25	161	67	72	1059	48	22 Oct
Paignton, Preston	SX 20 894623	43	...	127	181	26	166	61	42	42	79	24	163	66	79	1054	45	22 Oct
Paignton	SX 20 890609	2	982	137	183	32	183	68	50	41	82	26	170	69	85	1126	49	22 Oct
Laywell Res.	SX 20 917549	32	...	110	177	38	177	75	67	54	125	32	170	75	88	1187	40	22 Oct
Brixham, Rea Barn Road	SX 20 927556	50	967	99	159	37	168	75	56	52	123	23	164	66	80	1103	42	22 Oct
Dart																		
Postbridge	SX 20 655787	361	...	213	299	71	269	117	98	81	186	56	209	124	296	2020	52	10 Jan
Bellever Bridge R.F.M. Sta. Area 21.5 km²	SX 20 657775	309	2103	...	...	...	...	...	...	...	...	...	...	...	...	...	...	...
West Dart (Dart)																		
Beardown Hill	SX 20 605767	475	2307	176	328	77	279	172	122	98	206	63	203	107	391	2222	...	...
Cowsic Valley	SX 20 595767	403	2061	180	345	118	297	166	134	116	216	73	230	136	336	2346	...	...
Princetown Prison	SX 20 586741	414	2226	164	369	152	338	192	132	125	210	83	218	150	325	2459	71	24 Feb
Fox Tor	SX 20 620704	354	1910	187	334	98	(381)	145	96	87	202	58	244	120	268	2218	...	...
Swincombe	SX 20 635718	319	1911	235	418	119	381	164	102	86	217	69	249	136	286	2462	71	24 Feb
Dart																		
Venford Res.	SX 20 684707	290	...	272	440	116	398	152	83	75	198	87	291	141	265	2518	69	24 Feb
Venford Dam	SX 20 688711	290	1663	250	382	95	359	129	71	59	185	70	283	109	201	2194	64	24 Jan
Widecombe	SX 20 718767	244	--	202	307	62	273	97	67	56	161	55	215	117	213	1825	58	24 Feb
Buckfastleigh	SX 20 749659	27	...	153	279	72	250	99	39	31	145	47	224	77	164	1579	43	1 Oct
Austins Bridge R.F.M. Sta. Area 248 km²	SX 20 751659	23	1821	...	...	...	...	...	...	...	...	...	...	...	...	...	...	...
Rattery	SX 20 744607	125	--	158	249	70	258	121	58	51	136	44	211	90	161	1608	50	22 Oct
Totnes	SX 20 788620	37	1195	153	215	51	222	91	54	35	127	33	185	71	111	1349	47	22 Oct
Broadhempston	SX 20 804660	58	...	126	198	45	202	85	45	35	96	33	184	65	88	1202	43	22 Oct
Halwell	SX 20 776533	114	...	175	210	58	229	119	77	65	150	41	193	104	165	1585	53	22 Oct
Dartmouth, Public Gardens	SX 20 879515	3	1060	127	187	44	203	93	64	44	147	32	186	69	110	1306	50	6 Aug
Kingswear	SX 20 886512	46	...	115	166	33	185	75	59	40	134	26	161	63	87	1145	44	6 Aug
Dartmouth, Dean's Lodge	SX 20 880505	24	1074	112	156	35	184	79	61	44	135	27	165	70	97	1165	45	22 Oct 6 Aug
Coastal																		
Slapton	SX 20 824449	32	1108	125	172	44	189	79	49	73	137	28	169	89	118	1272	47	6 Aug
Kingsbridge	SX 20 735441	3	...	124	157	49	191	86	54	48	133	26	144	86	112	1209	47	6 Aug
Avon																		
Avon Res. R.F.M. Sta. Area 12.0 km²	SX 20 680651	312	2261	...	...	...	...	...	...	...	...	...	...	...	...	...	...	...
Avon Dam	SX 20 679649	311	...	229	395	113	380	183	95	83	227	70	295	135	304	2509	55	18 Feb
Yalland	SX 20 690628	264	1910	221	365	109	358	164	89	65	211	55	274	108	250	2268	60	6 Aug
Bala Brook	SX 20 675627	232	2124	207	363	112	341	169	93	75	207	59	274	118	290	2309	54	18 Feb
Badworthy	SX 20 685617	168	1872	200	317	103	318	157	92	63	212	57	259	115	245	2138	57	6 Aug
Avonwick, Black Hall	SX 20 712576	91	...	177	229	72	250	128	78	68	155	49	206	93	186	1688	52	22 Oct
Holsome	SX 20 732558	105	...	160	214	63	233	121	78	65	157	44	191	90	161	1579	48	22 Oct
Hazelwood House	SX 20 727522	85	1263	141	186	58	211	100	65	75	149	37	183	97	156	1459	53	6 Aug
Loddiswell	SX 20 718482	82	...	129	169	52	194	99	60	63	133	38	171	91	129	1329	49	6 Aug
Coastal																		
Mount Folly	SX 20 661446	94	...	107	137	39	136	69	39	42	110	25	155	81	120	1062	35	6 Aug
Erme																		
Erme Intake	SX 20 639631	299	...	185	324	111	(330)	164	98	69	203	78	279	91	167	2098	...	...
Harford	SX 20 638595	207	...	173	255	72	283	124	74	64	175	40	261	86	172	1777	60	1 Oct
Wrangaton	SX 20 666575	198	1561	171	246	85	248	138	84	82	172	56	228	108	200	1817	53	22 Oct
Kingston	SX 20 643490	91	...	137	189	46	169	79	53	50	138	31	172	82	111	1256	47	6 Aug

CORNWALL RIVER AUTHORITY AREA, 4644 SQUARE KILOMETRES

HYDROMETRIC AREAS 47 TO 49

Station	Grid reference	Altitude m	Av. 1916–1950 mm	Jan	Feb	Mar	Apr	May	Jun	Jul	Aug	Sep	Oct	Nov	Dec	Year	Highest recorded fall in a day mm	date
Yealm																		
Watercombe	SX 20 629612	271	...	189	260	91	301	153	88	88	191	59	253	100	230	2003	60	6 Aug
Slade	SX 20 598585	85	...	146	201	66	193	118	58	71	149	45	246	88	163	1542	65	1 Oct
Yealmpton, Torr	SX 20 582513	15	...	119	171	52	164	95	59	58	145	35	180	75	124	1276	50	22 Oct
Puslinch R.F.M. Sta. Area 54.9 km²	SX 20 574511	5	1448	...	...	...	...	...	...	...	...	...	...	...	...	...	...	...
Coastal																		
Mount Batten	SX 20 492529	27	963	117	160	39	151	77	43	45	123	26	201	61	106	1147	49	22 Oct
Meavy (Plym)																		
Leather Tor Farm	SX 20 566698	277	1582	155	272	97	251	151	83	86	191	55	212	136	219	1909	...	...
Siward's Cross	SX 20 595699	327	1829	206	354	108	316	163	136	96	203	70	245	158	256	2311	...	...
Deancombe Farm	SX 20 578687	246	1672	177	298	95	260	149	85	83	192	61	227	138	213	1978	...	...
Sheepstor, Head Weir	SX 20 553687	226	1582	131	237	85	219	139	82	81	180	47	197	103	201	1703	40	24 Feb
Burrator Lawn	SX 20 552685	229	...	131	242	91	219	139	78	80	175	47	199	102	201	1704	44	2 Mar
Burrator	SX 20 553680	230	...	141	259	83	215	136	76	74	184	48	210	113	186	1724	...	...
Plym																		
Lee Moor House	SX 20 573628	268	1560	163	242	81	243	151	80	74	179	52	221	87	187	1759	47	22 Oct
Little Woodford	SX 20 531571	24	1129	132	176	51	176	103	57	57	145	35	185	71	117	1307	52	22 Oct
Hartley Res.	SX 20 486569	98	1094	113	163	50	163	101	41	49	134	31	195	61	112	1213	45	22 Oct
Coastal																		
Maristow Gardens	SX 20 472640	15	1345	137	204	60	214	116	53	61	140	47	196	81	153	1459	44	22 Oct
Tavy																		
Tavistock	SX 20 482748	139	1337	111	198	77	180	119	63	67	171	49	179	81	148	1444	38	24 Feb
Walkham (Tavy)																		
North Hessary Tor	SX 20 577742	510	...	161	253	106	276	137	117	107	169	61	209	148	260	2004	38	28 Dec
Horrabridge	SX 20 511700	88	1355	120	212	84	177	126	63	62	143	42	196	93	168	1485	36	22 Oct
Tavy																		
Yelverton	SX 20 513676	189	...	133	217	70	190	117	61	64	159	45	188	92	163	1497	38	22 Oct
Tamar																		
Davies	SX 20 314982	119	1114	83	134	38	130	94	62	89	118	43	129	78	145	1144	33	28 Dec
Holsworthy	SS 21 343040	145	1151	91	131	51	138	102	65	77	136	48	140	81	124	1184	33	10 Aug
Ditchen	SX 20 309947	139	1096	88	163	41	150	88	59	83	122	42	144	75	149	1205	29	28 Dec
Ottery (Tamar)																		
Werrington Park R.F.M. Sta. Area 121 km²	SX 20 336866	55	1166	...	...	...	...	...	...	...	...	...	...	...	...	...	...	...
Werrington Park	SX 20 333872	98	1110	92	182	39	127	71	41	62	122	53	146	71	143	1149	28	24 Feb

GENERAL TABLE, ENGLAND AND WALES

Station	Grid reference	Altitude m	Av. 1916–1950 mm	Rainfall in millimetres Jan	Feb	Mar	Apr	May	Jun	Jul	Aug	Sep	Oct	Nov	Dec	Year	Highest recorded fall in a day mm	date
CORNWALL—*(contd)*																		
Tamar																		
Tregeare	SX 20 242864	165	1235	104	175	37	149	116	65	85	136	50	177	70	175	1340	...	...
Lyd (Tamar)																		
Prewley	SX 20 548910	322	1564	117	245	75	188	119	71	90	186	55	166	107	212	1631	49	24 Feb
Lifton Park R.F.M. Sta. Area 218 km²	SX 20 388842	48	1356	...	...	...	...	...	...	...	...	...	...	...	...	...	...	...
Tamar																		
Launceston, Pennygillow School	SX 20 324836	143	1087	111	188	38	136	75	57	56	135	51	150	64	149	1211	33	24 Feb
Landue	SX 20 349797	76	1132	110	193	44	151	71	51	57	143	40	149	66	134	1210	31	3 Aug
Gunnislake R.F.M. Sta. Area 917 km²	SX 20 426725	8	1229	...	...	...	...	...	...	...	...	...	...	...	...	...	...	...
Gunnislake	SX 20 422707	174	1296	131	217	73	166	101	71	58	157	48	189	73	159	1441	43	22 Oct
Pentillie Castle	SX 20 406647	46	1268	128	204	75	159	101	58	51	136	45	183	66	149	1355	41	22 Oct
Coastal																		
Ellbridge	SX 20 401632	58	1047	121	177	58	148	88	46	49	119	43	186	61	135	1231	43	22 Oct
Lynher																		
Launceston, Bastreet	SX 20 243764	244	1581	165	260	65	203	107	81	85	192	61	212	84	238	1753	43	28 Dec
Pensilva	SX 20 290701	213	...	153	228	67	195	99	66	73	178	56	197	77	194	1583	38	24 Jan
Pillaton Mill R.F.M. Sta. Area 135 km²	SX 20 368624	9	1436	...	...	...	...	...	...	...	...	...	...	...	...	...	...	...
Eglarooze Res.	SX 20 341546	64	...	115	161	49	148	93	43	54	127	40	182	64	141	1216	49	22 Oct
Coastal																		
Torpoint	SX 20 421557	58	...	123	159	48	163	90	44	47	126	38	186	64	116	1203	48	22 Oct
Seaton																		
St. Cleer Down	SX 20 253675	183	1369	149	222	67	179	118	70	64	152	55	211	88	221	1595	40	24 Jan
Little Trethew	SX 20 281602	98	...	134	213	69	187	98	46	64	162	45	179	57	165	1419	40	22 Oct
Coastal																		
Looe																		
Fowey																		
Trekeivesteps	SX 20 227698	189	...	155	239	57	176	109	79	86	181	62	183	83	207	1617	...	...
Trekeivesteps R.F.M. Sta. Area 36.8 km²	SX 20 227698	188	1678	...	...	...	...	...	...	...	...	...	...	...	...	...	...	...
Restormel R.F.M. Sta. Area 171 km²	SX 20 108613	4	1518	...	...	...	...	...	...	...	...	...	...	...	...	...	...	...
Coastal																		
Fowey	SX 20 120513	16	...	122	153	36	125	64	47	41	126	42	174	47	125	1102	41	14 Oct
Stenalees	SX 20 006572	290	...	166	254	56	203	108	93	62	189	42	198	57	144	1572	61	24 Jan
St. Austell	SX 20 018525	94	1232	135	192	42	158	82	59	42	141	40	211	75	174	1350	43	14 Oct
Fal																		
Hendra P. Sta.	SW 10 966518	71	1222	138	233	43	173	81	75	47	143	49	230	66	144	1421	50	14 Oct
Truro (Fal)																		
Truro, Broad Street	SW 10 834456	72	...	97	215	41	142	66	66	32	134	45	212	60	138	1247	47	14 Oct
Tresillian (Truro)																		
Ladock P. Sta.	SW 10 891511	30	1082	111	198	42	145	75	71	46	125	41	210	65	147	1276	49	14 Oct
Truro W. Wks.	SW 10 847464	18	1133	101	215	45	141	68	63	34	125	43	210	58	138	1242	45	14 Oct
Coastal																		
Trevince	SW 10 737403	73	1213	116	226	56	196	100	97	38	106	48	219	84	133	1419	49	22 Oct
Penryn Res.	SW 10 778336	72	1142	118	213	51	182	85	85	29	150	65	230	80	141	1428	45	14 Oct
Falmouth	SW 10 802325	51	1100	109	210	44	147	75	68	23	125	63	199	76	129	1268	45	22 Oct
Wendron	SW 10 677307	140	1138	117	195	45	173	82	80	25	145	64	219	88	138	1369	41	22 Oct
Culdrose	SW 10 669264	82	903	90	168	34	151	66	80	24	120	48	206	69	107	1161	49	14 Oct
Lizard	SW 10 701119	72	837	85	136	37	147	63	79	25	92	54	176	47	95	1035	34	14 Apr
St. Michael's Mount	SW 10 516300	5	938	79	153	33	152	74	(89)	(25)	127	80	224	50	123	1210	61	14 Oct
Gulval	SW 10 486317	15	1007	105	175	43	149	99	92	27	137	67	242	61	141	1340	56	14 Oct
Penzance	SW 10 473300	19	1055	106	186	49	168	109	95	35	150	66	247	63	132	1405	55	14 Oct
Trengwainton House	SW 10 445313	113	1156	111	184	50	176	101	110	29	139	57	268	70	129	1425	62	14 Oct
Drift W. Wks.	SW 10 438287	67	...	116	(178)	46	190	105	101	29	159	70	245	68	143	1449	57	14 Oct
Trewidden	SW 10 443296	84	1137	105	193	47	176	104	115	41	154	69	267	67	145	1481	61	14 Oct
Porthcurno	SW 10 385227	55	862	107	148	38	163	67	85	29	114	41	189	48	111	1140	48	14 Oct
St. Mary's, Scilly	SW 00 913120	48	831	80	120	40	123	65	116	24	88	57	184	46	98	1042	39	8 Jun
Tresco Abbey	SW 00 894141	5	858	77	115	31	129	56	106	28	76	47	190	43	90	990	41	14 Oct
Bussow Moor	SW 10 501392	119	...	100	206	44	177	88	77	29	141	67	267	46	135	1377	61	14 Oct
St. Ives, Halsetown Vicarage	SW 10 507401	99	1080	94	205	42	163	86	93	32	154	60	277	46	117	1368	65	14 Oct
Hayle																		
St. Erth	SW 10 541358	15	1009	96	171	39	145	91	82	20	125	77	232	50	119	1249	48	14 Oct
Coastal																		
Rosewarne	SW 10 643412	76	969	98	213	42	165	100	79	27	114	56	271	55	121	1344	60	14 Oct
Boswyn Res.	SW 10 659362	156	...	111	220	52	181	94	87	38	135	68	233	81	143	1443	54	14 Oct
St. Agnes	SW 10 721505	97	1007	102	178	36	138	68	71	36	115	45	216	63	112	1180	55	14 Oct
Gannel																		
Coastal																		
Newquay	SW 10 812614	53	868	82	128	31	108	51	63	36	107	42	148	54	103	954	32	22 Oct
St. Mawgan	SW 10 871642	103	970	81	143	37	103	61	56	51	129	44	170	64	116	1057	35	14 Oct
Constantine Bay	SW 10 867745	35	...	55	121	38	111	65	56	47	100	52	147	57	97	945	26	10 Aug
Padstow, The Nook	SW 10 914755	34	948	65	125	42	112	71	60	52	101	49	154	67	115	1014	29	14 Oct
Padstow, Barry's Lane	SW 10 917754	8	...	67	129	44	118	72	58	52	111	49	161	65	111	1035	30	14 Oct
Trevanson	SW 10 978729	30	1033	99	138	41	115	65	65	47	113	42	166	73	125	1089	31	28 Dec
Camel																		
Camelford, Roughtor View	SX 20 104832	213	1489	110	171	66	172	112	106	109	169	63	213	101	216	1608	38	28 Dec
Lower Moor	SX 20 128831	271	1526	132	211	74	209	125	105	131	183	67	198	104	228	1764	46	28 Dec
Stannon Wks.	SX 20 125805	232	1506	134	182	69	193	115	94	115	176	78	189	95	207	1647	33	10 Jan
De Lank Filters	SX 20 132765	221	...	133	199	63	184	98	97	113	188	70	229	97	219	1691	40	28 Dec
Denby R.F.M. Sta. Area 209 km²	SX 20 018682	5	1378	...	...	...	...	...	...	...	...	...	...	...	...	...	...	...
Allen (Camel)																		
Delabole No.1	SX 20 075835	198	1163	85	120	55	133	92	89	87	138	50	181	83	158	1272	34	28 Dec
Coastal																		
Trebetherick	SW 10 937772	46	922	71	120	40	113	70	61	51	102	51	137	58	102	973	26	28 Dec
Trelights	SW 10 990789	112	...	75	108	40	128	70	77	62	114	45	154	55	103	1030	29	15 Apr
Bossiney	SX 20 064889	88	1006	77	113	44	129	72	68	91	107	42	173	63	125	1103	33	10 Jul
Condolden Farm	SX 20 093869	290	...	76	125	63	145	88	95	132	150	58	199	90	194	1414	41	28 Dec
Bude	SS 21 208063	15	868	72	81	31	114	59	83	60	97	47	110	63	95	911	31	17 Jun

Station	Grid reference	Altitude m	Av. 1916–1950 mm	Jan	Feb	Mar	Apr	May	Jun	Jul	Aug	Sep	Oct	Nov	Dec	Year	Highest recorded fall in a day mm	date

DEVON RIVER AUTHORITY AREA cont'd

Devon, north area, 2357 square kilometres

Hydrometric areas 50 and 51D

Station	Grid reference	Altitude m	Av. 1916–1950 mm	Jan	Feb	Mar	Apr	May	Jun	Jul	Aug	Sep	Oct	Nov	Dec	Year	mm	date
Coastal																		
Hartland Point	SS 21 231277	93	923	80	88	45	117	75	63	51	88	45	88	49	99	889	29	1 Jan
Clovelly Court	SS 21 310250	107	1316	97	133	61	152	100	89	63	136	56	149	81	164	1279	25	10 Jan
Westward Ho	SS 21 443297	9	995	66	90	26	120	64	81	38	102	38	127	57	95	903	22	20 Oct
Torridge																		
West Putford	SS 21 361158	146	1360	96	135	59	160	95	97	64	115	56	123	91	199	1291	40	28 Dec
Shebbear	SS 21 447094	165	1218	107	141	56	164	103	63	56	127	52	119	80	159	1225	36	28 Dec
Ashbury Sta.	SX 20 483962	253	...	87	193	61	161	119	70	91	173	69	147	101	217	1489	41	28 Dec
Chilla	SS 21 445016	181	...	85	143	47	154	99	73	72	137	45	139	96	174	1265	36	28 Dec
Langabeer, Hatherleigh	SS 21 563008	125	...	109	121	54	128	97	50	42	151	51	118	83	(140)	1144	...	...
Highampton	SS 21 499048	126	...	96	151	45	152	91	64	68	153	48	115	95	166	1243	32	28 Aug 28 Dec
Okement (Torridge)																		
Vellake Intake	SX 20 560898	344	...	146	257	78	202	123	80	98	221	(63)	163	110	210	1751	...	...
Okehampton, Battle Camp	SX 20 585928	372	...	132	258	73	205	123	78	83	178	65	176	107	230	1709	53	24 Feb
Okehampton W. Wks.	SX 20 587938	262	1384	122	220	65	192	116	69	76	158	61	168	102	210	1557	37	28 Dec
Okement Hill	SX 20 606879	518	...	191	401	116	297	175	129	121	253	79	245	183	348	2537	...	...
Newbridge	SX 20 596904	442	1762	127	305	86	227	127	109	105	171	56	168	142	251	1876	...	...
Okehampton, Pleasure Gardens	SX 20 591948	162	1403	122	199	62	169	111	63	73	160	55	175	113	212	1515	39	28 Dec
Hawksland	SS 21 612069	125	...	101	122	39	120	69	58	59	132	46	111	81	145	1082	31	28 Dec
Torridge																		
Roborough	SS 21 579171	171	1099	90	131	44	135	87	67	51	104	53	132	82	134	1109	26	22 Oct
Newbridge R.F.M.Sta. Area 663 km²	SS 21 500185	14	1186	...	...	...	...	...	...	...	...	...	...	...	...	...	...	...
Torridge Vale S. Wks, New site	SS 21 482191	13	...	79	117	36	135	71	70	52	96	57	123	78	128	1043	25	28 Dec
Melbury Res.	SS 21 387202	146	1403	97	131	55	153	93	95	54	121	56	134	91	182	1264	39	28 Dec
Jennett's Res.	SS 21 444247	19	971	68	100	34	145	75	80	36	86	41	123	66	91	946	23	22 Oct
Gammaton Res.	SS 21 483253	107	931	89	125	36	149	77	96	39	86	44	134	71	112	1057	22	22 Oct 28 Dec
Taw																		
Taw Marsh	SX 20 621912	357	1558	138	243	77	203	125	68	77	172	53	186	121	248	1713	46	24 Feb
South Tawton	SX 20 654944	200	...	128	217	55	144	85	66	51	157	44	150	102	198	1396	45	24 Feb
North Wyke	SX 20 658983	177	...	122	164	39	153	68	80	50	117	41	158	89	116	1197	...	...
North Tawton	SS 21 672022	164	948	117	120	29	145	58	63	51	111	46	136	77	127	1081	34	10 Jan
Bow	SS 21 725017	131	934	118	138	33	147	63	68	50	127	40	129	73	133	1119	37	10 Jan
Lapford Cross	SS 21 729075	82	...	105	114	39	150	58	67	45	121	45	174	83	149	1151	53	4 Oct
Eggesford	SS 21 680118	64	1013	98	113	38	126	68	64	45	92	49	136	84	153	1066	32	28 Dec
Little Dart (Taw)																		
East Burrow	SS 21 775175	216	...	93	134	47	145	89	81	56	126	65	142	81	184	1244	32	28 Dec
Horseford Lodge	SS 21 773161	187	...	87	127	41	134	95	73	58	125	54	144	74	187	1198	28	22 Oct
Cheldon Barton	SS 21 734134	163	...	99	113	40	130	69	74	54	103	47	125	75	162	1090	28	28 Dec
Middle Garland	SS 21 711182	202	...	68	113	37	127	77	62	51	112	66	145	95	154	1108	26	18 Apr
Taw																		
Mole (Taw)																		
North Molton	SS 21 738298	197	...	88	143	49	154	112	105	73	123	75	178	105	264	1470	39	14 Oct
South Molton	SS 21 715255	131	1151	76	139	44	148	103	72	55	126	51	148	107	207	1274	33	12 Dec
East Anstey, Barton House	SS 21 866265	235	1261	102	128	65	161	97	89	(63)	(127)	62	150	110	246	1401	40	12 Dec
Yeo Mill	SS 21 842264	160	...	95	124	52	171	95	86	63	136	52	139	95	196	1305	31	28 Dec
Molland Sta.	SS 21 787264	130	...	83	140	58	158	115	83	67	115	57	136	101	241	1354	36	28 Dec
Ash Mill	SS 21 792229	146	1209	81	133	61	138	107	76	53	117	50	136	88	201	1241	30	28 Dec
Bray (Mole)																		
Challacombe	SS 21 692411	256	1704	109	171	72	189	140	167	96	155	96	152	147	311	1804	48	28 Dec
Fullaford	SS 21 682380	245	...	95	157	67	177	127	135	89	152	88	169	131	290	1677	47	28 Dec
Brayford	SS 21 685347	146	1420	79	155	68	161	134	115	87	128	77	130	123	280	1537	42	28 Dec
Hole Water (Bray)																		
Muxworthy R.F.M.Sta. Area 5.39 km²	SS 21 705373	223	1767	...	...	...	...	...	...	...	...	...	...	...	...	...	...	...
Bray (Mole)																		
Filleigh	SS 21 672283	98	...	75	141	49	140	93	85	67	118	68	136	93	203	1265	32	12 Dec
Taw																		
Umberleigh R.F.M.Sta. Area 826 km²	SS 21 608237	14	1158	...	...	...	...	...	...	...	...	...	...	...	...	...	...	...
Stowford House	SS 21 632266	79	1010	65	122	41	133	91	71	55	89	56	131	80	146	1078	24	28 Dec
Swimbridge	SS 21 622305	73	...	61	108	40	138	97	79	61	96	53	133	91	165	1122	27	22 Oct
Gunn	SS 21 634334	203	...	68	123	41	157	106	104	82	112	55	122	93	201	1265	29	28 Dec
Coastal																		
Barnstaple, Athenaeum	SS 21 558330	8	...	61	96	33	136	79	65	56	100	41	113	81	119	980	22	28 Dec
Wistlandpound Res.	SS 21 641415	251	...	90	161	87	187	122	137	95	129	102	146	110	292	1656	...	...
Stoke Rivers	SS 21 633354	209	...	68	124	44	158	113	105	74	112	50	141	91	186	1265	31	28 Dec
Chivenor	SS 21 494347	6	911	64	81	30	127	60	79	40	107	35	120	66	89	899	25	10 Aug
Narracott	SS 21 546432	232	...	74	126	36	170	97	143	69	122	68	128	93	175	1300	32	28 Dec
Spreacombe	SS 21 482419	125	...	72	94	48	143	83	110	62	116	48	143	74	139	1131	29	28 Dec
Halsinger	SS 21 513388	125	...	70	93	45	143	79	97	68	114	47	125	79	138	1099	28	28 Dec
Ilfracombe Res.	SS 21 505453	120	1232	75	110	49	153	97	121	57	116	61	140	77	157	1211	33	28 Dec
Ilfracombe	SS 21 520478	8	974	65	105	47	143	87	95	48	95	55	130	83	134	1088	29	28 Dec
Beera Farm	SS 21 539471	96	...	68	105	48	163	86	105	51	96	57	122	75	134	1110	32	10 Apr
Dean	SS 21 624455	258	...	83	128	57	176	109	110	69	121	77	140	145	179	1393	34	30 Dec
Caffyns Heanton	SS 21 699475	262	...	92	153	53	160	115	121	66	127	70	133	116	205	1411	36	30 Dec
Chains Barrow	SS 21 734420	484	1996	120	219	80	207	152	175	123	185	127	147	165	391	2091	...	...
Longstone Barrow	SS 21 708428	472	1853	119	195	76	203	148	166	127	173	115	147	158	349	1975	...	...

SOMERSET RIVER AUTHORITY AREA, 3086 SQUARE KILOMETRES

HYDROMETRIC AREAS 51S AND 52

Station	Grid reference	Altitude m	Av. 1916–1950 mm	Jan	Feb	Mar	Apr	May	Jun	Jul	Aug	Sep	Oct	Nov	Dec	Year	mm	date
Coastal																		
Nutscale Res.	SS 21 862432	305	1464	81	193	60	175	112	107	76	109	71	170	99	285	1538	...	...
West Dunkery	SS 21 880418	482	1814	116	266	89	174	135	114	81	123	60	188	79	(317)	1742	17	28 Jul
Minehead, Lower Park	SS 21 960462	24	...	65	111	28	119	51	72	51	81	32	104	63	88	865	29	10 Jun

Station	Grid reference	Altitude m	Av. 1916–1950 mm	Jan	Feb	Mar	Apr	May	Jun	Jul	Aug	Sep	Oct	Nov	Dec	Year	Highest recorded fall in a day mm	date
SOMERSET—*(contd)*																		
Avill																		
Wheddon Cross	SS 21 925389	287	...	111	188	63	175	91	105	64	109	52	163	108	256	1486	49	12 Dec
Timberscombe...	SS 21 956421	79	...	79	138	44	144	65	85	55	75	36	117	79	152	1069	36	12 Dec
Minehead, Recreation Ground	SS 21 972460	9	848	64	114	29	110	50	77	49	83	31	108	62	92	868	29	10 Jun
Coastal																		
Lype Hill	SS 21 953365	402	1491	115	182	72	184	89	118	(70)	129	56	163	114	266	1558	...	...
Luxborough	SS 21 973380	229	1256	107	137	56	159	81	109	56	95	52	144	89	177	1261	45	10 Jun
Sminhayes Farm	ST 31 014349	351	1336	125	167	58	193	104	117	(63)	145	71	166	86	206	1502	...	...
Weacombe	ST 31 109403	91	835	55	87	23	118	63	74	56	69	22	117	54	74	813	30	14 Oct
Sampford Brett	ST 31 085404	55	...	62	88	22	115	60	77	48	60	22	115	56	81	803	25	14 Oct
St. Audrie's	ST 31 111417	126	848	55	100	23	120	65	74	55	77	20	118	56	74	837	27	14 Oct
Dodington...	ST 31 175404	101	958	68	107	27	131	81	100	54	97	32	157	67	96	1015	41	10 Jun
Fairfield House	ST 31 186429	46	821	55	93	19	124	67	95	47	75	29	131	53	74	862	41	10 Jun
Whitewick Farm	ST 31 236453	3	789	55	96	15	120	94	82	53	65	33	113	57	71	854	32	10 Jun
Parrett																		
Chedington	ST 31 489057	175	945	70	172	40	130	91	65	48	119	44	192	90	93	1154	58	4 Nov
Misterton	ST 31 458083	61	863	71	153	34	116	84	74	42	103	32	197	76	86	1068	44	4 Nov
Crewkerne Res.	ST 31 436087	130	881	67	186	32	122	83	79	39	105	29	200	91	84	1118	59	4 Nov
Crewkerne, Lyme Road	ST 31 437089	122	894	73	179	34	127	78	74	36	100	24	203	93	88	1108	58	4 Nov
Crewkerne School	ST 31 447099	98	804	63	129	24	106	67	70	39	100	23	183	86	77	968	51	4 Nov
Bincombe House	ST 31 436098	89	823	62	143	27	117	72	73	43	104	21	188	93	83	1026	...	
Over Stratton	ST 31 435149	46	776	48	129	26	92	63	66	39	84	(17)	146	89	81	880	48	4 Nov
Little Norton	ST 31 480157	52	757	56	114	24	94	61	49	63	100	16	162	81	71	891	53	4 Nov
Norton-Sub-Hamdon	ST 31 468159	38	732	54	120	23	97	53	52	53	95	15	155	77	79	873	47	4 Nov
Compton Durville W. Wks, New site...	ST 31 417170	61	...	66	116	22	117	50	63	44	81	16	164	84	79	903	47	4 Nov
Compton Durville W. Wks.	ST 31 417170	61	763	63	113	21	112	49	63	45	81	16	161	83	78	885	45	4 Nov
New Cross Farm	ST 31 416191	49	...	63	90	15	117	46	65	48	71	16	157	83	80	851	43	4 Nov
Martock	ST 31 460190	18	751	55	103	20	103	49	48	49	90	15	151	84	69	835	48	4 Nov
Isle (Parrett)																		
Donyatt	ST 31 338140	43	...	89	152	24	130	62	67	42	83	18	143	72	92	974	38	4 Nov
Ashwell	ST 31 362160	30	...	74	127	21	121	63	65	39	75	17	166	74	84	924	39	4 Nov
Puckington	ST 31 376182	40	...	68	111	16	124	58	67	45	74	17	145	75	71	873	38	4 Nov
Ashford Mill R.F.M. Sta. Area 90.1 km²	ST 31 361188	15	904	...	...	...	...	...	...	...	...	...	...	...	...	...	...	...
Drayton	ST 31 406240	14	720	54	97	17	82	46	77	55	71	15	145	81	77	816	37	4 Nov
Parrett																		
Muchelney	ST 31 428249	8	...	61	99	18	87	47	71	53	74	16	142	91	76	836	44	4 Nov
Curry Rivel	ST 31 387250	55	...	58	107	16	94	51	81	59	79	16	154	82	82	879	36	4 Nov
Yeo (Parrett)																		
Sherborne, Leafield Cottage...	ST 31 643168	65	806	61	134	19	122	53	47	65	120	29	154	103	92	999	62	4 Nov
Sherborne, Manor House	ST 31 639170	75	...	60	126	13	128	47	43	59	120	25	153	95	86	956	61	4 Nov
Sherborne S. Wks.	ST 31 632154	50	754	58	124	17	101	47	40	59	113	26	153	96	86	917	59	4 Nov
Yetminster	ST 31 595106	53	...	68	147	22	106	59	39	55	99	31	163	95	87	972	63	4 Nov
Halstock (Yeo)																		
Sutton Bingham Res.	ST 31 556114	55	817	63	141	25	102	58	51	50	97	28	172	90	81	956	59	4 Nov
Sutton Bingham Res.	ST 31 556116	46	--	60	142	24	96	59	48	51	97	28	173	89	77	944	59	4 Nov
Sutton Bingham Res. R.F.M. Sta. Area 30.3 km²	ST 31 556116	41	909	...	...	...	...	...	...	...	...	...	...	...	...	...	...	...
Yeo (Parrett)																		
Yeovil, Preston Grove	ST 31 544161	53	793	43	117	17	110	57	51	52	108	25	154	86	72	891	55	4 Nov
Yeovil, Summerleaze Park School ...	ST 31 547162	49	...	47	115	17	103	53	48	50	109	25	150	84	66	866	54	4 Nov
Pen Mill R.F.M. Sta. Area 213 km²	ST 31 572159	24	846	...	...	...	...	...	...	...	...	...	...	...	...	...	...	...
Pen Mill S. Wks.	ST 31 373162	30	832	44	118	19	95	49	56	45	118	19	152	81	69	865	52	4 Nov
Rimpton	ST 31 606218	37	724	50	89	18	109	58	74	73	106	17	146	87	69	896	49	4 Nov
Woolston	ST 31 652279	107	...	45	113	20	122	66	54	80	115	19	134	95	87	949	56	4 Nov
West Camel	ST 31 574247	27	...	(51)	(102)	16	95	82	66	60	105	19	126	89	77	887	53	4 Nov
Yeovilton	ST 31 551237	18	...	54	100	20	95	74	49	55	107	21	127	91	84	875	50	4 Nov
Langport, Picts Hill	ST 31 432276	34	...	57	83	11	83	37	62	63	69	11	129	89	71	765	45	4 Nov
Parrett																		
Huish Episcopi	ST 31 427268	20	733	56	83	9	79	40	61	59	71	12	135	87	73	764	44	4 Nov
West Sedgemoor P. Sta.	ST 31 376286	8	...	50	91	12	92	68	62	67	78	13	141	83	72	830	34	4 Nov
Tone (Parrett)																		
Sedgeborough Farm	ST 31 054327	329	1203	117	143	63	220	93	132	58	114	42	166	95	180	1424	...	...
Syndercombe No.1...	ST 31 040308	244	...	100	150	42	224	94	126	50	134	58	171	88	143	1381	53	10 Jun
Milltown	ST 31 046311	201	...	129	129	53	206	88	131	55	140	40	168	103	182	1422	...	...
Clatworthy Res. R.F.M. Sta. Area 18.2 km²	ST 31 044312	197	1273	...	...	...	...	...	...	...	...	...	...	...	...	...	...	...
Huish Champflower	ST 31 046287	201	1158	89	141	50	207	93	117	54	128	50	168	84	178	1359	43	10 Jun
Bulland Lodge	ST 31 054273	229	1134	90	136	53	189	97	107	57	120	44	137	79	182	1292	35	10 Jun
Sampford Arundel	ST 31 112180	128	960	80	132	43	127	75	92	53	93	36	148	63	120	1061	37	10 Jun
Langford Budville	ST 31 114235	76	889	67	123	45	141	71	93	46	82	29	127	61	121	1008	37	10 Jun
Blagdon Res.	ST 31 202172	175	...	109	154	38	140	63	99	63	119	32	165	71	112	1166	...	...
Luxhay	ST 31 200178	142	...	87	155	41	132	73	99	58	102	31	162	62	113	1114	37	10 Jun
Bishops Hull R.F.M. Sta. Area 202 km²	ST 31 206250	16	993	...	...	...	...	...	...	...	...	...	...	...	...	...	...	...
Bagborough	ST 31 166334	137	900	67	105	31	130	63	77	52	91	27	121	75	106	946	31	14 Oct
Watts House	ST 31 163303	78	...	63	94	24	128	59	63	44	81	23	114	74	97	865	36	22 Oct
Yarford	ST 31 200296	61	...	54	89	23	112	55	65	52	67	21	106	83	97	825	...	...
Halse Water (Tone)																		
Bishops Hull R.F.M. Sta. Area 87.8 km²	ST 31 206253	17	909	...	...	...	...	...	...	...	...	...	...	...	...	...	...	...
Tone (Parrett)																		
Taunton, Vivary Park	ST 31 229238	22	744	63	114	23	111	52	68	41	67	19	124	61	73	815	30	22 Oct
Taunton, Cheddon Road	ST 31 228265	32	...	61	102	24	120	59	61	44	76	22	125	72	82	848	28	22 Oct
Orchard Portman	ST 31 244216	34	...	70	126	25	109	58	79	51	70	18	127	58	75	866	31	10 Jun
West Monkton, Glebe Court	ST 31 264284	67	...	64	98	24	115	56	59	47	81	20	134	73	93	865	26	22 Oct
West Newton Farm	ST 31 286291	37	...	56	96	20	118	56	61	44	67	16	129	67	90	820	25	22 Oct
West Newton	ST 31 289291	30	...	57	98	20	129	56	64	45	70	17	135	70	87	848	25	22 Oct
East Lyng...	ST 31 333289	13	...	58	93	17	105	57	69	44	80	20	134	75	81	835	28	22 Oct
Currymoor P. Sta.	ST 31 344288	6	...	55	89	15	89	60	64	43	76	18	144	72	73	799	28	22 Oct
Parrett																		
Allermoor P.Sta.	ST 31 357305	7	...	57	90	15	96	73	68	45	71	17	145	72	69	821	27	4 Nov
Northmoor P.Sta.	ST 31 331330	7	...	52	81	14	95	56	79	47	70	19	125	67	68	773	27	10 Jun
Clavelshay	ST 31 264316	130	...	60	101	25	135	68	70	60	91	24	135	75	104	948	(26	14 Apr)
Bridgwater, Blake Gardens	ST 31 300368	8	783	61	95	20	110	83	89	52	69	20	128	63	81	870	51	10 Jun
Spaxton	ST 31 224367	72	873	67	114	29	139	84	82	49	76	25	120	67	97	949	29	4 Nov
Durleigh	ST 31 274363	14	790	58	91	22	119	65	94	53	71	21	123	62	83	861	39	10 Jun
Bridgwater, Antony	ST 31 277378	34	...	58	84	18	108	55	97	55	71	20	113	54	78	811	47	10 Jun

Station	Grid reference	Altitude m	Av. 1916–1950 mm	Jan	Feb	Mar	Apr	May	Jun	Jul	Aug	Sep	Oct	Nov	Dec	Year	Highest recorded fall in a day mm	Highest recorded fall in a day date
SOMERSET—*(contd)*																		
Wembdon Hill	ST 31 280378	34	...	57	82	19	112	55	102	56	69	19	117	57	76	821	24	14 Oct
Wembdon	ST 31 283377	14	...	59	88	20	109	59	106	57	68	19	117	57	78	838	46	10 Jun
Cary (Parrett)																		
Castle Cary	ST 31 641319	88	834	59	120	35	128	80	79	92	113	20	126	97	83	1032	55	4 Nov
Ashcott, Bradley Cottage	ST 31 439365	35	722	55	84	18	95	62	80	86	70	17	137	85	73	862	44	4 Nov
Ashcott, Mellifera	ST 31 434372	64	...	55	81	20	94	62	79	74	70	21	128	84	77	845	44	4 Nov
Langport, Garden City	ST 31 425273	18	...	53	84	12	79	40	63	63	71	13	132	89	75	775	43	4 Nov
Parrett																		
Gaunts Farm	ST 31 287428	10	...	51	76	14	105	70	74	50	73	24	110	57	72	775	23	13 Apr
Cannington (Parrett)																		
Hawkridge Res.	ST 31 213362	114	...	69	136	32	138	97	97	57	87	35	134	78	111	1070	29	4 Nov
Ashford W. Wks.	ST 31 236385	31	786	63	99	27	127	83	94	48	89	28	125	59	86	928	46	10 Jun
Brymore	ST 31 244394	23	...	61	93	23	130	75	92	48	84	24	120	56	81	885	52	10 Jun
Cannington	ST 31 258400	29	760	49	88	19	110	70	108	40	81	23	111	53	75	826	59	10 Jun
Parrett																		
Nether Stowey	ST 31 192396	76	...	67	97	24	141	78	99	53	87	29	121	68	84	949	45	10 Jun
Huntspill																		
Leigh Holt	ST 31 485352	43	724	50	90	(24)	94	66	71	70	97	16	122	81	75	856	40	4 Nov
Millfield	ST 31 492362	30	764	53	95	25	101	70	74	78	105	17	(127)	82	83	911	42	4 Nov
Street Res.	ST 31 488363	30	...	54	95	26	88	85	76	77	100	16	131	82	74	905	43	4 Nov
Street, Hindhayes	ST 31 485364	24	...	49	92	25	90	78	77	76	101	17	(122)	84	82	893	42	4 Nov
Street S. Wks.	ST 31 486382	8	749	54	90	27	97	82	63	67	97	15	115	83	74	864	46	4 Nov
Gold Corner	ST 31 367431	4	...	40	86	19	93	67	69	58	85	26	111	69	75	797	31	10 Jun 4 Nov
Cossington	ST 31 357405	37	...	46	86	20	86	69	81	57	71	26	117	68	71	798	31	4 Nov
Puriton	ST 31 321415	15	...	53	86	20	107	67	77	67	72	30	127	59	83	848	26	14 Oct
Brue																		
Folly Farm	ST 31 704385	186	965	61	128	31	111	76	85	101	173	37	130	122	141	1196	73	4 Nov
Hardway	ST 31 723344	101	...	57	142	35	122	68	52	79	120	32	139	113	112	1071	69	4 Nov
Lovington R.F.M.Sta. Area 135 km²	ST 31 590318	20	909	...	...	...	...	...	...	...	...	...	...	...	...	...	...	...
Sheppey (Brue)																		
Shepton Mallet, Whitstone School	ST 31 623435	140	...	61	117	35	127	85	106	88	130	33	122	109	113	1125	65	4 Nov
Wells S. Wks.	ST 31 535451	23	852	43	103	33	120	88	89	91	106	31	102	91	109	1005	50	10 Jun
Fenny Castle R.F.M. Sta. Area 59.6 km²	ST 31 498439	6	996	...	...	...	...	...	...	...	...	...	...	...	...	...	...	...
Brue																		
North Drain P. Sta.	ST 31 399448	3	...	42	93	22	95	64	62	69	87	32	116	74	80	837	36	4 Nov
Bason Bridge	ST 31 349458	5	764	44	83	17	97	70	73	53	90	31	115	64	76	812	29	10 Jun 4 Nov
Coastal																		
Burnham, Hall Nurseries	ST 31 304495	7	761	55	85	17	127	62	78	55	78	27	120	56	74	835	27	10 Jun
Burnham, Berrow Road	ST 31 304506	3	759	(55)	83	13	123	67	79	59	76	24	108	61	77	822	25	4 Nov
Axe																		
Wookey Hole	ST 31 528477	69	963	44	103	38	127	86	76	70	96	38	117	95	111	1001	49	4 Nov
Wookey P. Sta.	ST 31 527459	30	895	47	106	33	117	88	79	83	106	33	113	101	107	1013	...	...
Wookey R.F.M. Sta. Area 18.2 km²	ST 31 527458	28	1117	...	...	...	...	...	...	...	...	...	...	...	...	...	...	...
Rodney Stoke	ST 31 488501	40	...	44	105	36	125	79	88	72	94	39	154	88	111	1036	42	4 Nov
Wedmore	ST 31 437487	30	...	51	95	26	88	65	84	65	98	37	109	78	82	878	44	4 Nov
Charterhouse	ST 31 488549	189	1120	63	(127)	43	160	92	97	100	108	47	161	106	152	1257	...	...
Cheddar Secondary School	ST 31 456532	12	...	(38)	(89)	33	109	73	91	71	78	40	121	92	107	942	46	4 Nov
Cheddar P. Sta.	ST 31 452534	15	820	44	97	34	113	68	92	74	81	38	118	89	105	953	44	4 Nov
Winscombe, Mendip View	ST 31 415572	30	930	44	94	24	120	67	84	74	82	30	109	79	105	912	36	4 Nov
Winscombe, Bore Hole	ST 31 407577	17	916	48	110	24	109	63	100	68	77	33	114	82	111	939	...	...
Bleadon	ST 31 340563	8	805	48	79	26	121	75	95	67	73	29	120	75	92	901	31	4 Nov
Brent Knoll	ST 31 327514	5	...	50	81	17	126	69	76	53	77	27	131	63	83	852	27	14 Oct
Hutton	ST 31 351588	15	...	41	84	24	120	81	88	57	80	35	113	73	100	896	31	4 Nov
Coastal																		
Weston-Super-Mare	ST 31 318602	9	823	45	78	23	111	73	79	42	82	38	142	71	106	892	28	4 Nov
Yeo																		
Blagdon, Street End	ST 31 498586	168	1098	65	140	38	145	87	70	72	107	48	150	100	138	1162	...	...
Blagdon P. Sta.	ST 31 505604	52	984	52	122	38	134	75	64	71	90	47	131	93	114	1033	43	4 Nov
Barley Wood	ST 31 479629	61	931	39	107	30	122	70	76	74	79	42	115	79	107	939	40	4 Nov
Yatton P. Sta.	ST 31 421658	5	876	36	72	21	106	77	92	48	102	52	143	(79)	101	928	36	6 Aug
Shipham	ST 31 447577	149	...	43	111	31	136	75	80	73	96	35	133	89	123	1026	42	4 Nov
Congresbury	ST 31 438632	11	...	45	93	25	111	74	90	58	80	35	128	77	113	930	39	4 Nov
Coastal																		
Yatton	ST 31 473680	18	891	49	88	25	111	61	68	52	70	48	122	79	93	865	24	14 Oct
Brockley Court Farm	ST 31 466670	15	...	56	88	26	116	60	65	56	71	53	122	78	90	879	42	4 Nov
Wraxall Court	ST 31 489721	61	884	52	90	27	116	60	(76)	56	75	40	124	84	95	896	44	4 Nov
Clevedon P. Sta.	ST 31 425715	15	830	49	88	30	112	67	73	37	81	43	136	67	91	873	33	19 Oct
Portishead, Bristol Road	ST 31 472754	6	830	54	93	32	125	63	58	34	74	39	121	79	97	869	36	4 Nov

BRISTOL AVON RIVER AUTHORITY AREA, 2220 SQUARE KILOMETRES

HYDROMETRIC AREA 53

Station	Grid reference	Altitude m	Av. 1916–1950 mm	Jan	Feb	Mar	Apr	May	Jun	Jul	Aug	Sep	Oct	Nov	Dec	Year	Highest recorded fall in a day mm	Highest recorded fall in a day date
Tetbury Avon																		
Shipton Moyne P.Sta.	ST 31 899887	95	849	54	102	26	118	79	40	65	88	41	136	65	88	904	37	29 Aug
Sherston Avon (Tetbury Avon)																		
Petty France Hotel	ST 31 786855	158	929	51	109	27	107	74	58	70	94	41	126	94	96	945	46	4 Nov
Pinkney Park	ST 31 863864	101	819	45	101	21	116	69	35	66	86	45	117	80	86	867	40	29 Aug
Avon																		
Great Somerford R.F.M.Sta. Area 303 km²	ST 31 966832	58	838	...	...	...	...	...	...	...	...	...	...	...	...	...	...	...
Seagry Mill	ST 31 955804	54	...	35	112	17	100	64	39	56	100	56	107	54	76	815	37	29 Aug
Bowood House	ST 31 974699	93	796	48	93	16	108	56	59	65	118	37	115	61	89	865	37	21 Aug
Lyneham	SU 41 012786	139	732	26	97	11	102	54	48	49	127	31	114	46	75	780	50	21 Aug
Hardenhuish Res.	ST 31 912751	105	...	34	102	15	103	61	49	56	80	31	104	60	81	775	32	29 Aug
Chippenham W.Wks.	ST 31 919727	44	791	38	100	15	103	49	60	51	90	27	98	57	81	770	37	29 Aug
Lacock	ST 31 923702	49	740	32	99	12	98	47	56	45	83	24	107	61	79	743	32	29 Aug
Thingley S.Wks.	ST 31 885694	66	...	37	106	15	93	59	58	47	64	29	102	68	84	760	30	4 Nov
Melksham, Forest Road	ST 31 909647	37	785	51	91	15	102	51	45	48	59	39	99	69	83	754	30	4 Nov
Melksham R.F.M.Sta. Area 666 km²	ST 31 903641	31	815	...	...	...	...	...	...	...	...	...	...	...	...	...	...	...

Station	Grid reference	Altitude m	Av. 1916–1950 mm	Jan	Feb	Mar	Apr	May	Jun	Jul	Aug	Sep	Oct	Nov	Dec	Year	Highest recorded fall in a day mm	date
BRISTOL AVON—*(contd)*																		
Great Chalfield	ST 31 858633	50	788	50	95	17	107	77	58	45	66	42	104	71	91	824	33	4 Nov
Semington (Avon)																		
West Lavington	SU 41 000535	99	774	53	100	15	88	55	46	70	90	26	157	58	74	832	30	1 Oct
Devizes, Grammar School	ST 31 999618	128	774	51	89	13	92	50	49	59	117	30	133	58	71	813	48	29 Aug
Semington R.F.M.Sta. Area 158 km²	ST 31 898609	32	785	...	...	...	...	...	...	...	...	...	...	...	...	...	...	...
Biss (Avon)																		
Upton Scudamore P.Sta.	ST 31 862482	93	847	51	119	24	109	77	57	70	109	37	124	77	100	954	39	21 Aug
Trowbridge, Bowling Green	ST 31 858577	40	744	47	94	18	106	57	45	50	75	33	100	65	75	763	32	4 Nov
Trowbridge, Upper Studley	ST 31 843562	55	687	42	100	15	102	62	45	51	74	30	96	68	80	766	35	4 Nov
Trowbridge, Victoria Road	ST 31 865588	48	...	45	89	18	107	58	46	45	77	31	104	65	76	761	18	9 Feb
Avon																		
Bradford-on-Avon, Frome Road	ST 31 825606	33	831	51	94	21	114	80	65	48	124	36	98	72	85	888	55	21 Aug
Frome (Avon)																		
Witham Friary Vicarage	ST 31 746414	107	...	56	111	34	105	81	60	74	125	36	128	95	121	1026	61	4 Nov
Maiden Bradley	ST 31 802398	152	...	65	159	37	137	95	61	79	231	34	171	98	129	1295	55	4 Nov
Frome S.Wks.	ST 31 772492	63	...	61	120	28	140	77	57	75	76	27	104	87	102	954	47	4 Nov
Wadbury (Frome)																		
Gurney Slade	ST 31 631494	179	...	73	158	55	163	97	96	106	130	52	129	124	167	1349	67	4 Nov
Wanstrow	ST 31 714418	136	...	66	135	37	132	89	71	113	129	48	125	111	135	1190	...	...
Egford P.Sta.	ST 31 756480	82	...	61	125	29	141	78	59	79	83	32	113	98	113	1010	54	4 Nov
Frome (Avon)																		
Orchardleigh Park	ST 31 774514	94	...	59	122	25	138	85	77	73	79	35	108	89	101	990	44	4 Nov
Tellisford R.F.M.Sta. Area 262 km²	ST 31 805564	35	983	...	...	...	...	...	...	...	...	...	...	...	...	...	...	...
Iford Manor	ST 31 801588	37	844	52	102	24	113	79	74	46	88	43	104	80	86	891	39	4 Nov
Midford (Avon)																		
Chilcompton	ST 31 647524	125	1103	67	155	47	147	86	71	91	123	48	120	121	150	1226	63	4 Nov
Downside Abbey	ST 31 656508	183	...	67	150	42	145	87	81	90	125	43	116	119	144	1208	63	4 Nov
Ammerdown House	ST 31 712527	149	1082	70	164	39	169	93	90	79	112	41	130	118	133	1237	62	4 Nov
Cam (Midford)																		
Hinton Blewett Manor	ST 31 594570	152	1051	63	159	51	132	89	47	62	102	51	144	123	127	1150	62	4 Nov
Midford (Avon)																		
Midford R.F.M.Sta. Area 147 km²	ST 31 763611	27	998	...	...	...	...	...	...	...	...	...	...	...	...	...	...	...
Southstoke	ST 31 750617	155	852	53	112	22	106	76	56	47	76	50	101	91	95	885	55	4 Nov
Tucking Mill P.Sta.	ST 31 763616	41	861	48	116	22	109	67	63	48	71	55	97	85	95	873	46	4 Nov
Avon																		
Monkswood Res., New site	ST 31 755711	111	...	36	115	26	105	72	45	60	67	49	121	99	91	885	52	4 Nov
Batheaston Res.	ST 31 769685	76	821	49	117	20	104	65	49	68	70	43	115	95	85	879	56	4 Nov
Charlcombe	ST 31 750668	88	830	49	107	19	106	65	54	56	74	42	113	97	90	872	57	4 Nov
Bath R.F.M.Sta. Area 1600 km²	ST 31 752651	17	861	...	...	...	...	...	...	...	...	...	...	...	...	...	...	...
Bath City Boys School	ST 31 751638	118	797	46	116	15	117	65	40	43	68	48	100	89	82	830	54	4 Nov
Bath, Priory Close	ST 31 759628	155	...	54	132	22	119	72	52	46	73	48	104	92	94	908	53	4 Nov
Weston Res.	ST 31 718676	112	...	38	112	20	102	65	45	62	73	40	97	104	81	839	57	4 Nov
Bitton	ST 31 667699	30	802	41	104	20	88	53	32	45	75	55	97	96	85	792	58	4 Nov
Chew (Avon)																		
Chewton House	ST 31 597528	161	1141	67	159	57	151	105	58	84	114	59	148	125	159	1287	66	4 Nov
Pitt Farm	ST 31 561544	215	1269	73	173	66	162	119	68	85	128	71	157	123	181	1406	...	...
North Widcombe	ST 31 580577	66	959	55	141	44	131	88	42	62	85	43	124	114	111	1042	...	...
Harptree Hill	ST 31 543559	239	1158	63	159	57	158	117	68	76	113	65	159	124	175	1336	...	...
Heron's Green	ST 31 557591	55	965	47	126	38	126	73	41	64	83	42	121	111	106	977	...	...
Clutton Res.	ST 31 615602	183	993	54	136	43	118	81	44	48	101	54	126	112	109	1025	...	...
Sutton Court	ST 31 597604	85	...	47	131	36	112	62	33	50	87	43	123	113	103	937	61	4 Nov
Ridgehill	ST 31 541625	122	980	44	115	28	115	68	55	56	86	44	132	99	112	955	...	...
Chew Stoke P.Sta.	ST 31 570617	43	956	51	112	35	110	61	44	54	80	44	118	97	107	912	46	4 Nov
Upper Town	ST 31 525660	168	995	54	117	29	116	78	58	61	87	62	127	99	105	991	...	...
Chew Magna	ST 31 580656	104	955	55	110	31	118	68	54	62	94	63	133	97	111	996	...	...
Compton Dando R.F.M.Sta. Area 130 km²	ST 31 648647	17	1013	...	...	...	...	...	...	...	...	...	...	...	...	...	...	...
Avon																		
Kingsfield School	ST 31 664737	52	...	46	92	22	94	61	54	45	72	52	131	107	89	865	53	5 Nov
Conham S.Wks.	ST 31 628720	15	838	43	108	29	115	68	62	42	76	58	134	106	103	945	55	4 Nov
Barrow Gurney Res.	ST 31 537679	91	940	45	106	27	112	74	75	51	73	58	121	93	91	926	49	4 Nov
Long Ashton	ST 31 537699	49	920	44	102	30	105	69	61	53	77	56	108	87	97	888	47	4 Nov
Frome (Avon)																		
Tytherington	ST 31 668884	73	883	57	99	29	105	69	51	52	68	44	118	88	86	866	51	4 Nov
Frampton Cotterell P.Sta.	ST 31 669819	47	757	50	99	28	96	56	65	58	68	41	116	93	86	857	50	4 Nov
Filton	ST 31 598802	59	791	47	97	27	27	58	52	37	73	35	123	93	85	754	56	4 Nov
Frenchay R.F.M.Sta. Area 149 km²	ST 31 637772	20	820	...	...	...	...	...	...	...	...	...	...	...	...	...	...	...
Bristol, Hopewell Hill	ST 31 645746	118	818	41	100	25	97	63	67	47	77	65	130	103	93	909	58	4 Nov
Horfield	ST 31 589763	76	832	47	109	23	107	64	59	39	71	38	131	97	88	874	52	4 Nov
Clifton, Oakfield Road	ST 31 577737	61	...	46	111	30	117	68	65	52	71	51	134	107	99	950	37	4 Nov
Avon																		
Failand House	ST 31 516732	131	967	63	127	29	129	76	66	61	78	48	135	89	119	1018	45	4 Nov

SEVERN RIVER AUTHORITY AREA, 11421 SQUARE KILOMETRES

HYDROMETRIC AREA 54

Station	Grid reference	Altitude m	Av. 1916–1950 mm	Jan	Feb	Mar	Apr	May	Jun	Jul	Aug	Sep	Oct	Nov	Dec	Year	Highest recorded fall in a day mm	date
Coastal																		
Lower Hazel	ST 31 623872	61	813	54	96	29	102	61	56	41	71	40	123	84	85	844	49	4 Nov
Little Avon																		
Alderley P.Sta.	ST 31 773914	63	825	59	103	27	120	78	48	79	89	46	108	102	86	945	...	...
Kingswood S.Wks.	ST 31 743929	42	...	45	90	21	93	65	48	69	90	38	120	79	79	837	49	4 Nov
Coastal																		
Sharpness	SO 32 670022	15	...	37	99	29	100	68	53	46	59	36	109	75	82	794	52	4 Nov
Upper Cam	ST 31 764988	107	858	53	97	31	95	65	55	55	99	31	122	84	83	869	47	4 Nov
Severn																		
Blaen Hafren	SN 22 835884	439	2438	114	274	205	156	157	257	173	119	152	179	263	580	2630	...	...
Hafren Lodge	SN 22 889853	244	1563	81	199	109	104	106	153	100	93	107	131	200	419	1802	91	9 Dec
Garth-Fawr	SN 22 996791	366	...	81	176	61	93	109	81	63	61	64	125	151	205	1269	47	9 Dec

Station	Grid reference	Altitude m	Av. 1916–1950 mm	Jan	Feb	Mar	Apr	May	Jun	Jul	Aug	Sep	Oct	Nov	Dec	Year	Highest recorded fall in a day mm	date
SEVERN—*(contd)*																		
Clywedog (Severn)																		
Dol Bachog	SN 22 884936	314	1583	86	209	104	112	104	133	95	83	109	147	187	324	1693	58	9 Dec
Hafod Cadwgan	SN 22 859905	351	1970	102	250	142	163	135	236	135	105	140	165	226	446	2244	...	...
Cwm Biga	SN 22 857890	309	1919	119	254	147	133	138	245	137	109	132	174	244	522	2356	...	...
Llanidloes R.F.M.Sta. Area 57.0 km²	SN 22 944855	175	1803	...	...	...	...	...	...	...	...	...	...	...	...	...	...	...
Severn																		
Llanidloes, Abernant	SN 22 955847	152	...	65	158	69	75	83	84	51	61	66	118	129	226	1187	56	9 Dec
Trefeglwys	SN 22 973912	146	...	60	127	65	77	70	77	54	72	65	107	121	191	1085	39	9 Dec
Carno	SN 22 960964	191	...	73	172	93	85	81	94	68	74	87	104	149	271	1351	62	9 Dec
Caersws	SO 32 036924	122	...	55	108	52	74	60	69	57	60	52	105	104	135	932	27	1 Dec
Gregynog Hall	SO 32 080969	198	900	64	133	56	92	62	77	72	68	52	102	116	144	1038	31	9 Dec
Mochdre Res.	SO 32 079871	198	981	65	159	49	85	89	69	55	65	58	124	126	148	1094	30	9 Dec
Kerry Forest	SO 32 131894	213	...	51	144	44	94	73	63	46	57	42	110	96	97	917	24	5 Nov
Abermule R.F.M.Sta. Area 580 km²	SO 32 165958	83	1257	...	...	...	...	...	...	...	...	...	...	...	...	...	...	...
Garthmyl Hall	SO 32 190990	101	...	52	111	41	87	77	61	59	67	42	96	100	109	901	23	1 Dec
Rhiw (Severn)																		
Cefn Coch	SJ 33 023007	261	...	79	165	71	104	83	89	71	85	66	111	145	248	1315	63	9 Dec
Camlad (Severn)																		
Church Stoke	SO 32 291933	140	792	41	111	26	78	67	74	42	47	38	(114)	105	87	827	27	4 Nov
Severn																		
Brithdir Hall	SJ 33 194026	107	829	44	109	39	71	69	58	51	62	39	90	89	97	818	24	7 Feb
Fron Llwyd	SJ 33 215077	152	885	36	111	48	81	63	60	48	68	39	96	86	96	835	24	5 Nov
Powis Castle	SJ 33 218065	102	...	38	108	44	81	67	61	49	57	39	93	96	108	841	26	5 Nov
Vyrnwy (Severn)																		
Rhiwargor	SH 23 961242	262	1865	117	279	130	146	151	106	94	98	130	156	246	281	1934	...	...
Carreg Goch	SH 23 974237	543	1874	125	304	141	186	164	111	105	111	129	159	238	275	2047	...	...
Gadfa	SH 23 928232	464	2301	146	293	152	174	164	142	132	126	145	174	251	339	2239	...	...
Eunant Bridge	SH 23 962224	253	1923	119	297	152	166	174	126	112	114	129	147	240	322	2098	...	...
Hirddu Fawr (Vyrnwy)																		
Nant Cawell	SH 23 936208	491	2194	152	304	141	201	165	164	131	125	149	158	255	384	2327	...	...
Llewst-Wen	SH 23 971202	494	2116	136	314	140	207	178	141	123	107	145	171	232	382	2276	...	...
Heol-y-Ffridd	SH 23 976215	253	1906	112	274	133	141	158	125	111	103	129	145	228	329	1987	...	...
Llwynrhiw	SH 23 975207	435	2125	125	301	155	184	175	142	126	111	133	160	239	428	2278	...	...
Vyrnwy (Severn)																		
Ty Uchaf	SH 23 998215	262	1707	108	253	(113)	142	139	105	94	95	108	148	197	282	1785	...	...
Erw Berfedd	SH 23 978264	549	2090	141	332	159	168	157	110	112	95	127	160	235	305	2102	...	...
Hen Eglwys	SH 23 993253	433	1869	126	279	126	177	142	100	91	99	128	164	240	278	1950	...	...
Fron Goch	SJ 33 000213	255	1651	103	249	(116)	136	130	103	100	90	102	145	204	275	1753	...	...
Bryn Gwyn	SH 23 999196	400	1781	103	275	127	177	138	118	111	93	104	131	199	315	1892	...	...
Fawnog	SJ 33 014219	401	1646	108	249	123	182	129	97	93	93	97	146	197	253	1765	...	...
Lake Vyrnwy	SJ 33 016192	303	1655	105	262	113	148	134	105	100	100	98	143	185	295	1788	58	9 Dec
Bank Cynon Isaf	SJ 33 018191	229	...	100	244	106	144	124	100	97	94	94	138	175	264	1678	52	9 Dec
Vyrnwy Res. R.F.M. Sta. Area 94.3 km²	SJ 33 019191	226	1908	...	...	...	...	...	...	...	...	...	...	...	...	...	...	...
Tyn-y-ffridd	SJ 33 040213	308	1520	87	226	86	134	118	85	111	94	84	125	170	225	1545	...	...
Hen Gerig	SH 23 968188	375	2076	131	312	160	185	166	146	131	108	132	157	236	400	2265	...	...
Cownwy Dam	SJ 33 000181	283	1625	111	262	107	155	125	122	107	91	99	141	194	276	1788	...	...
Banwy (Vyrnwy)																		
Afon Twrch	SH 23 941180	424	1993	147	305	152	175	178	(140)	127	114	127	160	244	368	2238	...	...
Cerniau	SH 23 951169	354	2025	116	279	134	164	152	151	107	101	113	144	207	331	2000	62	9 Dec
Pennant Twrch	SH 23 965155	309	...	117	268	137	157	148	156	105	84	100	116	214	309	1910	55	9 Dec
Abercannon Farm	SH 23 962069	244	1734	106	175	147	120	114	145	141	68	98	145	225	430	1914	94	9 Dec
Llanerfyl	SJ 33 014094	223	...	(89)	(152)	(114)	136	98	89	83	82	76	142	183	266	1510	49	9 Dec
Llanfair Caereinion S.Wks.	SJ 33 109071	113	...	56	137	66	101	64	60	50	63	55	93	122	128	995	25	9 Dec
Vyrnwy (Severn)																		
Llansantffraid	SJ 33 209178	76	862	52	113	41	106	66	59	57	75	44	102	92	107	915	20	1 Dec
Llanfyllyn P.Sta.	SJ 33 139197	130	...	70	150	58	106	77	79	63	82	80	117	147	187	1219	24	7 Feb
Tanat (Vyrnwy)																		
Pennant-Melangell	SJ 33 023265	201	1569	106	306	114	157	123	86	85	91	89	128	187	212	1682	...	...
Hirnant	SJ 33 048225	233	1499	95	296	102	145	115	83	83	93	85	123	163	209	1592	...	...
Rhiwlas	SJ 33 114260	228	1206	83	179	56	116	97	61	70	99	63	120	151	197	1293	39	1 Dec
Llangedwyn	SJ 33 174243	116	...	70	125	45	111	74	58	65	106	57	131	105	137	1083	31	1 Dec
Penygwely Res.No.1	SJ 33 220324	340	1057	90	156	54	150	109	72	72	128	63	142	125	182	1345	44	1 Dec
Penygwely Res.No.2	SJ 33 222322	322	...	89	156	53	150	107	74	73	120	64	141	125	185	1339	45	1 Dec
Llanyblodwel	SJ 33 233236	91	912	57	108	45	96	60	51	55	81	52	104	89	112	908	28	14 Oct
Morda (Vyrnwy)																		
Llanforda Res.	SJ 33 272297	214	928	63	125	41	113	79	49	61	105	51	123	85	125	1019	37	14 Oct
Oswestry	SJ 33 278296	174	906	58	103	37	100	71	45	59	100	49	125	87	130	963	39	14 Oct
Mount Res.	SJ 33 275304	213	927	65	122	40	116	77	62	67	116	60	132	95	151	1103	42	14 Oct
Oswestry, Yorkfields Depot	SJ 33 289302	136	...	54	108	37	102	71	59	61	95	47	128	93	133	989	40	14 Oct
Vyrnwy (Severn)																		
Four Crosses	SJ 33 272180	64	767	45	91	27	84	47	59	51	58	34	85	(81)	85	748	22	10 Jun
Severn																		
Montford R.F.M.Sta. Area 2030 km²	SJ 33 413145	52	1179	...	...	...	...	...	...	...	...	...	...	...	...	...	...	...
Ford	SJ 33 420129	77	678	41	84	26	92	65	85	59	79	30	100	79	95	835	33	10 Jun
Perry (Severn)																		
Ellesmere Dairy	SJ 33 399343	91	...	41	79	24	87	72	49	74	71	27	95	69	105	792	26	14 Oct
Ruyton of the Eleven Towns	SJ 33 392224	76	...	44	87	31	81	68	53	52	64	28	94	77	97	779	20	11 May
Yeaton R.F.M.Sta. Area 181 km²	SJ 33 435193	61	792	...	...	...	...	...	...	...	...	...	...	...	...	...	...	...
Severn																		
Onslow	SJ 33 437129	94	...	41	87	31	91	59	85	60	92	32	109	81	103	872	37	20 Aug
Rea (Severn)																		
Hampton	SJ 33 308055	201	828	44	83	41	88	70	64	65	63	43	103	101	95	860	28	5 Nov
Portesbury	SJ 33 404059	113	...	43	104	33	111	57	(76)	(63)	65	32	92	85	93	854	25	5 Nov
Hookagate R.F.M.Sta. Area 178 km²	SJ 33 465092	65	787	...	...	...	...	...	...	...	...	...	...	...	...	...	...	...
Severn																		
Shrewsbury	SJ 33 514138	56	654	35	88	33	75	57	108	61	61	39	91	71	99	817	31	17 Jun
Tern (Severn)																		
Sidway Hall Farm	SJ 33 757398	120	...	46	85	36	93	63	74	108	93	47	112	54	166	977	32	9 Dec
Market Drayton, Devon House	SJ 33 675341	98	729	42	75	32	77	73	78	87	90	43	104	73	125	899	22	9 Dec
Market Drayton S.Wks	SJ 33 670333	76	739	42	76	38	84	74	71	87	86	44	98	63	123	885	22	9 Dec

GENERAL TABLE, ENGLAND AND WALES

Station	Grid reference	Altitude m	Av. 1916–1950 mm	Jan	Feb	Mar	Apr	May	Jun	Jul	Aug	Sep	Oct	Nov	Dec	Year	Highest recorded fall in a day mm	date
SEVERN—*(contd)*																		
Ternhill	SJ 33 642309	82	711	38	80	29	75	64	62	78	70	45	101	68	117	825	15	1 Apr
Hawkstone Park	SJ 33 583301	131	...	39	88	31	79	63	83	69	96	34	99	68	120	869	26	20 Aug
Meese (Tern)																		
Norbury Junction	SJ 33 794228	101	728	37	83	39	104	60	82	97	97	39	99	66	136	938	29	20 Aug
Chetwynd Park	SJ 33 737215	70	739	43	78	35	87	59	84	75	87	38	103	62	127	879	25	22 Jun
Tern (Severn)																		
Crudgington	SJ 33 629181	55	...	37	71	29	83	54	70	71	97	34	121	66	102	835	25	20 Aug
Newport	SJ 33 712204	64	663	39	73	29	87	56	63	71	93	36	97	61	117	822	38	20 Aug
Wellington, Wrekin College	SJ 33 655117	104	681	42	88	37	99	71	83	79	112	36	100	73	107	927	43	20 Aug
Woodfield P.Sta.	SJ 33 630163	57	...	37	85	28	86	62	83	77	92	38	122	71	98	881	23	4 Nov
Allscott	SJ 33 606126	58	658	35	77	36	86	68	81	77	83	31	106	82	106	867	...	...
Roden (Tern)																		
Wem, Secondary Modern School	SJ 33 508293	81	721	40	75	29	76	67	66	68	72	28	95	65	104	785	19	11 May
Shawbury	SJ 33 553220	72	660	39	84	31	78	58	95	71	84	29	96	69	100	832	48	17 Jun
Rodington R.F.M.Sta. Area 259 km^2	SJ 33 589141	48	721	...	...	...	...	...	...	...	...	...	...	...	...	...	...	...
Walcot R.F.M.Sta. Area 852 km^2	SJ 33 592123	45	729	...	...	...	...	...	...	...	...	...	...	...	...	...	...	...
Cound (Severn)																		
Dorrington	SJ 33 472019	108	...	47	108	38	106	79	60	57	54	31	108	(76)	105	868	27	1 Dec
Harnage Grange	SJ 33 569021	117	725	32	85	30	93	61	81	60	96	33	105	82	90	849	29	4 Nov
Worfe (Severn)																		
Weston Park	SJ 33 806108	113	721	44	87	44	92	79	93	91	93	36	108	65	129	960	24	1 Dec
Cosford W. Wks, New site	SJ 33 780046	59	...	36	90	41	80	71	109	69	97	44	103	68	109	916	31	20 Aug
Hilton P.Sta.	SO 32 776959	51	...	35	96	33	80	62	107	55	109	26	90	61	83	837	41	10 Jun
Severn																		
Morville Hall	SO 32 667940	91	...	42	122	41	97	69	91	62	104	39	109	81	89	947	25	1 Dec
Alveley	SO 32 754842	72	...	41	90	33	90	71	82	47	86	26	95	(63)	69	793	29	8 May
Elan Aqueduct	SO 32 776784	28	741	41	109	29	88	76	81	49	78	29	101	71	82	832	29	8 May
Bewdley R.F.M.Sta. Area 4330 km^2	SO 32 782762	17	945	...	...	...	...	...	...	...	...	...	...	...	...	...	...	...
Ribbesford	SO 32 768737	157	726	40	109	32	83	69	60	51	71	27	94	68	77	780	26	4 Nov
Stour (Severn)																		
Hayley Green P.Sta.	SO 32 941815	180	790	38	104	41	79	70	85	77	104	29	101	65	92	884	42	21 Aug
Halesowen	SO 32 948822	154	761	35	101	38	78	65	79	74	97	27	99	65	90	849	29	21 Aug
Lye P.Sta.	SO 32 918849	75	695	31	86	36	71	61	80	64	101	26	82	63	81	782	26	21 Aug
Stourbridge P.Sta.	SO 32 893844	65	769	30	83	40	65	70	75	60	109	25	79	67	84	788	34	20 Aug
Coalbourn Brook P.Sta.	SO 32 894855	69	706	29	78	38	69	66	74	62	104	25	83	69	90	789	31	20 Aug
Smestow (Stour)																		
Wolverhampton, West Park	SO 32 906990	131	723	28	91	33	72	69	100	82	94	34	99	64	101	866	24	8 May
Tettenhall W.Wks.	SJ 33 885001	143	724	31	104	35	76	69	117	83	73	35	93	64	101	881	24	8 May
Dimmingsdale W. Wks, New site	SO 32 859961	98	...	28	95	34	77	72	102	72	83	30	93	61	92	840	33	10 Jun
Bratch W.Wks.	SO 32 868937	82	666	31	91	31	72	76	93	80	99	29	93	60	85	840	37	29 Aug
Ashwood P.Sta.	SO 32 866878	63	687	27	79	34	64	63	96	50	92	27	90	65	78	766	24	10 Jun
Enville	SO 32 824864	99	...	36	93	36	82	78	76	56	72	29	109	62	88	816	28	8 May
Stour (Severn)																		
Kinver	SO 32 848835	46	682	30	90	33	73	69	72	49	83	24	90	59	84	757	26	8 May
Park Hall	SO 32 864778	61	699	35	93	30	72	63	79	53	98	24	87	59	75	768	36	20 Aug
Kidderminster, Wolverley Road	SO 32 843786	75	666	33	85	34	76	61	75	49	83	25	79	61	77	737	25	20 Aug
Kidderminster R.F.M.Sta. Area 324 km^2	SO 32 828769	31	721	...	...	...	...	...	...	...	...	...	...	...	...	...	...	...
Severn																		
Lincombe Lock	SO 32 821693	19	682	41	98	27	77	56	54	51	71	25	96	56	75	727	28	14 Oct
Abberley	SO 32 746662	168	...	49	101	31	80	76	64	41	75	25	83	72	75	773	32	4 Nov
Holt Lock	SO 32 821634	17	645	37	88	24	77	53	54	47	73	24	88	57	68	690	19	4 Nov
Salwarpe (Severn)																		
Wildmoor P.Sta.	SO 32 954752	140	...	31	93	32	72	81	63	57	83	26	96	61	81	775	24	21 Aug
Washing Stocks P.Sta.	SO 32 959732	112	711	31	89	38	74	75	76	70	89	27	101	63	85	819	25	21 Aug
Burcot P.Sta.	SO 32 984716	132	734	35	103	37	91	69	67	74	97	26	106	67	76	848	23	29 Aug
Bromsgrove S.Wks.	SO 32 959685	67	702	35	86	32	79	62	54	69	83	22	107	59	74	763	20	29 Aug
Tardebigge Res.	SO 32 995693	139	753	40	97	37	90	68	61	70	86	24	127	63	86	847	23	29 Aug
Randan Wood	SO 32 921727	128	...	38	96	35	75	72	70	64	99	24	105	60	84	821	36	21 Aug
Harford Hill R.F.M.Sta. Area 184 km^2	SO 32 868619	19	691	...	...	...	...	...	...	...	...	...	...	...	...	...	...	...
Severn																		
Bevere Lock	SO 32 836594	16	629	39	87	18	68	54	53	56	77	20	95	53	51	670	19	4 Nov
Worcester, Samuel Southall School	SO 32 853564	45	...	36	87	20	71	49	59	59	77	24	97	54	59	692	22	20 Aug
Worcester W.Wks.	SO 32 841565	17	643	37	87	21	68	49	52	51	71	24	98	54	60	675	19	4 Nov
Worcester, Fort Royal Hill	SO 32 855543	41	...	39	82	18	61	44	49	51	66	22	109	44	52	636	17	10 Jun
Teme (Severn)																		
Anchor	SO 32 172857	381	1226	61	165	58	100	100	59	61	77	57	129	144	137	1147	34	5 Nov
Warren Hill	SO 32 151778	387	...	58	153	55	107	95	73	64	87	72	135	143	129	1170	31	4 Nov
Knighton, Town Hall	SO 32 287725	183	967	66	137	52	119	87	76	59	105	56	138	149	107	1150	52	4 Nov
Brampton Bryan	SO 32 371726	131	...	59	116	45	108	82	63	49	77	48	120	110	83	958	34	4 Nov
Clun (Teme)																		
Two Crosses	SO 32 238866	434	1087	61	162	57	111	89	68	55	69	57	166	119	132	1145	33	3 Oct
Clunbury	SO 32 372805	158	826	49	129	33	102	91	58	42	69	55	139	91	88	945	31	8 May
Totterton Hall	SO 32 360874	213	811	49	113	33	94	85	60	45	78	45	134	93	96	925	30	14 Oct
Lower Hodre	SO 32 325769	305	...	61	134	44	117	88	70	51	82	64	133	112	89	1045	36	4 Nov
Bucknell	SO 32 353735	146	815	60	123	42	107	76	65	46	73	46	131	102	76	947	34	4 Nov
Teme (Severn)																		
Downton Castle	SO 32 444747	139	759	44	99	32	117	89	65	39	81	40	107	91	73	877	31	8 May
Onny (Teme)																		
Shelve	SO 32 337989	343	...	56	103	41	94	107	66	63	54	48	127	98	116	972	27	14 Oct 4 Nov
Plowden Hall	SO 32 375866	191	...	58	115	32	95	87	59	51	64	48	142	90	93	933	34	14 Oct
Church Stretton	SO 32 438911	187	896	53	123	47	98	93	71	70	60	47	134	106	116	1017	27	8 May
Acton Scott	SO 32 452890	183	...	44	114	45	90	84	58	68	77	53	126	95	101	954	28	8 May
Teme (Severn)																		
Oakly Park	SO 32 491762	91	770	41	100	40	81	79	75	43	97	37	98	77	71	837	32	20 Aug
Corve (Teme)																		
Bourton	SO 32 596963	194	...	48	103	51	110	80	86	71	116	45	117	92	105	1023	28	20 Aug
Wilderhope	SO 32 544927	215	...	43	99	42	95	70	62	67	89	53	109	83	100	911	31	20 Aug
Westhope Cottage	SO 32 475863	186	761	42	113	37	87	93	59	63	93	50	121	77	91	927	33	8 May

Station	Grid reference	Altitude m	Av. 1916–1950 mm	Jan	Feb	Mar	Apr	May	Jun	Jul	Aug	Sep	Oct	Nov	Dec	Year	Highest recorded fall in a day mm	Highest recorded fall in a day date
SEVERN—*(contd)*																		
Teme (Severn)																		
Stoke St. Milborough	SO 32 568808	189	856	47	111	55	101	78	70	61	89	39	117	116	93	975	44	4 Nov
Tenbury Wells R.F.M.Sta. Area 1130 km²	SO 32 598685	49	871	...	...	...	...	...	...	...	...	...	...	...	...	...	...	...
Kyre	SO 32 638620	99	...	47	94	33	77	75	71	43	63	33	88	95	72	792	40	4 Nov
Tenbury, Belle Vue	SO 32 592666	96	...	39	81	32	62	69	75	41	65	30	106	68	72	739	31	22 Jun
Rea (Teme)																		
Dingle Leys	SO 32 623850	223	879	55	135	49	113	81	87	73	89	36	130	100	114	1062	26	4 Nov
Hopton Wafers	SO 32 642766	183	...	55	119	47	105	85	90	65	93	38	139	104	112	1052	32	4 Nov
Teme (Severn)																		
Shelsley Beauchamp	SO 32 739634	94	695	51	88	30	76	77	63	47	72	32	85	74	77	773	34	4 Nov
Kingswood	SO 32 743598	53	...	45	95	23	77	66	62	39	84	27	88	71	63	738	35	4 Nov
Brockhampton	SO 32 687550	165	...	45	95	32	89	73	90	45	87	29	106	83	75	847	38	4 Nov
Cradley	SO 32 734471	76	...	46	88	29	87	57	73	48	53	29	100	100	59	768	57	4 Nov
Old Storridge	SO 32 750514	84	...	47	90	29	95	62	66	47	77	32	111	80	61	796	44	4 Nov
Severn																		
Malvern, Lansdowne Crescent	SO 32 782461	84	...	48	112	26	90	61	69	49	72	35	117	68	65	812	30	4 Nov
Malvern, Hall Green	SO 32 800457	38	...	39	100	23	82	50	56	42	65	34	107	59	61	717	26	4 Nov
British Camp Res.	SO 32 764399	206	704	44	93	24	90	60	58	52	77	27	99	76	64	765	35	4 Nov
Tewkesbury P.Sta.	SO 32 890337	14	650	24	82	19	68	56	44	63	56	23	92	34	63	625	18	29 Aug
Tewkesbury, Upper Lode	SO 32 881328	12	649	26	92	22	67	62	49	61	65	25	97	40	65	671	17	19 Oct
Avon (Severn)																		
Kilworth Wharf	SP 42 627836	128	...	36	99	17	93	77	56	86	108	39	104	57	101	873	41	29 Aug
Stanford Res.	SP 42 596804	112	659	31	89	18	83	76	60	84	113	29	99	45	85	813	57	29 Aug
Avon W.Wks.	SP 42 512763	87	646	26	82	27	81	61	66	87	102	31	119	41	90	814	44	29 Aug
Swift (Avon)																		
Lutterworth W.Wks.	SP 42 548840	111	679	33	109	19	87	61	63	79	93	32	106	46	90	817	36	29 Aug
Avon (Severn)																		
Rugby	SP 42 507749	116	648	24	77	22	87	64	61	85	91	27	115	45	89	789	41	29 Aug
Stareton R.F.M.Sta. Area 347 km²	SP 42 333715	55	676	...	...	...	...	...	...	...	...	...	...	...	...	...	...	...
Sowe (Avon)																		
Bedworth W.Wks.	SP 42 354866	105	681	30	100	30	89	61	69	84	102	31	103	35	84	819	36	29 Aug
Doe Bank P.Sta.	SP 42 323792	84	727	30	106	33	79	67	70	59	87	28	113	56	87	816	37	29 Aug
Whitley P.Sta.	SP 42 349771	73	...	29	109	35	86	75	71	56	107	37	109	40	91	844	45	29 Aug
Coventry, Belvedere Road	SP 42 324778	104	724	32	113	36	81	75	77	58	90	40	118	41	94	854	36	29 Aug
Kenilworth	SP 42 289721	90	710	35	108	41	83	69	81	55	138	40	110	40	90	892	71	29 Aug
Coventry	SP 42 290780	107	...	(25)	(108)	(32)	78	65	93	53	93	25	119	52	97	840	37	29 Aug
Finham S.Wks.	SP 42 334740	61	693	29	101	33	87	73	78	59	106	33	108	40	90	838	35	29 Aug
Stoneleigh R.F.M.Sta. Area 264 km²	SP 42 332731	55	709	...	...	...	...	...	...	...	...	...	...	...	...	...	...	...
Leam (Avon)																		
Staverton	SP 42 537611	168	...	27	88	21	87	49	71	67	109	30	119	55	101	824	57	29 Aug
Braunston	SP 42 540661	122	666	25	81	21	88	51	74	68	103	25	111	42	88	777	46	29 Aug
Bilton Grange	SP 42 489716	122	651	24	75	29	81	53	68	90	84	24	115	43	94	782	35	29 Aug
Dunchurch, Winton Hill	SP 42 484711	119	686	24	79	26	81	51	59	95	91	25	108	40	89	768	52	29 Aug
Itchen (Leam)																		
Southam, Banbury Road	SP 42 416612	76	650	23	77	24	82	41	74	60	94	21	118	46	82	742	53	29 Aug
Southam, Council Office	SP 42 417622	101	...	25	81	22	83	39	72	58	101	23	121	47	84	757	58	29 Aug
Leam (Avon)																		
Avon (Severn)																		
Warwick, St. Nicholas Park	SP 42 287649	47	651	25	86	25	87	61	74	67	149	33	110	41	72	831	81	29 Aug
Heathcote S.Wks.	SP 42 309627	67	...	24	90	21	95	62	78	81	133	32	115	44	74	849	79	29 Aug
Hatton	SP 42 243668	94	705	28	98	29	79	57	79	46	136	29	97	40	72	789	64	29 Aug
Ashorne	SP 42 306577	46	623	18	71	24	80	46	63	71	70	25	100	44	70	683	27	29 Aug
Moreton Morrell	SP 42 306553	85	640	19	76	20	81	47	58	71	68	26	102	37	68	673	30	29 Aug
Wellesbourne	SP 42 271565	47	636	19	69	23	73	47	55	68	67	32	105	36	68	661	31	29 Aug
Dene (Avon)																		
Kineton High School	SP 42 341510	87	650	21	69	21	73	42	59	74	94	21	91	37	65	667	58	29 Aug
Gaydon	SP 42 342541	13	...	21	72	19	81	43	61	66	67	22	97	36	67	651	31	29 Aug
Avon (Severn)																		
Snitterfield	SP 42 207584	84	628	27	76	27	83	60	68	84	97	25	92	35	66	740	44	29 Aug
Stratford-on-Avon, Thurlaston	SP 42 206540	41	632	25	63	20	81	50	64	75	91	29	100	39	63	699	47	29 Aug
Stratford-on-Avon P.Sta.	SP 42 195540	37	628	23	63	20	79	49	62	80	87	31	93	36	73	695	43	29 Aug
Stour (Avon)																		
Swalcliffe	SP 42 372369	175	...	29	82	25	79	52	65	69	92	31	112	42	93	770	50	29 Aug
Sibford Ferris	SP 42 356373	174	711	26	78	24	77	57	61	67	89	25	113	38	71	727	44	29 Aug
Weston Park	SP 42 287345	155	703	28	88	21	83	50	49	59	74	20	117	38	80	707	36	29 Aug
Chipping Campden, Bedfont House	SP 42 152392	143	774	31	89	29	93	58	58	63	74	24	112	48	95	777	31	29 Aug
Blockley, Greenway Res.	SP 42 160351	191	...	32	97	27	90	59	52	64	73	25	107	45	120	789	26	29 Aug
Shipston-on-Stour	SP 42 213407	111	674	25	86	19	90	54	51	57	82	22	100	35	87	710	40	29 Aug
Alscot Park R.F.M.Sta. Area 319 km²	SP 42 208507	38	711	...	...	...	...	...	...	...	...	...	...	...	...	...	...	...
Avon (Severn)																		
Luddington	SP 42 154528	42	600	22	71	22	90	48	72	77	68	28	102	31	68	700	28	29 Aug
Stratford-on-Avon	SP 42 164549	49	609	24	81	23	92	51	70	84	85	28	101	37	77	752	37	29 Aug
Arrow (Avon)																		
Lower Bittell Res.	SP 42 021743	139	779	43	119	49	98	87	69	81	97	34	129	70	111	987	25	1 Dec
Barnt Green	SO 32 996732	198	820	42	111	42	103	90	76	80	106	32	118	69	89	959	25	21 Aug 29 Aug
Brockhill P.Sta.	SP 42 001701	134	752	35	100	38	93	68	67	73	85	30	125	60	83	855	25	29 Aug
Ipsley S.Wks.	SP 42 062656	73	...	36	95	31	82	59	56	96	111	35	139	55	79	874	35	21 Aug
Studley College	SP 42 088640	78	668	39	93	33	83	66	55	106	93	25	113	52	82	842	41	5 Jul
Alcester, The Pleck	SP 42 078589	76	659	33	85	29	79	63	53	86	93	30	116	43	79	789	26	29 Aug
Alcester, Council Office	SP 42 092579	43	649	31	91	27	88	65	49	87	91	34	104	40	74	782	23	29 Aug
Alne (Arrow)																		
Henley-in-Arden P.Sta.	SP 42 154679	78	...	39	101	39	77	69	71	76	113	27	97	52	80	841	61	29 Aug
Wroxall	SP 42 220708	122	713	36	105	37	81	69	74	65	158	33	104	43	82	888	105	29 Aug
Arrow (Avon)																		
Wixford	SP 42 093543	40	...	31	88	31	89	60	58	79	77	32	108	40	79	772	23	29 Aug
Broom R.F.M.Sta. Area 319 km²	SP 42 087532	29	709	...	...	...	...	...	...	...	...	...	...	...	...	...	...	...
Avon (Severn)																		
Bretforton Manor	SP 42 092438	40	657	28	68	24	82	51	63	67	62	23	102	42	79	692	26	29 Aug
Evesham, High Street	SP 42 041444	37	622	22	70	26	66	52	70	65	70	23	101	32	73	671	28	29 Aug

GENERAL TABLE, ENGLAND AND WALES

Station	Grid reference	Altitude m	Av. 1916–1950 mm	Jan	Feb	Mar	Apr	May	Jun	Jul	Aug	Sep	Oct	Nov	Dec	Year	Highest recorded fall in a day mm	date
SEVERN—*(contd)*																		
Bengeworth	SP 42 048442	39	617	22	69	25	71	58	69	71	75	27	105	39	79	710	28	29 Aug
Evesham R.F.M.Sta. Area 2210 km²	SP 42 034431	20	683	...	...	...	...	...	...	...	...	...	...	...	...	...	...	...
Abbey Road P.Sta....	SP 42 034433	26	616	25	74	27	68	52	69	65	71	23	113	31	78	695	32	29 Aug
Isborne (Avon)																		
Sudeley Castle...	SP 42 030282	94	754	28	79	23	71	61	51	76	74	20	115	56	96	751	31	18 Oct
Stanway House	SP 42 062324	94	...	32	73	25	77	54	44	65	71	19	102	50	93	704	24	29 Aug
Avon (Severn)																		
Hampton Park S.Wks.	SP 42 030446	30	...	23	74	27	71	53	70	66	75	22	110	32	75	699	34	29 Aug
Pershore	SO 32 973495	30	...	36	71	19	63	51	48	78	74	29	116	35	55	675	22	29 Aug
Pershore, College of Horticulture ...	SO 32 960450	40	647	35	84	20	68	48	56	73	80	28	102	35	63	693	23	29 Aug
Bredon Hill Res.	SO 32 954405	216	...	28	85	29	80	67	55	78	85	22	102	39	69	738	...	...
Bow (Avon)																		
Bradley Green...	SO 32 990614	71	...	39	80	22	78	55	68	69	103	22	121	47	72	775	25	14 Oct
Crowle	SO 32 919571	80	...	38	76	19	70	47	44	62	59	27	115	40	66	663	25	14 Oct
Worcester...	SO 32 873535	58	645	40	80	19	63	51	44	59	61	32	109	45	55	659	18	5 Jul
Avon (Severn)																		
Overbury Court	SO 32 956373	65	656	27	84	26	62	63	73	69	69	27	100	46	73	719	24	17 Jun
Hewletts Res.	SO 32 973222	128	720	34	94	29	69	66	45	61	73	22	102	63	91	749	25	4 Nov
Cheltenham Cemetery	SO 32 971233	82	...	33	83	26	55	64	44	64	72	27	110	61	81	721	25	29 Aug
Severn																		
Forthampton Court...	SO 32 870318	15	682	26	90	22	70	61	48	57	58	26	97	42	66	664	18	17 Jun
Chelt (Severn)																		
Dowdeswell Res.	SO 32 984198	109	764	38	101	31	72	70	33	63	81	25	101	56	95	767	26	29 Aug
Cheltenham, Sandford Mead	SO 32 953216	64	693	27	80	24	68	68	45	63	61	24	106	59	82	708	26	4 Nov
Cheltenham	SO 32 946218	65	668	32	80	29	68	66	45	67	59	27	103	57	86	721	24	4 Nov
Severn																		
Witcombe Res.	SO 32 903147	91	738	41	91	31	79	79	41	74	113	33	113	61	79	834	38	20 Aug
Gloucester	SO 32 851177	27	...	24	81	26	75	63	39	67	111	21	88	55	70	719	40	20 Aug
Leadon (Severn)																		
Bosbury	SO 32 691433	69	709	37	90	31	73	58	63	48	55	25	106	92	61	740	52	4 Nov
Much Marcle, Hall Court	SO 32 644353	61	...	45	91	23	75	53	61	44	60	22	120	87	61	743	42	4 Nov
Upton Bishop	SO 32 658282	98	718	42	79	23	90	49	50	37	77	33	105	89	53	726	52	4 Nov
Bromsberrow Heath P.Sta.	SO 32 738332	55	724	44	88	28	72	65	48	44	68	32	118	77	63	744	39	4 Nov
Eastnor, Bronsil House...	SO 32 748372	114	...	52	88	25	81	67	59	52	69	23	109	75	59	762	36	4 Nov
Staunton	SO 32 791292	21	709	41	101	26	73	62	43	44	63	29	107	56	64	710	26	4 Nov
Kilcot...	SO 32 681255	64	779	49	116	26	100	58	43	43	72	37	119	85	62	810	47	4 Nov
Ford House	SO 32 735273	38	...	42	114	27	76	58	50	45	71	32	111	71	70	767	34	4 Nov
Wedderburn Bridge R.F.M.Sta. Area 293 km²	SO 32 777234	9	737	...	...	...	...	...	...	...	...	...	...	...	...	...	...	...
Taynton	SO 32 742216	24	757	42	127	31	95	76	48	72	79	27	112	61	79	848	31	4 Nov
Severn																		
Harescombe Grange	SO 32 843105	170	...	41	93	33	90	74	51	61	180	35	107	72	72	909	84	20 Aug
Hempsted...	SO 32 811157	12	732	26	85	28	81	65	42	67	67	23	91	62	70	709	33	4 Nov
Frome (Severn)																		
Miserden Park...	SO 32 938088	232	948	57	109	35	102	85	42	73	95	31	121	69	103	922	24	4 Nov
Chalford	SO 32 906033	175	--	63	105	31	113	90	48	67	106	32	118	69	91	932	27	29 Aug 4 Nov
Chalford W.Wks.	SO 32 891024	67	...	54	102	26	108	79	44	61	98	29	112	68	93	875	28	4 Nov
Pitchcombe	SO 32 854084	91	...	42	89	32	87	72	55	57	148	37	113	74	75	880	56	20 Aug
Minchinhampton, Rectory Cottage, New site	SO 32 873008	186	906	61	104	25	103	78	44	58	102	30	110	69	84	869	40	29 Aug
Dudbridge...	SO 32 831045	40	...	44	99	28	93	67	50	53	138	38	105	74	81	872	65	20 Aug
Stonehouse	SO 32 808051	41	738	41	97	26	84	60	47	47	98	32	118	74	70	794	40	4 Nov
Whitminster	SO 32 756093	10	783	40	105	36	96	72	58	59	66	31	113	70	83	829	34	4 Nov
Coastal																		
Blakeney Hill Lodge	SO 32 661086	134	834	58	162	46	128	78	66	55	160	36	139	77	105	1111	63	29 Aug
Whitemead Park	SO 32 612077	76	...	75	173	46	134	84	61	58	100	32	135	75	110	1082	43	4 Nov
Lydney Recreation Ground	SO 32 638028	12	...	(70)	(171)	(46)	126	73	62	52	87	34	126	66	103	1016	36	4 Nov

WYE RIVER AUTHORITY AREA, 4184 SQUARE KILOMETRES

HYDROMETRIC AREA 55

Station	Grid reference	Altitude m	Av. 1916–1950 mm	Jan	Feb	Mar	Apr	May	Jun	Jul	Aug	Sep	Oct	Nov	Dec	Year	Highest recorded fall in a day mm	date
Wye																		
Llugnant	SN 22 812858	442	...	116	255	183	172	200	307	236	123	173	208	274	601	2849	...	...
Nant Iago	SN 22 826858	408	2375	107	252	171	154	158	233	188	109	148	193	248	576	2537	...	...
Cefn-brwyn R.F.M. Sta. Area 10.4 km²	SN 22 829838	341	2532	...	...	...	...	...	...	...	...	...	...	...	...	...	...	...
Cefn-brwyn	SN 22 830837	343	2350	100	245	154	147	170	256	161	102	155	170	255	576	2491	...	...
Manod	SN 22 840827	320	2198	94	207	147	118	167	235	151	100	147	164	229	541	2300	121	9 Dec
Plynlimon	SN 22 791839	488	2363	107	229	155	127	175	305	216	102	155	203	254	574	2601	...	...
Pant Mawr	SN 22 831825	335	...	99	230	137	135	160	242	157	100	145	163	235	526	2330	122	9 Dec
Pant Mawr R.F.M. Sta. Area 27.2 km²	SN 22 843825	310	2461	...	...	...	...	...	...	...	...	...	...	...	...	...	...	...
Dernol	SN 22 914748	265	1435	89	191	89	89	114	127	89	102	89	127	178	368	1651	...	...
St. Harmon	SN 22 990726	280	1196	89	177	75	83	103	88	72	74	79	137	137	219	1334	44	9 Dec
Marteg	SN 22 947710	239	...	84	212	96	107	109	138	93	91	105	137	158	300	1628	...	...
Rhayader R.F.M. Sta. Area 167 km²	SN 22 969676	195	1664	...	...	...	...	...	...	...	...	...	...	...	...	...	...	...
Rhayader	SN 22 972680	213	1246	80	183	76	108	92	(94)	94	90	77	120	130	256	1399	...	...
Elan (Wye)																		
Glan Fedwen	SN 22 871759	419	1850	116	221	131	139	134	189	137	101	123	169	239	507	2207	...	...
Bod-talog	SN 22 868749	381	1820	116	216	127	143	143	187	136	108	121	166	232	504	2199	103	9 Dec
Abergwngu	SN 22 870734	351	1844	125	228	127	146	147	176	150	117	125	154	239	499	2232	111	9 Dec
Aberhenllan	SN 22 898717	326	1714	140	227	109	118	126	135	116	114	125	129	178	491	2008	...	...
Calettwr	SN 22 875701	454	1887	149	244	137	152	144	168	132	128	153	155	178	438	2177	...	...
Abercaethon	SN 22 870690	360	1791	128	217	112	143	131	153	117	117	131	150	180	452	2031	...	...
Tynant	SN 22 899673	311	1757	120	262	125	149	134	159	105	113	125	182	217	433	2124	99	9 Dec
Pen-y-Gareg	SN 22 925681	466	1502	95	230	(97)	132	104	119	94	102	117	135	150	343	1717	...	...
Pen-y-Gareg No.2	SN 22 926677	469	...	97	229	(97)	135	110	125	97	103	119	129	145	348	1732	...	...
Ty'n-y-llidiart	SN 22 907659	307	1688	108	242	119	147	125	132	114	114	140	170	216	394	2021	...	...
Nantgwillt...	SN 22 909639	256	1601	107	263	103	143	128	110	97	111	103	179	189	365	1899	79	9 Dec
Claerwen (Elan)																		
Figyn	SN 22 828715	494	2065	118	221	104	135	137	137	127	102	140	147	206	407	1981	...	...
Claerwen	SN 22 820673	381	1783	104	201	102	129	125	127	125	97	122	150	208	419	1909	...	...
Claerddu	SN 22 792686	434	1889	107	198	94	140	123	117	115	91	117	145	204	396	1847	...	...

Station	Grid reference	Altitude m	Av. 1916–1950 mm	Jan	Feb	Mar	Apr	May	Jun	Jul	Aug	Sep	Oct	Nov	Dec	Year	Highest recorded fall in a day mm	date
WYE—*(contd)*																		
Bwlchyrhendre...	SN 22 807664	483	1807	109	204	97	129	129	131	117	115	129	155	221	407	1944	...	...
Llyn Gynon	SN 22 801651	442	1824	108	209	109	119	131	122	109	86	129	153	211	401	1888	...	...
Cerigllwydion	SN 22 843692	521	1959	108	213	107	128	129	132	122	104	135	160	229	414	1981	...	...
Bryn Hir	SN 22 843681	448	1969	115	226	104	131	127	119	117	102	125	150	213	427	1956	...	...
Esgair-y-Gader	SN 22 859665	504	1821	112	230	102	147	119	(137)	104	129	109	175	203	407	1976	...	...
Arban...	SN 22 830629	457	2069	128	255	153	178	154	137	135	140	103	173	204	432	2192	...	...
Esgair Garthen	SN 22 842643	485	1983	141	255	145	176	140	135	128	153	102	178	216	458	2227	...	...
Ciloerwynt	SN 22 882629	319	1570	120	245	110	144	124	116	102	111	104	175	187	356	1895	66	9 Dec
Nant-y-car	SN 22 877617	475	2005	142	306	155	159	147	135	102	141	122	188	201	407	2203	...	...
Cors yr Hwch	SN 22 859600	550	2306	135	328	165	178	153	140	142	168	128	201	216	483	2436	...	...
Llyn Carw	SN 22 851614	541	2136	142	305	153	188	153	128	114	165	107	191	206	432	2284	...	...
Esgair Ceiliog	SN 22 898601	475	2106	122	295	160	175	150	(132)	114	140	114	191	206	427	2225	...	...
Tyrmynydd	SN 22 914625	254	1595	102	262	103	148	126	111	96	108	100	176	184	348	1863	76	9 Dec
Elan (Wye)																		
Tyn-y-Pant	SN 22 926626	402	1659	109	290	104	157	135	112	99	120	99	175	191	345	1936	...	...
Caban Coch R.F.M. Sta. Area 184 km^2	SN 22 924645	251	1872	...	...	...	...	...	...	...	...	...	...	...	...	...	...	...
Elan Village	SN 22 934651	213	...	94	239	91	129	100	108	83	93	90	156	169	333	1685	79	9 Dec
Wye																		
Llysdinam, New site	SO 32 007586	206	1308	77	159	88	99	99	79	80	86	76	147	125	225	1339	41	1 Dec
Eithon (Wye)																		
Llanbadarn-fynydd	SO 32 097777	297	1129	64	203	65	111	107	69	70	127	69	133	144	166	1328	...	...
Llanddewi-Ystradenny	SO 32 107685	234	979	58	141	63	88	101	86	71	76	65	137	123	126	1136	...	...
Llanddewi R.F.M. Sta. Area 111 km^2	SO 32 105683	230	1166	...	...	...	...	...	...	...	...	...	...	...	...	...	...	...
Bwlch-y-sarnau	SO 32 029746	393	...	67	179	62	85	107	83	77	83	75	137	125	155	1235	...	...
Abbey-Cwmhir	SO 32 044719	290	...	64	190	63	88	101	95	72	96	79	155	124	177	1304	38	24 Feb
Llandrindod Wells	SO 32 061605	235	1032	59	128	60	77	92	80	66	90	64	139	118	135	1109	26	8 May
Irfon (Wye)																		
Llanerch-yrfa	SN 22 836555	326	2044	126	272	129	209	171	121	140	139	127	198	215	395	2242	...	...
Nant-y-brain	SN 22 849518	251	1946	112	293	155	191	185	122	141	153	122	198	191	393	2256	...	...
Cefn Fannog	SN 22 826513	439	2083	127	254	165	203	193	145	160	155	127	203	193	457	2382	...	...
Abernant	SN 22 888465	195	1486	84	198	107	135	145	99	115	103	97	156	144	318	1700	...	...
Abernant R.F.M. Sta. Area 72.8 km^2	SN 22 892460	185	1953	...	...	...	...	...	...	...	...	...	...	...	...	...	...	...
Llandulas	SN 22 902428	238	1568	88	221	115	157	148	109	131	110	97	171	146	350	1843	...	...
Llwyn Madoc	SN 22 903524	223	1589	99	233	115	156	141	113	115	120	117	186	150	322	1867	63	1 Dec
Garth	SN 22 954498	165	1313	79	183	77	125	102	84	97	94	81	153	123	242	1439	44	9 Dec
Wye																		
Builth Wells	SO 32 038530	229	1071	57	126	43	71	86	80	69	86	66	147	(102)	148	1082	28	8 May
Cwrt-Gwenddwr	SO 32 072456	124	1113	73	138	57	106	93	79	53	97	69	141	105	173	1183	33	1 Dec
Erwood R.F.M. Sta. Area 1280 km^2	SO 32 076445	107	1425	...	...	...	...	...	...	...	...	...	...	...	...	...	...	...
Afon Llynfi (Wye)																		
Felin-fach	SO 32 088333	168	928	78	130	44	93	79	85	45	78	53	144	101	174	1105	28	1 Dec
Talgarth	SO 32 136354	152	925	75	134	52	94	87	69	39	78	50	141	93	153	1065	33	8 May
Wye																		
Glasbury	SO 32 181393	84	913	67	96	47	115	83	60	45	64	37	129	90	116	947	36	8 May
Clifford	SO 32 268445	146	871	76	102	38	113	98	(58)	(43)	83	34	127	(102)	95	969	19	2 Mar
Sarnesfield	SO 32 368509	122	739	49	81	37	101	82	59	44	91	31	121	92	60	847	39	4 Nov
Lulham House	SO 32 405413	67	733	49	88	27	91	57	40	29	85	32	131	76	56	761	35	4 Nov
Breinton Court	SO 32 470396	73	671	56	84	21	88	51	53	40	83	29	149	73	43	769	31	4 Nov
Belmont R.F.M. Sta. Area 1900 km^2	SO 32 485388	46	1267	...	...	...	...	...	...	...	...	...	...	...	...	...	...	...
Broomy Hill	SO 32 496395	73	...	49	90	20	78	54	49	34	58	26	119	65	44	687	30	4 Nov
Hereford S. Wks.	SO 32 519388	51	633	47	89	17	78	55	42	31	57	23	108	66	50	665	29	4 Nov
Tillington	SO 32 467448	82	663	47	80	24	87	61	51	40	83	27	118	72	51	740	27	4 Nov
Hereford, St. Mary's Hospital	SO 32 482433	82	654	50	81	20	81	56	51	40	85	27	134	70	49	745	28	4 Nov
Lugg (Wye)																		
Cascob	SO 32 228658	305	1184	74	155	58	125	93	69	74	109	58	145	133	121	1215	32	4 Nov
Summergill (Lugg)																		
Llanfihangel-Nant-Melan	SO 32 167590	366	...	67	141	68	120	92	80	65	89	51	148	142	164	1228	...	...
New Radnor	SO 32 213606	300	1111	73	139	54	132	88	92	61	92	58	147	133	116	1185	31	8 May
Evancoyd	SO 32 261630	227	948	65	127	50	116	70	66	46	91	51	129	131	107	1049	44	4 Nov
Lugg (Wye)																		
Shobdon	SO 32 395619	145	...	42	97	47	101	77	67	45	101	34	98	103	71	884	31	8 May
Grantsfield	SO 32 526603	129	719	47	85	32	77	82	69	32	71	39	97	90	65	785	35	8 May
Leominster	SO 32 501587	72	...	43	89	31	78	76	71	35	69	38	91	81	68	769	33	8 May
Arrow (Lugg)																		
Newchurch	SO 32 217508	244	979	59	126	64	122	103	87	69	89	56	133	133	143	1184	...	...
Lyonshall, The Croft	SO 32 335558	146	811	44	97	43	105	72	60	38	113	40	122	106	87	928	38	4 Nov
Lyonshall	SO 32 339576	155	832	44	105	42	107	71	54	41	99	37	119	105	86	910	38	4 Nov
Monkland	SO 32 449573	88	686	45	87	40	93	90	59	31	81	39	116	86	68	834	35	4 Nov
Lugg (Wye)																		
Hope under Dinmore	SO 32 513538	75	679	43	86	28	85	69	52	35	81	36	106	72	63	756	27	4 Nov
Stoke Prior	SO 32 533557	113	...	41	85	29	81	71	58	40	77	33	107	84	65	771	29	4 Nov
Preston Wynne	SO 32 565475	84	642	37	77	22	69	62	58	45	65	21	104	73	58	692	28	4 Nov
Lugwardine R.F.M. Sta. Area 886 km^2	SO 32 548405	45	836	...	...	...	...	...	...	...	...	...	...	...	...	...	...	...
Frome (Lugg)																		
Down House	SO 32 669545	148	708	46	88	28	90	81	81	43	77	28	101	85	72	821	42	4 Nov
Wye																		
Sollers Hope	SO 32 616325	94	...	43	97	23	83	59	52	37	77	25	118	78	56	746	39	4 Nov
Perrystone Court	SO 32 625295	158	...	45	91	26	91	63	51	39	88	22	115	77	59	768	41	4 Nov
Ross-on-Wye	SO 32 598237	68	708	44	106	24	94	54	30	34	80	30	107	79	54	737	45	4 Nov
Alton Court P. Sta.	SO 32 606235	50	...	44	102	22	95	54	31	39	73	33	105	79	52	727	44	4 Nov
Ross-on-Wye S. Wks.	SO 32 584234	34	...	46	105	23	102	48	31	35	75	(36)	117	74	(63)	756	42	4 Nov
Walford, Hill Court	SO 32 575215	38	739	53	110	21	98	51	33	35	79	34	118	78	60	771	42	4 Nov
Coleford, Marions Lodge	SO 32 571127	198	...	88	161	36	156	89	49	51	97	34	163	91	89	1104	55	4 Nov
Monnow (Wye)																		
Longtown	SO 32 323291	172	...	116	169	54	159	73	71	40	93	46	165	127	159	1273	50	4 Nov
Honddu (Monnow)																		
Blaen-Bwch	SO 32 241322	411	1484	150	215	85	214	112	99	79	133	73	182	177	189	1709	...	...
Tafolog	SO 32 277294	259	1272	130	198	74	194	95	103	70	113	63	177	175	151	1544	...	...
Dore (Monnow)																		
Vowchurch	SO 32 349362	125	913	87	127	38	130	66	54	31	97	36	166	102	83	1015	45	4 Nov
The Mynde	SO 32 470298	128	736	54	115	15	87	53	49	36	71	27	116	60	44	729	25	4 Nov

Station	Grid reference	Altitude m	Av. 1916–1950 mm	Jan	Feb	Mar	Apr	May	Jun	Jul	Aug	Sep	Oct	Nov	Dec	Year	Highest recorded fall in a day mm	date
WYE—*(contd)*																		
Monnow (Wye)																		
Kentchurch Court	SO 32 423258	72	825	90	137	23	116	59	51	33	93	36	141	95	65	940	46	4 Nov
Kentchurch R.F.M. Sta. Area 357 km²	SO 32 419251	58	993	...	...	...	...	...	...	...	...	...	...	...	...	...	...	...
Skenfrith	SO 32 459204	46	831	65	134	33	112	68	47	33	90	28	148	84	65	905	...	
Monmouth, Cornwall House	SO 32 507126	16	789	66	135	28	106	58	40	39	89	30	138	79	67	875	53	4 Nov
Troddi (Wye)																		
Llantilio Crossenny	SO 32 393172	162	913	91	164	53	130	70	66	51	95	(38)	154	103	85	1100	52	4 Nov
Wye																		
Redbrook P. Sta.	SO 32 539108	67	938	81	167	38	126	81	42	48	89	30	149	82	88	1020	52	4 Nov
Cadora R.F.M. Sta. Area 4040 km²	SO 32 535090	8	1039	...	...	...	...	...	...	...	...	...	...	...	...	...	...	...
Chepstow, Bethdour	ST 31 527936	94	949	69	130	37	146	75	85	56	91	48	150	77	116	1079	44	19 Oct
Coastal																		
Shirenewton Filters	ST 31 484944	84	1033	84	147	44	168	84	96	69	98	50	153	83	117	1192	47	4 Nov

USK RIVER AUTHORITY AREA, 1741 SQUARE KILOMETRES

HYDROMETRIC AREA 56

Station	Grid reference	Altitude m	Av. 1916–1950 mm	Jan	Feb	Mar	Apr	May	Jun	Jul	Aug	Sep	Oct	Nov	Dec	Year	Highest recorded fall in a day mm	date
Coastal																		
Newchurch	ST 31 437964	149	1139	105	179	54	186	97	84	60	122	54	168	91	114	1314	...	...
Nantypridd	ST 31 436944	152	1106	101	168	46	175	91	81	49	125	52	173	89	111	1261	...	...
Llanvaches	ST 31 428929	139	1031	71	153	42	156	89	79	53	127	49	163	76	118	1177	43	4 Nov
Redwick	ST 31 413841	6	...	66	117	38	143	68	70	33	77	52	164	69	107	1004	43	19 Oct
Usk																		
Upper Usk 6	SN 22 812249	393	1975	143	307	97	216	156	127	131	143	118	240	154	267	2099	...	...
Upper Usk 5	SN 22 815264	332	1741	77	260	75	170	140	117	126	125	105	201	133	296	1825	...	...
Upper Usk 2	SN 22 808299	418	1412	67	179	76	147	129	103	125	102	81	187	114	276	1586	...	...
Blaenau-isaf	SN 22 845258	287	1669	92	321	88	191	157	129	119	147	116	188	159	327	2031	54	24 Feb
Cray Res. No.3	SN 22 880209	320	...	131	372	95	265	171	145	149	184	137	234	174	291	2349	...	...
Cray Res. No.2	SN 22 880210	311	2403	130	372	94	265	171	143	149	184	136	233	173	289	2339	...	...
Cray Res. No.1, New site	SN 22 887220	305	2112	128	361	84	259	164	128	116	178	131	227	161	280	2216	46	18 Oct
Sennybridge	SN 22 918285	189	...	83	225	69	168	117	97	87	129	102	177	137	290	1681	45	9 Dec
Llwyn Llydan Farm	SN 22 920198	351	...	146	392	139	244	226	152	136	189	172	242	221	454	2713	...	...
Garn Wen Farm	SN 22 927216	256	...	112	325	102	206	191	110	115	159	136	233	186	351	2226	...	...
Cray	SN 22 896241	287	1808	107	307	92	199	152	98	120	154	113	202	151	326	2021	...	...
Berth Ddu Farm	SN 22 917322	259	...	70	184	66	136	109	92	94	109	87	188	121	305	1561	...	...
Trallong R.F.M.Sta. Area 184 km²	SN 22 947295	166	1708	...	...	...	...	...	...	...	...	...	...	...	...	...	...	...
Llanfihangel, Nant Bran School	SN 22 943342	213	...	66	167	72	132	106	85	84	105	71	188	112	284	1472	...	...
Aberbran Mill	SN 22 986294	154	...	63	186	61	113	100	73	68	91	72	156	98	225	1306	44	9 Dec
Merthyr Cynog	SN 22 987377	311	...	74	177	91	128	123	99	106	92	95	190	137	308	1622	...	...
Llwyn-y-Celyn Farm	SN 22 973225	312	...	123	331	104	204	189	110	105	153	120	215	165	328	2147	...	...
Tairbull	SN 22 996261	198	1360	89	192	75	143	117	90	67	108	87	147	111	247	1473	44	9 Dec
Brecon, North Road	SO 32 042289	148	...	68	169	45	107	84	82	52	90	59	146	89	192	1183	41	1 Dec
Honddu (Usk)																		
Drovers Arms	SN 22 986452	439	...	94	200	93	147	121	112	111	117	100	173	136	260	1662	...	...
Castle Madoc	SO 32 025369	244	...	55	151	61	121	97	99	73	99	79	172	116	212	1337	...	...
Llanerch Wen	SO 32 024321	280	...	59	161	59	97	98	88	59	85	72	177	111	228	1294	...	...
Pytin Gwyn Farm	SO 32 042314	198	...	59	137	46	103	77	85	53	95	61	174	107	189	1187	...	...
The Forge, Brecon R.F.M.Sta. Area 62.2 km²	SO 32 051297	143	1194	...	...	...	...	...	...	...	...	...	...	...	...	...	...	...
Usk																		
Caerfanell (Usk)																		
Darenfawr	SO 32 073164	357	1847	136	305	124	206	152	134	79	151	98	206	170	331	2092	...	...
Ynys-poste	SO 32 072169	226	1779	119	361	126	201	145	117	69	151	103	191	179	326	2088	...	...
Abercynafon No.2	SO 32 085167	357	1872	129	295	116	180	147	126	75	146	96	195	155	304	1965	...	...
Rhyd Bine	SO 32 091195	373	1814	118	298	115	181	131	119	70	136	102	189	154	303	1917	...	...
Talybont Filters	SO 32 106207	169	1506	97	237	88	159	106	90	48	103	67	165	120	195	1475	36	24 Feb
Graig Pwllfa	SO 32 074205	533	1841	122	339	120	213	146	125	81	141	113	200	158	312	2069	...	...
Cefnbach	SO 32 080210	385	...	116	333	124	180	135	118	76	168	102	188	128	309	1977	...	...
Nant-y-Llanerch	SO 32 085212	307	1797	127	300	115	209	140	109	66	141	98	191	163	298	1956	...	...
Tyle-Clydach	SO 32 099212	216	1536	112	239	99	164	119	95	56	111	81	186	138	230	1630	50	1 Dec
Talybont-on-Usk	SO 32 119229	116	...	89	188	67	122	89	58	44	94	55	145	113	165	1229	40	1 Dec
Usk																		
Ashford	SO 32 123218	119	1222	98	207	69	132	93	67	44	95	60	149	115	177	1305	44	1 Dec
Tretower House	SO 32 185213	88	...	87	163	49	121	68	77	47	84	51	128	118	120	1114	47	4 Nov
Grwyne Fawr (Usk)																		
Grwyne Fawr	SO 32 233307	545	1477	140	183	113	163	125	113	82	138	74	170	144	167	1611	51	2 Mar
Blaen-y-cwm	SO 32 246295	434	1426	137	175	109	157	122	108	78	132	72	165	140	160	1555	50	2 Mar
Coed Dias	SO 32 275244	274	1162	122	181	67	179	81	101	64	123	59	181	139	174	1473	47	4 Nov
Tal-y-maes	SO 32 225259	381	...	107	166	63	125	86	112	70	127	46	140	136	163	1340	...	...
Usk																		
Gilwern	SO 32 230138	175	1281	129	197	61	177	79	77	48	92	62	185	128	175	1410	49	4 Nov
Chain Bridge R.F.M.Sta. Area 912 km²	SO 32 345056	23	1379	...	...	...	...	...	...	...	...	...	...	...	...	...	...	...
Usk	SO 32 368018	21	1008	91	174	34	134	70	39	35	95	45	163	85	103	1066	46	4 Nov
Crumbland Plantation	SO 32 474024	245	1034	83	187	48	146	90	83	52	88	36	176	90	92	1170	56	4 Nov
Sluvad Treatment Wks.	ST 31 318992	143	...	105	207	48	166	88	61	33	112	50	169	91	121	1251	47	4 Nov
Lwyd (Usk)																		
Upper Cwmbran	ST 31 273968	194	...	145	253	81	203	124	91	51	153	74	195	109	189	1670	42	19 Oct 4 Nov
Cwmbran No.1	ST 31 273953	165	1312	114	211	61	180	113	77	48	150	64	184	109	175	1486	40	4 Nov
Caerleon, The Croft	ST 31 338908	16	...	89	155	45	163	82	61	39	98	44	174	79	126	1154	34	4 Nov
Usk																		
Pant-yr-Eos Res.	ST 31 254915	133	1280	104	172	54	171	107	103	53	127	73	156	90	182	1393	41	19 Oct
Ynys-y-fro Farm	ST 31 285892	43	...	91	155	43	170	77	89	39	105	44	168	84	143	1209	43	19 Oct
Newport, Frobisher Road	ST 31 336875	11	...	80	143	42	149	74	73	36	99	51	183	89	124	1143	48	19 Oct
Ebbw																		
Llangynidr No.2	SO 32 151143	448	1679	156	265	95	193	131	121	73	129	85	174	153	287	1862	48	19 Oct
Llangynidr No.3	SO 32 150141	447	1722	162	256	92	186	133	122	70	127	81	170	148	275	1822	47	19 Oct
Llangynidr No.1	SO 32 155139	418	1658	164	271	96	189	131	122	72	127	90	177	152	303	1894	48	19 Oct
Carno North	SO 32 160139	433	1628	149	259	91	176	126	116	69	122	81	171	143	269	1773	46	19 Oct
Carno West	SO 32 156132	430	1678	150	253	88	178	128	113	69	119	79	171	143	267	1758	47	19 Oct
Carno East	SO 32 165134	411	1606	133	234	79	162	117	102	61	111	76	161	133	248	1614	46	19 Oct

Station	Grid reference	Altitude m	Av. 1916–1950 mm	Rainfall in millimetres Jan	Feb	Mar	Apr	May	Jun	Jul	Aug	Sep	Oct	Nov	Dec	Year	Highest recorded fall in a day mm	date
USK—*(contd)*																		
Carno South	SO 32 163127	385	1641	145	251	87	182	130	117	70	124	89	182	145	269	1791	47	19 Oct
Ebbw Vale Iron Wks.	SO 32 170098	274	...	120	243	77	165	99	110	59	110	73	168	134	232	1591	51	19 Oct
Abertillery Park	SO 32 207049	213	1490	131	235	72	184	(112)	(102)	75	126	68	187	116	181	1588	51	19 Oct
Cwmtillery Res.	SO 32 221069	344	1491	139	260	86	193	112	97	71	114	67	198	136	213	1686	53	4 Nov
Llanhilleth...	SO 32 198012	312	1458	134	222	76	176	114	110	62	111	68	182	122	198	1575	41	19 Oct
Sirhowy (Ebbw)																		
Nant-y-Bwch W.Wks.	SO 32 132114	351	1620	127	247	89	169	119	121	68	122	85	171	144	284	1745	47	1 Dec
Tredegar	SO 32 141086	313	1636	125	265	95	168	123	136	71	119	80	162	129	259	1731	46	1 Dec
Georgetown W.Wks.	SO 32 149085	344	...	125	244	89	155	116	125	68	116	73	165	127	252	1653	47	1 Dec
Blackwood	ST 31 167973	219	1390	142	224	70	186	112	85	49	111	76	168	109	183	1515	46	19 Oct
Nantydraenog Res....	ST 31 188934	210	1434	127	211	60	177	103	90	57	113	78	166	108	145	1435	41	4 Nov
Ebbw																		
Rhiwderyn R.F.M.Sta. Area 217 km²	ST 31 258888	31	1491	...	...	...	...	...	...	...	...	...	...	...	...	...	...	...
Rogerstone Golf Club	ST 31 268893	93	1207	108	175	49	169	91	87	43	113	56	166	93	164	1316	41	19 Oct
Newport	ST 31 279849	23	...	87	149	47	145	86	75	36	98	58	149	81	139	1150	39	19 Oct
Coastal																		
Peterstone Wentllooge	ST 31 270802	6	...	89	131	39	127	71	53	33	88	44	147	74	120	1016	...	

GLAMORGAN RIVER AUTHORITY AREA, 1954 SQUARE KILOMETRES

HYDROMETRIC AREAS 57 AND 58

Rhymney																		
Trethomas...	ST 31 181893	107	...	122	180	53	156	106	89	54	135	77	155	109	180	1415	41	4 Nov
Machen	ST 31 210893	76	...	104	168	59	150	112	86	46	117	64	147	96	171	1321	39	19 Oct
Cardiff	ST 31 195793	62	1069	96	133	37	138	80	82	44	95	54	158	75	145	1135	41	19 Oct
Llanishen Res....	ST 31 188815	47	1051	111	140	46	158	85	83	46	111	60	163	94	178	1275	52	19 Oct
Cardiff, Heath Filters	ST 31 181799	40	1112	104	135	41	137	77	95	44	100	56	151	81	173	1195	46	19 Oct
Cardiff, Roath Park	ST 31 185789	16	1100	93	136	39	145	77	86	39	104	57	149	71	156	1152	39	19 Oct
Taff																		
Blaen Taf Fawr	SN 22 992207	511	2225	125	301	126	199	191	199	135	175	154	228	190	383	2406	...	...
Storey Arms Field	SN 22 986200	436	2170	137	313	135	216	192	171	117	189	150	225	194	411	2449	54	30 Dec
Beacon Res.	SN 22 987192	414	2132	130	323	138	207	200	168	117	184	138	221	198	402	2424	53	30 Dec
Nant Penig	SN 22 977191	610	2406	122	323	133	203	194	209	146	175	(129)	226	208	398	2468	...	...
Cantref Res.	SN 22 995155	341	1871	111	263	109	184	175	140	103	173	123	216	176	377	2149	53	9 Dec
Llysiog	SO 32 015149	329	1780	101	251	107	173	166	136	109	164	95	225	185	325	2037	...	...
Garw-nant...	SN 22 994132	335	1924	110	274	110	175	168	158	104	180	117	211	175	369	2150	...	...
Cefn Gougarn	SN 22 986150	415	1931	107	261	117	180	179	178	114	175	141	219	201	399	2272	...	...
Nant Ffynnon	SN 22 983119	412	1902	113	275	119	178	159	160	105	188	123	216	171	375	2183	...	...
Nant Gwinell	SO 32 017132	366	1762	105	245	117	162	164	161	103	148	98	200	177	344	2023	...	...
Garn Ddu	SO 32 027129	408	1836	108	235	97	171	164	173	83	178	113	189	179	337	2027	...	...
Llwynon	SO 32 012115	262	...	113	248	96	157	143	131	92	163	103	188	149	344	1926	53	9 Dec
Llwynon Res.R.F.M.Sta. Area 43.0 km²	SO 32 010112	236	2055	...	...	...	...	...	...	...	...	...	...	...	...	...	...	...
Cefn-Coed-y-Cymmer	SO 32 024086	241	1461	111	231	85	150	123	120	88	144	98	161	125	298	1734	53	9 Dec
Taf Fechan (Taff)																		
Neuadd Res.No.6	SO 32 011209	640	2512	129	363	119	218	180	191	132	206	135	239	206	408	2526	...	...
Neuadd Res.No.5	SO 32 026195	470	...	125	307	122	218	157	152	129	183	147	223	226	431	2422	...	...
Neuadd Res.No.10	SO 32 032193	518	1959	119	333	140	223	175	160	112	196	129	226	213	408	2435	...	...
Neuadd Res.No.2	SO 32 020192	604	...	132	312	125	208	188	175	140	203	150	226	223	461	2544	...	...
Neuadd Res No.11A	SO 32 032185	463	2156	131	297	129	227	168	169	115	190	124	213	189	348	2299	45	30 Dec
Pentwyn Res.No.16	SO 32 049158	332	1764	116	269	121	180	156	133	85	152	100	191	170	351	2024	...	...
Pentwyn Res.No.15	SO 32 056145	341	1804	127	263	123	193	159	134	83	170	101	195	167	347	2060	52	19 Oct
Pontsticill No.12	SO 32 059116	315	1604	101	213	94	155	149	123	80	150	95	173	145	294	1773	48	9 Dec
Taf Fechan Res.R.F.M.Sta. Area 33.7 km²	SO 32 060117	298	2007	...	...	...	...	...	...	...	...	...	...	...	...	...	...	...
Pontsticill West No.14	SO 32 043116	382	...	94	215	97	143	151	125	82	152	95	178	150	296	1778	...	...
Taff																		
Merthyr Tydfil, Bronygarth	SO 32 048071	235	1404	108	234	77	149	108	107	81	144	83	169	127	289	1675	50	9 Dec
Dowlais Filters	SO 32 066096	379	1535	101	219	75	155	133	118	73	151	81	173	128	238	1646	39	19 Oct
Dowlais, Gwernllwyn, New site	SO 32 068081	273	1458	101	225	76	139	119	115	73	141	74	159	118	259	1599	45	9 Dec
Penybryn Filters	SO 32 054077	286	1453	102	212	70	147	128	111	74	134	78	180	116	251	1603	41	9 Dec
Treharris Res.	ST 31 096974	233	1422	127	235	79	185	128	102	59	124	84	182	125	228	1659	46	19 Oct
Cynon (Taff)																		
Darranlas Res....	ST 31 044991	165	1650	137	281	83	191	141	145	57	165	97	196	117	253	1863	51	24 Feb
Abercynon R.F.M. Sta. Area 109 km²	ST 31 079956	81	1801	...	...	...	...	...	...	...	...	...	...	...	...	...	...	...
Taff																		
Lanwood Res.	ST 31 073908	91	1589	113	214	80	190	138	134	67	135	116	191	118	259	1755	51	30 Dec
Rhondda (Taff)																		
Treherbert, Tyn-y-Waun, No.1	SS 21 931993	244	2377	164	346	108	239	197	183	117	207	121	237	164	423	2505	60	24 Feb
Treherbert, Tyn-y-Waun, No.2	SS 21 931993	244	2379	164	347	109	242	200	184	118	209	120	237	164	425	2520	59	24 Feb
Treherbert Park	SS 21 938983	183	2345	167	319	113	228	196	193	110	210	127	253	165	389	2469	61	30 Dec
Treorchy P.Sta.	SS 21 963964	157	2201	158	295	109	230	187	170	102	196	125	247	164	403	2385	65	30 Dec
Gelli-galed	SS 21 990947	145	1878	134	274	96	210	173	131	77	175	109	234	150	307	2068	54	30 Dec
Lluest Wen Res.	SN 22 950015	390	2388	178	368	124	256	196	188	136	215	139	233	189	432	2654	62	28 Dec
Lluest Wen Filters	SN 22 957014	366	2247	172	335	117	248	189	181	125	211	138	223	179	431	2549	65	28 Dec
Castell Nos Res.	SN 22 964001	323	2207	161	337	108	232	180	192	133	197	127	239	180	418	2504	68	24 Feb
Penrhys Cemetery	ST 31 003947	287	...	133	278	99	216	172	138	78	179	112	218	158	342	2124	56	30 Dec
Porth Park	ST 31 023911	122	1861	113	232	100	213	158	132	63	151	115	202	142	331	1952	60	30 Dec
Taff																		
Nantgawr	ST 31 118856	36	...	129	164	74	173	113	115	54	139	83	136	97	223	1499	39	30 Dec
Rhiwbina Res.	ST 31 150824	102	1274	130	148	56	163	95	115	53	118	74	151	99	224	1426	53	19 Oct
Ely																		
Llantrisant	ST 31 028849	85	...	115	180	84	175	133	150	67	143	94	149	98	283	1671	45	28 Dec
St.Fagans Castle	ST 31 119773	30	1226	118	140	51	172	95	135	47	108	63	161	77	171	1339	55	19 Oct
Cogan P.Sta.	ST 31 171729	37	1049	92	118	40	145	75	137	49	94	79	151	81	158	1220	52	19 Oct
Coastal																		
Penarth, Alexandra Park	ST 31 187714	37	...	75	98	29	133	64	132	48	88	68	163	76	138	1111	49	19 Oct
Biglis P.Sta.	ST 31 147698	8	...	81	90	28	133	58	122	36	93	49	134	63	125	1014	...	...
Penarth Cemetery	ST 31 180702	30	927	84	97	29	135	62	119	44	90	59	145	70	122	1054	42	19 Oct
Lower Penarth...	ST 31 182695	20	915	70	82	29	113	56	105	37	77	57	137	64	122	949	42	19 Oct
Rhoose	ST 31 060670	6	922	60	88	34	128	58	107	40	86	39	132	71	119	962	45	19 Oct
Thaw																		
Cowbridge, Pwllwy...	SS 21 991766	28	1142	93	129	58	129	104	128	49	113	61	173	72	182	1290	57	19 Oct
Tair Onen	ST 31 034742	128	1123	80	109	55	135	90	128	44	105	67	165	73	165	1216	53	19 Oct

Station	Grid reference	Altitude m	Av. 1916–1950 mm	Rainfall in millimetres Jan	Feb	Mar	Apr	May	Jun	Jul	Aug	Sep	Oct	Nov	Dec	Year	Highest recorded fall in a day mm	date
GLAMORGAN—*(contd)*																		
St.Mary Church	ST 31 019716	14	...	67	100	54	134	83	115	39	93	52	181	79	157	1153	52	19 Oct
St.Athan	ST 31 026671	15	948	65	91	36	119	61	112	38	103	32	136	73	116	981	22	21 Apr
Ogmore																		
Ogmore Vale	SS 21 933936	260	2197	154	254	122	232	201	197	138	212	125	228	174	463	2500	69	30 Dec
Pontycymmer	SS 21 903916	171	2063	145	226	119	209	203	189	118	188	117	209	151	421	2295	61	19 Oct 30 Dec
Maesteg Park	SS 21 847913	168	1833	133	205	110	197	161	161	97	164	86	191	131	310	1944	54	19 Oct
Bridgend R.F.M.Sta. Area 158 km²	SS 21 904794	14	1765	...	...	...	...	...	...	...	...	...	...	...	...	...	...	...
Ewenny (Ogmore)																		
Llanharan S. Wks	SS 21 974821	46	1269	92	149	69	140	120	141	57	127	65	153	85	236	1436	40	19 Oct
Tremains House	SS 21 918796	43	1140	80	108	52	146	104	125	50	103	60	138	72	165	1205	33	19 Oct
Ewenny Priory R.F.M. Sta. Area 62.9 km²	SS 21 914780	8	1328	...	...	...	...	...	...	...	...	...	...	...	...	...	...	...
Merthyr Mawr, Schwyll	SS 21 887770	14	1055	61	99	44	122	97	118	46	115	49	161	59	142	1113	54	19 Oct
Coastal																		
Port Talbot	SS 21 789867	7	1088	72	124	59	151	87	91	54	102	51	141	75	143	1150	46	19 Oct
Avan																		
Neath																		
Ystradfellte Res.	SN 22 945173	343	2050	114	263	117	227	175	135	132	167	137	216	203	434	2320	59	9 Dec
Twyn Gareg	SN 22 984166	458	2108	113	255	118	185	171	148	124	162	151	204	193	403	2226	...	...
Llyn Fawr Res.No.2	SN 22 919035	341	2438	188	355	133	268	222	183	156	235	155	242	183	415	2734	61	30 Dec
Llyn Fawr Res.No.1	SN 22 919035	341	2384	188	357	133	270	223	183	157	237	156	243	184	416	2745	60	30 Dec
Maesgwyn...	SN 22 869079	287	...	96	213	104	178	154	122	101	167	105	186	141	314	1882	49	19 Oct
Gnoll Res.	SS 21 764975	62	1338	87	166	74	157	139	120	81	129	86	145	104	216	1505	37	2 Mar
Cimla Res.	SS 21 767962	110	1343	95	167	78	164	147	134	85	127	88	145	104	230	1563	38	2 Mar
Neath...	SS 21 718957	44	1257	81	131	59	144	118	103	61	118	85	134	99	188	1322	34	30 Dec
Briton Ferry	SS 21 753940	153	1283	103	151	76	181	134	137	73	115	87	143	108	226	1533	37	2 Mar 30 Dec

SOUTH WEST WALES RIVER AUTHORITY AREA, 6263 SQUARE KILOMETRES

HYDROMETRIC AREAS 59 TO 63

Station	Grid reference	Altitude m	Av. 1916–1950 mm	Jan	Feb	Mar	Apr	May	Jun	Jul	Aug	Sep	Oct	Nov	Dec	Year	mm	date
Tawe																		
Nantyrwydd	SN 22 859181	277	2019	114	295	131	216	179	145	145	164	139	197	190	444	2359	58	12 Dec
Ynyscedwyn House	SN 22 778096	72	1737	(95)	225	97	185	169	128	97	182	124	195	129	299	1926	45	19 Oct
Trebanos S.Wks.	SN 22 711022	23	...	86	191	74	159	142	134	92	152	89	174	130	231	1653	33	18 Feb
Clydach	SN 22 696014	24	1374	81	178	70	155	131	113	88	163	96	151	116	228	1571	35	30 Dec
Ynys Tanglws R.F.M.Sta. Area 228 km²	SS 21 685998	10	1841	...	...	...	...	...	...	...	...	...	...	...	...	...	...	...
Wernfadog	SN 22 672001	122	1415	73	174	69	173	129	113	89	145	94	159	118	205	1539	36	21 Apr
Coastal																		
Swansea	SS 21 642923	10	1092	88	124	50	169	98	111	61	96	82	156	90	159	1283	34	21 Apr
Penmaen	SS 21 532888	82	...	83	123	57	152	88	89	70	110	86	183	87	170	1297	44	19 Oct
Lliw																		
Upper Lliw Res.	SN 22 662058	183	1689	104	200	78	188	148	140	112	151	108	201	136	239	1805	45	30 Dec
Lower Lliw Res.	SN 22 649034	127	1578	74	181	70	189	148	139	103	157	97	206	137	236	1737	42	30 Dec
Loughor																		
Upper Brynamman...	SN 22 721152	247	...	96	212	83	188	153	129	119	160	132	218	151	305	1947	52	12 Dec
Garnswllt S.Wks.	SN 22 621100	21	...	91	171	66	183	129	103	83	138	90	189	139	252	1634	34	30 Dec
Llannon Res.	SN 22 552102	262	1624	100	224	77	181	155	132	93	154	99	244	135	226	1819	...	...
Coastal																		
Upper Lliedi Res.	SN 22 514043	88	1329	100	183	59	205	111	103	76	121	72	199	118	171	1516	36	21 Apr
Cwmlliedi W.Wks.	SN 22 519031	73	1359	96	156	58	193	103	113	88	123	91	183	118	185	1509	33	21 Apr
Llanelli Filters...	SN 22 516024	72	1337	91	147	50	173	93	103	77	115	79	178	109	173	1386	29	21 Apr 19 Oct
Penyfai House...	SN 22 497015	52	...	88	140	42	170	81	83	69	108	67	166	94	(171)	1278	29	19 Oct
Pembrey	SN 22 415022	6	1071	75	141	41	159	95	91	60	95	70	243	73	138	1280	24	30 Dec
Gwendraeth Fawr																		
Gorslas Res.	SN 22 563146	229	...	96	214	67	183	147	107	105	150	94	270	141	247	1821	49	19 Oct
Gwendraeth Fach																		
Kidwelly P.Sta.	SN 22 420087	18	1276	74	140	47	175	88	79	67	105	65	223	73	146	1282	26	21 Apr
Tywi																		
Nant-y-Maen	SN 22 760583	418	1855	101	220	103	171	179	152	173	147	95	233	202	363	2139	58	12 Dec
Nant Rhyd Ifor	SN 22 764467	405	1570	114	229	89	146	140	114	89	152	95	165	152	356	1841	...	...
Bran (Tywi)																		
Cynghordy S. Wks...	SN 22 803399	122	1327	75	148	81	116	118	103	99	108	113	175	127	257	1521	...	...
Llandovery P. Sta....	SN 22 765353	69	1270	61	153	67	123	116	94	95	89	76	191	117	262	1442	36	12 Dec
Tywi																		
Troedyrhiw Res.	SN 22 686367	256	...	81	213	81	140	137	114	94	140	91	221	137	246	1697	...	...
Sawdde (Tywi)																		
Llyn-y-fan Fach	SN 22 802219	503	2750	163	393	156	250	238	188	168	263	180	261	175	364	2800	...	...
Waen Sychlwch	SN 22 804220	502	2697	164	390	145	253	227	177	164	229	167	252	185	399	2752	...	...
Nant Coch	SN 22 804233	357	2100	120	336	101	219	169	125	111	161	138	242	159	332	2214	62	24 Feb
Troed-y-Foel	SN 22 798238	264	1960	112	311	87	205	162	106	98	137	111	229	147	302	2005	65	24 Feb
Tywi																		
Golden Grove Farm	SN 22 597203	30	1377	95	209	61	182	128	89	72	126	78	240	127	220	1627	38	4 Nov
Cothi (Tywi)																		
Llwyn-Diried	SN 22 687452	198	...	104	211	79	155	125	117	112	117	89	198	150	287	1742	...	...
Rhydcymmerau	SN 22 578389	192	...	122	247	86	167	151	123	138	124	97	244	131	295	1924	56	12 Dec
Abergorlech	SN 22 585336	75	1547	91	243	89	171	149	122	113	125	88	242	139	295	1868	64	12 Dec
Brechfa, New site	SN 22 523299	88	1700	98	235	102	192	158	134	110	148	98	229	146	276	1925	61	12 Dec
Felin-Mynachdy R.F.M.Sta. Area 298 km²	SN 22 508225	16	1646	...	...	...	...	...	...	...	...	...	...	...	...	...	...	...
Tywi																		
Ty-Castell Farm R.F.M.Sta. Area 1090 km²	SN 22 491204	8	1572	...	...	...	...	...	...	...	...	...	...	...	...	...	...	...
Gwilli (Tywi)																		
Cwmtawel Res.	SN 22 386257	91	1717	101	220	77	226	135	116	107	157	97	245	121	260	1862	52	12 Dec
Tywi																		
Cwmoernant Res.	SN 22 419212	41	1321	74	184	60	169	120	78	82	139	79	219	(95)	185	1484	32	12 Dec
Trevaughan	SN 22 400218	76	...	71	192	63	162	117	83	90	128	87	216	102	183	1493	37	21 Apr

Station	Grid reference	Altitude m	Av. 1916–1950 mm	Jan	Feb	Mar	Apr	May	Jun	Jul	Aug	Sep	Oct	Nov	Dec	Year	Highest recorded fall in a day mm	Highest recorded fall in a day date
SOUTH WEST WALES—*(contd)*																		
Carmarthen	SN 22 413183	15	...	83	183	47	170	102	71	81	123	75	255	89	164	1441	38	19 Oct
Taf																		
Whitland P.Sta.	SN 22 201160	24	...	101	219	50	202	107	90	90	131	72	248	92	171	1573	36	2 Oct
Llanboidy	SN 22 215233	104	1468	105	205	48	219	103	99	101	123	90	220	102	231	1647	44	12 Dec
Whitland	SN 22 201164	18	1309	91	205	50	199	103	87	93	133	70	273	89	175	1569	50	19 Oct
Willow Glyn	SN 22 222143	128	...	97	216	62	188	116	85	80	139	79	230	86	183	1560	...	...
Coastal																		
Pendine	SN 22 262085	5	...	65	163	45	161	82	65	60	97	74	173	76	144	1203	33	18 Apr
Stackpole Court	SR 11 972960	18	1064	75	139	29	156	69	83	63	101	58	182	69	99	1122	28	12 Aug
Orielton	SR 11 953992	61	...	77	155	34	146	76	81	73	125	62	221	81	111	1242	41	19 Oct
Eastern Cleddau																		
Allt-y-Gog	SN 22 141287	163	1628	113	212	63	175	119	124	129	111	94	217	114	(203)	1676	43	31 Jul
Maenclochog	SN 22 069300	226	1600	117	208	63	196	118	133	145	114	81	223	118	204	1721	60	31 Jul
Clarbeston	SN 22 035219	96	1348	97	189	43	149	90	83	106	95	65	203	95	182	1398	37	31 Jul
Llandyssilio	SN 22 122213	101	1294	92	190	46	143	88	116	99	92	71	240	95	202	1473	44	12 Dec
Canaston Bridge R.F.M.Sta. Area 183 km^2	SN 22 072153	5	1461	...	...	...	...	...	...	...	...	...	...	...	...	...	...	...
Blackpool	SN 22 058137	64	1261	89	184	38	163	87	83	102	115	61	187	88	145	1343	41	31 Jul
Western Cleddau (Eastern Cleddau)																		
Trecwn House	SM 12 975325	119	1335	122	201	47	182	106	106	120	94	62	243	91	150	1524	49	31 Jul
Hotwells Res.	SM 12 977305	187	1292	100	189	45	146	100	114	121	94	58	203	98	169	1439	48	31 Jul
Coastal																		
Milford Haven, Grammar School	SM 12 914072	52	...	83	144	31	157	53	(89)	74	100	53	196	(63)	(104)	1147	35	12 Aug
Milford Haven, Conservancy Board	SM 12 892054	37	...	81	139	29	152	69	91	55	98	54	191	65	104	1130	35	12 Aug
Brawdy	SM 12 851249	110	1165	94	145	43	166	82	95	133	79	56	207	64	92	1257	72	31 Jul
Trenewydd	SM 12 913396	91	1003	94	139	30	159	81	86	97	83	55	210	73	91	1197	46	31 Jul
Teifi																		
Swyddffynnon	SN 22 696655	179	1310	53	117	85	65	97	92	110	51	80	157	116	295	1318	55	12 Dec
Tregaron	SN 22 677595	171	...	73	148	72	91	98	92	93	(63)	65	161	121	262	1339	45	12 Dec
Lampeter, Grammar School	SN 22 574479	116	...	90	153	46	101	81	91	78	70	65	170	104	229	1278	63	12 Dec
Gorsgoch Res.	SN 22 460501	323	...	88	185	63	156	119	122	128	90	98	231	103	300	1684	...	...
Glan Teifi R.F.M.Sta. Area 894 km^2	SN 22 244416	5	1389	...	...	...	...	...	...	...	...	...	...	...	...	...	...	...
Cardigan, Cedrwydd Glas	SN 22 190459	12	1055	71	154	24	113	63	87	96	63	51	206	91	128	1147	45	31 Jul
Coastal																		
Aeron																		
Brynamlwg, Blaenpenal	SN 22 608632	305	...	58	150	69	101	88	123	109	(75)	81	143	127	279	1404	...	...
Felinfach	SN 22 518575	67	...	54	161	54	94	84	94	102	79	67	177	81	177	1225	50	12 Dec
Aberayron County School	SN 22 461626	24	...	52	136	32	90	69	95	75	39	50	199	68	133	1037	48	3 Oct
Ystwyth																		
Cwmystwyth	SN 22 773749	304	1769	69	152	103	81	133	168	133	59	109	156	179	420	1762	79	9 Dec
Trawscoed	SN 22 674736	61	1200	42	136	57	62	85	113	89	47	72	141	111	214	1169	37	26 Jun
Llanilar	SN 22 621751	38	...	49	101	44	81	88	93	95	38	61	147	103	152	1053	26	26 Jun
Pont Llolwyn R.F.M.Sta. Area 170 km^2	SN 22 590772	12	1488	...	...	...	...	...	...	...	...	...	...	...	...	...	...	...
Rheidol (Ystwyth)																		
Cwmrheidol	SN 22 709792	53	...	46	113	67	72	104	134	95	82	74	147	120	233	1286	45	26 Jun
Aberystwyth, Plant Breeding Sta.	SN 22 605825	137	1057	33	113	41	83	70	98	80	52	47	171	75	137	1001	34	19 Oct
Coastal																		
Lletty-Evan-hen	SN 22 693851	290	1430	45	130	75	106	106	152	121	78	79	169	126	235	1421	61	26 Jun

GWYNEDD RIVER AUTHORITY AREA, 4047 SQUARE KILOMETRES

HYDROMETRIC AREAS 64 TO 66G AND 102

Station	Grid reference	Altitude m	Av. 1916–1950 mm	Jan	Feb	Mar	Apr	May	Jun	Jul	Aug	Sep	Oct	Nov	Dec	Year	Highest recorded fall in a day mm	Highest recorded fall in a day date
Coastal																		
Dovey																		
Ceinws	SH 23 759061	61	1959	87	198	168	153	150	235	141	91	126	152	231	408	2141	83	26 Jun
Dovey Bridge R.F.M. Sta. Area 471 km^2	SH 23 745019	7	–	...	...	...	...	...	...	...	...	...	...	...	...	...	...	...
Coastal																		
Aberdovey	SN 22 621963	52	1134	36	115	59	92	75	125	79	51	49	140	82	165	1068	49	26 Jun
Dysynni																		
Abergynolwyn	SH 23 675069	36	...	65	168	90	145	138	191	133	89	84	159	163	302	1728	81	26 Jun
Coastal																		
Arthog	SH 23 632148	2	1244	48	167	89	102	87	137	72	65	62	186	119	239	1374	40	26 Jun
Mawddach																		
Cefn Gallt-y-Cwm	SH 23 705345	203	...	114	226	122	152	151	225	118	104	117	184	202	281	1997	...	...
Graig-ddu-Ganol	SH 23 685292	247	...	109	290	180	157	180	333	180	127	137	234	236	307	2471	...	...
Dwyryd																		
Minafon	SH 23 717434	229	...	123	229	203	125	167	262	187	122	113	(254)	275	390	2451	...	...
Oakeley Quarry	SH 23 693469	335	2939	134	276	233	199	220	283	207	164	195	219	364	424	2919	98	26 Jun
Llechwedd Quarry	SH 23 697468	213	2794	112	244	207	191	218	297	186	161	168	224	324	371	2703	95	4 Jun
Conglog	SH 23 670473	610	...	180	361	279	290	276	348	211	188	249	274	305	544	3504	...	...
Cwmorthin	SH 23 668464	341	...	173	314	229	236	239	340	165	178	203	246	292	477	3093	...	...
Cynfal	SH 23 751416	378	2604	93	361	170	152	127	273	127	121	145	267	231	349	2415	...	...
Pont Newydd	SH 23 713408	186	1938	100	204	123	131	137	224	129	102	102	179	191	322	1944	91	26 Jun
Trawsfynydd	SH 23 691388	186	...	84	177	99	133	123	215	121	109	83	183	166	263	1756	89	26 Jun
Darn Gae	SH 23 762368	290	2001	102	222	170	123	175	169	155	125	75	235	254	263	2067	...	...
Tyddyn-y-Garreg	SH 23 700383	201	1811	91	202	131	165	136	198	129	136	99	184	181	258	1910	...	...
Trawsfynydd Dam	SH 23 677377	198	1808	93	197	125	154	144	239	125	136	87	194	166	285	1944	...	...
Glaslyn																		
Glaslyn	SH 23 618545	625	...	376	541	312	389	411	432	333	234	249	282	450	574	4582	...	...
Crib Goch	SH 23 623550	713	4361	282	536	282	381	373	432	323	269	305	330	483	523	4519	...	...
Delta	SH 23 624544	437	4319	302	566	305	373	381	429	320	254	307	310	490	566	4605	...	...
Llydaw Intake	SH 23 636547	451	3829	274	478	258	301	311	369	262	222	282	296	403	512	3968	...	...
Llydaw Mine	SH 23 631548	451	4011	255	503	272	328	323	429	295	221	279	305	437	498	4144	...	...
Teyrn	SH 23 644544	329	3962	254	490	285	279	287	356	241	246	300	257	434	544	3973	...	...
Cwm Dyli	SH 23 654542	94	3568	165	357	256	257	266	337	222	198	244	244	367	505	3418	103	26 Jun
Bryn Gwynant	SH 23 642513	95	...	139	309	207	220	227	274	175	176	190	227	307	440	2890	90	26 Jun
Beddgelert R.F.M. Sta. Area 68.6 km^2	SH 23 592478	33	3279	...	...	...	...	...	...	...	...	...	...	...	...	...	...	...

Station	Grid reference	Altitude m	Av. 1916–1950 mm	Jan	Feb	Mar	Apr	May	Jun	Jul	Aug	Sep	Oct	Nov	Dec	Year	Highest recorded fall in a day mm	Highest recorded fall in a day date
GWYNEDD—*(contd)*																		
Coastal																		
Cwmystradllyn Dam	SH 23 558439	198	...	99	221	136	165	166	195	145	140	114	175	211	286	2054	67	26 Jun
West End	SH 23 366341	3	1080	62	115	66	114	80	96	64	79	56	152	100	125	1108	25	11 Aug
Botwnnog	SH 23 263313	34	...	62	125	63	120	73	100	65	78	61	139	94	153	1131	27	9 Dec
Bardsey Island	SH 23 111206	15	...	65	108	37	118	61	65	52	75	46	145	64	121	957	26	19 Apr
Cefnamwlch	SH 23 234354	70	1122	75	147	58	132	85	91	70	92	62	163	84	149	1208	31	7 Feb
Nantlle	SH 23 497532	122	1610	90	205	102	152	140	103	95	81	76	171	128	203	1548	...	...
Glynllifon Park	SH 23 455554	30	1204	78	138	54	141	95	96	76	76	58	151	75	148	1186	37	1 Apr
Vaynol Park	SH 23 538693	43	...	38	103	73	95	75	132	118	77	93	133	104	154	1195	43	1 Apr
Braint																		
Gadlys Cottage	SH 23 548758	104	1113	46	99	76	91	82	153	88	69	106	127	110	155	1202	35	1 Apr
Treffos	SH 23 547753	88	1111	47	108	75	90	90	145	92	76	106	136	116	153	1235	36	26 Jun 3 Sep
Cefni																		
Cefni Res.	SH 23 444772	27	...	55	112	73	87	82	103	63	65	81	106	80	150	1058	28	1 Oct
Glanaber	SH 23 457752	21	1025	43	116	69	97	78	112	67	63	79	103	81	150	1059	27	1 Apr
Newborough Forest	SH 23 414669	6	987	41	96	58	84	63	99	64	62	70	113	67	134	950	27	1 Apr
Coastal																		
Valley	SH 23 310758	10	855	52	109	51	93	55	81	47	88	76	98	57	113	921	24	11 Aug
Llwydiarth Esgob	SH 23 436843	61	1024	60	117	69	92	72	97	71	80	85	90	79	145	1057	28	7 Feb
Plas Cadnant	SH 23 557731	37	1104	44	99	75	104	85	149	94	75	101	140	115	152	1231	42	1 Apr
Pen-y-Ffridd	SH 23 563705	84	1193	39	96	79	101	75	148	109	81	103	135	109	154	1228	49	1 Apr
Bangor, Town Hall	SH 23 580721	9	1074	41	95	71	100	74	143	89	81	112	132	111	157	1204	47	1 Apr
Afon Berthen	SH 23 643641	427	...	136	262	226	174	150	315	155	119	194	184	222	225	2362	...	...
Gerlan	SH 23 638658	250	...	113	221	105	189	129	177	145	87	156	188	185	186	1881	63	1 Apr
Aber	SH 23 650728	18	1057	40	103	64	107	61	124	84	65	84	130	114	144	1120	41	1 Apr
Llyn Anafon	SH 23 700699	497	...	112	262	262	174	100	223	159	125	223	203	244	241	2328	...	...
Conway Mussel Tanks	SH 23 785772	15	774	38	78	34	63	28	73	58	45	60	113	61	106	757	28	3 Oct
Conway																		
Afon Ddu	SH 23 770435	457	2711	104	353	175	180	165	317	137	114	133	258	251	353	2543	...	...
Afon Serw	SH 23 814452	335	2473	125	375	197	166	132	222	152	132	189	198	248	326	2462	...	...
Voelas Hall	SH 23 851515	195	1424	74	149	84	122	90	102	96	79	86	147	166	180	1374	45	1 Apr
Afon Machno	SH 23 753476	213	2504	105	279	170	170	198	198	155	119	181	208	234	328	2346	...	...
Ffynnon Llugwy	SH 23 694622	536	2561	140	403	243	248	159	241	178	178	356	248	241	394	3026	...	...
Llwyn Cowlyd	SH 23 715608	408	2310	129	419	221	239	192	182	125	140	317	241	196	395	2795	...	...
Gwydyr Forest	SH 23 748574	137	2070	95	251	142	177	158	160	130	97	184	186	216	289	2085	61	4 Jun
Cwm Llanerch R.F.M. Sta. Area 344 km^2	SH 23 802581	7	2284	...	...	...	...	...	...	...	...	...	...	...	...	...	...	...
Ffrith Garreg Wen	SH 23 739635	347	2270	178	363	175	214	176	226	143	152	262	223	161	381	2654	...	...
Cowlyd	SH 23 737637	366	2296	151	354	176	224	156	212	150	150	238	237	(173)	394	2615	...	...
Penstock	SH 23 756668	274	1852	101	263	132	180	120	149	137	108	184	214	177	318	2083	...	...
Melynllyn	SH 23 703657	629	3022	200	367	256	177	189	229	190	208	228	252	200	462	2957	...	...
Llyndulyn	SH 23 706663	497	2674	191	353	231	180	170	229	167	199	225	241	189	405	2779	...	...
Dulyn No.1	SH 23 726675	381	...	132	272	234	180	156	185	140	129	221	240	174	381	2445	...	...
Afon Tafolog	SH 23 733711	344	1644	86	220	126	151	81	125	128	116	175	142	166	199	1717	...	...
Llanbedr-y-Cennin	SH 23 754698	155	1405	77	149	77	131	72	96	106	111	115	143	146	196	1420	53	1 Apr
Bodnant	SH 23 800725	52	976	41	131	47	104	50	79	68	60	67	156	116	141	1060	52	3 Oct
Coastal																		

DEE AND CLWYD RIVER AUTHORITY AREA, 2948 SQUARE KILOMETRES

HYDROMETRIC AREAS 66D AND 67

Station	Grid reference	Altitude m	Av. 1916–1950 mm	Jan	Feb	Mar	Apr	May	Jun	Jul	Aug	Sep	Oct	Nov	Dec	Year	Highest recorded fall in a day mm	Highest recorded fall in a day date
Clwyd																		
Clawdd-newydd	SJ 33 078521	300	1013	61	106	57	87	69	47	87	69	65	110	129	189	1076	36	1 Dec
Rhoslydan	SJ 33 161498	312	...	46	76	36	112	51	55	93	73	40	123	113	147	965	32	4 Nov
Llysfasi	SJ 33 147523	125	867	46	72	43	97	53	45	84	51	50	119	118	156	934	35	9 Dec
Ruthin, Corwen Road	SJ 33 125556	71	834	29	69	44	69	44	36	84	43	45	85	109	129	786	...	...
Bwlch Tunnel, New site	SJ 33 163581	277	857	30	61	41	92	40	49	89	66	42	97	79	126	813	33	1 Apr
Ruthin	SJ 33 133584	76	822	21	59	41	70	42	40	65	45	45	97	111	125	763	31	1 Apr
Gyffylliog	SJ 33 058577	168	...	42	105	63	121	68	47	78	61	56	115	(121)	183	1059	38	10 Apr
Llangwyfan	SJ 33 122661	140	788	30	52	38	121	48	73	84	60	45	118	94	140	904	32	1 Apr
Denbigh Hospital	SJ 33 051653	96	844	17	59	54	109	52	63	86	68	51	118	94	155	927	37	1 Dec
Eglwyswen S. Wks.	SJ 33 071664	41	...	27	55	42	83	46	54	73	58	47	95	84	130	793	31	1 Apr
Nantlys	SJ 33 081711	54	754	22	46	23	87	28	96	76	58	40	94	68	124	763	31	1 Apr
Pont-y-Cambwll R.F.M. Sta. Area 404 km^2	SJ 33 069709	15	922	...	...	...	...	...	...	...	...	...	...	...	...	...	...	...
Tremeirchion College	SJ 33 080742	137	796	31	57	33	95	34	96	81	59	47	101	84	98	817	25	1 Apr
Elwy (Clwyd)																		
Gwytherin	SH 23 875615	238	...	48	108	65	114	66	78	113	78	85	150	140	156	1201	53	1 Apr
Haffotty Bennett	SH 23 835692	328	...	45	144	58	141	61	87	98	85	76	141	120	155	1209	45	1 Apr
Llyn Aled Res.	SH 23 915579	372	1281	88	146	82	106	84	89	120	92	122	150	149	195	1423	...	...
Moel-y-Bryniau	SH 23 913591	375	1275	95	154	90	111	90	93	129	90	112	154	153	200	1470	...	...
Bylchau	SH 23 969626	302	...	67	88	75	128	79	62	100	93	72	160	145	188	1256	...	...
Aled (Elwy)																		
Bryn Aled R.F.M. Sta. Area 69.9 km^2	SJ 33 957703	105	1171	...	...	...	...	...	...	...	...	...	...	...	...	...	...	...
Elwy (Clwyd)																		
Dolwen Res. South	SH 23 972702	177	884	21	76	41	94	46	78	98	69	57	108	98	144	928	...	...
Llannefydd	SH 23 974706	177	882	21	74	41	96	46	72	92	68	52	111	96	141	910	...	...
Plas Uchaf Res.	SH 23 968712	148	829	30	77	42	93	44	76	89	63	52	106	97	128	897	29	1 Apr
Pant-yr-Onen R.F.M. Sta. Area 220 km^2	SJ 33 021704	36	1175	...	...	...	...	...	...	...	...	...	...	...	...	...	...	...
Groes Hall	SJ 33 009648	198	898	39	81	56	111	58	57	85	73	55	130	109	145	999	19	12 Aug
Clwyd																		
Dyserth	SJ 33 058783	79	741	21	58	27	86	35	79	81	66	41	93	62	102	752	29	1 Apr
Glascoed Filters	SH 23 996736	110	820	26	67	34	90	34	69	80	68	51	98	78	127	823	42	1 Apr
Rhyl	SJ 33 002807	9	676	24	60	24	68	35	60	68	51	41	84	61	116	692	31	9 Dec
Coastal																		
Prestatyn	SJ 33 065831	4	726	27	62	27	88	40	77	77	55	42	87	68	104	755	29	1 Apr
Dee																		
Blaen Lliw Isaf	SH 23 804328	390	2353	178	254	178	132	178	178	208	132	155	(264)	229	305	2390	...	...
Llanuwchllyn, Post Office	SH 23 877299	173	...	102	205	111	132	122	122	142	100	108	211	191	253	1799	54	9 Dec
Tryweryn (Dee)																		
Arenig	SH 23 839391	411	2057	114	292	152	125	175	173	140	117	165	191	170	287	2101	...	...
Afon Mynach	SH 23 898436	365	1817	111	255	112	138	140	133	122	101	133	175	205	298	1923	47	14 Sep
Bala Sluices	SH 23 935357	164	1337	79	149	84	87	84	94	85	63	94	165	150	218	1352	43	9 Dec

Station	Grid reference	Alti-tude	Av. 1916–1950	Rainfall in millimetres													Highest recorded fall in a day	
		m	mm	Jan	Feb	Mar	Apr	May	Jun	Jul	Aug	Sep	Oct	Nov	Dec	Year	mm	date
DEE AND CLWYD—*(contd)*																		
Dee																		
Bala R.F.M. Sta. Area 262 km^2	SH 23 942357	158	1933	...	...	...	...	...	...	...	...	...	...	...	...	...	...	...
Maesafallen	SH 23 948308	292	1642	114	(191)	114	140	140	140	114	89	114	197	191	(241)	1784	...	...
Crogen	SJ 33 008367	168	...	77	134	74	107	92	77	84	90	88	132	126	149	1230	33	14 Sep
Corwen	SJ 33 048399	144	1082	65	109	57	87	64	49	73	56	62	115	141	169	1049	41	1 Dec
Alwen (Dee)																		
Cottage Bridge...	SH 23 920562	366	1309	65	136	82	113	68	77	100	80	96	154	163	191	1323	...	...
Tan-y-graig	SH 23 935563	335	1332	65	123	80	100	67	74	124	77	94	148	152	233	1337	...	...
Nant Heilyn	SH 23 925554	373	1361	59	140	93	86	(76)	88	79	74	98	148	159	202	1303	...	...
Coed Tai Isaf	SH 23 944544	370	...	72	143	77	79	72	83	110	79	78	150	157	217	1319	...	...
Mwdwl-eithen	SH 23 935535	401	1419	71	152	93	96	77	82	127	76	73	145	157	222	1371	...	...
Craig-yr-Iyrchen	SH 23 938523	457	1416	67	153	85	103	88	83	114	77	(76)	148	157	228	1379	...	...
Alwen Dam	SH 23 955528	362	1342	63	143	75	105	74	77	113	82	69	141	147	209	1300	37	3 Oct
Alwen Res. R.F.M. Sta. Area 25.5 km^2	SH 23 957528	332	1326	...	...	...	...	...	...	...	...	...	...	...	...	...	...	...
Alwen	SH 23 957529	335	1264	61	140	75	101	73	77	117	79	72	144	158	231	1329	46	1 Dec
Brenig (Alwen)																		
Llyn Bran	SH 23 960593	435	1427	64	142	98	110	81	77	123	(89)	110	143	167	245	1450	...	...
Pant-y-maen	SH 23 977587	402	1397	66	125	93	109	76	78	122	85	95	144	170	226	1389	...	...
Wirral Water Board's Weir R.F.M. Sta. Area 20.2 km^2	SH 23 974539	326	1308	...	...	...	...	...	...	...	...	...	...	...	...	...	...	...
Elor-garreg-uchaf	SH 23 964539	399	...	62	142	87	98	79	82	125	76	79	146	177	246	1399	...	...
Pentre-draw	SH 23 972528	335	1152	48	115	61	81	65	78	116	73	59	147	147	219	1211	...	...
Alwen (Dee)																		
Llanfihangel-glyn-myfyr	SH 23 986491	274	1108	47	139	55	82	75	70	93	87	71	131	119	140	1108	31	3 Oct
Minffordd	SJ 33 030483	296	1031	49	120	61	90	65	56	81	71	63	120	133	183	1092	...	...
Llangwm, Cwm Llan	SH 23 962430	366	...	49	173	65	103	74	87	89	106	(89)	163	(152)	(241)	1389	29	1 Apr
Druid R.F.M. Sta. Area 185 km^2	SJ 33 042436	146	1333	...	...	...	...	...	...	...	...	...	...	...	...	...	...	...
Dee																		
Glyndyfrdwy R.F.M. Sta. Area 728 km^2	SJ 33 155428	118	1562	...	...	...	...	...	...	...	...	...	...	...	...	...	...	...
Vivod...	SJ 33 191424	195	972	61	133	39	121	79	40	67	103	48	119	122	164	1096	41	1 Dec
Worlds End	SJ 33 242481	479	1121	79	107	59	128	73	43	107	105	61	121	135	221	1240	...	...
Ruabon Mountain	SJ 33 237457	498	1107	81	107	60	107	85	44	97	101	58	137	139	221	1238	...	...
Ruabon	SJ 33 275421	95	877	63	109	45	104	69	37	75	90	48	97	106	208	1052	56	1 Dec
Mountain Lodge	SJ 33 258474	335	...	73	111	63	134	78	38	90	97	58	117	134	242	1236	...	...
Bryn Adda Flat	SJ 33 250453	370	1067	79	109	58	112	78	40	93	98	56	128	137	227	1213	...	...
Penycae	SJ 33 268453	207	1032	65	112	57	105	76	48	83	89	52	106	116	239	1148	54	1 Dec
Ceiriog (Dee)																		
Llanarmon Dyffryn Ceiriog	SJ 33 156327	267	1228	91	188	61	130	108	50	74	112	60	126	142	221	1362	52	1 Dec
Chirk	SJ 33 293376	107	912	63	125	41	109	85	39	73	100	47	118	95	161	1053	44	1 Dec
Brynkinalt Weir	SJ 33 295373	72	908	60	120	40	107	82	39	68	99	40	123	105	165	1049	28	14 Oct
Brynkinalt Weir R.F.M. Sta. Area 114 km^2	SJ 33 295373	64	1473	...	...	...	...	...	...	...	...	...	...	...	...	...	...	...
Dee																		
Erbistock Rectory	SJ 33 356413	28	--	46	95	34	95	77	61	54	97	38	97	77	136	908	33	1 Dec
Erbistock Rectory R.F.M. Sta. Area 1040 km^2	SJ 33 357413	23	1445	...	...	...	...	...	...	...	...	...	...	...	...	...	...	...
Clywedog (Dee)																		
Abersychnant	SJ 33 232498	418	1099	79	91	67	128	66	52	105	104	63	129	125	218	1228	...	...
Minera	SJ 33 262508	427	1023	79	107	66	127	81	31	91	96	61	119	130	194	1182	...	...
Erwau	SJ 33 262488	354	1137	72	117	69	135	79	43	102	102	58	128	139	251	1294	...	...
Ty-Mawr Res.	SJ 33 274480	245	977	73	109	56	119	81	46	89	93	54	121	125	236	1205	56	1 Dec
Pack Saddle Res.	SJ 33 305479	113	887	63	91	46	117	79	43	69	73	47	102	93	177	1001	40	1 Dec
Wrexham	SJ 33 338507	84	774	53	89	38	108	77	39	60	90	44	104	86	127	915	33	9 Dec
Wrexham, Camberley Drive...	SJ 33 342512	82	769	49	84	35	106	70	47	60	92	42	100	75	141	900	33	9 Dec
Worthenbury (Dee)																		
Gredington	SJ 33 445385	91	772	38	92	27	75	79	40	85	82	29	85	72	125	831	25	1 Dec
Wych (Worthenbury)																		
Whitchurch	SJ 33 544410	83	749	36	70	36	86	78	73	84	84	36	90	82	127	880	24	9 Dec
Alyn (Dee)																		
Pendinas Res.	SJ 33 232518	381	1069	76	95	61	137	88	75	102	103	69	147	147	216	1316	43	1 Dec
Loggerheads	SJ 33 201622	210	...	44	72	42	128	70	77	97	95	46	99	118	185	1073	38	9 Dec
Cilcain Res. No.4	SJ 33 154640	287	925	33	66	47	111	54	77	98	87	51	122	99	153	997	...	...
Cilcain Res. No.2	SJ 33 163645	239	917	42	64	47	116	58	84	94	87	55	126	102	173	1047	33	9 Dec
Glan-yr-Afon	SJ 33 202677	134	...	41	69	43	119	70	97	109	67	59	100	109	173	1057	35	1 Apr
Llyn Cyfynwy	SJ 33 223545	345	1064	68	83	59	135	67	51	101	114	61	145	151	209	1245	...	...
Leeswood Farm	SJ 33 254605	168	...	57	69	50	121	63	71	74	85	46	116	106	162	1019	38	9 Dec
Bwlchgwyn	SJ 33 236520	386	1092	77	95	63	134	91	76	106	103	68	151	147	227	1337	43	1 Dec
Nantyfrith Res.	SJ 33 243534	317	...	72	101	56	144	87	83	95	104	60	147	136	219	1302	48	1 Dec
Ithel's Bridge	SJ 33 390562	7	...	43	65	31	83	86	65	85	81	37	79	70	112	837	27	9 Dec
Dee																		
Tower Wood P. Sta.	SJ 33 503540	99	...	38	71	38	77	73	80	86	102	47	70	62	128	871	28	20 Aug
Eaton Hall	SJ 33 414609	18	739	37	(63)	40	81	76	85	82	100	38	93	82	117	893	31	20 Aug
Chester W. Wks.	SJ 33 418666	20	693	30	61	33	67	74	96	79	102	43	93	72	133	883	32	20 Aug
Chester, Dicksons Drive	SJ 33 409677	20	783	35	58	33	72	69	79	80	99	43	98	70	131	866	30	20 Aug
Chester S. Wks.	SJ 33 393667	5	697	34	51	24	71	68	82	74	86	41	100	68	112	813	27	9 Dec
Aston Park	SJ 33 306675	30	...	47	59	34	102	67	90	73	79	46	100	87	118	901	29	11 Jun
Hawarden Bridge	SJ 33 322693	5	698	39	55	29	89	54	77	79	80	42	89	74	103	810	28	9 Dec
Soughton Hall	SJ 33 243671	131	873	46	66	40	117	66	110	90	93	49	92	100	167	1036	36	1 Apr
Mold	SJ 33 256663	153	834	45	68	42	111	71	100	84	87	41	100	88	150	987	31	9 Dec
Coastal																		
Ness Gardens	SJ 33 303754	38	...	33	54	29	91	69	105	77	83	43	88	73	101	849	30	1 Apr
West Kirby Park	SJ 33 216865	7	775	27	61	27	108	53	93	85	51	46	97	73	111	831	28	10 Apr

MERSEY AND WEAVER RIVER AUTHORITY AREA, 4573 SQUARE KILOMETRES

HYDROMETRIC AREAS 68 AND 69

Station	Grid reference	Altitude m	Av. 1916–1950 mm	Jan	Feb	Mar	Apr	May	Jun	Jul	Aug	Sep	Oct	Nov	Dec	Year	mm	date
Coastal																		
Crosshill Res.	SJ 33 280842	65	792	33	65	27	103	52	101	95	86	47	96	67	113	886	32	1 Apr
Prenton W. Wks., New site	SJ 33 295863	28	...	29	63	29	100	58	95	93	92	50	97	67	118	889	36	1 Apr
Bidston, The Woodlands	SJ 33 284893	30	772	32	70	28	98	63	87	81	90	48	90	68	109	864	34	1 Apr
Bidston Observatory	SJ 33 287897	60	737	29	60	27	86	58	89	78	84	45	87	62	92	798	28	1 Apr
Hooton W. Wks.	SJ 33 351779	30	791	33	58	27	81	69	111	70	74	45	93	61	116	840	...	...

Station	Grid reference	Altitude m	Av. 1916–1950 mm	Jan	Feb	Mar	Apr	May	Jun	Jul	Aug	Sep	Oct	Nov	Dec	Year	Highest recorded fall in a day mm	Highest recorded fall in a day date
MERSEY AND WEAVER—*(contd)*																		
Port Sunlight, Oval	SJ 33 330851	19	...	35	67	27	96	81	100	102	77	49	87	57	122	900	34	1 Apr
Port Sunlight, Estate Nursery	SJ 33 336842	15	830	35	69	28	96	84	100	107	78	47	85	61	126	916	34	1 Apr
Sutton Hall P. Sta.	SJ 33 371757	38	...	34	55	29	77	67	103	79	76	44	93	58	119	834	22	1 Apr 3 Oct
Gowy																		
Barrowmore	SJ 33 474691	35	724	24	57	32	70	71	103	83	74	39	87	67	119	825	20	1 Apr
Delamere	SJ 33 548704	85	837	33	75	40	99	72	108	105	76	44	102	79	141	973	26	1 Apr
Eddisbury Fruit Farm	SJ 33 534702	99	...	30	73	37	100	66	105	99	74	47	97	68	125	922	23	26 Jun
Mouldsworth P. Sta.	SJ 33 503704	40	778	26	58	30	90	82	109	84	65	45	86	72	120	868	25	1 Apr
Picton R.F.M. Sta. Area 156 km^2	SJ 33 444714	3	754	...	...	...	...	...	...	...	...	...	...	...	...	...	...	...
Coastal																		
Weaver																		
Peckforton P. Sta.	SJ 33 530553	99	793	35	74	40	78	81	94	101	84	52	78	66	136	918	24	9 Dec
Cholmondeley	SJ 33 553505	89	766	41	63	40	78	76	88	99	74	35	72	67	136	871	(27	11 Dec)
Mere Farm	SJ 33 691409	128	...	40	68	32	93	68	59	100	87	40	98	61	122	867	...	...
Audlem	SJ 33 656432	65	758	38	68	41	81	74	82	104	97	39	86	73	133	916	23	20 Aug
Audlem R.F.M. Sta. Area 262 km^2	SJ 33 652432	44	772	...	...	...	...	...	...	...	...	...	...	...	...	...	...	...
Madeley School	SJ 33 772443	107	...	46	82	42	77	70	80	90	75	45	103	57	122	889	26	1 Apr
Nantwich, Brookfield House	SJ 33 650515	40	740	33	66	35	85	60	98	109	84	37	87	63	136	893	25	1 Dec
Dorfold Hall	SJ 33 634524	58	775	35	68	36	82	64	105	98	81	39	84	66	(140)	897	24	26 Jun
Nantwich, Beam Bridge	SJ 33 651533	37	760	31	66	36	86	60	112	89	80	41	88	63	141	893	25	1 Dec
Reaseheath Hall	SJ 33 048542	40	735	29	63	36	77	58	101	86	82	44	87	55	137	855	23	1 Dec
Hurleston	SJ 33 621552	61	...	31	70	39	77	70	105	96	72	42	81	65	148	895	25	1 Dec
Valley (Weaver)																		
Radway Green	SJ 33 776542	78	775	39	79	33	(81)	61	99	99	(81)	45	96	68	106	887	25	11 May
Crewe S. Wks.	SJ 33 684554	38	747	32	69	38	81	54	111	90	77	43	99	59	117	869	24	1 Apr
Marshfield Bridge R.F.M. Sta. Area 88.1 km^2	SJ 33 673552	30	785	...	...	...	...	...	...	...	...	...	...	...	...	...	...	...
Weaver																		
Sandyford W. Wks.	SJ 33 567667	76	822	33	72	40	96	75	89	100	85	52	101	73	139	955	28	1 Apr
Tiresford	SJ 33 556612	74	742	33	63	35	83	69	96	87	80	35	81	60	119	840	24	20 Aug
Tilstone Lodge	SJ 33 568611	70	...	33	69	35	76	70	98	88	79	35	75	68	137	865	24	11 May
Eaton P. Sta.	SJ 33 569635	87	825	36	65	44	97	77	103	101	82	40	82	70	142	939	24	1 Apr
Below Ashbrook R.F.M. Sta. Area 609 km^2	SJ 33 670633	17	772	...	...	...	...	...	...	...	...	...	...	...	...	...	...	...
Northwich, Hunts Lock	SJ 33 656729	14	787	34	76	46	95	63	98	68	80	62	92	72	123	910	35	1 Apr
Dane (Weaver)																		
Wincle, Gighall	SJ 33 955641	168	...	61	103	53	92	79	120	123	95	80	100	109	157	1173	33	28 Jul
Bosley Res.	SJ 33 927657	180	864	50	87	45	88	85	102	123	93	68	98	81	129	1049	36	1 Apr
Bosley Lock	SJ 33 905669	160	866	51	90	56	74	70	116	107	100	64	88	73	129	1019	28	1 Apr
Buglawton Hall	SJ 33 886645	146	853	46	72	44	78	53	105	109	88	54	87	69	125	929	25	28 Jul
Biddulph Grange	SJ 33 894591	171	896	53	102	41	87	73	106	94	89	66	104	65	134	1014	28	1 Apr
Congleton S. Wks.	SJ 33 853635	75	790	42	67	36	79	54	96	91	90	49	85	65	121	874	25	21 Aug
Hulme Walfield R.F.M. Sta. Area 152 km^2	SJ 33 844644	66	1034	...	...	...	...	...	...	...	...	...	...	...	...	...	...	...
Cranage, Hall Hospital	SJ 33 750683	55	723	47	83	49	76	68	(102)	(89)	(83)	45	79	69	111	899	21	14 Sep
Wheelock (Dane)																		
Mow Cop P. Sta.	SJ 33 843577	140	800	44	84	38	80	62	99	87	87	49	99	61	124	912	25	1 Apr
Red Bull S. Wks.	SJ 33 829551	114	778	39	90	39	83	64	94	110	83	46	99	56	111	913	29	5 Jul
Dane (Weaver)																		
Bostock Hall	SJ 33 675683	43	778	37	77	43	90	67	118	90	91	60	102	69	119	962	28	22 Jun
Rudheath R.F.M. Sta. Area 414 km^2	SJ 33 668718	13	874	...	...	...	...	...	...	...	...	...	...	...	...	...	...	...
Wincham (Weaver)																		
Chelford	SJ 33 814742	75	791	47	97	49	87	76	122	115	89	67	83	79	127	1038	29	11 May
Lostock Gralam R.F.M. Sta. Area 148 km^2	SJ 33 697757	17	833	...	...	...	...	...	...	...	...	...	...	...	...	...	...	...
Northwich, Marbury Road	SJ 33 650759	24	777	33	74	40	92	66	111	83	77	51	99	68	121	917	36	1 Apr
Weaver																		
Winnington	SJ 33 643747	20	770	32	73	44	90	27	113	76	75	59	94	65	119	866	36	1 Apr
Weaverham Grange	SJ 33 634727	43	810	35	79	44	99	71	111	74	86	61	97	72	126	954	33	1 Apr
Crabtree Green	SJ 33 584711	50	829	31	80	47	98	92	97	100	95	57	106	83	143	1029	30	1 Apr
Coastal																		
Weston Point	SJ 33 495814	6	809	29	70	30	100	63	103	83	95	55	101	76	123	926	18	15 Jul
Moore	SJ 33 582841	37	...	33	78	27	83	70	104	74	83	52	87	59	106	856	29	20 Aug
Appleton Res.	SJ 33 602845	37	838	35	89	28	102	67	121	72	97	55	100	63	114	943	34	20 Aug
Goyt																		
Goyts Moss	SK 43 019717	430	...	89	157	86	141	102	166	136	140	88	141	154	239	1638	...	...
Berry Clough	SK 43 029727	472	1387	81	178	102	138	107	163	137	129	86	137	152	211	1621	...	...
Cat and Fiddle Inn	SK 43 005717	491	1379	84	159	102	122	104	166	157	127	89	132	137	241	1620	...	...
Goytsclough	SK 43 011733	358	1418	87	178	89	120	101	165	152	129	113	131	153	240	1659	45	26 Jun
Goytshead	SK 43 012748	282	1335	77	164	87	118	97	153	136	114	106	123	144	210	1529	40	14 Sep
Embridge	SJ 33 999764	436	1329	65	138	83	103	85	150	127	108	84	106	107	183	1339	...	...
Longhill	SK 43 031752	424	1305	76	183	84	(108)	100	145	112	129	86	126	128	184	1462	...	...
Upper Hall	SK 43 017771	287	1175	65	151	77	104	90	142	121	104	98	109	129	223	1413	42	14 Sep
Oldfield	SK 43 012777	250	1210	68	154	81	104	95	145	127	103	101	112	137	235	1463	43	14 Sep
Blackhillgate	SJ 33 997787	363	1124	63	135	76	93	76	133	132	102	83	106	112	180	1290	...	...
Sparrow Pit No.102	SK 43 091804	366	1131	58	157	81	86	76	122	127	122	79	112	127	137	1285	...	...
Chapel-en-le-Frith	SK 43 056809	223	1098	57	141	65	76	(76)	124	116	93	84	83	(127)	(140)	1182	33	14 Sep
Sett (Goyt)																		
Upper Sett	SK 43 074855	433	1271	73	162	66	127	84	163	133	126	101	113	157	192	1498	...	...
Dimpus Clough	SK 43 065851	381	...	70	143	70	86	93	175	113	109	104	88	141	207	1399	...	...
Oaken Clough	SK 43 064860	381	...	74	136	53	133	(94)	(159)	122	111	88	95	137	136	1337	...	...
Tunstead Clough	SK 43 064867	411	...	79	165	59	150	94	148	119	116	93	95	138	170	1427	...	...
Kinder Source	SK 43 087878	625	1547	81	196	76	155	108	173	132	127	104	108	165	205	1629	...	...
Hollin Head	SK 43 064883	290	1298	70	139	59	101	98	149	120	104	95	100	151	178	1365	41	26 Jun
Upper Moor	SK 43 067879	415	1287	74	177	71	142	101	163	116	111	99	105	158	198	1513	...	...
Kinder Filters	SK 43 054880	244	1173	65	132	54	94	92	141	114	99	91	95	143	169	1290	39	26 Jun
Leygatehead Moor	SK 43 054894	466	1227	79	156	69	136	105	171	136	111	102	95	163	197	1519	...	...
Middle Moor	SK 43 049887	357	1191	70	152	65	105	102	166	119	108	94	107	156	194	1439	...	...
Etherow (Goyt)																		
Longside Moss	SE 44 127010	432	1549	77	214	123	145	141	129	122	147	121	136	161	207	1722	...	...
Far Black Clough	SK 43 127984	513	1503	84	213	116	155	136	117	115	145	107	134	185	241	1749	...	...
Pikenaze Moor	SE 44 114004	350	1491	70	208	117	140	111	119	102	131	114	129	193	241	1676	...	...
Pikenaze	SE 44 098001	282	1292	60	169	100	146	112	122	99	114	90	119	153	202	1486	...	...
Bleaklow Hill	SK 43 093979	524	1516	66	216	(114)	107	98	142	126	150	114	132	178	225	1669	...	...

Station	Grid reference	Altitude	Av. 1916–1950	Rainfall in millimetres													Highest recorded fall in a day	
		m	*mm*	Jan	Feb	Mar	Apr	May	Jun	Jul	Aug	Sep	Oct	Nov	Dec	Year	*mm*	*date*
MERSEY AND WEAVER—*(contd)*																		
Upper Heyden...	SE 44 098035	523	1774	102	227	116	161	141	142	111	152	145	129	204	281	1911	...	...
West Withens	SE 44 114022	421	1618	77	237	121	152	114	131	128	128	112	127	194	241	1762	...	...
Westend Moor...	SE 44 086020	515	1567	59	179	96	110	114	138	111	122	102	131	176	248	1586	...	...
Woodhead Res.	SK 43 078993	203	1260	62	156	80	126	94	131	113	119	105	116	181	208	1490	47	19 Dec
Grains Moss	SE 44 068039	479	1669	99	231	118	152	135	145	122	147	137	136	199	263	1884	...	...
Near Broadslate	SE 44 055027	507	1673	102	234	136	140	140	147	146	159	135	143	207	263	1951	...	...
Span Gutter	SK 43 058999	466	1429	68	139	95	102	112	132	113	132	135	117	163	211	1519	...	...
Crowden, Meadow Clough	SE 44 074025	434	1648	83	207	111	140	134	166	142	137	152	121	203	254	1850	...	...
Clough Edge	SK 43 066970	443	1431	75	147	90	99	106	147	125	133	127	121	157	227	1553	...	...
Torside Res.	SK 43 054985	204	1196	61	137	83	129	95	123	122	126	118	115	170	200	1479	...	...
Bottoms Res.	SK 43 023972	153	1074	50	126	61	105	86	111	128	121	102	86	146	154	1276	...	...
Bottoms Res. Longdendale R.F.M. Sta. Area 78.2 km²	SK 43 022971	151	1481	...	...	...	...	...	...	...	...	...	...	...	...	...	...	...
Ogden (Etherow)																		
Boar Flat	SE 44 020003	363	1324	54	145	79	(114)	93	107	145	136	119	112	168	191	1462	...	...
Hollingworth Hall Moor	SK 43 002984	364	1214	57	144	83	84	95	120	134	131	(119)	100	164	186	1416	...	...
Arnfield Moor	SK 43 033996	393	1359	54	140	84	89	94	106	145	135	101	118	148	201	1414	...	...
Arnfield Res.	SK 43 012972	156	1057	49	122	64	89	87	109	133	120	103	84	144	170	1275	42	14 Sep
Etherow (Goyt)																		
Swineshaw W. Wks.	SK 43 044958	223	1113	53	137	63	86	88	105	123	124	100	94	146	167	1287	38	14 Sep
Hurst Res.	SK 43 053939	195	...	64	160	68	112	93	119	105	116	96	97	160	177	1368	...	...
Glossop S. Wks.	SK 43 003946	117	...	52	123	57	63	85	111	117	107	99	82	130	163	1188	36	26 Jun
Goyt																		
Bredbury S. Wks.	SJ 33 917920	50	...	39	112	39	72	76	115	132	101	75	79	94	136	1068	34	26 Jun
Tame (Goyt)																		
Ox Hey, New site	SD 34 986104	311	...	73	231	106	164	141	120	136	143	106	129	159	211	1718	51	7 Feb
Broadhead Noddle	SD 34 996102	261	1386	65	230	96	157	128	111	123	139	103	134	154	198	1638	51	7 Feb
Saddleworth	SD 34 998067	175	1228	55	182	78	120	122	112	136	125	117	119	153	174	1492	(49	26 Jun)
Greenfield S. Wks....	SD 34 992042	152	1151	47	161	77	99	117	111	115	131	111	105	157	172	1403	43	7 Feb
Greenfield (Tame)																		
Birchen Clough	SE 44 031054	310	...	71	221	101	155	133	145	137	150	110	141	220	225	1808	...	...
Upperwood Moor	SE 44 031063	463	1377	75	231	103	158	130	141	136	153	113	151	222	226	1839	...	...
Yeoman Hey	SE 44 022046	239	1329	69	217	99	160	125	144	143	146	111	147	223	216	1802	51	7 Feb
Ashway Moss	SE 44 032046	482	1430	74	227	111	157	131	143	139	154	115	150	222	234	1857	...	...
Chew (Greenfield)																		
Chew Res.	SE 44 046020	502	...	73	209	116	127	133	145	150	143	126	145	216	241	1824	...	...
Chew, Blindstones	SE 44 038014	532	1536	69	208	117	109	125	138	151	150	122	145	173	230	1736	...	...
Chew Sluice	SE 44 037029	509	1512	64	196	87	95	115	132	150	149	118	146	186	249	1685	...	...
Chew Mount	SE 44 035020	481	1532	70	212	94	115	123	145	153	148	112	146	204	234	1757	54	7 Feb
Chew-Great-Gruff	SE 44 032013	503	1532	72	205	103	113	125	144	152	149	111	145	177	225	1719	...	...
Tame (Goyt)																		
Swineshaw Moor	SE 44 010008	408	1234	59	151	72	116	106	123	138	134	120	104	155	170	1447	...	...
Higher Swineshaw	SK 43 006996	294	1177	57	150	70	113	105	124	138	133	118	96	155	168	1426	...	...
Lower Swineshaw Res....	SK 43 004994	270	1138	51	149	70	107	95	118	136	128	120	96	151	165	1386	39	7 Feb
Brushes Filters	SJ 33 993990	188	1088	51	152	65	107	86	113	130	129	112	95	149	165	1353	41	7 Feb
Stalybridge, Market Ground...	SJ 33 964983	116	1010	(51)	131	47	78	72	95	109	110	90	75	106	127	1092	37	7 Feb
Ashton-Under-Lyne, Henrietta Street	SJ 33 940995	120	995	51	151	55	86	94	105	102	136	87	83	112	145	1207	43	7 Feb
Dukinfield Res.	SJ 33 955972	133	...	42	131	47	78	82	99	109	115	81	83	94	111	1072	35	7 Feb
Godley Res.	SJ 33 967954	149	937	42	117	45	79	75	100	109	112	81	79	99	123	1060	...	...
Denton S. Wks.	SJ 33 921935	58	830	38	119	44	65	73	116	113	104	75	77	94	128	1048	36	26 Jun
Denton Res.	SJ 33 913956	93	937	39	123	43	70	77	102	100	110	81	75	92	125	1035	...	...
Mersey																		
Disley, Higher Stoneridge	SJ 33 975837	229	945	41	103	43	72	69	120	113	103	77	89	94	144	1067	36	14 Sep
Lyme Park	SJ 33 971836	205	892	41	100	43	72	67	123	114	101	72	86	98	153	1070	38	26 Jun
High Lane...	SJ 33 964847	206	898	37	100	41	65	76	118	112	100	69	74	92	134	1017	34	14 Sep
Stepping Hill Hospital	SJ 33 911876	85	...	31	104	39	66	71	112	105	102	75	77	100	131	1013	31	26 Jun
Audenshaw Res.	SJ 33 913973	91	954	41	133	46	72	84	117	92	114	83	76	95	136	1088	...	...
Fallowfield, Platt Fields Park	SJ 33 852943	36	843	34	117	38	76	77	124	102	106	74	82	82	99	1011	31	1 Apr
Chorlton-cum-Hardy S. Wks.	SJ 33 808932	23	809	33	116	41	91	82	113	91	109	70	89	88	122	1045	34	1 Apr
Sale S. Wks.	SJ 33 768930	18	853	35	121	37	97	75	109	86	112	69	105	81	124	1051	34	1 Apr
Sale, Carrington Lane	SJ 33 765928	22	808	31	116	35	93	73	104	87	109	68	101	77	122	1016	44	20 Aug
Irlam Weir R.F.M. Sta. Area 679 km²	SJ 33 726937	10	1118	...	...	...	...	...	...	...	...	...	...	...	...	...	...	...
Irwell (Mersey)																		
Stake Brook	SD 34 872247	343	...	58	206	77	106	148	142	121	155	110	137	115	181	1556	46	26 Jun
Sheephouse Res.	SD 34 876219	297	1432	72	260	105	134	173	123	121	167	119	142	128	214	1759	55	7 Feb
Cragg Res.	SD 34 847198	404	1472	81	259	103	133	175	133	121	162	123	151	121	201	1763	...	...
Cowpe Res.	SD 34 843204	282	1497	77	248	104	135	175	139	120	166	119	149	139	215	1786	54	7 Feb
Cloughbottom	SD 34 845269	312	1305	78	186	78	107	131	139	128	145	116	139	118	180	1546	44	3 Oct
Cloughbottom No.2	SD 34 837271	328	...	73	173	81	117	123	131	132	139	115	128	115	198	1525	...	...
Limy (Irwell)																		
Clowbridge Res. No.2	SD 34 835285	311	...	71	182	83	107	111	126	129	121	105	134	102	172	1442	...	...
Clowbridge Res. No.1	SD 34 824281	283	1333	66	178	81	83	126	131	134	130	117	125	111	200	1481	41	26 Jun
Greenfold Res. No.2	SD 34 829264	312	...	76	196	70	108	135	121	127	143	106	132	102	175	1492	...	...
Greenfold Res. No.1	SD 34 823261	305	...	77	192	75	104	138	125	127	141	111	131	100	181	1502	43	3 Oct
Rawtenstall Cemetery	SD 34 813233	174	1333	67	198	87	111	158	120	113	168	113	136	117	192	1579	44	11 May
Ogden (Irwell)																		
Calf Hey	SD 34 742225	244	1455	62	189	92	87	145	128	112	129	109	121	118	212	1503	...	...
Ogden Res. No.1	SD 34 754234	226	...	61	169	94	89	136	111	110	125	110	120	120	210	1456	...	...
Holden Wood No.2	SD 34 767224	195	1425	72	194	96	99	147	125	114	128	108	121	116	216	1536	44	3 Oct
Uglow	SD 34 758206	332	1424	62	187	98	91	129	127	125	123	115	138	137	219	1551	...	...
Musbury Valley	SD 34 770213	229	1423	69	194	87	102	146	109	117	127	106	122	116	203	1499	...	...
Helmshore	SD 34 774205	260	1539	77	213	107	111	162	129	129	136	118	119	130	248	1680	50	7 Feb
Irwell (Mersey)																		
Scout Moor No.1	SD 34 827191	434	...	89	198	102	90	177	152	115	148	121	140	127	198	1657	...	...
Scout Moor No.3	SD 34 828198	404	...	78	221	98	115	185	149	134	153	137	155	149	249	1822	51	3 Oct
New Hall Filter Sta.	SD 34 807197	244	1280	65	206	91	94	153	121	112	119	113	106	120	196	1497	44	7 Feb
Ramsbottom	SD 34 793168	127	1303	65	197	92	102	150	122	110	107	106	120	(121)	(203)	1495	48	7 Feb
Gin Hall	SD 34 805144	145	1216	60	196	84	111	139	113	100	117	99	(114)	(102)	(178)	1413	53	7 Feb
Bury S. Wks.	SD 34 804079	67	...	45	167	49	101	113	101	79	110	70	116	88	141	1180	49	7 Feb
Roch (Irwell)																		
Upper Chelburn	SD 34 952185	221	1267	63	228	93	143	144	120	112	169	116	121	138	214	1661	46	7 Feb
Lower Chelburn, New site	SD 34 949187	1301	1324	63	226	86	155	140	120	109	180	107	124	136	206	1653	49	7 Feb
Blackstone Edge	SD 34 966182	387	1264	68	223	99	170	139	123	115	173	102	133	144	211	1700	51	7 Feb
Broadhead...	SD 34 969171	404	...	65	227	87	175	132	113	109	172	99	134	141	161	1615	55	7 Feb
Longden End	SD 34 957140	229	1285	61	211	84	143	126	115	101	130	98	117	129	185	1500	45	7 Feb
Hollingworth Lake, New site	SD 34 942150	182	...	57	199	77	131	123	108	101	141	95	102	131	173	1435	48	7 Feb

GENERAL TABLE, ENGLAND AND WALES

Station	Grid reference	Altitude m	Av. 1916–1950 mm	Jan	Feb	Mar	Apr	May	Jun	Jul	Aug	Sep	Oct	Nov	Dec	Year	Highest recorded fall in a day mm	date
MERSEY AND WEAVER—*(contd)*																		
Beal (Roch)																		
Brushes Clough	SD 34 956097	286	1257	63	214	81	140	134	115	134	148	100	118	137	182	1567	54	7 Feb
Cold Greave	SD 34 967124	255	1260	61	205	77	155	130	110	112	140	97	132	143	178	1541	46	7 Feb
Kitcliffe	SD 34 960123	233	1245	61	211	80	158	137	114	120	130	100	125	143	177	1556	47	7 Feb
Newhey S. Wks.	SD 34 933111	156	1150	53	177	68	112	127	108	103	115	86	98	133	171	1351	45	7 Feb
Milnrow, Butterworth Hall	SD 34 931124	155	1218	47	175	70	105	121	98	93	115	84	102	132	150	1292	...	...
Roch (Irwell)																		
Watergrove Res.	SD 34 910183	245	...	65	201	80	89	148	118	111	185	120	117	135	196	1564	49	20 Aug
Cowm Res.	SD 34 881185	232	1473	65	233	105	110	165	140	127	173	127	131	147	257	1780	54	7 Feb
Spring Mill Res.	SD 34 875168	212	1409	61	211	98	106	163	142	124	158	122	115	149	234	1684	51	7 Feb
Rochdale	SD 34 882124	110	1148	46	172	67	85	144	152	100	128	82	95	111	166	1349	49	7 Feb
Whitefield Park	SD 34 807061	104	...	46	167	54	106	122	119	89	113	82	118	107	174	1298	50	7 Feb
Bradshaw (Irwell)																		
Brown Lowe	SD 34 682193	366	1448	61	194	99	99	129	157	133	111	105	142	111	201	1541	...	...
Clough	SD 34 687187	328	1488	59	179	96	102	149	164	152	113	113	143	126	217	1614	...	...
Top o'th' Brow	SD 34 709171	285	1526	69	206	110	98	156	143	151	117	131	150	147	232	1710	...	...
Entwistle	SD 34 724173	213	1433	66	202	101	111	148	135	133	120	114	135	124	218	1606	50	7 Feb
Longshoot...	SD 34 736202	283	1439	63	179	102	94	135	129	127	116	116	126	132	211	1529	...	...
Moorside	SD 34 738184	259	1423	67	179	95	99	134	131	122	116	112	131	131	215	1532	...	...
Wayoh	SD 34 734162	185	1420	60	198	105	99	135	129	135	114	111	127	129	223	1565	51	7 Feb
Turton Tower	SD 34 730151	178	1392	68	208	92	111	132	120	131	117	107	131	120	198	1534	52	7 Feb
Eagley (Bradshaw)																		
Hulton Pasture	SD 34 672183	332	1402	70	163	75	98	122	150	141	112	93	140	109	189	1461	...	...
Bromiley Heys...	SD 34 665174	282	1479	68	165	83	103	123	157	143	115	107	130	120	224	1538	...	...
Hordern	SD 34 661153	354	1480	69	165	77	109	142	147	139	112	91	148	113	229	1541	...	...
Eagley	SD 34 700148	166	1500	73	212	102	118	139	145	129	117	105	137	132	240	1649	...	...
Holden's Farm...	SD 34 686171	291	1423	71	210	99	120	141	147	127	121	102	137	136	222	1633	...	...
Longworth	SD 34 692161	235	1447	69	209	97	116	140	145	131	117	101	134	133	221	1613	...	...
Delph...	SD 34 704156	216	1483	67	211	100	116	135	132	131	113	101	136	133	230	1604	55	7 Feb
Springs	SD 34 692142	247	1505	67	197	96	93	136	170	156	120	109	147	147	247	1685	67	26 Jun
Smithills Hall	SD 34 699118	157	1258	60	179	81	100	116	128	117	98	93	127	114	190	1402	57	26 Jun
Croal (Irwell)																		
Heaton Res.	SD 34 687096	152	1190	52	163	68	101	103	114	101	93	83	125	101	191	1296	45	7 Feb
Bolton, Leverhulme Avenue...	SD 34 724070	99	...	50	164	67	102	100	102	104	92	83	123	96	175	1259	...	...
Irwell (Mersey)																		
Kearsley, Mossfield Road	SD 34 746045	90	...	(44)	144	47	102	100	102	91	93	79	113	90	157	1163	44	7 Feb
Ringley S. Wks.	SD 34 768048	46	...	42	145	47	101	99	97	91	100	71	113	83	137	1124	44	7 Feb
Prestwich S. Wks.	SD 34 801032	46	...	43	143	45	93	106	103	88	112	86	117	90	137	1164	43	7 Feb
Prestwich Res....	SD 34 812032	104	1060	44	145	48	86	108	111	81	110	94	112	99	147	1186	...	...
Adelphi Weir R.F.M. Sta. Area 559 km²	SJ 33 824988	24	1265	...	...	...	...	...	...	...	...	...	...	...	...	...	...	...
Irk (Irwell)																		
Royton S. Wks.	SD 34 907072	131	...	45	156	58	88	114	96	104	125	85	111	109	165	1257	46	7 Feb
Trub S. Wks.	SD 34 880095	134	...	49	167	63	97	135	113	95	116	85	95	109	165	1289	49	7 Feb
Heaton Park Res.	SD 34 824047	108	...	45	142	47	83	119	113	88	114	80	98	95	151	1175	...	...
Heaton Park	SD 34 826043	99	1077	46	149	51	84	120	102	88	123	84	106	101	143	1197	44	7 Feb
Scotland Weir R.F.M. Sta. Area 74.3 km²	SJ 33 843992	26	1064	...	...	...	...	...	...	...	...	...	...	...	...	...	...	...
Medlock (Irwell)																		
Strinesdale	SD 34 957067	240	...	50	180	72	109	111	105	120	137	101	107	129	158	1379	48	7 Feb
Strinesdale	SD 34 957066	241	1128	49	178	69	102	113	105	121	139	102	109	130	160	1377	48	7 Feb
Oldham, Glodwick	SD 34 939046	194	...	43	166	60	107	109	103	115	129	99	95	124	144	1294	43	7 Feb
Oldham, Alexandra Park	SD 34 931039	167	...	42	163	58	104	119	104	123	129	104	99	120	142	1307	44	1 Apr
Knott Hill Res.	SD 34 958014	184	1070	44	152	59	89	95	103	110	139	90	95	120	142	1239	42	7 Feb
Irwell (Mersey)																		
Weaste S. Wks.	SJ 33 795980	21	888	36	121	43	88	85	110	69	93	73	91	84	121	1014	35	7 Feb
Eccles S. Wks....	SJ 33 758976	22	864	38	116	36	83	78	99	73	112	69	101	85	126	1017	42	20 Aug
Swinton S. Wks.	SD 34 773003	44	...	39	125	46	90	93	97	76	112	85	107	85	134	1089	41	20 Aug
Davyhulme S. Wks.	SJ 33 756969	22	853	36	117	33	90	76	100	73	113	64	104	78	114	997	43	20 Aug
Glaze (Mersey)																		
Platt Bridge S. Wks.	SD 34 609025	28	955	41	135	39	100	83	124	79	87	74	103	87	151	1103	38	7 Feb
Leigh S. Wks.	SJ 33 661992	20	874	37	122	30	89	71	96	78	91	78	95	81	128	997	38	7 Feb
Atherton Cemetery...	SD 34 666029	50	984	42	137	38	89	79	87	84	93	79	97	93	151	1070	39	7 Feb
Leigh, Bedford P. Sta.	SD 34 668000	22	...	39	125	37	91	79	90	81	91	78	100	87	136	1034	37	7 Feb
Little Woolden Hall R.F.M. Sta. Area 152 km²	SJ 33 684935	9	950	...	...	...	...	...	...	...	...	...	...	...	...	...	...	...
Mersey																		
Ringway	SJ 33 818850	76	802	29	100	38	80	76	123	122	90	60	76	74	124	992	32	5 Jul
Altrincham, Norman's Place	SJ 33 765877	46	828	36	111	38	97	70	132	82	123	75	87	82	130	1062	39	1 Apr
Bollin (Mersey)																		
Standing Stone...	SJ 33 976715	372	1123	77	129	70	129	93	139	133	104	84	109	125	182	1373	...	...
Twelve Ashes	SJ 33 967712	275	1075	56	112	63	108	82	121	115	102	79	98	111	159	1207	36	1 Apr
Trentabank Res.	SJ 33 962712	263	1069	55	111	63	104	81	121	114	101	78	97	110	157	1193	35	1 Apr
Nessitt Hill	SJ 33 963706	384	1034	71	123	65	113	81	120	118	103	80	104	82	147	1206	...	...
Bottoms Res.	SJ 33 944715	201	953	50	99	56	94	70	115	106	101	72	87	105	143	1097	32	1 Apr
Macclesfield	SJ 33 911741	152	923	43	88	52	67	71	123	114	92	74	77	80	140	1022	41	26 Jun
Prestbury S. Wks.	SJ 33 897783	98	830	33	95	47	70	75	131	119	90	79	71	81	135	1026	32	14 Sep
Dean (Bollin)																		
Lamaload Res....	SJ 33 968755	267	...	51	111	59	99	78	127	119	109	90	91	104	150	1187	41	14 Sep
Rainow	SJ 33 947751	248	944	42	95	55	90	72	124	121	96	84	81	94	139	1092	36	14 Sep
Bollin (Mersey)																		
Nether Alderley	SJ 33 843764	100	828	38	97	45	72	75	109	96	98	68	88	89	157	1033	29	11 May
Snelson House...	SJ 33 803746	76	...	41	87	40	79	75	121	96	87	62	75	68	114	945	27	1 Apr
Knutsford	SJ 33 757783	65	840	33	79	45	80	72	107	92	84	62	77	78	122	930	26	1 Apr
Dunham Massey R.F.M. Sta. Area 256 km²	SJ 33 726876	13	871	...	...	...	...	...	...	...	...	...	...	...	...	...	...	...
Dunham Massey S. Wks.	SJ 33 722877	17	809	31	107	36	80	71	136	85	103	72	90	71	124	1005	44	20 Aug
Mersey																		
Warrington, Highways Yard...	SJ 33 613885	13	...	37	85	30	95	86	133	79	86	52	87	61	134	964	32	1 Apr
Sankey (Mersey)																		
Rookery S. Wks.	SJ 33 485999	33	894	47	104	36	107	76	99	99	81	55	103	81	131	1018	37	1 Apr
Prescot	SJ 33 472935	86	903	39	90	42	104	83	113	109	95	57	95	77	140	1044	...	...
Eccleston Hill P. Sta., New site	SJ 33 488941	81	...	45	94	40	101	79	115	89	93	56	92	75	140	1019	35	1 Apr
Carr Mill P. Sta.	SJ 33 526976	37	...	39	108	38	98	93	100	84	83	68	110	76	141	1037	34	1 Apr
Great Sankey	SJ 33 552889	20	837	33	77	36	94	81	134	86	90	62	87	70	125	977	29	20 Aug
Penketh	SJ 33 560884	16	...	35	76	34	88	79	115	86	95	70	99	71	139	987	31	20 Aug

Station	Grid reference	Altitude m	Av. 1916–1950 mm	Jan	Feb	Mar	Apr	May	Jun	Jul	Aug	Sep	Oct	Nov	Dec	Year	Highest recorded fall in a day mm	date
				Rainfall in millimetres														
MERSEY AND WEAVER—*(contd)*																		
Coastal																		
Widnes, Laboratory	SJ 33 515847	8	...	31	65	31	93	63	128	91	84	49	85	66	124	910	32	20 Jun
Netherly W. Wks.	SJ 33 446882	12	...	32	78	37	93	69	102	84	65	39	91	70	140	901	30	1 Apr
Halewood S. Wks.	SJ 33 460859	8	867	30	69	33	95	68	(102)	84	69	54	89	66	140	900	28	1 Apr
Pex Hill Res.	SJ 33 500888	69	875	32	78	33	97	71	107	102	103	53	91	69	132	968	32	20 Aug
Widnes, Corporation Depot	SJ 33 507852	7	...	34	73	33	98	61	119	86	82	54	87	66	128	923	27	1 Apr
Speke...	SJ 33 415839	23	773	29	63	29	97	75	112	95	27	22	102	57	128	837	31	1 Apr
Garston Depot...	SJ 33 395853	25	802	33	76	33	97	81	99	97	68	48	99	68	134	932	32	1 Apr
Liverpool, Smithdown Road...	SJ 33 391885	41	863	38	75	35	101	73	101	96	70	56	89	65	134	932	34	1 Apr
Liverpool, Lark Lane	SJ 33 370874	27	817	35	69	32	92	73	110	103	81	56	91	64	126	932	32	1 Apr
Liverpool, Rathbone Road	SJ 33 391907	60	881	39	82	35	98	75	101	93	77	55	95	64	124	939	34	1 Apr
Liverpool, Monument Place...	SJ 33 354907	39	837	38	74	37	95	84	92	89	96	62	99	69	120	953	31	1 Apr
Liverpool, Queen's Drive	SJ 33 361950	30	891	41	81	37	101	75	91	73	89	44	107	69	111	921	34	1 Apr
Litherland, Hatton Hill Park	SJ 33 335976	18	791	44	93	42	97	79	93	77	84	49	111	77	119	966	37	1 Apr
Alt																		
Liverpool, Knotty Ash	SJ 33 411915	32	898	39	83	40	108	72	101	93	93	59	100	70	135	992	35	1 Apr
Kirkby	SJ 33 399980	15	876	49	90	45	113	93	106	79	82	48	113	79	136	1033	39	1 Apr
Melling P. Sta....	SD 34 393009	24	...	41	89	38	90	86	92	79	74	45	109	72	112	927	32	26 Jun
Bickerstaffe P. Sta....	SD 34 419036	39	901	45	103	43	93	106	90	86	80	56	123	86	127	1041	35	7 Feb
Springfield P. Sta.	SD 34 388046	21	894	40	95	54	88	94	76	86	82	56	121	89	133	1014	33	7 Feb
Ainsdale S. Wks.	SD 34 318112	6	...	31	100	41	87	61	87	64	86	75	119	87	117	956	38	7 Feb
Formby, Hightown	SD 34 294044	8	815	34	91	54	88	63	94	55	71	56	112	79	127	925	35	7 Feb
Coastal																		
Formby Golf Club	SD 34 289084	3	845	36	107	55	99	67	96	62	74	69	121	86	129	1002	38	7 Feb

LANCASHIRE RIVER AUTHORITY AREA, 5260 SQUARE KILOMETRES

HYDROMETRIC AREAS 70 TO 74L

Station	Grid reference	Altitude m	Av. 1916–1950 mm	Jan	Feb	Mar	Apr	May	Jun	Jul	Aug	Sep	Oct	Nov	Dec	Year	Highest recorded fall in a day mm	date
Fine Janes																		
Crossens P.Sta.	SD 34 378206	5	849	30	101	40	78	71	85	56	90	83	121	85	126	966	33	26 Jun
Rufford	SD 34 456164	12	864	36	113	49	91	80	81	83	92	65	124	87	138	1041	41	7 Feb
Southport	SD 34 371208	5	818	32	104	40	75	76	86	52	90	81	123	85	121	965	31	7 Feb
Douglas																		
Scotchmans Post	SD 34 660146	434	...	71	175	79	114	137	146	131	118	95	155	112	184	1517	...	...
Lower Rivington Res.	SD 34 631126	134	1049	56	159	75	112	119	124	109	91	84	118	93	165	1304	47	7 Feb
Rivington Group R.F.M.Sta. Area 39.4 km²	SD 34 631119	113	1237	...	...	...	...	...	...	...	...	...	...	...	...	...	...	...
Worthington W.Wks.	SD 34 580102	69	1010	48	142	51	91	93	112	91	83	57	105	79	146	1100	42	26 Jun
Wigan, Mesnes Park	SD 34 577066	39	...	43	145	50	88	102	102	93	91	56	112	84	143	1109	39	7 Feb
Billinge W.Wks.	SD 34 525030	114	1013	53	129	42	95	87	90	97	76	55	94	68	144	1031	37	7 Feb
Mawdesley	SD 34 493150	21	...	42	116	52	91	95	99	91	91	58	118	86	138	1078	41	26 Jun
Dark Lane P, Sta.	SD 34 428088	41	...	38	110	58	89	103	93	83	89	59	124	79	127	1053	39	7 Feb
Lathom Park	SD 34 458086	52	...	45	115	51	91	101	83	87	84	58	127	81	135	1057	40	7 Feb
Yarrow (Douglas)																		
Stones House	SD 34 621163	195	...	70	146	66	105	113	125	119	86	79	114	101	168	1291	47	26 Jun
Horden Stoops...	SD 34 650154	325	...	63	164	67	118	118	121	109	105	85	135	103	174	1362	...	...
Great Hill...	SD 34 646190	378	...	67	129	80	119	116	132	143	89	85	122	94	165	1341	...	...
Round Loaf	SD 34 632184	296	...	51	146	63	89	109	133	127	100	83	127	97	159	1284	...	...
Coppice Stile	SD 34 627191	268	1142	51	145	63	88	98	138	114	99	81	109	(89)	(143)	1221	...	...
Common BankS.Wks.	SD 34 562174	46	964	43	120	45	90	106	102	102	98	65	96	86	137	1088	44	26 Jun
Ribble																		
Ribblehead	SD 34 766789	312	1998	107	265	169	106	207	201	121	162	177	128	220	320	2184	57	1 Dec
Holden Moor No.27	SD 34 765589	316	1498	67	227	108	96	131	111	86	152	102	112	173	224	1589	...	...
Barnoldswick	SD 34 878458	223	...	55	187	69	89	109	117	75	145	92	136	118	180	1371	41	3 Oct
Greenber Field	SD 34 887481	149	1034	55	175	59	77	102	106	65	145	80	120	93	157	1234	34	3 Oct
Tosside No.33	SD 34 766563	268	...	63	211	93	112	127	116	90	149	108	103	126	183	1480	...	...
Chatburn	SD 34 761433	94	1131	60	169	75	87	106	123	101	159	113	135	114	189	1432	44	19 Dec
Hodder (Ribble)																		
White Hill No.30	SD 34 677585	521	...	85	264	151	124	186	211	148	161	172	161	216	324	2201	...	...
Lamb Hill No.3	SD 34 688592	405	1841	81	283	143	102	187	211	145	163	160	151	225	292	2143	...	...
Greet Bridge No.24	SD 34 702589	251	1846	83	283	113	112	170	160	127	161	170	134	229	259	2000	...	...
Saddle Hill No.16A	SD 34 690579	427	2023	83	272	149	101	182	211	142	151	176	158	241	323	2189	...	...
Catlow No.4A	SD 34 705591	251	...	80	269	116	117	149	145	115	156	145	137	215	277	1919	...	...
Parks Clough No.40	SD 34 720587	277	...	74	278	103	98	127	145	98	149	148	104	170	204	1697	...	...
Collyholme No.20	SD 34 716578	198	1661	66	254	105	112	155	137	108	152	138	122	185	258	1794	...	...
Merrybent No.39A	SD 34 710571	305	...	81	229	157	121	160	159	118	163	136	125	213	225	1887	...	...
Merrybent No.39	SD 34 710571	305	...	85	215	146	125	164	161	118	165	142	129	219	234	1903	...	...
Swire Clough No.25A	SD 34 725594	363	1655	84	221	116	95	151	131	102	152	147	119	196	245	1759	...	...
Holme House No.19	SD 34 725571	183	1603	69	231	121	98	114	113	103	145	126	121	187	248	1674	...	...
Dalehead No.1...	SD 34 733564	186	1500	71	216	116	86	142	119	95	143	114	122	175	244	1644	...	...
Dob Dale (Hodder)																		
Brown Hill No.43	SD 34 756602	277	...	81	228	109	108	141	125	103	146	128	115	167	237	1688	...	...
Knotteranum No.26	SD 34 737606	366	1729	83	260	127	127	158	143	102	163	145	137	186	279	1910	...	...
Dobdale No.28	SD 34 747592	244	1451	76	206	103	108	136	117	98	141	128	121	166	237	1636	...	...
Tenters No.42	SD 34 757588	328	...	81	211	109	97	145	124	108	145	127	123	169	247	1685	...	...
Bottoms No.6	SD 34 748585	263	1469	79	225	95	107	154	128	94	148	129	113	175	253	1700	...	...
Far Barn No.22A	SD 34 746572	206	1513	70	216	109	113	148	121	95	147	118	118	173	232	1660	...	...
Bottoms Beck Flume R.F.M.Sta. Area 10.6 km²	SD 34 745565	186	1524	...	...	...	...	...	...	...	...	...	...	...	...	...	...	...
Hodder (Ribble)																		
Stocks No.10	SD 34 719550	201	...	67	241	123	105	146	153	101	139	125	127	173	266	1768	71	19 Dec
Stocks No.21D	SD 34 717548	192	...	68	235	114	108	138	146	96	143	122	131	169	237	1707	52	19 Dec
Stocks Res.R.F.M.Sta. Area 37.6 km²	SD 34 720546	182	1656	...	...	...	...	...	...	...	...	...	...	...	...	...	...	...
Slaidburn	SD 34 715549	192	1579	65	223	100	104	132	138	91	143	115	130	149	227	1617	52	19 Dec
Stocks No.11A	SD 34 715549	192	...	66	225	106	104	137	141	92	142	115	132	152	238	1651	58	19 Dec
Phynis No.18	SD 34 712545	174	1531	66	221	106	102	142	145	86	141	116	128	159	235	1647	53	19 Dec
Rushton Hill No.41	SD 34 733552	191	...	63	220	105	90	140	127	104	138	118	112	143	225	1585	...	...
Croasdale (Hodder)																		
Croasdale Top No.29	SD 34 668574	399	...	90	281	149	123	176	206	142	164	159	116	208	307	2120	...	...
Upper Croasdale No.37...	SD 34 684569	265	...	80	278	142	103	169	193	131	168	147	122	205	296	2034	...	...
Baxton Fell	SD 34 673561	465	...	85	275	161	111	151	212	147	154	141	132	202	312	2082	...	...
Croasdale Valley No.15...	SD 34 694557	219	1841	66	256	107	108	153	168	120	160	133	154	195	303	1922	...	...
Croasdale House No.14...	SD 34 704550	183	1556	71	247	109	110	137	145	99	151	126	129	169	260	1753	62	19 Dec
Croasdale Flume R.F.M.Sta. Area 10.4 km²	SD 34 706546	177	1839	...	...	...	...	...	...	...	...	...	...	...	...	...	...	...

Station	Grid reference	Altitude m	Av. 1916–1950 mm	Jan	Feb	Mar	Apr	May	Jun	Jul	Aug	Sep	Oct	Nov	Dec	Year	Highest recorded fall in a day mm	date
LANCASHIRE—*(contd)*																		
Dunsop (Hodder)																		
Hornby Gate No.45	SD 34 650587	389	...	93	290	174	126	189	232	166	164	185	171	221	331	2341	...	...
Whitendale No.4	SD 34 653559	253	1838	67	259	127	113	142	183	132	136	139	117	191	281	1887	...	...
Whitendale Bridge	SD 34 660547	207	...	71	274	121	126	138	191	117	149	137	118	188	276	1907	...	...
Footholme No.2	SD 34 653533	171	1675	72	261	126	118	159	184	128	157	145	125	184	274	1932	...	...
Cabin Hill	SD 34 621563	475	1848	63	274	118	113	133	205	164	132	148	128	166	241	1886	...	...
Brennand	SD 34 631548	250	1731	81	263	127	123	152	185	140	141	145	126	154	250	1889	...	...
Swine Clough No.5	SD 34 643543	201	...	74	285	122	125	149	182	136	145	143	123	181	271	1934	...	...
Dunsop House No.1	SD 34 655515	136	...	70	249	133	114	160	185	124	151	138	128	188	257	1896	...	...
Hodder (Ribble)																		
Birch Bank	SD 34 618493	396	1954	84	228	143	99	172	202	154	147	170	148	207	303	2058	...	...
Langden Brook	SD 34 628510	170	1875	79	255	134	109	175	197	134	150	144	141	194	287	2000	57	26 Jun
Hareden Fell	SD 34 636502	172	1826	79	248	137	111	158	186	119	141	141	132	192	280	1925	61	19 Dec
Calder (Ribble)																		
Causeway Side	SD 34 890292	338	1367	70	236	77	137	137	120	104	124	111	139	125	179	1559	...	...
Habergham Eaves	SD 34 838308	238	1186	78	168	70	97	116	129	114	123	120	127	116	164	1422	37	14 Sep
Don (Calder)																		
Red Spa Moor	SD 34 910353	358	1274	74	173	72	100	123	154	109	116	117	144	106	178	1466	...	...
Entwistle Moor Top	SD 34 907335	379	1284	68	199	81	116	104	144	111	115	108	137	129	192	1505	...	...
Rams Clough Head	SD 34 903326	377	1292	76	215	80	125	124	142	108	125	111	144	127	193	1570	...	...
Swinden Upper	SD 34 893331	247	1234	72	171	70	98	109	137	97	114	103	132	110	173	1386	44	3 Oct
Swinden Lower	SD 34 885334	227	1203	69	168	67	93	105	134	97	109	104	127	106	163	1342	42	3 Oct
Shedden	SD 34 893301	293	1344	70	241	76	145	132	122	101	122	115	139	128	179	1568	...	...
Cant Clough, South	SD 34 894306	275	1350	(69)	199	83	108	128	125	100	121	109	134	125	197	1497	42	3 Oct
Cant Clough, North	SD 34 890313	273	1283	68	211	81	117	123	129	92	119	109	141	116	196	1501	47	3 Oct
Calder (Ribble)																		
Burnley	SD 34 855353	200	...	61	145	74	81	105	137	101	115	101	117	103	167	1307	39	19 Dec
Pendle (Calder)																		
Ogden Res.	SD 34 817398	277	1419	69	166	106	81	125	157	127	138	145	135	140	241	1630	56	19 Dec
Blackmoss Res.	SD 34 826414	251	1346	72	198	111	89	128	142	108	140	126	145	153	222	1634	...	...
Buttock	SD 34 807401	358	1504	65	162	104	82	141	166	124	161	146	142	158	260	1709	70	19 Dec
Colne (Pendle)																		
Laneshaw Res.	SD 34 943414	242	1100	58	160	58	87	95	109	90	145	93	130	99	146	1270	39	3 Oct
Swinden S.Wks.	SD 34 871394	129	1132	55	139	83	64	104	128	102	109	101	113	121	200	1320	42	19 Dec
Foulridge, Lower Res.	SD 34 880418	160	1220	58	162	82	77	117	118	93	128	99	131	129	201	1394	43	3 Oct
Barrowford Res.	SD 34 868403	144	1185	63	151	86	71	117	122	96	119	105	122	123	194	1369	41	3 Oct
Pendle (Calder)																		
Nelson	SD 34 872384	165	1133	57	141	73	78	100	126	94	115	106	114	103	169	1275	40	3 Oct
Coldwell Res.No.3	SD 34 901356	323	1334	75	160	74	97	112	152	118	126	115	142	107	182	1460	44	3 Oct
Coldwell Res.No.2	SD 34 900362	293	1266	74	160	74	97	112	150	118	124	114	141	108	180	1452	43	3 Oct
Nelson S.Wks.	SD 34 845375	109	1147	51	122	75	64	103	145	102	117	103	112	100	164	1259	42	19 Dec
Calder (Ribble)																		
Padiham Park	SD 34 789337	93	1116	56	129	67	86	103	113	106	109	99	109	93	145	1216	37	26 Jun
Altham S.Wks.	SD 34 778329	69	1071	53	124	65	72	104	105	104	102	95	106	88	142	1159	39	11 May
Burnley Road Res.	SD 34 773296	187	...	59	166	77	95	118	117	139	111	99	111	97	171	1360	38	26 Jun
Mitchell's House Res.	SD 34 787274	283	1229	66	181	77	118	126	119	135	137	110	132	86	165	1453	49	3 Oct
Jackhouse Res.	SD 34 741259	198	...	59	177	83	83	136	138	100	138	109	118	113	210	1465	47	26 Jun
Coppy Clough S.Wks.	SD 34 742295	106	1127	49	150	68	80	106	114	101	110	98	111	90	160	1237	43	26 Jun
Rishton S.Wks.	SD 34 738311	91	1093	46	128	76	75	95	120	94	96	97	102	94	154	1177	47	26 Jun
Churn Clough Res.	SD 34 785382	241	1209	58	146	80	87	111	137	107	134	121	120	110	194	1405	47	19 Dec
Churn Clough Res., New Site	SD 34 785382	241	...	58	145	80	85	112	137	106	134	120	119	111	194	1403	47	19 Dec
Great Harwood, Kebb Farm	SD 34 722327	204	1230	53	136	83	78	109	133	102	123	109	127	101	164	1316	45	26 Jun
Whalley R.F.M.Sta. Area 316 km²	SD 34 730360	40	1227	...	...	...	...	...	...	...	...	...	...	...	...	...	...	...
Whalley	SD 34 725371	60	1117	42	134	75	63	99	126	109	129	107	117	99	160	1259	45	26 Jun
Ribble																		
Parsonage Res.	SD 34 699321	192	1047	60	122	62	90	87	99	97	134	103	125	80	136	1195	38	1 Apr
Stonyhurst	SD 34 692388	115	1213	51	142	84	76	112	134	124	125	112	114	111	197	1380	51	26 Jun
Jeffrey Hill, New site	SD 34 647402	283	...	(51)	(191)	(89)	86	115	162	142	129	135	137	127	202	1566	...	...
Cowley Brook	SD 34 641392	188	1220	60	162	76	87	119	164	124	127	119	139	119	188	1486	47	26 Jun
Longridge	SD 34 620372	110	1102	45	149	65	77	105	148	113	126	108	125	90	143	1294	43	7 Feb
Ribchester Institution	SD 34 627365	84	1132	41	137	65	79	(108)	(152)	(114)	120	95	122	93	152	1280	37	7 Feb
Alston Res.	SD 34 603366	101	1065	44	135	67	79	103	153	102	127	98	136	86	138	1267	42	7 Feb
Samlesbury R.F.M.Sta. Area 1140 km²	SD 34 589304	7	1323	...	...	...	...	...	...	...	...	...	...	...	...	...	...	...
Darwen (Ribble)																		
Cranberry Moss	SD 34 710192	299	1435	69	194	101	100	147	131	134	109	104	140	111	199	1537	...	...
Earnsdale Res.	SD 34 670220	215	1323	68	173	82	94	139	145	128	130	88	122	105	181	1455	58	26 Jun
Sunnyhurst Hey Res.	SD 34 677221	279	1304	70	178	78	90	137	143	128	127	85	122	106	185	1449	52	26 Jun
Pickup Bank	SD 34 721237	222	1308	70	180	77	114	129	128	113	122	88	137	101	198	1457	46	26 Jun
Guide Res.	SD 34 698264	186	1114	65	155	67	99	121	111	101	115	81	107	96	163	1282	42	26 Jun
Rishton Res.	SD 34 715301	142	...	48	(127)	76	86	92	113	102	130	91	123	81	128	1196	27	11 May
Blackburn, Witton	SD 34 668276	96	...	44	129	63	99	89	117	110	118	73	104	86	136	1167	47	26 Jun
Roddlesworth (Darwen)																		
Pimms	SD 34 652195	301	1360	74	154	79	(117)	118	149	155	109	97	135	108	(175)	1469	...	...
Slater Lane	SD 34 662192	300	1421	69	169	81	110	123	161	157	111	113	135	123	223	1575	...	...
Tockholes	SD 34 663201	296	1347	69	179	79	118	122	151	157	105	93	138	109	196	1515	...	...
Roddlesworth	SD 34 655216	221	...	65	157	63	77	114	(138)	140	107	(61)	129	90	(157)	1300	...	...
Rivington, Gamekeeper's	SD 34 640209	259	1230	66	145	(64)	79	112	140	141	102	62	123	(95)	(156)	1283	...	...
Darwen (Ribble)																		
Ribble																		
Preston, West Cliff	SD 34 532288	8	937	38	119	57	81	85	108	80	113	90	135	82	125	1112	38	26 Jun
Hutton	SD 34 512266	25	935	47	109	54	75	85	97	66	119	95	132	80	123	1082	...	...
Haighton Res.	SD 34 573352	80	1027	42	133	63	77	95	128	108	111	98	137	84	126	1202	40	26 Jun
Preston, Moor Park	SD 34 537311	33	1003	38	119	56	79	88	107	77	118	79	143	83	122	1111	40	26 Jun
Coastal																		
Squires Gate	SD 34 316317	10	852	43	101	41	73	59	67	51	85	81	115	78	121	916	26	7 Feb
Thornton	SD 34 336421	5	...	57	117	46	64	72	99	41	97	87	118	91	110	1000	41	3 Oct
Poulton le Fylde	SD 34 346384	12	934	51	114	65	82	75	88	51	88	80	127	84	114	1019	35	3 Oct
Wyre																		
Hare Syke	SD 34 599575	463	1863	76	180	160	97	140	246	183	140	160	119	163	239	1903	...	...
Abbeystead Res.	SD 34 563543	112	1344	67	172	79	107	120	147	110	128	113	106	(112)	(165)	1425	46	26 Jun
Old Gauge Basin	SD 34 541576	294	1326	63	137	97	81	102	145	122	127	112	97	112	145	1339	...	...
Damas Ghyll Res.	SD 34 527573	175	1263	69	138	58	71	92	123	103	123	111	96	103	135	1221	34	26 Jun
Scorton	SD 34 502487	28	1061	53	114	51	85	92	105	74	108	95	84	85	134	1078	33	26 Jun
Grizedale Intake	SD 34 541496	196	1200	61	157	55	85	102	104	91	112	111	97	91	134	1199	...	...
Barnacre Res.	SD 34 528480	188	1236	66	145	61	89	105	139	105	111	108	95	93	153	1270	34	26 Jun

Station	Grid reference	Altitude m	Av. 1916–1950 mm	Jan	Feb	Mar	Apr	May	Jun	Jul	Aug	Sep	Oct	Nov	Dec	Year	Highest recorded fall in a day mm	date
LANCASHIRE—*(contd)*																		
Calder Intake	SD 34 544484	193	1225	62	140	49	66	104	105	97	107	107	99	102	132	1169	...	...
Barnsfold Res.	SD 34 577416	155	1218	56	149	65	81	101	136	123	136	114	128	94	157	1340	...	...
St.Michaels-on-Wyre	SD 34 465411	8	962	47	120	52	72	82	98	73	103	95	106	84	109	1042	31	26 Jun
St.Michael's R.F.M.Sta. Area 275 km²	SD 34 465411	4	1257	...	...	...	...	...	...	...	...	...	...	...	...	...	...	...
Kirkham	SD 34 414346	24	961	42	109	44	71	77	95	70	102	90	115	75	117	1006	33	26 Jun
Coastal																		
Preesal P.Sta.	SD 34 365465	9	889	55	123	51	69	71	95	42	105	86	108	73	116	995	32	26 Jun
Hampson-in-Ellel	SD 34 490541	44	1020	49	119	53	60	80	100	68	107	88	101	87	121	1032	32	26 Jun
Garstang	SD 34 469480	21	961	46	111	46	70	76	93	64	96	87	80	81	119	969	27	26 Jun
Blea Tarn Res.	SD 34 495583	101	1157	54	153	54	76	91	135	83	109	99	79	86	128	1147	36	26 Jun
Rawthey (Lune)																		
Sedbergh	SD 34 653919	123	1399	54	191	97	68	130	135	93	117	145	122	120	132	1404	46	4 Jun
Wenning (Lune)																		
Austwick, Town Head	SD 34 769687	149	1205	59	170	75	61	90	114	95	118	104	84	111	178	1259	35	23 Feb
Great Harlow	SD 34 696612	454	...	86	303	143	125	191	180	117	150	165	118	215	323	2115	...	...
Clapham Common	SD 34 729615	360	...	83	277	121	79	152	143	103	165	143	124	185	248	1824	...	...
Bentham, Summerhill	SD 34 666682	116	...	56	175	80	71	77	132	88	137	91	83	101	156	1247	33	19 Dec
High Salter	SD 34 616620	302	1568	93	235	102	140	128	133	100	146	128	122	160	174	1661	...	...
Lune																		
Snab	SD 34 564682	30	1161	40	132	50	50	71	129	83	106	93	97	75	111	1037	35	26 Jun
Halton R.F.M.Sta. Area 995 km²	SD 34 502647	5	1577	...	...	...	...	...	...	...	...	...	...	...	...	...	...	...
Lancaster, Edenbreck	SD 34 469612	15	...	53	141	57	66	80	120	79	122	97	94	96	139	1143	33	26 Jun
Coastal																		
Heysham Harbour	SD 34 404601	11	964	49	139	54	75	73	97	51	111	90	127	76	127	1070	57	3 Oct
Morecambe	SD 34 431645	7	975	47	142	61	73	86	93	68	124	83	120	83	127	1106	46	3 Oct
Bare	SD 34 454644	6	979	45	138	57	61	83	114	71	119	89	102	77	124	1079	31	26 Jun
Peddar Pots Res.	SD 34 533700	82	1187	44	161	64	68	72	145	81	127	100	110	95	142	1208	37	3 Oct
Beela																		
Crooklands	SD 34 534835	48	1201	57	171	73	79	121	129	81	99	114	(119)	109	154	1306	37	4 Jun
Lupton Res.	SD 34 574834	168	1324	60	182	85	73	129	163	84	105	128	113	111	151	1385	...	...
Fisher Tarn Res.	SD 34 549927	226	1406	73	190	109	78	157	175	85	100	159	125	133	164	1548	50	4 Jun
Kent																		
Dubbs Res.	NY 35 421018	227	1794	121	232	176	100	173	137	72	152	237	159	186	255	2001	...	...
Kendal	SD 34 519914	41	1303	64	198	99	73	139	146	67	85	138	111	120	166	1408	36	4 Jun
Crook	SD 34 461948	79	1447	94	220	123	102	143	154	66	112	165	117	140	206	1643	44	4 Feb
Levens, Lowgate House	SD 34 487862	34	1310	78	194	91	77	147	133	68	127	136	123	131	189	1493	40	4 Feb
Levens, Bridge End	SD 34 474857	7	1261	77	180	85	74	131	136	67	112	128	107	113	177	1387	39	4 Feb
Coastal																		
High Newton Res.	SD 34 399839	178	1353	86	216	88	97	141	173	91	122	126	131	136	217	1625	45	17 Jun
Leven																		
Long House	NY 35 306067	108	2759	208	392	285	141	277	227	101	190	334	178	335	434	3101	77	29 Nov
Langdale	NY 35 315058	113	2593	193	358	234	140	266	214	103	174	310	167	290	363	2811	79	3 Sep
Grasmere	NY 35 330083	84	...	175	343	216	122	237	201	86	174	322	160	280	335	2651	72	3 Sep
Rydal Hall	NY 35 364064	108	2186	178	264	171	108	192	164	73	173	267	159	219	327	2297	65	3 Sep
Ambleside	NY 35 372040	46	1890	149	265	165	116	189	181	69	157	224	154	180	289	2138	65	3 Sep
High Wray	NY 35 367019	53	1756	148	277	154	120	221	189	75	150	237	160	200	311	2241	67	3 Sep
Briery Close	NY 35 391019	107	1757	143	248	145	117	191	145	67	151	205	165	186	262	2024	51	4 Feb
Holehird	NY 35 411007	149	1807	130	251	138	119	184	133	72	151	248	166	162	255	2009	62	3 Sep
Windermere Nurseries	SD 34 410982	99	1698	122	248	144	109	175	128	67	139	231	148	161	251	1923	56	4 Feb
Ferry House	SD 34 390956	44	1563	120	244	143	108	156	145	60	119	208	132	154	241	1830	59	4 Feb
Bryerswood	SD 34 383953	110	1689	124	237	158	115	176	152	73	123	219	140	172	281	1969	56	4 Feb
Hawkshead	SD 34 362973	84	1789	127	265	170	125	204	166	75	141	225	149	195	293	2134	59	19 Dec
Ghyll Head Res.	SD 34 397922	128	1615	114	244	143	115	166	157	69	113	214	137	166	233	1870	...	...
Tower Wood	SD 34 385912	46	1560	104	237	137	120	168	156	70	132	205	139	146	237	1851	54	3 Sep
Fell Foot	SD 34 382871	53	1493	98	247	118	137	155	166	89	125	164	147	143	242	1832	52	3 Sep
Newby Bridge R.F.M.Sta. Area 241 km²	SD 34 371863	38	2195	...	...	...	...	...	...	...	...	...	...	...	...	...	...	...
Haverthwaite	SD 34 345839	13	...	95	234	109	120	136	164	88	107	155	139	142	237	1725	51	3 Sep
Grizedale	SD 34 337941	91	1897	144	266	162	131	189	199	88	139	235	146	194	303	2196	62	4 Feb 19 Dec
Coniston (Leven)																		
Coniston	SD 34 299978	76	2462	169	350	269	154	214	238	115	170	310	168	226	420	2803	93	19 Dec
Low Nibthwaite R.F.M.Sta. Area 73.0 km²	SD 34 294882	39	--	...	...	...	...	...	...	...	...	...	...	...	...	...	...	...
Coastal																		
Urswick	SD 34 268742	40	1179	70	188	77	101	87	119	61	103	123	168	82	172	1350	52	3 Oct
Little Urswick	SD 34 264738	40	1234	71	196	83	105	96	128	70	108	130	165	90	191	1434	58	3 Oct
Poaka Beck	SD 34 240781	156	1339	83	188	90	94	109	149	97	106	159	124	101	196	1495	46	26 Jun
Dalton-in-Furness, Eynhallow	SD 34 226742	46	1176	75	191	77	102	99	116	66	113	136	133	89	184	1379	42	26 Jun
Barrow-in-Furness, Gasworks	SD 34 217688	5	1030	60	175	70	84	78	117	64	110	107	124	72	147	1207	42	26 Jun 3 Oct
Barrow-in-Furness S.Wks.	SD 34 216689	7	1043	64	180	71	93	76	117	64	114	110	129	74	143	1237	46	3 Oct
Barrow-in-Furness, Hill Road	SD 34 207718	58	...	71	184	75	94	83	114	69	108	124	130	76	168	1296	52	3 Sep
Duddon																		
Seathwaite Tarn	SD 34 249984	378	2548	191	347	191	137	264	235	142	194	310	169	230	373	2783	...	...
Ulpha	SD 34 209947	78	2044	164	316	194	117	190	187	115	131	258	132	206	342	2351	75	3 Sep
Coastal																		
Baystone Bank Res.	SD 34 171859	88	1623	103	219	139	94	133	159	109	141	195	123	151	297	1863	...	...

CUMBERLAND RIVER AUTHORITY AREA, 4590 SQUARE KILOMETRES

HYDROMETRIC AREAS 74C TO 76

Coastal																		
Fellside	SD 34 117882	81	1275	91	191	76	81	109	118	79	127	136	101	87	165	1361	26	25 Jan
Esk																		
Eskmeals	SD 34 085931	9	...	87	134	65	76	92	101	78	116	133	88	81	135	1187	39	3 Sep
Coastal																		
Sellafield	NY 35 027032	13	...	64	118	77	67	93	101	47	108	127	85	70	144	1101	42	3 Sep
Ehen																		
Gillerthwaite	NY 35 143140	128	2363	137	277	249	133	279	285	150	165	320	180	216	282	2673	...	...
Ennerdale	NY 35 085153	117	1778	116	197	162	91	188	180	81	133	256	108	152	227	1891	75	11 Sep
Summergrove	NX 25 999159	86	...	78	140	99	63	117	123	52	133	178	102	95	155	1334	63	13 Aug

Station	Grid reference	Altitude m	Av. 1916–1950 mm	Jan	Feb	Mar	Apr	May	Jun	Jul	Aug	Sep	Oct	Nov	Dec	Year	Highest recorded fall in a day mm	date
CUMBERLAND—*(contd)*																		
Braystones	NY 35 008058	11	1056	66	119	63	68	86	104	49	122	127	89	70	134	1095	...	...
Coastal																		
Stanley Pond	NX 25 982143	27	1143	62	105	73	56	82	106	(46)	130	151	92	92	(127)	1122	38	3 Sep
St.Bees Head	NX 25 941143	94	975	56	100	56	43	83	76	32	125	123	84	67	105	952	65	13 Aug
Workington, Coke Ovens	NX 25 989280	9	964	47	109	71	58	115	77	33	112	140	80	79	105	1026	57	13 Aug
Derwent																		
Styehead No.1	NY 35 231106	335	4015	305	457	381	178	267	305	216	229	508	178	533	508	4064	...	...
Stockley Bridge	NY 35 233110	178	3672	279	470	343	165	254	292	178	356	521	191	457	445	3950	...	...
Styehead No.2	NY 35 229107	326	...	305	546	381	184	279	343	203	241	559	191	597	571	4401	...	...
Seathwaite	NY 35 235121	129	3342	241	368	305	152	241	267	165	267	(432)	178	457	425	3499	...	...
Borrowdale	NY 35 256143	101	2677	147	331	231	139	220	261	125	264	266	178	292	273	2726	95	13 Aug
Keswick, Braeside	NY 35 274230	123	...	78	175	112	76	113	130	78	150	179	113	157	183	1544	62	3 Sep
Greta (Derwent)																		
Hill Top Gill	NY 35 317121	290	2650	176	338	223	121	240	196	105	156	324	167	278	351	2674	...	...
Thirlmere, The Nook	NY 35 319129	189	2511	158	354	213	143	239	198	102	175	331	178	275	347	2713	78	3 Sep
Birkside	NY 35 337133	579	2500	166	333	207	150	234	196	111	267	300	202	243	288	2695	...	...
Wythburn	NY 35 318139	186	2590	174	365	219	152	249	207	102	183	305	180	273	361	2769	...	...
Ullscarf	NY 35 298123	640	3158	195	415	263	149	317	280	163	243	386	217	331	439	3399	...	...
Brown Rigg	NY 35 309148	427	3009	178	395	248	161	277	251	106	209	325	186	282	379	2998	...	...
Armboth Fells	NY 35 292156	448	2573	188	403	256	158	284	271	117	218	350	175	280	392	3091	...	...
Dale Head Hall	NY 35 313175	189	2303	169	347	182	148	196	186	85	182	224	164	255	295	2433	64	3 Sep
Thirlmere Res.R.F.M.Sta. Area 40.9 km²	NY 35 309191	179	2710	...	...	...	...	...	...	...	...	...	...	...	...	...	...	...
Brown Cove	NY 35 333159	549	2478	204	340	155	149	249	227	98	278	269	216	236	306	2727	...	...
Mungrisdale	NY 35 364302	226	--	(76)	191	161	95	113	125	87	159	207	125	199	(203)	1744	57	3 Sep
Derwent																		
Braithwaite	NY 35 232238	85	...	76	195	147	73	124	150	100	170	191	132	178	225	1762	75	13 Aug
Hause Gill W.Wks....	NY 35 263324	267	...	78	153	(146)	88	87	169	114	134	234	89	157	173	1622	...	...
Cocker (Derwent)																		
Fleetwith	NY 35 209132	442	...	220	400	375	171	375	331	188	241	451	191	317	533	3793	...	...
Bleaberry Tarn	NY 35 167154	506	...	160	339	250	154	317	178	144	202	333	169	274	375	2895	...	...
Casgale Gill	NY 35 187210	604	...	132	274	196	119	193	272	168	257	257	182	262	234	2544	...	...
Cockermouth Moor Res.	NY 35 117300	79	...	48	99	82	45	80	114	47	104	180	65	74	132	1072	53	13 Aug
Derwent																		
Great Broughton	NY 35 072315	76	...	41	79	64	51	74	96	52	98	164	85	83	105	990	39	13 Aug
Camerton R.F.M.Sta. Area 663 km²	NY 35 037305	17	1821	...	...	...	...	...	...	...	...	...	...	...	...	...	...	...
Stainburn Res....	NY 35 022293	56	1015	51	100	74	55	108	99	54	113	174	85	83	119	1114	57	13 Aug
Ellen																		
Quarry Hill Filter Wks....	NY 35 219412	143	...	44	97	79	51	79	105	71	107	152	65	79	107	1033	54	13 Aug
Coastal																		
Waver																		
Wampool																		
Brookfield...	NY 35 242478	34	865	(43)	106	59	(51)	57	109	44	114	123	76	78	89	950	51	13 Aug
Coastal																		
Anthorn	NY 35 185577	11	...	49	92	46	75	73	93	39	122	136	71	72	86	953	51	3 Sep
Eden																		
Mallerstang	NY 35 782006	275	...	(56)	203	83	69	(102)	131	59	85	117	110	125	164	1302	27	4 Sep
Eden Place	NY 35 776097	166	1064	47	140	49	60	79	84	53	112	82	102	81	125	1014	34	1 Dec
South Stainmore	NY 35 837124	271	...	36	135	47	69	69	98	79	132	89	97	105	126	1081	37	13 Aug
Appleby Castle	NY 35 684198	146	928	47	121	56	60	67	84	57	98	84	116	63	103	956	34	3 Oct
Appleby, Highfield	NY 35 688201	146	905	45	115	52	56	63	77	50	92	81	119	50	112	912	41	1 Dec
Temple Sowerby R.F.M.Sta. Area 617 km²	NY 35 604283	92	1207	...	...	...	...	...	...	...	...	...	...	...	...	...	...	...
Temple Sowerby	NY 35 609273	108	828	43	95	36	56	52	93	41	100	104	112	59	76	869	30	20 Jun
Eamont (Eden)																		
Patterdale Hall...	NY 35 391163	146	2356	157	313	181	145	195	135	65	184	232	166	239	263	2275	61	3 Sep
Hutton John	NY 35 439270	210	1221	60	170	92	84	75	93	65	139	139	127	148	158	1350	51	13 Aug
Penrith, Castle Park	NY 35 513299	149	847	39	87	43	53	42	77	49	117	111	105	76	84	884	37	13 Aug
Lowther (Eamont)																		
Wasdale Pike	NY 35 537085	564	2561	200	339	234	186	223	191	104	188	(305)	180	264	282	2695	...	...
Wet Sleddale	NY 35 537107	293	2232	164	333	173	147	204	138	69	169	254	155	240	259	2305	...	...
Birbeck Fell	NY 35 564115	282	1550	103	235	99	111	134	101	55	142	174	145	150	170	1620	...	...
Thornship Gill	NY 35 551132	257	1570	120	232	91	111	149	101	(56)	141	182	138	166	173	1658	...	...
Swindale (Lowther)																		
Artle Crag...	NY 35 482102	650	2722	216	369	263	193	235	202	86	245	382	197	292	328	3008	...	...
Mosedale Cottage	NY 35 496095	440	2589	225	367	241	159	253	199	107	212	357	185	279	310	2893	...	...
High Wether Howe	NY 35 513111	494	2208	184	329	129	195	202	128	63	221	276	169	216	243	2355	...	...
Tailbert Head	NY 35 530145	294	...	153	315	114	133	170	104	61	148	220	143	198	213	1972	...	...
Haweswater (Lowther)																		
Blea Water	NY 35 456108	427	2986	285	431	281	199	293	255	127	244	427	200	298	343	3384	...	...
Riggindale	NY 35 466118	272	2613	235	432	260	184	255	215	97	229	293	164	261	287	2910	...	...
Whelter	NY 35 469132	250	2400	231	398	209	173	229	156	75	221	264	175	253	264	2649	...	...
Guerness	NY 35 483135	276	2185	191	335	157	164	189	132	69	214	253	171	213	241	2330	...	...
Sandhills	NY 35 485153	251	...	184	335	183	147	180	131	59	189	212	154	212	237	2224	...	...
Low Raise	NY 35 454137	751	2838	323	486	283	277	268	175	88	268	324	(198)	291	324	3305	...	...
Measand Beck	NY 35 461151	549	2350	223	351	193	227	227	148	80	248	272	199	241	270	2680	...	...
Wallow Crag	NY 35 496149	430	2022	186	335	143	153	185	123	60	202	224	159	213	229	2213	...	...
Burn Banks	NY 35 503159	213	1818	140	281	126	127	143	90	55	181	169	142	160	205	1819	59	13 Aug
Naddle House	NY 35 509153	251	1817	148	301	129	148	153	95	53	179	200	146	183	201	1936	...	...
Thornthwaite R.F.M.Sta. Area 33.9 km²	NY 35 515161	189	2497	...	...	...	...	...	...	...	...	...	...	...	...	...	...	...
Atkinson's Grain	NY 35 457164	663	...	(229)	328	204	163	195	132	77	236	217	183	249	272	2484	...	...
Sealhole Grain	NY 35 470165	541	2053	213	306	180	145	209	128	76	206	254	177	216	241	2352	...	...
Cawdale Beck	NY 35 478176	360	1880	158	282	175	143	139	94	71	198	180	165	193	(207)	2006	...	...
Stonegarth...	NY 35 497177	278	1634	142	241	120	126	117	100	63	172	180	141	165	191	1756	...	...
Lowther (Eamont)																		
Kidmoor	NY 35 482197	318	1810	149	306	127	124	131	102	71	151	154	163	185	203	1865	...	...
Askham Hall	NY 35 515239	186	1052	63	150	57	72	59	82	46	118	107	118	103	116	1091	31	3 Oct
Eamont Bridge R.F.M.Sta. Area 159 km²	NY 35 525285	113	1890	...	...	...	...	...	...	...	...	...	...	...	...	...	...	...
Eamont (Eden)																		
Udford R.F.M.Sta. Area 396 km²	NY 35 575305	91	1882	...	...	...	...	...	...	...	...	...	...	...	...	...	...	...

Station	Grid reference	Altitude m	Av. 1916–1950 mm	Jan	Feb	Mar	Apr	May	Jun	Jul	Aug	Sep	Oct	Nov	Dec	Year	Highest recorded fall in a day mm	date
CUMBERLAND—*(contd)*																		
Eden																		
Nunwick Hall	NY 35 553358	88	814	34	73	34	48	50	74	45	107	112	93	67	86	824	24	12 Sep
Blencarn	NY 35 637312	170	...	36	72	37	49	53	89	51	105	114	107	67	91	871	31	12 Sep
Great Salkeld	NY 35 551365	101	814	30	71	32	45	53	77	47	109	113	97	67	84	825	24	12 Sep
Renwick	NY 35 598433	217	983	36	81	44	57	61	98	63	120	116	112	78	104	970	38	13 Aug
Irthing (Eden)																		
Spadeadam	NY 35 599720	274	...	41	90	74	68	94	137	71	123	147	96	121	150	1211	32	19 Dec
Gelt (Irthing)																		
·New Water	NY 35 580513	299	1024	38	80	50	57	65	127	96	(152)	(152)	(127)	91	125	1161	...	...
Geltsdale	NY 35 575537	229	1071	40	84	56	60	77	128	78	154	151	113	93	141	1175	62	13 Aug
Old Water...	NY 35 590533	302	1029	42	122	52	57	65	125	83	170	152	131	103	126	1230	...	...
Castle Carrock...	NY 35 545551	146	...	33	79	44	51	58	97	69	133	124	102	68	113	971	48	13 Aug
Eden																		
Petteril (Eden)																		
Newton Rigg	NY 35 493310	171	887	45	92	48	63	46	88	45	116	112	109	90	109	963	33	13 Aug
Broadfield House	NY 35 440443	155	...	27	57	40	49	45	108	59	119	91	91	61	79	827	39	13 Aug
Caldew (Eden)																		
Skelton	NY 35 426378	195	...	36	93	66	61	58	96	40	132	108	101	82	111	983	41	13 Aug
Carlisle Cemetery	NY 35 390544	35	816	21	51	39	44	45	118	63	105	110	72	61	88	816	35	11 Sep
Eden																		
Carlisle S.Wks.	NY 35 388566	12	795	25	57	45	46	56	123	56	95	120	71	70	95	859	45	11 Sep
Carlisle	NY 35 384603	26	...	38	70	41	54	52	140	44	111	113	79	66	97	906	47	11 Sep

ISLE OF MAN, 572 SQUARE KILOMETRES

HYDROMETRIC AREA 103

Station	Grid reference	Altitude m	Av. 1916–1950 mm	Jan	Feb	Mar	Apr	May	Jun	Jul	Aug	Sep	Oct	Nov	Dec	Year	Highest recorded fall in a day mm	date
Coastal																		
Block Eary Dam	SC 24 388900	189	1600	113	358	114	169	169	104	41	152	131	155	168	206	1880	...	...
Ballure Res.	SC 24 455930	75	1300	97	299	93	149	144	93	31	136	111	136	135	195	1619	52	21 Apr
Corrany Weir	DC 24 439894	168	...	(116)	327	114	155	149	115	45	153	142	159	168	260	1903	...	...
Clypse Res.West	SC 24 402807	137	1295	90	203	95	109	123	116	31	101	117	123	120	182	1409	49	13 Aug
West Baldwin Res....	SC 24 358831	145	1521	109	279	104	171	143	126	41	127	129	168	171	222	1790	69	13 Aug
Douglas	SC 24 382776	87	1147	83	183	81	119	118	98	24	90	118	90	103	159	1265	38	3 Sep
Ronaldsway	SC 24 279688	17	871	62	137	54	108	83	85	21	80	87	77	61	127	983	33	13 Aug
Cringle Res.	SC 24 253743	152	1217	95	207	79	160	119	100	45	84	127	133	84	184	1418	...	...
Ballagawne Res.	SC 24 237715	111	1008	87	181	51	139	103	89	29	74	103	100	63	133	1152	41	7 Feb
Port St.Mary	SC 24 209671	8	...	70	142	56	115	85	75	19	68	92	92	55	119	988	28	3 Sep
Point of Ayre	NX 25 467043	9	786	42	122	51	84	69	64	33	102	73	67	84	107	899	54	13 Aug

Scotland

SOLWAY RIVER PURIFICATION BOARD AREA, 7404 SQUARE KILOMETRES

HYDROMETRIC AREAS 77 TO 81

Station	Grid reference	Altitude m	Av. 1916–1950 mm	Rainfall in millimetres Jan	Feb	Mar	Apr	May	Jun	Jul	Aug	Sep	Oct	Nov	Dec	Year	Highest recorded fall in a day mm	date
Esk																		
Eskdalemuir	NT 36 235026	242	1582	109	201	122	84	121	112	55	122	182	117	109	214	1546	44	3 Sep
Black Esk (Esk)																		
Cairn Knowe	NY 35 199971	323	1633	81	157	116	87	111	142	46	118	187	106	109	207	1467	...	...
Sandyford	NY 35 205939	204	...	84	172	93	79	96	124	52	103	178	88	77	181	1326	38	11 Sep
Fauldbrae	NY 35 230959	283	...	95	174	115	90	108	152	53	97	185	109	108	197	1484	...	...
Esk																		
Megdale	NY 35 300954	166	1402	94	151	79	131	90	108	53	159	165	151	115	166	1462	52	13 Aug
Irvine House	NY 35 374806	61	1259	80	130	84	98	103	117	63	126	177	94	105	162	1339	41	3 Sep
Canonbie R.F.M. Sta. Area 495 km²	NY 35 397751	23	1550	...	...	...	...	...	...	...	...	...	...	...	...	...	...	...
Liddel (Esk)																		
Newcastleton Forest	NY 35 504877	191	1326	105	156	85	116	122	166	67	159	173	97	120	169	1534	60	3 Sep
Esk																		
Netherby R.F.M. Sta. Area 842 km²	NY 35 390718	15	1502	...	...	...	...	...	...	...	...	...	...	...	...	...	...	...
Lyne (Esk)																		
Ashycroft	NY 35 567769	226	1158	65	100	72	83	92	113	78	144	155	88	139	158	1285	22	22 May
Esk																		
Longtown	NY 35 347673	15	853	41	78	47	72	66	108	54	103	155	73	56	101	955	31	11 Sep
Coastal																		
Solway Bank	NY 35 308774	139	1274	66	129	80	104	113	105	61	129	180	87	84	141	1278	44	3 Sep
Kirtle																		
Winterhope	NY 35 273823	174	1422	86	163	94	119	124	135	63	157	172	97	97	179	1485	41	4 Feb
Springkell	NY 35 255751	91	1150	69	125	69	99	107	130	51	129	157	74	70	137	1218	42	3 Sep
Coastal																		
Annan	NY 35 193651	11	...	55	105	49	67	82	103	49	82	137	79	69	101	977	40	4 Sep
Annan																		
Corehead	NT 36 073124	197	...	98	134	90	75	102	86	39	117	161	100	125	195	1325	38	13 Aug
Moffat, St. Margarets'	NT 36 068038	141	1459	89	187	111	82	113	80	45	117	175	90	114	208	1412	40	4 Feb
Moffat (Annan)																		
Polmoody	NT 36 166121	183	...	118	191	123	117	148	123	61	152	202	138	148	241	1761	48	13 Aug
Kinnel (Annan)																		
Ae (Kinnel)																		
Cran Loch	NY 35 005938	335	1561	90	196	98	123	136	69	51	211	193	125	107	188	1585	...	...
Ae	NX 25 984895	122	--	82	168	87	103	129	66	48	138	183	116	101	183	1405	65	13 Aug
Elshieshields R.F.M. Sta. Area 143 km²	NY 35 068852	55	1494	...	...	...	...	...	...	...	...	...	...	...	...	...	...	...
Annan																		
Lochmaben Hospital	NY 35 074827	65	...	61	115	67	69	102	63	36	110	158	93	75	135	1085	47	13 Aug
Kirkwood	NY 35 120752	79	1129	58	110	63	75	89	82	42	130	152	81	75	101	1057	...	...
Water of Milk (Annan)																		
Coastal																		
Lochar Water																		
Coastal																		
Nith																		
Bank Colliery	NS 26 595114	226	1341	74	141	138	65	110	100	33	138	124	131	147	254	1457	64	13 Aug
Afton (Nith)																		
Afton Water	NS 26 637024	462	2300	163	202	206	112	171	149	98	183	222	209	217	422	2353	...	...
Alwhat Burn No.2	NS 26 642029	437	2242	174	236	188	119	175	140	86	177	206	209	221	400	2332	...	...
Montraw	NS 26 639035	418	2120	157	203	177	114	165	114	72	160	173	208	220	353	2117	...	...
Montraw Burn	NS 26 649032	474	...	179	196	199	117	168	137	90	175	206	207	218	382	2275	...	...
Green Burn	NS 26 633035	431	2320	156	197	191	104	169	132	86	170	187	206	213	432	2243	...	...
Afton Res. R.F.M. Sta. Area 8.5 km²	NS 26 632047	386	2334	...	...	...	...	...	...	...	...	...	...	...	...	...	...	...
Glen Afton Hospital	NS 26 616101	254	1459	87	164	126	88	121	88	40	157	125	136	179	267	1578	...	...
Nith																		
Hall Bridge R.F.M. Sta. Area 155 km²	NS 26 684129	173	1692	...	...	...	...	...	...	...	...	...	...	...	...	...	...	...
Kirkconnel	NS 26 735123	162	1259	63	136	104	79	84	79	47	110	125	101	150	207	1284	55	13 Aug
Euchanhead	NS 26 683058	418	2050	186	211	206	111	175	138	81	180	172	179	223	353	2216	...	...
Poltallan Burn	NS 26 692075	472	1866	122	152	144	91	133	105	74	185	173	139	166	276	1763	...	...
Benzien Craig	NS 26 723067	302	1695	125	174	145	97	132	88	53	175	143	156	149	279	1715	...	...
Euchan Filters	NS 26 729070	288	1645	110	175	126	102	125	91	53	178	150	132	164	248	1653	78	13 Aug
Eliock	NS 26 797074	149	1247	78	144	89	72	108	104	39	178	134	128	104	218	1396	102	13 Aug
Drumlanrig Castle	NX 25 852992	58	...	97	129	86	77	106	95	43	132	176	106	113	172	1332	63	1 Sep
Kettleton Filters	NX 25 901995	195	...	89	111	71	92	98	91	54	143	168	141	111	158	1326	78	13 Aug
Scar (Nith)																		
Appin	NX 25 753973	219	...	102	163	105	97	120	101	35	161	166	145	113	214	1522	...	...
Capenoch R.F.M.Sta. Area 142 km²	NX 25 845940	48	1700	...	...	...	...	...	...	...	...	...	...	...	...	...	...	...
Nith																		
Blackwood	NX 25 907870	30	1176	82	137	63	84	86	65	34	94	127	104	84	151	1111	38	13 Aug
Friar's Carse R.F.M. Sta. Area 799 km²	NX 25 923851	20	1598	...	...	...	...	...	...	...	...	...	...	...	...	...	...	...
Cluden (Nith)																		
Dalmacallan	NX 25 799890	146	...	99	174	95	109	111	89	39	154	154	138	117	203	1483	81	13 Aug
Maxwelton House	NX 25 821896	107	...	79	153	82	93	91	69	32	153	127	115	106	163	1263	76	13 Aug
Glenkiln Res.	NX 25 849776	137	...	100	182	102	124	134	91	44	131	151	122	110	195	1485	63	13 Aug
Fiddlers Ford R.F.M.Sta. Area 238 km²	NX 25 928795	23	1407	...	...	...	...	...	...	...	...	...	...	...	...	...	...	...
Nith																		
Dumfries	NX 25 982747	49	1087	63	119	54	86	85	54	36	104	131	93	87	109	1019	40	13 Aug
Lochrutton W. Wks.	NX 25 901743	83	1258	85	125	77	104	101	77	44	121	127	102	98	145	1205	60	13 Aug

Station	Grid reference	Altitude m	Av. 1916–1950 mm	Jan	Feb	Mar	Apr	May	Jun	Jul	Aug	Sep	Oct	Nov	Dec	Year	Highest recorded fall in a day mm	date
SOLWAY—*(contd)*																		
Glencaple	NX 25 995682	9	...	65	146	47	112	89	56	32	104	132	79	73	109	1043	...	...
Coastal																		
Fairgirth	NX 25 879566	43	...	77	142	78	99	101	87	44	107	140	92	115	153	1235	...	...
Urr																		
Corsock	NX 25 758755	154	...	106	195	83	103	125	82	35	112	143	116	118	182	1398	46	13 Aug
Chipperkyle	NX 25 784688	94	1273	81	145	73	79	102	81	29	94	133	90	113	149	1169	37	13 Aug
Corbieton House	NX 25 796658	53	1196	74	145	62	85	98	78	27	100	125	84	113	133	1123	...	...
Dalbeattie R.F.M. Sta. Area 199 km²	NX 25 822610	4	–	...	...	...	...	...	...	...	...	...	...	...	...	...	...	...
Coastal																		
St. Mary's Isle	NX 25 674494	12	...	100	136	78	85	88	95	38	86	120	95	105	144	1167	45	13 Aug
Dee																		
Loch Dee	NX 25 479792	244	2219	155	210	258	117	169	134	61	170	202	202	224	396	2298	...	...
Craigencallie	NX 25 504779	226	2089	141	235	226	119	166	140	75	153	240	190	245	377	2306	...	...
Clatteringshaws	NX 25 554780	178	2022	149	234	163	142	158	113	47	169	181	191	212	333	2091	82	13 Aug
Pullaugh (Dee)																		
Loch Grennoch No. 1	NX 25 542681	218	...	166	234	178	118	194	122	65	165	311	194	197	362	2306	...	...
Loch Grennoch No. 2	NX 25 514692	480	...	251	264	188	133	234	191	122	174	330	197	253	394	2731	...	...
Loch Grennoch No. 3	NX 25 537698	213	...	207	239	183	118	194	132	62	152	283	184	191	352	2297	...	...
Loch Grennoch No. 4	NX 25 531712	315	...	187	277	175	127	206	127	66	171	187	187	213	394	2317	...	...
Loch Grennoch No. 5	NX 25 548718	213	...	177	256	164	147	182	120	59	174	193	190	218	393	2274	...	...
Loch Grennoch No. 6	NX 25 543736	216	...	187	270	158	145	181	111	65	173	168	204	197	427	2285	...	...
Pullaugh Burn	NX 25 544741	183	...	182	260	170	141	180	117	70	175	183	193	213	397	2279	83	13 Aug
Dee																		
Laggan of Dee	NX 25 572745	168	1813	144	199	156	117	156	111	53	175	171	161	171	319	1933	...	...
White Burn	NX 25 564713	399	...	211	254	180	157	188	146	63	183	208	191	191	362	2334	...	...
Ken (Dee)																		
Cornharrow	NX 25 667928	247	1509	121	158	109	111	117	110	44	138	172	171	171	244	1666	...	...
Fingland	NX 25 670902	287	...	110	152	101	107	138	100	38	128	171	174	165	248	1631	...	...
Kendoon	NX 25 615884	158	1371	109	171	103	97	103	76	26	116	173	139	134	230	1475	52	13 Aug
Deugh (Ken)																		
Moor	NS 26 570036	241	1721	104	179	123	90	171	121	65	166	135	161	203	340	1857	...	...
Knockengarroch	NX 25 556971	226	1481	122	148	112	103	105	113	34	109	160	173	159	326	1665	...	...
Drumjohn	NX 25 524976	227	1530	93	181	104	94	131	107	30	178	141	176	144	271	1651	96	13 Aug
Carsphairn	NX 25 561934	180	1518	117	150	113	79	113	104	23	111	143	168	145	214	1482	...	...
Shiel of Castlemaddy	NX 25 537901	274	...	193	226	170	111	185	112	35	145	188	185	185	335	2070	...	...
Dundeugh	NX 25 598879	119	1500	109	168	116	92	108	91	23	119	167	132	147	257	1530	51	13 Aug
Ken (Dee)																		
Forrest Lodge	NX 25 555866	152	1893	131	201	166	95	133	103	36	129	179	160	212	320	1865	53	3 Sep
Creaganfois	NX 25 629817	125	1287	90	144	93	87	106	79	33	104	140	128	115	195	1313	52	13 Aug
Glenlee	NX 25 606805	55	1526	113	180	122	101	127	80	34	119	149	146	145	261	1577	51	13 Aug
Dee																		
Glenlochar Barrage	NX 25 732646	47	1217	87	153	84	86	97	81	42	106	143	84	111	167	1240	51	13 Aug
Threave	NX 25 751607	73	1282	97	157	86	98	108	81	45	117	125	86	109	153	1261	56	13 Aug
Kirkcudbright	NX 25 683509	12	...	99	132	71	81	79	91	41	110	110	88	91	134	1125	56	13 Aug
Coastal																		
Roberton	NX 25 600486	15	1031	80	121	63	66	86	80	41	96	113	76	99	141	1062	49	13 Aug
Water of Fleet																		
Loch Fleet	NX 25 556698	358	...	180	218	197	114	193	145	61	169	197	152	203	381	2211	...	...
Cally Market Gardens	NX 25 604550	32	...	90	181	100	73	117	99	47	125	117	94	129	176	1347	...	...
Cree																		
Bargrennan	NX 25 361789	110	1511	109	159	125	77	120	86	59	111	185	159	170	295	1657	50	13 Aug
Penninghame Prison	NX 25 382699	46	...	83	139	118	71	119	74	49	132	166	111	134	250	1445	64	13 Aug
Minnigaff	NX 25 401674	50	...	85	115	84	63	104	73	39	105	141	103	105	205	1223	47	13 Aug
Newton Stewart	NX 25 409657	26	...	80	114	87	56	92	84	39	104	128	97	107	200	1187	46	13 Aug
Newton Stewart R.F.M.Sta. Area 370 km²	NX 25 412653	5	–	...	...	...	...	...	...	...	...	...	...	...	...	...	...	...
Palnure	NX 25 452642	15	1321	93	137	95	77	88	81	43	110	139	104	118	205	1289	59	13 Aug
Bladnoch																		
Fordbank	NX 25 425548	15	...	95	129	89	65	106	77	45	102	124	97	107	201	1238	46	13 Aug
Coastal																		
Culderry	NX 25 476467	8	1062	68	113	76	56	92	65	30	89	95	82	102	162	1030	44	13 Aug
Galloway House	NX 25 474455	6	1085	79	131	79	67	92	65	25	101	103	90	98	167	1096	50	13 Aug
Monreith	NX 25 361433	44	1042	94	127	82	77	87	55	29	90	104	110	101	189	1145	...	...
Luce																		
Lagafater Lodge	NX 25 139761	244	1631	103	153	131	79	129	102	77	113	116	179	195	313	1689	34	21 Apr 9 Sep
Penwhirn (Luce)																		
Penwhirn	NX 25 127693	166	1547	95	163	147	102	151	97	66	108	159	150	159	278	1674	41	1 Mar
Penwhirn Res. R.F.M.Sta. Area 18.2 km²	NX 25 128694	151	1567	...	...	...	...	...	...	...	...	...	...	...	...	...	...	...
Luce																		
Camrie Farm	NX 25 201600	107	1283	78	121	101	72	135	79	38	91	135	133	128	246	1356	45	13 Aug
Piltanton																		
Coastal																		
West Freugh	NX 25 109546	15	1015	55	101	81	63	103	59	33	79	105	117	99	174	1070	29	3 Oct
Stranraer Filters	NX 25 036619	64	1275	62	124	106	78	126	61	42	108	122	124	125	256	1334	34	3 Oct
Castle Kennedy	NX 25 112609	21	1196	67	114	103	77	116	70	43	111	116	143	137	207	1306	33	1 Sep

AYRSHIRE RIVER PURIFICATION BOARD AREA, 2601 SQUARE KILOMETRES

HYDROMETRIC AREAS 82 AND 83

Station	Grid reference	Altitude m	Av. 1916–1950 mm	Jan	Feb	Mar	Apr	May	Jun	Jul	Aug	Sep	Oct	Nov	Dec	Year	Highest recorded fall in a day mm	date
Stinchar																		
Barr	NX 25 286941	122	...	81	134	127	62	123	123	54	138	119	155	179	282	1577	62	13 Aug
Ballantrae	NX 25 084825	12	1111	66	117	85	54	80	60	42	88	86	105	145	179	1107	...	...
Girvan																		
Loch Bradan	NX 25 434978	296	1758	134	132	148	90	117	122	49	152	156	146	166	340	1751	...	...
Loch Bradan, New site	NX 25 436984	311	...	122	124	142	85	100	117	46	149	139	134	155	315	1627	...	...
Kirkbride House	NS 26 334047	114	...	56	91	125	33	99	132	43	98	114	127	147	244	1309	55	13 Aug
Penwhapple Res.	NX 25 257969	202	1377	70	101	103	53	101	111	53	113	119	146	147	225	1341	55	13 Aug
Robston R.F.M.Sta. Area 246 km²	NX 25 217997	9	1435	...	...	...	...	...	...	...	...	...	...	...	...	...	...	...

GENERAL TABLE, SCOTLAND

Station	Grid reference	Altitude m	Av. 1916–1950 mm	Rainfall in millimetres Jan	Feb	Mar	Apr	May	Jun	Jul	Aug	Sep	Oct	Nov	Dec	Year	Highest recorded fall in a day mm	date
AYRSHIRE—*(contd)*																		
Girvan	NX 25 184979	8	...	46	72	87	30	73	83	39	102	83	95	123	165	997	46	13 Aug
Coastal																		
Culzean Castle	NS 26 231098	42	1101	50	75	97	26	82	106	40	100	100	88	141	178	1085	...	...
Doon																		
Starr	NX 25 481936	244	1628	86	172	160	72	111	93	27	142	151	122	187	321	1645	...	...
Loch Recawr	NX 25 441938	284	1907	132	132	206	112	127	123	47	168	152	175	210	351	1934	...	...
Loch Finlas	NX 25 469981	256	1565	101	156	148	81	90	127	31	184	149	152	149	310	1678	...	...
Garpel Burn	NX 25 482985	216	1462	89	138	119	80	98	110	31	161	141	145	141	249	1503	84	13 Aug
Ardoon	NS 26 441086	209	1286	51	76	103	32	78	113	47	147	137	121	153	230	1288	...	...
Kerse Filters	NS 26 425131	226	1415	57	89	108	42	93	147	53	143	139	105	170	235	1381	65	13 Aug
Knockjarder W. Wks.	NS 26 353150	84	...	46	74	93	31	70	159	37	116	117	96	152	187	1178	...	...
Otterden House	NS 26 307146	107	1150	51	77	94	31	81	140	42	112	115	90	137	179	1146	53	13 Aug
Doonholm	NS 26 338175	27	1124	49	80	88	32	69	169	37	104	102	98	144	177	1149	56	13 Aug
Alloway, Doonbrae	NS 26 331180	23	...	44	75	79	33	66	154	34	99	95	92	124	146	1041	49	13 Aug
Ayr																		
Kaimeshill	NS 26 696266	229	1221	56	107	112	37	89	126	55	141	129	97	131	198	1277	...	...
Blackside Water Tank	NS 26 575285	256	...	49	76	109	32	74	137	61	132	129	77	129	191	1197	...	...
Lugar (Ayr)																		
Cumnock	NS 26 558204	104	1089	41	102	87	23	80	114	39	114	112	81	114	183	1088	63	13 Aug
Ayr																		
Hillhead Water Tank	NS 26 494283	172	1134	43	79	97	30	75	122	49	118	118	71	123	172	1099	54	13 Aug
Auchincruive	NS 26 389236	45	1002	39	70	70	26	65	141	36	113	100	78	111	142	990	64	13 Aug
Ayr Cemetery	NS 26 348211	12	912	41	71	67	32	58	137	29	101	85	79	94	128	921	56	13 Aug
Coastal																		
Prestwick	NS 26 369261	16	938	46	76	72	33	63	124	34	117	103	81	105	149	1006	67	13 Aug
Prestwick, Sherwood Road	NS 26 363258	15	967	45	69	57	37	58	127	28	114	94	74	99	138	940	...	...
Collenan Res.	NS 26 349330	46	...	59	96	75	55	79	116	51	106	100	68	102	161	1069	51	13 Aug
Gailes	NS 26 323354	4	990	59	107	64	59	74	103	39	105	101	79	103	163	1056	44	13 Aug
Irvine																		
Darvel, Burgh Yard	NS 26 563376	106	1250	64	105	118	41	91	150	61	129	131	82	156	208	1337	58	13 Aug
Lanfine	NS 26 547364	152	1294	65	103	115	42	89	132	54	123	127	77	149	202	1279	...	...
Whiteleehill	NS 26 546428	279	1397	65	102	125	42	92	152	74	141	146	75	170	214	1398	...	...
Carnell	NS 26 467322	76	...	55	88	94	34	77	126	42	111	116	63	122	183	1110	...	...
Fenwick (Irvine)																		
Moor Farm	NS 26 510479	256	1468	85	164	142	68	106	157	62	155	167	86	166	242	1600	...	...
Raithill	NS 26 482468	195	1397	77	128	138	68	119	131	67	155	181	73	167	247	1552	...	...
Amlaird Filters	NS 26 483444	180	1284	68	111	115	68	100	126	59	138	155	75	150	201	1366	60	13 Aug
North Craig Res.	NS 26 438412	97	1182	64	98	101	57	99	113	49	106	140	78	129	197	1232	47	13 Aug
Lochgoin Res.	NS 26 534474	264	1512	86	134	165	70	102	173	63	126	161	82	185	295	1644	...	...
Craigendunton Res.	NS 26 522455	245	1431	80	132	164	63	108	165	68	142	174	77	178	269	1620	...	...
Croilburn	NS 26 537440	221	1312	69	115	119	56	101	144	62	133	149	76	162	270	1455	...	...
Kilmarnock	NS 26 414378	34	...	57	97	87	42	69	115	40	100	108	67	125	171	1079	37	9 Sep
Annick (Irvine)																		
Stewarton Filters	NS 26 439478	162	...	69	111	137	70	101	99	52	130	159	83	143	248	1402	(58	17 Dec)
Long Loch Res. South	NS 26 470518	235	...	95	142	143	86	107	140	62	150	182	98	184	248	1637	...	...
Dunlop	NS 26 412489	146	1259	81	137	134	78	107	96	51	117	150	85	147	249	1430	61	17 Dec
Garnock (Irvine)																		
Pundeavon Res.	NS 26 294577	276	...	178	205	237	136	164	152	92	126	255	147	214	369	2275	...	...
Hole House Filters	NS 26 309560	115	1785	120	195	216	106	132	127	75	129	205	123	186	302	1915	57	3 Sep
Beith Filters	NS 26 371538	126	1452	87	139	117	83	104	95	47	109	161	94	151	247	1435	66	17 Dec
Rye (Garnock)																		
Rye Water	NS 26 253585	315	1850	152	211	165	140	165	109	58	104	191	133	137	292	1858	...	...
Blairpark	NS 26 248574	220	1733	154	207	147	135	152	140	71	112	160	132	140	290	1839	...	...
Catburn	NS 26 255559	309	1810	157	231	170	135	164	160	79	125	211	147	135	338	2051	...	...
Whitehill	NS 26 263563	191	1785	155	203	180	125	157	152	81	119	188	142	175	320	1999	...	...
Camphill Res.	NS 26 275549	186	1792	120	197	181	116	134	136	78	127	198	131	152	284	1853	...	...
Haining	NS 26 264545	273	1676	135	189	183	112	155	145	74	107	234	135	163	361	1990	...	...
Garnock (Irvine)																		
Thirdpart	NS 26 264506	171	...	80	165	134	93	123	100	81	108	182	117	146	217	1546	...	...
Dalry Filters	NS 26 277501	98	1405	95	158	141	95	120	117	79	98	185	103	153	257	1602	57	17 Dec
Dalry R.F.M.Sta. Area 88.8 km²	NS 26 293488	19	1742	...	...	...	...	...	...	...	...	...	...	...	...	...	...	...
Caaf (Garnock)																		
Bradshaw	NS 26 239539	323	1500	105	155	114	132	133	122	83	116	168	137	136	246	1647	...	...
Knockendon	NS 26 238526	229	1528	99	175	117	125	127	131	75	105	175	133	132	243	1637	...	...
South Bradshaw	NS 26 247527	284	...	104	185	121	138	129	125	79	117	174	138	136	246	1693	...	...
South Knockendon	NS 26 246519	216	...	109	180	123	141	133	132	86	118	174	135	141	248	1721	...	...
Knockendon Res. R.F.M.Sta. Area 6.03 km²	NS 26 246514	208	1643	...	...	...	...	...	...	...	...	...	...	...	...	...	...	...
Drum Burn	NS 26 237513	208	...	90	175	109	116	114	123	67	105	171	131	133	243	1579	...	...
Birkheadsteel	NS 26 245503	158	1524	86	160	116	98	117	117	72	101	184	112	129	241	1533	46	9 Sep
Caaf Res.	NS 26 255500	168	...	86	160	116	98	117	117	73	101	184	111	129	240	1531	46	9 Sep
Munnoch Res.	NS 26 254481	99	1332	78	143	106	92	111	107	69	95	177	108	136	218	1441	49	9 Sep
Garnock (Irvine)																		
Blair House	NS 26 302481	47	1296	83	152	123	84	105	94	65	101	162	97	148	226	1439	46	9 Sep
Lugton (Garnock)																		
Montgreenan	NS 26 344443	55	1202	64	112	95	74	99	97	60	118	161	88	132	202	1301	52	9 Sep
Garnock (Irvine)																		
Ravenspark Hospital	NS 26 309407	7	1038	63	115	87	79	105	99	50	111	124	89	121	175	1220	46	13 Aug
Coastal																		
Greenhead Filters	NS 26 273437	56	1090	70	119	73	80	95	101	53	85	125	77	118	188	1184	38	13 Aug
Ardrossan Filters	NS 26 235444	51	1106	67	115	65	95	78	88	44	87	124	92	105	169	1127	41	13 Aug
Busbie Res.	NS 26 240463	113	1210	72	143	71	93	97	86	47	99	147	102	128	192	1278	...	...
Crosbie Res.	NS 26 218507	114	...	87	129	82	101	89	99	45	84	159	106	104	196	1281	49	9 Sep
Hunterston	NS 26 183514	8	...	104	116	79	134	75	95	39	65	136	100	111	160	1214	63	5 Apr
Fairlie W. Wks.	NS 26 214555	82	1351	129	163	117	137	104	109	54	89	146	114	109	248	1519	...	...
Largs	NS 26 204599	6	1276	105	126	120	121	86	116	46	65	125	101	127	185	1324	51	5 Apr
Middleton Filters	NS 26 215624	70	1325	116	118	125	115	104	121	49	81	121	101	127	202	1380	55	25 Jan

Station	Grid reference	Altitude m	Av. 1916–1950 mm	Rainfall in millimetres Jan	Feb	Mar	Apr	May	Jun	Jul	Aug	Sep	Oct	Nov	Dec	Year	Highest recorded fall in a day mm	date

CLYDE RIVER PURIFICATION BOARD AREA, 5565 SQUARE KILOMETRES

HYDROMETRIC AREAS 84 TO 87

Station	Grid reference	Altitude m	Av. 1916–1950 mm	Jan	Feb	Mar	Apr	May	Jun	Jul	Aug	Sep	Oct	Nov	Dec	Year	Highest mm	date
Coastal																		
Largs, Woodbank	NS 26 199607	8	1240	115	124	120	115	95	118	45	68	126	93	133	184	1335	...	...
Kelly Dam	NS 26 223693	191	...	145	145	168	128	128	146	56	112	174	119	160	269	1750	...	...
Kip																		
Knockencorsan	NS 26 246676	317	...	165	184	185	150	140	109	69	109	191	118	194	297	1911	...	...
Largs Road	NS 26 249685	277	1798	168	216	193	152	127	132	67	114	175	119	188	287	1939	...	...
Loch Thom	NS 26 256725	196	1788	146	180	180	117	121	148	58	102	175	116	175	293	1811	(44	19 Dec)
Compensation Res.	NS 26 248725	183	...	137	165	159	113	112	142	49	91	159	111	160	266	1664	43	19 Dec
Shielhill	NS 26 237727	278	1837	143	157	152	132	123	143	56	108	163	122	164	262	1726	...	...
Larkfield Filters	NS 26 245761	103	1590	113	161	155	100	104	146	49	82	162	110	154	246	1581	40	17 Dec
Spango Burn	NS 26 243736	231	1802	147	175	165	129	119	159	55	102	179	116	187	279	1812	...	...
Daff Burn Res.	NS 26 228712	147	...	153	159	166	147	133	138	56	87	157	120	157	271	1744	...	...
Coastal																		
Greenock, Newton Street	NS 26 264771	53	...	100	150	132	80	100	129	29	75	131	109	141	217	1392	39	17 Dec
Hole Glen	NS 26 259745	189	1768	121	183	152	104	114	145	52	93	158	112	172	268	1674	54	17 Dec
Greenock	NS 26 274757	61	1588	112	174	156	84	100	130	34	85	143	108	154	242	1521	57	17 Dec
New Yetts Dam	NS 26 264738	241	1708	125	188	152	106	119	151	53	98	164	118	175	266	1716	53	17 Dec
Dougliehill Filters	NS 26 317738	169	...	116	172	109	68	49	158	24	98	111	108	139	249	1401	55	17 Dec
Clyde																		
Thick Cleuch	NS 26 952021	457	2093	152	237	159	143	156	121	95	159	246	169	136	273	2047	...	...
Daerhead	NS 26 964030	397	2166	159	230	177	135	179	113	75	191	213	160	150	271	2051	...	...
Lower Carsehope	NS 26 963048	373	2054	154	257	170	142	163	104	74	180	213	155	142	265	2019	...	...
Upper Carsehope	NS 26 939040	518	2125	154	213	143	133	178	122	95	159	236	156	125	271	1985	...	...
New Kirkhope	NS 26 963055	357	...	170	232	177	135	175	98	79	182	204	158	156	266	2034	95	13 Aug
New Crookburn	NS 26 966056	354	2141	151	224	169	141	169	104	77	180	203	152	144	260	1972	91	13 Aug
Whiteholm	NS 26 984036	427	2106	165	285	184	145	183	118	76	221	211	146	151	292	2177	...	...
Lamb Sike	NS 26 977043	408	2044	163	254	177	145	171	108	75	197	217	145	143	307	2101	...	...
Kirkhope Cleuch	NS 26 954069	427	2133	164	230	165	133	182	121	86	171	234	146	155	293	2080	...	...
White Burn	NS 26 997060	381	2017	160	276	136	128	152	95	66	213	182	112	127	263	1910	...	...
Garls Craig	NS 26 976092	300	1812	140	197	140	111	138	87	65	167	170	125	150	241	1731	83	13 Aug
Hitteril	NS 26 963099	297	1728	122	203	129	107	109	85	62	165	191	126	133	243	1675	...	...
Potrenick Burn	NS 26 934101	306	1707	119	184	143	97	136	100	75	187	186	111	149	241	1730	...	...
Peden Res.	NS 26 942123	294	1711	114	171	131	92	126	96	70	169	154	123	143	242	1631	...	...
Camps (Clyde)																		
Risingclaw	NT 36 036200	480	1637	112	229	94	86	102	135	61	140	211	122	140	216	1646	...	...
Peat Rig	NT 36 013209	427	1578	91	191	89	97	89	122	51	127	140	127	121	211	1454	...	...
Whitecamp	NT 36 029214	358	1388	90	171	77	63	89	127	51	140	135	122	127	178	1370	...	...
Kneesend	NT 36 013225	366	1343	81	108	76	60	84	128	53	140	156	112	121	173	1291	...	...
Hillshaw	NT 36 042252	549	1472	83	152	79	52	79	135	58	198	147	114	160	229	1486	...	...
Linnburn	NT 36 022253	381	1353	77	132	74	51	84	125	56	157	151	114	141	188	1350	...	...
Camps Res.	NS 26 997226	295	1272	78	103	74	51	79	104	48	140	127	108	116	160	1187	66	13 Aug
Clyde																		
Leadhills	NS 26 888153	387	1660	105	153	104	100	155	118	81	176	159	121	129	213	1615	85	13 Aug
Abington	NS 26 932230	247	1225	61	109	76	45	79	90	43	143	118	100	116	159	1139	68	13 Aug
Duneaton (Clyde)																		
Cairn Table	NS 26 741234	381	1548	79	141	125	69	106	155	85	160	115	107	160	240	1541	...	...
Auchandaff	NS 26 782236	358	1482	74	134	123	62	104	128	78	152	105	103	150	225	1438	...	...
Glentewing	NS 26 877222	262	1232	67	103	94	54	69	54	39	164	111	91	106	168	1121	...	...
Clyde																		
Coulter Res.	NT 36 036276	347	1353	79	100	79	66	83	127	58	143	127	123	137	163	1284	...	...
Cowgill Res.	NT 36 009288	308	1113	55	99	70	42	68	131	49	147	101	90	91	112	1055	...	...
Huntfield	NT 36 012402	271	894	85	94	28	32	53	125	51	140	112	86	76	79	960	...	...
Easton	NT 36 085493	255	921	39	68	36	33	61	139	47	130	95	82	83	102	913	...	...
Carnwath	NS 26 974464	208	877	36	80	40	38	56	115	52	113	91	61	74	104	860	40	13 Aug
Sills R.F.M. Sta. Area 742 km²	NS 26 927424	183	1339	...	...	...	...	...	...	...	...	...	...	...	...	...	...	...
Loch Lyoch Res.	NS 26 932357	273	1082	56	93	62	36	63	103	48	143	116	87	92	119	1019	57	13 Aug
Douglas (Clyde)																		
Monksfoot	NS 26 787285	238	1271	55	129	111	53	67	101	66	163	124	110	144	208	1331	90	13 Aug
Bonnington Power Sta.	NS 26 884417	119	875	40	80	50	31	49	104	45	119	84	83	78	109	872	46	13 Aug
Lanark	NS 26 875434	152	811	36	74	39	31	42	102	35	104	73	79	63	89	765	45	13 Aug
Mouse (Clyde)																		
Springfield Res.	NS 26 907520	262	...	44	82	59	43	65	159	55	119	102	98	73	107	1007	...	...
Woolfords	NT 36 004567	259	1034	47	104	58	39	60	164	66	102	114	66	102	119	1042	...	...
Clyde																		
Hazelbank R.F.M. Sta. Area 1090 km²	NS 26 835452	52	1262	...	...	...	...	...	...	...	...	...	...	...	...	...	...	...
Coldstream Res.	NS 26 873499	244	...	41	75	43	43	63	125	50	113	94	96	68	104	916	45	13 Aug
Nethan (Clyde)																		
Dunside Res.	NS 26 748372	305	1365	63	114	128	58	89	143	69	154	124	93	135	201	1371	64	13 Aug
Clyde																		
Mauldslie	NS 26 808502	53	882	44	73	47	36	51	135	25	109	82	80	72	107	860	45	13 Aug
Gair Res.	NS 26 853535	200	838	49	69	36	52	58	129	45	108	93	87	61	104	890	45	13 Aug
Avon (Clyde)																		
Glengavel Res.	NS 26 663351	256	1406	65	109	148	46	91	147	67	154	148	85	173	218	1453	81	13 Aug
West Browncastle	NS 26 622432	238	1453	70	124	149	53	95	173	75	157	137	90	187	247	1559	83	13 Aug
Kype Res.	NS 26 735387	287	1359	65	115	124	64	86	137	66	163	129	93	123	202	1369	74	13 Aug
Glassford Filters	NS 26 709478	236	1234	59	99	95	51	70	168	57	124	109	78	122	159	1190	52	13 Aug
Fairholm R.F.M. Sta. Area 258 km²	NS 26 755518	53	1295	...	...	...	...	...	...	...	...	...	...	...	...	...	...	...
South Calder (Clyde)																		
Shotts Res.	NS 26 880613	247	1042	61	115	86	80	83	164	57	121	108	95	115	144	1230	43	13 Aug
Clyde																		
Townhill Filters	NS 26 694546	131	997	49	97	70	51	52	125	46	124	79	80	90	125	989	44	13 Aug
Blairston R.F.M. Sta. Area 1710 km²	NS 26 704579	18	1222	...	...	...	...	...	...	...	...	...	...	...	...	...	...	...
North Calder (Clyde)																		
Roughrigg Res.	NS 26 804639	202	992	53	85	58	62	71	143	54	117	84	85	78	136	1025	...	...
Luggie (North Calder)																		
Coatbridge, Dunbeth Park	NS 26 741655	99	...	40	79	57	43	67	129	28	94	82	78	229	122	1049	91	29 Nov
Airdrie, Broomknoll Street	NS 26 762653	123	933	47	83	61	53	71	130	44	128	85	81	83	133	999	57	13 Aug
Coatbridge, Whifflet Park	NS 26 736638	94	923	39	83	53	49	74	156	44	115	92	86	84	147	1021	51	13 Aug
Coatbridge, West End Park	NS 26 722649	73	...	39	83	51	53	69	144	20	112	91	87	77	130	957	46	13 Aug
Coatbridge	NS 26 712643	78	898	39	77	47	49	63	131	31	113	90	79	70	122	911	49	13 Aug

GENERAL TABLE, SCOTLAND

Station	Grid reference	Altitude m	Av. 1916–1950 mm	Jan	Feb	Mar	Apr	May	Jun	Jul	Aug	Sep	Oct	Nov	Dec	Year	Highest recorded fall in a day mm	Highest recorded fall in a day date
CLYDE—*(contd)*																		
Clyde																		
Greenlees Filters	NS 26 640590	98	...	51	95	65	61	58	129	40	111	95	77	92	111	986	42	13 Aug
Alexandra Park	NS 26 618657	43	995	47	96	65	73	71	149	44	110	91	80	87	131	1044	39	13 Aug
Glasgow Green	NS 26 599645	11	988	49	99	64	71	73	132	43	108	92	76	87	131	1027	(32	4 Sep)
Bellahouston Park	NS 26 553637	24	1036	59	106	83	77	85	162	45	111	107	73	86	150	1144	46	22 Jun
Kelvin (Clyde)																		
Kilsyth, Colzium House	NS 26 728787	93	1353	73	139	140	98	95	146	48	158	142	106	134	201	1478	63	13 Aug
Bridgend R.F.M. Sta. Area 63.7 km²	NS 26 672749	35	1331	...	...	...	...	...	...	...	...	...	...	...	...	...	...	...
Luggie (Kelvin)																		
Condorrat	NS 26 728726	78	...	(63)	97	104	67	77	129	40	128	107	93	101	159	1165	55	13 Aug
Gartloch Hospital	NS 26 680669	89	1014	43	103	74	58	67	139	39	117	103	81	95	142	1062	44	13 Aug
Kelvin (Clyde)																		
Bishopbriggs	NS 26 617700	91	...	54	108	91	73	79	143	38	105	106	92	98	150	1137	36	17 Dec
Allander (Kelvin)																		
Burncrooks Filters	NS 26 501793	216	...	108	148	146	90	129	144	66	112	150	106	120	248	1568	43	17 Dec
Mugdock Res.	NS 26 558755	98	1294	70	126	121	89	99	142	53	93	141	93	115	207	1348	46	17 Dec
Kelvin (Clyde)																		
Killermont R.F.M. Sta. Area 335 km²	NS 26 558705	28	1245	...	...	...	...	...	...	...	...	...	...	...	...	...	...	...
Glasgow, Springburn	NS 26 608686	107	1012	54	104	78	78	78	146	42	109	104	88	91	135	1107	33	13 Aug
Ruchill Park	NS 26 578680	69	973	53	102	71	72	81	138	46	98	109	86	89	141	1087	38	22 Jun
Glasgow, Botanic Gardens	NS 26 568675	40	1018	54	107	70	70	85	139	47	97	99	86	87	140	1081	37	13 Aug
Kelvingrove Park	NS 26 572666	15	...	48	102	67	71	86	131	44	99	91	83	83	132	1037	(36	6 Feb)
Clyde																		
Glasgow, Elder Park	NS 26 543658	8	1056	56	112	80	76	84	146	46	105	103	87	86	149	1132	40	17 Dec
Victoria Park	NS 26 541673	9	986	56	106	72	73	84	131	46	98	99	83	90	137	1076	37	13 Aug
Black Cart (Clyde)																		
Upper Calder	NS 26 265647	408	2400	173	268	217	175	202	174	97	141	234	133	229	337	2379	...	...
Calder Water, Barytes Mine	NS 26 283650	307	2367	174	282	236	158	212	155	117	139	208	140	245	343	2409	51	11 Nov
Calder Dam	NS 26 295653	280	2420	178	258	267	155	205	196	95	146	222	117	260	386	2484	...	...
Queenside Loch	NS 26 294643	384	2479	178	254	298	160	202	159	104	151	236	136	267	381	2526	...	...
Muirshiel R.F.M. Sta. Area 12.4 km²	NS 26 309638	229	2329	...	...	...	...	...	...	...	...	...	...	...	...	...	...	...
Calder Water, Muirshiel Mill	NS 26 310633	238	2150	173	276	219	149	173	123	102	148	215	135	222	326	2260	57	5 Oct
White Grain Burn	NS 26 279633	438	2411	175	254	226	157	207	160	109	151	229	138	216	310	2333	...	...
Kaim Dam	NS 26 346622	188	1999	139	215	207	108	136	127	79	134	216	125	201	329	2015	62	3 Sep
Springside	NS 26 404578	162	1499	95	167	121	104	118	109	57	116	169	105	138	231	1528	...	...
Muirhead	NS 26 390576	147	1500	91	160	115	97	111	101	55	112	162	102	132	221	1460	...	...
Lochwinnoch S.Wks.	NS 26 356588	29	1626	98	161	209	97	114	100	67	125	178	116	145	272	1681	67	17 Dec
Glentyan House	NS 26 395631	91	1561	98	154	155	112	109	143	56	120	175	112	153	268	1655	80	17 Dec
Gryfe (Black Cart)																		
Darndaff Moor	NS 26 276731	259	1725	125	163	165	107	117	163	53	98	147	119	178	262	1695	...	...
Kilmacolm	NS 26 358691	74	1533	127	212	168	90	119	119	40	110	141	98	162	241	1627	53	5 Oct
Creuch Hill	NS 26 267688	411	...	163	221	216	155	196	183	86	145	198	132	196	297	2187	...	...
Green Water	NS 26 288700	240	...	142	201	188	132	142	152	71	129	160	125	203	282	1928	...	...
Kilmacolm W.Wks.	NS 26 317674	183	...	154	252	226	129	159	139	53	137	173	125	218	335	2101	54	17 Dec
Duchal Moor	NS 26 282667	351	2120	185	226	213	183	178	191	84	137	191	150	201	302	2241	...	...
Craigend R.F.M. Sta. Area 71.0 km²	NS 26 415664	10	1729	...	...	...	...	...	...	...	...	...	...	...	...	...	...	...
White Cart (Black Cart)																		
Dunwan Res. No.3	NS 26 552486	247	...	74	126	138	75	83	164	54	138	135	91	174	223	1474	...	...
Dunwan Res. No.4	NS 26 542494	253	...	74	126	138	75	83	163	54	139	135	91	174	222	1474	...	...
Dunwan Res. No.2	NS 26 554496	250	...	73	126	138	75	83	163	54	139	134	91	175	222	1473	...	...
Picketlaw Res. No.1	NS 26 567516	210	1448	72	125	137	75	82	167	53	150	134	90	173	222	1481	56	17 Dec
South Moorhouse	NS 26 529508	249	1653	99	177	185	85	133	180	61	155	191	113	212	269	1861	70	17 Dec
South Moorhouse, New site	NS 26 527509	236	...	88	166	148	80	109	166	55	153	167	111	176	252	1671	63	17 Dec
Thorntonhall	NS 26 593552	138	1171	65	114	90	71	59	129	48	121	104	80	124	159	1163	51	13 Aug
Linn Park	NS 26 583591	50	1066	59	100	68	70	62	140	40	109	103	73	96	131	1050	42	13 Aug
Queens Park	NS 26 581619	45	–	61	89	71	67	75	139	44	106	99	77	81	128	1037	41	13 Aug
Rouken Glen Park	NS 26 547585	47	1113	71	111	90	72	78	143	41	116	108	82	118	157	1187	45	13 Aug
Crossmyloof	NS 26 572627	23	...	55	95	75	72	71	127	37	106	79	73	77	134	1001	42	17 Dec
Pollok House	NS 26 549619	15	1009	63	96	75	71	74	127	41	105	95	75	87	148	1057	42	13 Aug
Levern (White Cart)																		
Neilston Filters	NS 26 475564	171	...	100	132	138	90	91	131	51	139	139	79	149	219	1459	61	17 Dec
Barrhead	NS 26 511601	30	1156	76	119	102	77	73	113	33	106	104	73	117	169	1162	52	17 Dec
Waulk Glen	NS 26 523580	91	1258	69	131	113	62	79	111	53	117	120	81	103	122	1161	44	3 Sep
White Cart (Black Cart)																		
Thornly Park School	NS 26 487613	59	1305	79	118	88	84	91	113	39	110	115	82	127	201	1249	...	...
Hawkhead R.F.M. Sta. Area 235 km²	NS 26 499629	5	1264	...	...	...	...	...	...	...	...	...	...	...	...	...	...	...
Back Thornlymuir	NS 26 463599	197	...	97	129	131	116	97	125	42	145	135	94	120	210	1440	...	...
Stanely Res.	NS 26 469616	58	1282	84	128	115	94	95	107	31	133	122	87	108	194	1297	...	...
Paisley	NS 26 478642	32	1123	72	131	93	79	88	111	33	105	115	82	100	175	1185	50	17 Dec
Clyde																		
Loch Cochno	NS 26 487763	279	...	87	143	124	82	107	135	59	97	157	98	107	228	1425	...	...
Jaw Res.	NS 26 496757	277	...	75	145	127	81	101	131	72	93	157	92	119	240	1433	...	...
Cochno Filter	NS 26 495737	126	1241	65	118	102	81	88	130	48	78	133	91	98	182	1214	36	3 Sep
Greenland Dam	NS 26 441757	259	...	87	138	99	100	91	163	36	114	136	127	115	151	1356	...	...
Greenland Dam	NS 26 441757	259	...	102	156	89	106	97	168	45	136	141	127	102	184	1453	...	...
Whitemoss Res.	NS 26 414719	58	...	86	137	98	71	111	125	26	82	120	106	94	193	1249	...	...
Dumbarton S.Wks.	NS 26 407745	3	1174	79	124	87	83	87	123	25	94	112	104	97	163	1178	41	5 Oct
Garshake Filters	NS 26 415762	70	...	92	150	86	95	85	131	38	96	122	100	107	173	1273	37	5 Oct
Falloch (Clyde)																		
Allt-Nan-Caorrunn	NN 27 278222	366	3021	203	251	258	113	168	200	93	107	244	131	269	321	2357	...	...
Dubh Eas	NN 27 291203	369	3108	197	297	389	167	227	246	97	117	289	158	319	506	3010	...	...
Lairig Arnan	NN 27 295182	340	3197	204	287	341	154	216	210	92	132	306	170	305	506	2923	...	...
Ardvorlich	NN 27 315116	342	3179	272	335	371	186	213	246	145	141	270	161	276	458	3074	...	...
Sloy Power Sta.	NN 27 324099	12	2699	227	286	288	141	175	184	65	93	264	143	255	413	2534	60	3 Sep
Inveruglas (Falloch)																		
Allt a' Chnoic	NN 27 255140	411	3292	233	350	331	218	225	226	121	137	308	171	264	510	3095	...	...
Sloy, Main Adit	NN 27 293105	204	...	250	347	355	149	226	211	84	113	266	184	265	465	2915	...	...
Beinn Ime, South	NN 27 270080	381	...	278	329	381	173	258	269	111	151	324	196	372	633	3476	...	...
Beinn Ime, North	NN 27 266093	467	3529	258	345	353	176	264	262	139	152	323	209	307	525	3314	...	...
Falloch (Clyde)																		
Corriehichan	NN 27 376085	198	...	197	254	182	133	171	165	51	88	203	149	179	334	2105	...	...
Ardess	NS 26 360994	18	2161	223	280	175	166	148	132	49	114	180	147	204	315	2133	...	...

Station	Grid reference	Altitude m	Av. 1916–1950 mm	Jan	Feb	Mar	Apr	May	Jun	Jul	Aug	Sep	Oct	Nov	Dec	Year	Highest recorded fall in a day mm	date
CLYDE—*(contd)*																		
Doune	NS 26 313981	98	...	174	264	247	138	211	149	63	107	212	147	231	384	2329	78	17 Dec
Sallochy	NS 26 387955	47	1925	192	201	130	138	141	158	49	121	179	119	173	310	1911	...	...
Cashell Burn	NS 26 418954	334	2096	221	211	140	145	155	183	48	132	175	133	157	340	2041	...	...
Fruin (Falloch)																		
Auchingaich Res. North	NS 26 276918	244	...	147	161	160	255	152	164	42	84	217	138	155	285	1960	...	...
Auchingaich Res. South	NS 26 275915	244	...	142	159	161	245	142	152	49	80	228	142	151	272	1922	...	...
Ballyvoulin	NS 26 295885	137	...	130	194	184	104	167	168	66	127	180	125	168	231	1844	...	...
Blairnairn No. 1	NS 26 302882	175	1884	142	203	195	110	159	174	70	147	184	141	177	233	1936	...	...
Endrick (Falloch)																		
Endrick Weir	NS 26 682867	238	1559	128	199	107	146	100	140	38	144	139	114	124	166	1545	...	...
Fintry	NS 26 616870	91	1522	109	157	103	101	111	148	35	116	103	99	97	175	1352	43	13 Aug
Old Ballikinrain	NS 26 559879	65	1314	91	129	90	73	118	143	41	116	104	104	104	165	1278	37	13 Aug
Blanefield	NS 26 555796	76	1404	95	126	131	91	114	143	49	120	141	95	119	218	1442	...	...
Killearn Hospital	NS 26 509850	23	...	95	126	131	91	114	143	49	120	141	95	119	218	1442	43	17 Dec
Killearn Hospital	NS 26 509850	23	...	91	136	95	95	100	125	42	99	107	102	106	186	1283	...	...
Burncrooks Res.	NS 26 483795	255	...	101	140	121	84	123	138	74	101	142	103	116	219	1463	...	...
Drymen, Muir Park	NS 26 488919	201	1567	117	165	113	91	116	157	53	109	180	109	125	241	1577	...	...
Gartocharn	NS 26 425857	45	1447	89	151	134	83	108	159	47	79	135	99	118	207	1407	43	17 Dec
Leven (Clyde)																		
Loch Lomond Park	NS 26 391832	56	1496	87	167	131	104	111	139	58	80	135	106	116	204	1439	37	17 Dec
Linnbrane R.F.M.Sta. Area 786 km²	NS 26 394803	4	2168	...	...	...	...	...	...	...	...	...	...	...	...	...	...	...
Clyde																		
Ardoch	NS 26 364763	14	...	38	135	152	173	167	233	93	150	212	168	131	231	1883	...	...
Ardoch	NS 26 364763	14	...	81	125	82	72	95	152	40	120	101	101	93	157	1221	...	...
Parkhill Filters	NS 26 353736	74	...	112	164	107	61	43	151	21	94	106	105	133	239	1336	55	17 Dec
Coastal																		
Helensburgh	NS 26 303838	96	1463	89	157	133	77	106	133	46	114	143	109	133	223	1465	37	22 Jun 17 Dec
Garelochhead	NS 26 240911	3	...	129	195	200	129	147	136	42	83	174	136	176	342	1889	52	17 Dec
Finnart	NS 26 241951	15	...	190	214	193	133	141	139	44	71	177	122	199	343	1967	67	25 Jan
Allt a Bhalachain	NN 27 279051	351	3389	241	259	311	158	223	229	108	123	270	162	312	504	2899	...	...
Coilessan	NN 27 264012	30	...	213	235	266	127	195	165	83	92	246	149	243	442	2456	73	17 Dec
Drimsyniebeg	NN 27 187043	24	...	197	272	269	129	161	195	77	87	250	159	247	461	2503	94	17 Dec
Eachaig																		
Glenbranter	NS 26 113978	24	...	163	240	231	117	157	214	71	78	204	154	241	359	2231	52	17 Dec
Younger Botanic Garden	NS 26 141856	12	2395	149	246	243	133	174	185	61	107	229	152	248	402	2331	61	17 Dec
Coastal																		
Ardnadam	NS 26 167803	8	2027	136	207	199	90	141	154	54	110	185	151	199	331	1958	51	19 Dec
Knockdow	NS 26 104707	31	1573	105	146	178	69	134	120	56	84	168	136	161	277	1632	44	9 Sep 19 Dec
Glen Tarsan	NS 26 070833	117	2221	165	241	191	147	183	153	65	91	218	166	223	345	2189	...	...
Loch Striven Power Sta.	NS 26 056839	4	2142	161	208	187	104	155	152	59	75	221	157	193	326	2000	...	...
Rothesay	NS 26 083649	43	1427	86	120	149	61	127	109	55	63	157	100	162	223	1411	47	9 Sep
Dhu Loch W.Wks.	NS 26 071624	83	1319	89	104	118	55	99	119	49	61	130	103	126	184	1237	37	9 Sep
Millport Cemetery	NS 26 158554	38	...	83	117	109	85	85	107	38	58	115	88	130	184	1198	36	5 Oct
Millport	NS 26 174544	5	1147	100	120	99	97	81	93	36	49	132	87	116	190	1199	42	25 Jan
Ruel																		
Garvie Farm	NS 26 037904	43	2211	141	188	229	111	154	157	77	63	207	131	238	339	2034	39	17 Dec
Ormidale	NS 26 002817	15	2053	110	133	141	82	134	145	69	52	193	167	167	300	1693	...	...
Coastal																		
Tighnabruaich	NR 16 973726	55	...	101	116	174	89	122	141	44	35	158	113	173	255	1521	(47	26 Mar)
Lephinmore	NR 16 985926	9	1790	113	126	180	82	123	176	59	42	173	128	194	299	1694	42	19 Dec
Barnacarry	NS 26 001943	13	1788	109	125	184	92	126	172	61	47	189	130	196	279	1711	...	...
Ardkinglas House	NN 27 175104	8	2183	150	217	168	102	131	202	47	56	175	110	196	291	1844	...	...
Kinglas																		
Beinn Chorranach	NN 27 264104	408	3709	272	383	371	250	263	241	125	165	291	216	298	512	3387	...	...
Allt Uaine	NN 27 263113	342	3388	244	345	342	214	285	241	113	153	271	200	248	507	3164	...	...
Fyne																		
Upper Glen Fyne	NN 27 236213	335	3062	210	291	351	151	205	215	89	107	291	161	307	484	2863	...	...
Allt Na Lairige	NN 27 252171	328	3248	234	310	333	179	215	232	114	127	290	180	315	489	3017	...	...
Glenfyne Lodge	NN 27 220155	38	2552	219	255	239	136	145	189	105	75	247	143	238	381	2373	94	25 Jan
Meall Bhig	NN 27 222134	488	...	223	259	279	138	173	207	115	98	279	171	244	548	2733	...	...
Clachan Power Sta.	NN 27 195133	8	...	161	225	209	106	177	205	71	66	221	123	223	366	2154	...	...
Shira																		
Shira Main Dam	NN 27 161203	351	2528	154	205	264	94	192	226	87	109	239	134	221	355	2280	...	...
Brannie Burn	NN 27 165150	415	2653	182	238	295	127	183	242	98	100	287	145	265	409	2572	...	...
Ellerigbeg	NN 27 135145	9	2225	139	169	234	83	141	187	70	71	230	113	210	347	1994	...	...
Coastal																		
Inveraray Castle	NN 27 090098	12	2236	151	170	205	73	157	185	57	67	227	113	225	378	2006	61	17 Dec
Beinn Bhreac	NN 27 024103	457	...	153	180	242	79	195	179	106	85	239	143	234	427	2264	...	...
Crarae Lodge	NR 16 986973	6	1947	123	129	216	81	139	165	81	40	215	127	208	344	1870	40	14 Sep 19 Dec
Glashan Dam	NR 16 921927	101	...	105	123	193	71	140	141	79	53	203	139	184	301	1731	...	...
Loch Gair	NR 16 924908	9	...	109	125	180	70	128	141	74	47	201	147	188	277	1688	...	...
Skipness House	NR 16 908576	24	...	83	100	137	82	99	81	54	53	139	102	148	222	1301	43	19 Dec

KINTYRE AND SOUTH WESTERN ISLANDS AREA, 2293 SQUARE KILOMETRES

HYDROMETRIC AREA 104

Coastal																		
Maol Donn	NS 26 011405	320	...	144	231	265	164	166	143	136	175	182	163	293	460	2523	...	...
Brodick Castle	NS 26 013377	15	...	122	227	205	150	135	96	82	125	165	149	221	348	2024	48	19 Dec
Kildonan	NS 26 035209	18	...	86	127	112	83	115	92	61	88	137	125	169	214	1408	34	9 Sep
Pladda Lighthouse	NS 26 028190	16	...	88	153	77	70	101	83	49	84	107	121	139	189	1261	41	5 Oct
Dougarie Lodge	NR 16 884373	15	1229	87	108	131	106	115	100	35	59	152	128	135	202	1357	33	3 Sep
Carradale Manse	NR 16 798380	9	...	87	152	199	72	123	118	47	78	179	151	207	245	1657	...	...
Bordadubh Water	NR 16 735315	175	...	116	155	123	89	156	112	70	101	177	154	209	290	1751	...	...
Corrylach	NR 16 707300	137	...	106	138	106	84	128	97	60	95	169	145	179	225	1531	...	...
Campbeltown	NR 16 716208	12	...	65	102	99	86	109	61	34	73	120	115	146	149	1161	31	30 Nov
Davaar Lighthouse	NR 16 760206	15	...	66	105	117	79	107	105	38	73	151	151	236	230	1459	...	...
Sanda Lighthouse	NR 16 724038	6	977	70	109	95	99	110	94	34	69	108	131	138	170	1226	35	9 Sep
Machrihanish	NR 16 663226	10	...	80	112	90	70	84	67	36	52	111	118	126	145	1089	23	25 Jan
Largie Castle	NR 16 699461	24	1268	83	132	135	85	100	95	43	47	113	120	142	158	1252	28	9 Sep
Gigha	NR 16 642478	20	1048	53	88	115	67	94	102	38	33	101	89	109	132	1021	34	6 Jun
McArthur's Head Lighthouse	NR 16 463597	53	...	85	123	207	65	109	76	59	58	137	130	137	235	1422	44	5 Oct

Station	Grid reference	Altitude *m*	Av. 1916–1950 *mm*	Jan	Feb	Mar	Apr	May	Jun	Jul	Aug	Sep	Oct	Nov	Dec	Year	Highest recorded fall in a day *mm*	*date*
KINTYRE AND SOUTHWESTERN ISLANDS—(*contd*)																		
Eallabus	NR 16 335634	21	1347	71	110	145	76	84	86	46	47	127	113	123	191	1221	23	9 Sep
Lochindaul Lighthouse	NR 16 257588	14	1382	49	97	87	55	76	83	39	37	109	100	106	151	989	26	24 Feb
Rhuvaal	NR 16 426792	20	1476	80	120	123	75	112	88	52	43	113	133	114	173	1226	24	22 Feb
Ardlussa	NR 16 651879	30	1627	80	103	182	78	128	111	66	35	156	108	162	275	1485	51	19 Dec
Craighouse	NR 16 528680	3	1629	87	111	181	61	124	109	73	51	131	132	163	266	1489	36	19 Dec
Colonsay	NR 16 393968	30	1349	77	97	107	59	93	97	73	37	134	92	106	170	1142	32	23 Jun
Seil, An Cala	NM 17 748173	9	1649	91	102	156	54	139	125	77	61	180	89	129	234	1438	(33	25 Mar)

ADD, AWE AND ETIVE AREA, 2227 SQUARE KILOMETRES

HYDROMETRIC AREAS 88 AND 89

Station	Grid reference	Altitude *m*	Av. 1916–1950 *mm*	Jan	Feb	Mar	Apr	May	Jun	Jul	Aug	Sep	Oct	Nov	Dec	Year	Highest recorded fall in a day *mm*	*date*
Coastal																		
Achaglachgach...	NR 16 813643	12	1557	107	161	189	113	125	138	67	64	167	134	219	288	1771	33	19 Dec
Ormsary House	NR 16 740718	15	1451	88	107	144	61	127	109	41	42	121	124	170	212	1345	38	5 Oct
Add																		
Tunns Farm, Old site	NR 16 924954	101	...	97	118	186	72	144	135	76	54	197	130	201	263	1672	...	...
Dunardry, Woodglen	NR 16 827910	34	...	84	130	157	68	161	138	61	51	139	130	175	240	1534	...	...
Coastal																		
Duntrune Castle	NR 16 795959	15	1448	71	95	138	59	140	151	61	37	143	98	139	224	1355	39	19 Dec
Blaran	NM 17 859172	123	...	114	143	228	79	174	214	110	100	237	124	200	395	2119	...	...
Kilchoan	NM 17 795134	9	1634	97	97	143	59	145	152	90	54	182	86	152	255	1513	33	21 May
Midmuir Aqueduct	NM 17 947232	259	2318	132	168	233	79	157	198	74	75	226	160	208	339	2049	...	...
Nant Dam...	NN 27 004250	201	2229	141	189	245	63	150	157	53	64	207	122	207	334	1931	...	...
Awe																		
Ford	NM 17 870035	46	1849	111	136	170	75	178	159	95	71	205	111	193	333	1838	71	31 Aug
Abhainn a'Bhealach	NM 17 976055	251	...	175	189	237	97	207	177	97	75	225	139	247	407	2271	...	...
Loch Avich	NM 17 934163	288	2320	132	185	251	75	184	226	97	88	239	133	207	403	2221	...	...
Inverinan	NM 17 998176	46	2225	117	162	223	72	158	160	72	57	191	110	235	391	1949	...	...
Orchy (Awe)																		
Inveroran Hotel	NN 27 273413	177	...	140	265	325	79	148	150	65	66	211	108	221	381	2159	...	...
Bridge of Orchy	NN 27 301397	186	...	144	214	303	96	152	166	79	78	196	117	261	376	2183	...	...
Airidh Castulaich	NN 27 256342	111	2557	150	196	287	99	135	178	80	65	211	121	296	397	2216	76	17 Dec
Glenlochy Crossing	NN 27 254293	200	...	129	279	236	117	158	176	99	72	226	142	224	419	2277	...	...
Craig Lodge	NN 27 178277	61	...	116	166	228	73	128	178	87	63	207	105	198	312	1859	40	17 Dec
Allt Mhoille	NN 27 107318	503	...	215	283	458	134	271	327	170	185	375	159	304	509	3390	...	...
Awe																		
Ben Cruachan	NN 27 085295	427	...	157	172	387	93	236	263	155	123	315	175	286	439	2802	...	...
Pass of Brander	NN 27 050284	61	...	136	151	225	94	157	212	85	85	198	130	182	304	1961	...	...
Inverawe Power Sta.	NN 27 016322	5	...	97	126	172	54	124	170	61	37	163	109	143	249	1505	...	...
Coastal																		
Glen Noe	NN 27 061341	30	...	136	185	258	93	215	165	71	60	168	136	171	378	2038	...	...
Etive																		
Black Corries	NN 27 299561	361	...	142	221	279	86	133	165	58	71	193	133	263	469	2215	...	...
Dalness	NN 27 169512	73	3081	156	230	484	120	196	218	130	106	282	176	378	673	3147	...	...
Coastal																		
Alltnacree	NM 17 933360	3	...	97	112	173	48	129	148	59	45	150	97	133	238	1429	26	29 Jan
Rhu Garbh	NM 17 933407	6	1681	120	108	184	55	160	141	53	59	170	91	151	243	1535	41	28 Jan
Barcaldine...	NM 17 958415	15	1756	120	132	213	61	173	153	59	58	180	91	174	294	1709	...	...

LOCHY AND LINNHE AREA, 2504 SQUARE KILOMETRES

HYDROMETRIC AREAS 90 AND 91

Station	Grid reference	Altitude *m*	Av. 1916–1950 *mm*	Jan	Feb	Mar	Apr	May	Jun	Jul	Aug	Sep	Oct	Nov	Dec	Year	Highest recorded fall in a day *mm*	*date*
Coastal																		
Appin, Ardveich	NM 17 919425	15	...	91	102	165	47	139	129	50	45	157	80	138	229	1372	28	29 Jan
Ardtur	NM 17 913464	8	1762	113	98	182	58	161	141	59	61	148	86	149	271	1526	...	...
Glenduror Forest	NN 27 048544	296	2921	178	268	410	109	254	189	105	88	241	194	365	559	2961	...	...
Leven																		
Glenleven, Invercam	NN 27 322608	326	...	127	191	298	74	100	147	51	53	152	111	236	417	1957	...	...
Loch Chiaran	NN 27 291634	366	2069	135	244	216	107	137	173	70	100	177	133	249	428	2168	...	...
Blackwater House	NN 27 282599	326	2189	137	244	271	100	131	166	65	67	185	131	257	419	2173	...	...
Blackwater Res.R.F.M.Sta. Area 171 km²	NN 27 247604	299	2146	...	...	...	...	...	...	...	...	...	...	...	...	...	...	...
Blackwater Dam South	NN 27 245601	335	2424	159	292	329	127	183	209	87	90	214	132	287	550	2661	172	17 Dec
Loch Eilde Beg	NN 27 251653	343	2607	165	282	418	107	169	170	88	72	215	149	329	610	2772	...	...
Loch Eilde Mhor	NN 27 217633	344	2534	152	265	409	95	150	179	94	76	217	143	303	588	2673	...	...
Coastal																		
Upper Glenrigh	NN 27 076647	146	...	149	168	315	72	155	150	107	76	193	113	262	391	2150	...	...
Lower Glenrigh	NN 27 061630	128	...	143	170	277	70	147	151	99	76	183	112	257	317	2003	...	...
Onich...	NN 27 030635	15	2067	106	155	263	78	149	165	94	90	179	115	223	344	1960	51	17 Dec
Fort William, Glasdrum...	NN 27 101735	66	1998	152	142	249	60	133	126	65	67	145	102	180	323	1744	...	...
Nevis																		
Achriabhach	NN 27 145684	52	...	143	239	299	91	170	152	76	72	166	107	234	472	2220	65	16 Dec
Lochy																		
Arkaig (Lochy)																		
Achnacarry House	NN 27 179879	39	2089	96	170	311	63	146	109	81	73	158	97	217	434	1955	89	17 Dec
Lochy																		
Gairlochy	NN 27 177842	30	1991	78	131	309	55	119	129	70	72	164	94	230	413	1862	77	17 Dec
Mucomir Generating Sta.	NN 27 182839	24	...	80	132	317	57	125	116	70	70	149	98	232	400	1847	77	17 Dec
Spean (Lochy)																		
Loch Pattack	NN 27 542798	433	1599	89	165	191	51	108	142	89	83	89	89	196	297	1587	...	...
Ardverikie...	NN 27 507874	251	1439	61	104	162	50	108	129	55	59	103	76	180	401	1487	...	...
Craigbeg	NN 27 406817	262	1521	51	108	196	44	89	127	48	55	112	90	168	382	1469	...	...
Treig (Spean)																		
Corrour	NN 27 353668	411	...	152	249	213	74	142	170	53	75	173	114	188	382	1986	...	...
Fersit	NN 27 351782	259	1704	115	180	227	56	124	145	52	62	123	102	195	443	1823	118	17 Dec
Spean (Lochy)																		
Allt Laire	NN 27 328776	320	1983	178	250	249	71	157	180	49	72	131	107	251	406	2103	...	...
Roy (Spean)																		
Braeroy Lodge...	NN 27 336914	223	...	89	170	245	47	131	147	57	64	141	88	209	361	1749	...	...

Station	Grid reference	Altitude m	Av. 1916–1950 mm	Jan	Feb	Mar	Apr	May	Jun	Jul	Aug	Sep	Oct	Nov	Dec	Year	Highest recorded fall in a day mm	date
LOCHY AND LINNHE—*(contd)*																		
Spean (Lochy)																		
Lairig Road	NN 27 279753	503	2587	203	269	407	97	198	203	72	88	150	146	231	497	2561	...	...
Allt Leachdach	NN 27 256789	198	1695	129	175	178	76	138	128	51	61	114	112	168	312	1643	...	...
The Cour	NN 27 222758	320	2580	159	305	382	94	188	129	77	75	170	160	277	461	2476	...	...
Spean Bridge	NN 27 221816	61	1708	84	120	236	48	111	120	65	61	119	97	173	333	1569	...	...
Lochy																		
Sgurr Finisgaig...	NN 27 188757	564	2427	191	291	272	117	199	145	74	93	151	159	215	392	2298	...	...
Allt a' Mhuilinn	NN 27 149746	305	2386	183	196	263	79	168	156	83	81	141	145	211	383	2088	...	...
Fort William, The Factory	NN 27 130751	23	...	113	134	190	53	122	123	54	58	126	100	146	308	1527	...	...
Fort William	NN 27 123759	8	1754	109	135	216	55	124	121	47	58	127	93	174	315	1573	55	16 Dec
Lochaber Water Power Area R.F.M.Sta. Area 780 km^2	NN 27 126752	23	2019	...	...	...	...	...	...	...	...	...	...	...	...	...	...	...
Coastal																		
Achdalieu	NN 27 060782	30	1955	111	167	213	58	121	119	52	66	115	123	177	321	1645	34	28 Jan
Corran Lighthouse	NN 27 016635	4	2167	119	131	283	83	142	147	80	94	209	105	223	366	1981	61	17 Dec
Ardgour House	NM 17 994639	18	2505	135	165	358	95	192	207	101	94	248	121	259	464	2440	59	17 Dec
Kingairloch House	NM 17 836531	12	2670	216	244	156	100	277	147	103	113	222	168	244	454	2443	...	...

SHIEL, ALSH AND MAREE AREA, 3905 SQUARE KILOMETRES

HYDROMETRIC AREAS 92 TO 94

Station	Grid reference	Altitude m	Av. 1916–1950 mm	Jan	Feb	Mar	Apr	May	Jun	Jul	Aug	Sep	Oct	Nov	Dec	Year	Highest recorded fall in a day mm	date
Coastal																		
Drimnin House	NM 17 553550	69	1549	106	127	184	63	133	109	95	78	159	134	160	258	1605	39	23 Jun
Shiel																		
Polloch House	NM 17 791689	15	...	139	174	235	95	200	144	119	110	186	141	241	396	2179	(46	22 May)
Coastal																		
Glenmoidart House	NM 17 745728	61	2010	100	104	171	68	145	147	139	99	210	160	222	376	1939	57	1 Dec
Inverailort...	NM 17 763815	2	2283	113	185	247	70	189	137	116	61	180	122	212	359	1991	63	1 Dec
Arisaig House	NM 17 692848	12	1620	58	83	152	43	108	159	139	67	153	113	165	256	1495	30	23 Jun
Traigh House	NM 17 657906	6	1527	89	85	147	51	121	139	116	78	156	88	135	260	1463	32	21 May
Morar																		
Rhubana	NM 17 688922	16	1801	76	97	190	55	146	141	125	81	165	102	157	310	1645	34	17 Dec
Coastal																		
Inverie House	NM 17 774994	5	2137	79	118	294	86	175	151	141	79	171	143	192	391	2020	53	17 Dec
Kinlochourn	NG 18 951066	5	...	95	221	383	105	180	169	116	95	204	143	279	557	2547	98	17 Dec
Druim Sgurr	NG 18 903198	335	...	77	183	513	111	203	153	114	87	161	137	269	491	2499	...	...
Ratagan	NG 18 919197	4	2042	69	165	408	87	159	152	101	69	148	131	258	480	2227	74	17 Dec
Glenshiel Forest	NH 28 017132	183	...	95	214	588	149	254	178	134	102	177	155	282	503	2831	...	...
Elchaig																		
Loch Mhoicean	NH 28 076319	436	...	(57)	181	449	81	171	199	142	102	192	113	304	516	2507	...	...
Ling																		
Loch na Maoile Buidhe	NH 28 046344	363	...	52	196	467	63	199	184	149	127	193	120	342	580	2673	...	...
Coastal																		
Nostie	NG 18 853272	7	1841	57	107	259	51	127	163	69	55	130	107	201	401	1726	...	...
Balmacara	NG 18 800276	4	...	65	126	222	41	129	135	61	46	124	105	148	310	1511	37	1 Dec
Plockton	NG 18 802332	12	1290	67	79	173	36	111	113	43	32	88	86	145	263	1237	47	1 Dec
Achmore, Dailfearn	NG 18 854337	18	...	69	105	203	62	149	138	80	40	111	105	187	301	1550	47	1 Dec
Am Meallan	NG 18 879341	229	...	76	105	206	52	152	165	63	51	152	76	223	324	1647	...	...
Attadale	NG 18 936383	15	...	75	110	294	49	168	131	93	45	116	115	215	398	1809	58	17 Dec
Carron																		
Achnashellach	NH 28 038492	67	...	52	167	381	73	159	149	157	78	188	162	312	537	2414	92	17 Dec
Coulin Pass	NH 28 025501	290	...	63	173	401	74	166	160	168	102	175	142	335	592	2552	...	...
Coastal																		
Couldoran	NG 18 843441	15	2202	81	135	384	50	189	140	79	77	181	161	324	493	2294	...	...
Applecross, Bealach	NG 18 757441	396	...	150	152	292	97	203	203	142	63	132	179	137	330	2081	...	...
Applecross House	NG 18 723456	21	1519	64	109	197	72	145	145	92	35	122	139	125	280	1525	63	1 Dec
Badachro	NG 18 778737	15	1329	49	65	195	41	123	138	97	33	102	108	163	269	1383	...	...
Loch na h' Oidhche	NG 18 899642	465	3486	112	151	388	86	197	243	195	112	199	289	301	593	2867	...	...
A' Garbh-Choire	NG 18 873701	244	2245	74	110	310	50	160	193	159	71	133	215	226	443	2145	...	...
Loch Bad na Sgalaig	NG 18 848719	108	1803	55	91	297	44	144	170	138	65	127	169	223	397	1922	...	...
Sithean Mor	NG 18 848741	259	...	86	93	297	58	159	184	173	61	133	195	243	431	2111	...	...
Rudh Re Lighthouse	NG 18 739918	16	...	52	71	173	43	90	131	90	40	118	123	134	214	1278	39	5 Mar
Ewe																		
Loch Gleann na Muice	NH 28 063681	316	...	80	107	330	114	184	200	121	107	151	191	361	601	2547	...	...
Heights of Kinlochewe	NH 28 069638	91	...	69	101	288	24	137	168	90	102	129	169	311	543	2131	...	...
Kinlochewe	NH 28 024630	23	1970	55	130	345	50	151	149	117	55	149	161	291	567	2220	119	16 Dec
Coastal																		
Poolewe	NG 18 861818	6	1497	36	59	261	32	121	150	102	55	123	129	183	332	1582	57	16 Dec

INNER AND OUTER HEBRIDES AREA, 6040 SQUARE KILOMETRES

HYDROMETRIC AREAS 105 AND 106

Station	Grid reference	Altitude m	Av. 1916–1950 mm	Jan	Feb	Mar	Apr	May	Jun	Jul	Aug	Sep	Oct	Nov	Dec	Year	Highest recorded fall in a day mm	date
Coastal																		
Lochdonhead	NM 17 728334	3	...	130	130	257	77	166	122	56	73	207	127	230	387	1961	55	17 Dec
Iona, Cnoc Mor	NM 17 282240	18	--	77	62	106	37	81	79	65	46	131	73	113	160	1029	27	23 Jun
Ulva House	NM 17 442389	15	1608	85	121	153	71	143	89	104	68	164	108	155	261	1522	29	7 Dec
White House of Aros	NM 17 564455	3	1958	112	142	211	61	132	107	80	84	165	117	196	331	1735	31	19 Dec
Tiree	NL 07 999446	9	1117	59	91	93	49	85	62	66	33	113	72	107	166	995	23	28 Jan 7 Dec
Isle of Eigg	NM 17 479836	37	1539	83	82	156	37	114	84	107	47	150	109	172	257	1397	29	28 Jan 7 Dec
Kinloch Glen	NG 18 370001	107	...	182	146	286	103	206	147	202	105	220	136	225	343	2300	...	...
Isle of Rhum	NM 17 402996	5	2520	192	175	310	93	209	129	117	74	225	146	228	349	2246	73	28 Jan
Coire Dubh	NM 17 387977	305	...	286	189	351	112	272	180	218	126	310	177	278	394	2892	...	...
Slugan Burn	NM 17 393983	152	...	277	189	377	114	259	170	192	112	281	161	281	419	2832	...	...
Harris	NM 17 337958	21	...	125	103	157	55	109	111	77	47	127	86	132	216	1345	...	...
Ard Nev	NM 17 354987	274	...	160	136	229	85	180	156	173	93	213	114	194	274	2008	...	...
Kilmory	NG 18 362038	15	...	161	143	194	102	156	105	99	76	152	100	166	265	1722	...	...
Staffin	NG 18 491673	46	...	82	86	214	48	117	129	81	41	113	131	175	246	1461	39	16 Dec
Storr Lochs No.2	NG 18 506526	152	...	98	99	322	75	181	159	141	95	169	174	264	421	2198	...	...
Storr Lochs No.1	NG 18 513523	140	...	98	100	324	75	183	159	141	96	173	176	269	429	2223	...	...
Portree	NG 18 479440	27	1813	99	103	265	73	152	126	92	70	143	93	208	341	1767	47	16 Dec

GENERAL TABLE, SCOTLAND

Station	Grid reference	Altitude m	Av. 1916–1950 mm	Rainfall in millimetres: Jan	Feb	Mar	Apr	May	Jun	Jul	Aug	Sep	Oct	Nov	Dec	Year	Highest recorded fall in a day: mm	date
INNER AND OUTER HEBRIDES—*(contd)*																		
Alltdearg House	NG 18 477293	55	...	186	285	499	179	265	191	163	117	232	141	330	527	3115	...	...
Broadford	NG 18 649228	30	2283	106	125	389	93	164	156	123	129	208	143	226	503	2366	151	17 Dec
Kyle House	NG 18 745265	9	...	56	99	195	31	102	161	51	31	74	89	147	242	1279	...	...
Upper Ostaig	NG 18 646060	46	...	84	79	263	58	136	149	140	74	207	101	185	265	1741	(39	16 Aug)
Bealach Brittle	NG 18 394231	232	...	172	127	356	98	234	155	175	129	251	112	258	329	2397	...	...
Grula	NG 18 384266	18	2124	161	125	332	96	207	142	152	75	197	133	251	310	2182	84	28 Jan
Treen	NG 18 399311	2	...	131	120	313	90	174	153	169	73	198	134	255	370	2181	61	28 Jan
Prabost	NG 18 418501	67	1825	107	96	263	74	135	136	108	65	148	116	210	319	1777	49	5 Mar
The Storr, New site	NG 18 489534	472	...	122	111	339	77	163	178	152	87	165	161	254	391	2199	...	...
Duntulm	NG 18 399718	90	1349	85	71	159	61	99	125	91	46	111	111	127	228	1313	30	1 Jan
Loch Ionadagro	NB 19 513464	82	...	63	67	150	45	83	117	70	49	103	123	153	189	1211	...	...
Back Filters	NB 19 466425	69	...	61	65	149	46	81	115	67	47	120	121	156	194	1223	...	...
Stornoway	NB 19 459332	3	...	57	68	148	37	84	92	68	41	86	100	132	186	1099	20	28 Jan
Loch Oichean	NB 19 389271	55	...	53	73	132	43	83	116	54	50	100	102	146	174	1125	...	...
Lag Garbh Aqueduct	NB 19 074095	178	...	120	93	232	74	128	193	129	66	234	146	178	282	1876	...	...
Chliostair Power Sta.	NB 19 058092	47	...	97	84	216	65	129	179	108	51	196	152	174	313	1765	...	...
Loch Ashavat	NB 19 073123	241	...	99	116	354	100	157	291	241	81	338	170	263	394	2605	...	...
Todale Intake	NB 19 104287	73	...	102	74	255	66	114	182	114	69	154	138	192	326	1787	...	...
Gisla Power Sta.	NB 19 131258	7	1579	91	69	253	67	111	132	109	68	148	137	203	333	1721	52	16 Dec
Loch Grunavat Weir	NB 19 102262	111	...	108	79	259	61	116	147	109	75	159	197	182	306	1797	...	...
Shawbost Filters	NB 19 261436	73	...	55	76	130	44	82	122	71	51	112	120	133	165	1163	...	...
Lochmaddy	NF 08 922696	10	...	71	68	124	41	85	76	65	39	130	88	104	159	1050	27	1 Sep
Benbecula	NF 08 782555	5	1140	78	63	149	44	99	69	66	32	123	96	123	192	1134	29	28 Jan
Glac a'Bhodaich	NF 08 800387	6	...	118	113	168	64	146	64	92	49	136	111	127	228	1417	...	...
Stilligarry	NF 08 770387	5	...	103	112	166	61	144	61	86	46	135	101	127	212	1354	...	...

LAXFORD AREA, 2220 SQUARE KILOMETRES

HYDROMETRIC AREA 95

Station	Grid reference	Altitude m	Av. 1916–1950 mm	Jan	Feb	Mar	Apr	May	Jun	Jul	Aug	Sep	Oct	Nov	Dec	Year	mm	date
Broom																		
Loch Droma	NH 28 258755	282	...	28	86	231	47	129	111	71	52	53	147	219	589	1763	...	...
Loch Droma	NH 28 257755	279	1641	24	77	243	36	130	112	58	45	75	143	213	482	1638	...	...
Glackour	NH 28 191820	20	...	29	74	225	22	107	159	100	40	92	159	234	412	1654	70	1 Dec
Coastal																		
Strone Nea	NH 28 204843	366	...	33	84	207	24	116	165	112	70	119	182	198	419	1729	...	...
Ullapool	NH 28 126940	12	1259	28	53	178	20	85	131	100	36	94	159	204	368	1457	55	1 Dec
Strath Kanaird Power Sta.	NC 29 149013	14	1418	44	65	239	22	124	145	133	71	124	143	180	346	1636	...	...
Dubh Loch	NH 28 150995	197	...	45	65	241	23	124	146	134	72	125	144	181	347	1646	...	...
Knockanrock	NC 29 187087	244	...	81	89	254	51	132	186	133	99	191	203	293	372	2085	...	...
Lochan Gainmheich	NC 29 137117	76	...	66	68	228	39	125	158	122	76	152	157	231	305	1727	...	...
Inverpolly	NC 29 074134	14	1307	39	50	180	31	101	134	99	53	115	143	166	266	1376	51	1 Dec
Kirkaig																		
Elphin	NC 29 212113	145	1672	56	61	299	43	142	150	122	71	173	160	294	409	1981	67	1 Dec
Falls of Kirkaig	NC 29 115180	137	...	41	45	203	21	99	106	84	27	143	108	202	279	1360	...	...
Coastal																		
Lochinver	NC 29 092222	6	1236	63	51	202	23	100	126	90	33	115	158	143	209	1312	46	1 Dec
Stoer	NC 29 040284	21	...	54	49	136	20	105	130	84	43	110	119	143	235	1228	46	16 Dec
Kylestrome	NC 29 220344	30	...	62	75	209	23	139	174	129	93	135	183	226	314	1760	63	1 Dec
Laxford																		
Lochmore Lodge	NC 29 300386	46	1933	66	76	335	44	155	219	175	94	179	211	311	400	2264	56	1 Dec
Coastal																		
Kinlochbervie, Garbet Hotel	NC 29 221565	24	...	78	56	157	22	87	132	76	59	85	159	146	319	1377	...	...
Cape Wrath	NC 29 259747	112	1203	55	71	156	15	98	126	67	101	112	132	133	181	1248	46	9 Aug

NAVER AND THURSO AREA, 2878 SQUARE KILOMETRES

HYDROMETRIC AREAS 96 AND 97

Station	Grid reference	Altitude m	Av. 1916–1950 mm	Jan	Feb	Mar	Apr	May	Jun	Jul	Aug	Sep	Oct	Nov	Dec	Year	mm	date
Coastal																		
Inshore	NC 29 327692	94	...	70	71	199	25	101	149	104	88	184	175	208	287	1661	...	...
Hope																		
Gobernuisgach Lodge	NC 29 437417	76	1994	62	92	315	39	148	196	108	75	174	212	327	464	2212	86	16 Dec
Ben Hee	NC 29 409368	523	2721	95	146	457	63	175	208	198	132	254	298	383	477	2889	...	...
Coastal																		
Tongue	NC 29 593588	6	1100	35	59	117	11	79	150	60	51	75	178	127	183	1123	33	22 May
Borgie																		
Borgie	NC 29 663579	34	1113	41	57	131	16	98	131	71	61	92	186	143	182	1208	...	...
Naver																		
Altnaharra	NC 29 569351	91	...	34	61	129	19	149	129	53	47	52	145	212	285	1313	...	...
Vagastie Aqueduct	NC 29 518272	235	...	38	66	204	34	131	170	75	58	108	185	239	385	1692	...	...
Dalvina Lodge	NC 29 696438	46	...	45	53	131	13	133	119	59	79	64	185	179	233	1293	...	...
Coastal																		
Strathy																		
Lochstrathy	NC 29 797493	160	...	50	66	121	13	75	118	75	63	80	193	139	184	1177	...	...
Strathy	NC 29 830609	38	1088	51	60	134	12	73	159	83	63	101	176	127	186	1224	36	1 Oct
Halladale																		
Croick	NC 29 894543	30	...	43	52	120	9	68	133	81	62	67	157	126	189	1107	43	1 Dec
Coastal																		
Shebster	ND 39 006644	70	...	36	60	98	13	83	143	87	83	103	140	109	173	1129	45	1 Oct
Dounreay	NC 29 990670	21	955	51	65	113	12	67	128	87	53	104	109	89	168	1047	32	1 Sep
Forss																		
Loch Shurrery	ND 39 041568	100	1075	37	61	102	14	88	149	90	85	105	145	112	180	1169	46	1 Oct
Coastal																		
Holborn Head Lighthouse	ND 39 108707	21	...	34	38	74	21	85	138	74	81	121	158	133	205	1159	47	16 Dec
Calder Burn (Thurso)																		
Loch Calder Res. R.F.M. Sta. Area 24.5 km^2	ND 39 085596	65	970	...	...	...	...	...	...	...	...	...	...	...	...	...	...	...
Thurso																		
Achavarn Lodge	ND 39 086595	67	958	49	67	99	14	88	119	76	80	79	141	102	168	1082	...	...
Hoy P. Sta.	ND 39 137607	23	984	49	77	116	18	83	113	73	81	87	139	106	202	1145	37	1 Oct
Buckies	ND 39 109633	73	989	53	65	110	19	93	126	84	92	85	145	105	181	1157	37	1 Oct

Station	Grid reference	Altitude m	Av. 1916–1950 mm	Jan	Feb	Mar	Apr	May	Jun	Jul	Aug	Sep	Oct	Nov	Dec	Year	Highest recorded fall in a day mm	Highest recorded fall in a day date
NAVER AND THURSO—*(contd)*																		
Coastal																		

ORKNEY AND SHETLAND ISLANDS AREA, 2485 SQUARE KILOMETRES

HYDROMETRIC AREAS 107 AND 108

Station	Grid reference	Altitude m	Av. 1916–1950 mm	Jan	Feb	Mar	Apr	May	Jun	Jul	Aug	Sep	Oct	Nov	Dec	Year	Highest recorded fall in a day mm	Highest recorded fall in a day date
Coastal																		
Baltasound	HP412 607089	24	1141	125	96	145	42	98	49	63	35	113	144	111	192	1213	43	22 May
Uyeasound	HP412 591009	5	...	121	111	126	33	118	39	47	35	103	154	81	180	1146	50	22 May
Saltness	HU411 541628	9	1030	98	92	111	29	117	58	44	24	85	124	103	184	1069	42	22 May
Vementry	HU411 309597	8	980	112	93	122	35	88	68	61	35	56	171	122	183	1146	46	22 May
Kergord	HU411 394543	46	...	109	114	161	36	139	82	79	41	85	179	120	204	1348	49	22 May
Lerwick	HU411 453397	82	1132	107	99	140	34	125	74	61	35	87	140	109	189	1201	50	22 May
Hamnavoe	HU411 372357	21	...	150	93	107	27	116	57	51	29	68	129	103	193	1124	52	28 Jan
Sumburgh	HU411 388126	79	...	113	105	82	25	95	94	47	31	64	122	109	149	1035	...	...
North Ronaldshay Lighthouse	HY310 784559	7	900	78	93	92	30	92	88	69	64	58	125	111	175	1076	39	1 Oct
Balfour Castle	HY310 476165	15	...	100	86	113	19	72	58	93	79	63	137	117	198	1136	...	...
Evie	HY310 378247	34	...	118	86	110	29	99	61	113	57	70	149	117	207	1217	31	1 Oct
Stenness	HY310 298112	23	1006	93	90	105	22	86	72	100	86	69	126	125	193	1166	27	27 Jul
Kirkwall	HY310 483076	26	1000	108	89	111	18	80	65	86	83	72	144	117	222	1194	34	1 Oct
Graemsay, High Lighthouse	HY310 268061	9	...	93	88	117	10	98	74	79	83	61	147	155	231	1236	...	...
Cantick Head Lighthouse	ND 39 345893	21	865	84	72	106	14	62	81	61	67	68	114	101	232	1062	28	15 Dec
Sule Skerry	HX210 623241	13	...	64	84	117	18	74	87	63	41	98	135	115	151	1048	27	1 Oct

WICK WATER TO CONON AREA, 6358 SQUARE KILOMETRES

HYDROMETRIC AREAS 1 TO 4

Station	Grid reference	Altitude m	Av. 1916–1950 mm	Jan	Feb	Mar	Apr	May	Jun	Jul	Aug	Sep	Oct	Nov	Dec	Year	Highest recorded fall in a day mm	Highest recorded fall in a day date
Wick																		
Acharole	ND 39 221510	55	...	56	74	92	19	77	96	87	97	88	151	108	163	1107	33	1 Dec
Wick	ND 39 364522	36	761	45	67	81	13	68	107	87	86	76	136	102	138	1006	35	1 Dec
Coastal																		
Rumster	ND 39 212371	116	...	58	65	91	17	85	109	85	89	67	141	105	157	1070	...	...
Berriedale																		
Langwell House	ND 39 100222	70	909	57	56	42	24	110	69	61	91	66	141	129	313	1158	...	...
Helmsdale																		
Rimsdale	NC 29 751388	168	...	41	42	116	17	122	105	83	48	60	182	185	225	1224	...	...
Achentoul	NC 29 873354	137	...	42	64	109	18	87	104	77	51	68	149	150	191	1109	47	1 Dec
Suisgill Lodge	NC 29 901231	69	...	70	77	97	33	91	109	69	69	81	132	139	254	1222	61	1 Dec
Brora																		
Dalnessie	NC 29 631152	177	...	61	48	166	28	109	136	63	54	73	118	180	313	1349	...	...
Coastal																		
Brora	NC 29 904036	15	...	38	59	100	30	85	103	31	72	56	76	119	163	931	34	1 Dec
Uppat	NC 29 870025	67	854	37	58	120	35	81	120	52	65	62	93	108	200	1031	51	1 Dec
Fleet																		
Coastal																		
Ospisdale House	NH 28 714897	30	798	26	80	97	29	64	159	94	49	47	91	110	186	1032	54	1 Dec
Oykel																		
Ben More Lodge	NC 29 323114	160	...	47	79	278	43	141	166	122	61	142	186	346	440	2052	...	...
Cassley (Oykel)																		
Ben More	NC 29 346194	386	3142	117	160	660	51	236	298	179	140	303	294	(470)	762	3670	...	...
Duchally	NC 29 388170	133	...	43	86	292	29	140	177	74	52	140	176	269	470	1950	...	...
Rosehall	NC 29 469022	15	...	41	56	160	21	86	166	48	42	85	122	167	308	1302	56	1 Dec
Shin (Oykel)																		
Garvault	NC 29 372305	335	2322	83	108	330	44	175	211	170	95	215	271	363	511	2577	...	...
Loch Merkland	NC 29 401297	118	...	60	101	304	35	163	174	134	83	188	223	324	462	2249	...	...
Corrykinloch	NC 29 365257	113	1947	69	95	330	31	177	198	126	106	190	240	362	419	2344	...	...
Cassley Power Sta.	NC 29 396232	99	1795	54	85	323	33	135	156	80	43	171	159	360	500	2100	105	16 Dec
Cnoc an Laoigh	NC 29 487199	137	...	39	43	152	42	83	173	53	41	131	149	194	301	1400	...	...
Cnoc Riabhach	NC 29 543077	259	...	57	78	174	24	95	133	56	51	83	134	166	279	1330	...	...
Tirry (Shin)																		
Cnoc na Doire	NC 29 548199	207	...	67	49	169	41	102	159	66	42	75	147	189	352	1457	...	...
Shin (Oykel)																		
Lairg Dam	NC 29 576071	94	...	53	76	149	23	92	128	43	59	67	120	143	271	1224	44	1 Dec
Achany	NC 29 569018	85	1054	41	64	168	28	87	128	49	64	83	99	149	260	1219	38	1 Dec
Carron (Oykel)																		
Gleann Beag	NH 28 337836	335	...	44	154	432	53	158	254	155	75	165	191	363	791	2835	...	...
Coastal																		
Morangie	NH 28 760832	46	...	27	60	68	22	64	145	72	77	39	84	75	140	871	41	1 Dec
Tain	NH 28 785823	2	...	31	65	47	19	52	102	65	61	35	46	74	131	727	...	...
Tarbatness	NH 28 947875	18	558	21	57	57	17	54	138	76	61	40	87	52	121	781	47	1 Dec
Geanies House	NH 28 895793	61	599	18	47	48	19	61	141	76	61	37	82	39	98	726	42	23 Jun
Rosskeen	NH 28 683695	15	...	26	67	63	15	68	120	45	69	33	49	95	171	822	53	1 Dec
Alness																		
Gleann Mhuire	NH 28 493786	251	...	44	93	227	53	102	170	99	76	102	109	295	432	1802	...	...
Loch Bad a Bhathaich	NH 28 542791	317	...	(51)	91	208	41	94	163	93	81	94	90	169	406	1581	...	...
Ardross, Glensax	NH 28 629737	171	948	33	72	77	28	104	143	63	76	31	84	114	207	1031	72	1 Dec
Glass																		
Loch Glass	NH 28 534701	219	...	49	83	177	32	104	97	47	63	69	109	191	312	1334	...	...
Assynt Filters	NH 28 586669	152	...	31	57	105	15	71	98	37	67	55	84	125	244	990	63	1 Dec
Conon																		
Allt na Feithe Baine	NH 28 086585	253	...	49	138	337	43	127	161	100	60	156	168	251	491	2081	...	...
Luib	NH 28 130547	191	...	40	122	285	61	119	150	85	68	145	164	245	452	1935	76	17 Dec
Druimdhu, New site	NH 28 189592	146	...	31	115	274	37	118	123	61	43	104	134	297	481	1817	...	...
Grudie (Conon)																		
Loch Fannich South	NH 28 178647	293	...	52	158	310	39	154	161	82	65	152	153	346	521	2192	...	...
Fannich Lodge	NH 28 219660	268	...	43	110	295	33	135	167	80	49	132	139	309	465	1958	74	1 Dec
Aultdearg	NH 28 288652	221	1374	26	91	245	33	109	126	55	55	109	132	247	412	1641	74	16 Dec
Conon																		
Loch Luichart Manse	NH 28 317625	94	1137	21	73	203	35	95	113	65	54	83	104	206	433	1485	82	17 Dec
Loch Luichart, East	NH 28 390575	50	1090	36	101	160	31	87	102	40	65	81	109	190	322	1324	...	...

Station	Grid reference	Altitude m	Av. 1916–1950 mm	Rainfall in millimetres Jan	Feb	Mar	Apr	May	Jun	Jul	Aug	Sep	Oct	Nov	Dec	Year	Highest recorded fall in a day mm	date
WICK WATER TO CONON—*(contd)*																		
Meig (Conon)																		
Gleann Fhiodhaig	NH 28 175490	223	...	63	127	273	57	171	119	63	61	154	117	229	394	1828	...	...
Corriefeol	NH 28 203509	168	...	56	105	237	47	92	117	69	34	109	91	201	346	1505	63	17 Dec
Loch na Caoidhe	NH 28 228473	442	...	61	151	315	66	140	161	131	49	163	234	172	211	1854	...	...
Bridgend	NH 28 323551	122	1272	34	134	179	33	84	122	63	77	81	114	187	308	1415	60	1 Dec
Black (Conon)																		
Luibachlaggan	NH 28 350782	259	...	32	145	296	40	146	167	114	55	114	147	281	438	1975	...	...
Vaich Dam	NH 28 347747	230	...	26	59	234	24	102	141	72	43	91	116	273	419	1601	...	...
Blackbridge	NH 28 374710	210	...	24	70	184	22	94	156	59	42	82	109	236	383	1462	...	...
Strathrannoch	NH 28 386745	282	...	31	102	311	37	133	161	82	42	98	120	203	456	1776	...	...
Dubh Choille	NH 28 397684	351	...	40	72	173	31	89	188	74	55	99	109	168	411	1508	...	...
Inchbae Lodge	NH 28 404692	160	...	24	75	148	27	97	156	77	49	78	90	198	370	1387	87	1 Dec
Conon																		
Moy Bridge R.F.M. Sta. Area 1170 km²	NH 28 483547	10	1758	...	...	...	...	...	...	...	...	...	...	...	...	...	...	...
Orrin (Conon)																		
Orrin Dam	NH 28 401504	221	...	40	112	142	38	93	119	61	72	63	106	176	291	1312	48	1 Dec
Fairburn House	NH 28 455528	152	953	36	78	119	25	78	123	49	72	59	90	131	236	1097	37	17 Dec
Conon																		
Brahan	NH 28 515548	53	...	24	72	87	28	66	104	48	63	53	71	108	196	919	47	1 Dec
Coastal																		
Cromarty Lighthouse	NH 28 786677	6	...	25	49	52	23	66	137	47	49	30	65	66	127	738	...	...
Blackstand	NH 28 716612	158	797	25	55	67	21	70	148	58	88	44	75	91	158	901	48	23 Jun
Fortrose	NH 28 749557	5	646	15	41	61	9	53	162	48	68	40	68	83	125	775	44	23 Jun
Allangrange House	NH 28 625516	21	...	15	47	87	27	43	183	55	71	70	50	88	159	894	41	1 Dec
Muir of Ord	NH 28 527500	46	756	20	53	77	20	68	124	53	53	41	72	97	165	843	...	...

BEAULY AND NESS AREA, 3065 SQUARE KILOMETRES

HYDROMETRIC AREAS 5 AND 6

Station	Grid reference	Altitude m	Av. 1916–1950 mm	Jan	Feb	Mar	Apr	May	Jun	Jul	Aug	Sep	Oct	Nov	Dec	Year	Highest mm	date
Beauly																		
Creag na Calman	NH 28 185218	320	1897	92	182	283	124	143	136	72	95	185	135	310	438	2195	...	...
Affric Lodge	NH 28 185230	233	1612	61	157	276	59	125	110	45	61	132	99	275	463	1864	92	17 Dec
Glen Fiadh	NH 28 180257	472	...	105	228	301	132	179	175	91	113	196	144	354	667	2683	...	...
Benevean Dam	NH 28 275280	232	1305	63	106	178	45	93	135	52	74	77	99	195	383	1499	...	...
Cannich	NH 28 297259	107	1188	53	117	139	47	93	127	53	83	71	75	141	297	1296	66	17 Dec
Fasnakyle	NH 28 314288	80	1071	44	103	150	39	84	125	47	72	69	74	159	315	1280	...	...
Cannich (Beauly)																		
Upper Mullardoch	NH 28 063286	320	...	58	207	428	89	170	203	138	97	199	129	328	555	2602	...	...
Mullardoch Dam	NH 28 223310	245	1744	65	159	263	57	125	119	59	63	122	100	252	520	1904	121	17 Dec
Muchrachd	NH 28 287337	165	...	44	119	171	40	90	122	48	51	81	123	190	426	1505	84	17 Dec
Beauly																		
Glassburn	NH 28 370344	52	1096	38	88	151	28	76	106	53	74	57	102	167	298	1238	47	17 Dec
Farrar (Beauly)																		
Upper Misgeach	NH 28 171382	247	...	51	174	317	48	122	123	84	44	124	120	278	693	2180	...	...
Misgeach Intake	NH 28 184381	206	...	48	163	322	67	121	151	81	52	134	160	247	407	1953	...	...
Loch Toll a Mhuic	NH 28 235419	572	...	83	216	368	150	166	239	160	94	202	208	187	395	2467	...	...
Inchvuilt	NH 28 229387	151	...	49	114	227	52	102	115	72	49	95	131	189	335	1529	...	...
Deanie Generating Sta.	NH 28 294388	128	...	47	101	207	49	97	128	57	59	84	129	173	364	1492	...	...
Upper Neaty Burn	NH 28 346436	549	...	111	223	300	75	157	155	112	97	156	114	203	381	2084	...	...
Struy	NH 28 394404	53	...	53	101	173	34	95	104	91	55	67	86	183	321	1362	...	...
Beauly																		
Aigas Dam	NH 28 475435	53	...	33	76	104	25	73	122	41	60	51	81	126	210	1002	...	...
Teanacoil	NH 28 492413	67	...	28	64	94	20	73	138	41	63	45	67	101	180	914	39	23 Jun
Ness																		
South Laggan	NN 27 295974	46	...	85	158	275	63	118	124	93	70	123	80	213	396	1797	73	17 Dec
Oich (Ness)																		
Garry (Ness)																		
Coire nan Gall	NM 17 926974	290	4193	155	346	728	170	276	293	154	203	343	222	457	926	4275	...	...
Cruadhach	NM 17 932994	216	4127	153	318	750	163	277	307	186	204	345	221	457	927	4309	...	...
Kinlochquoich	NG 18 961007	206	3708	135	293	625	142	241	239	141	153	296	190	447	889	3792	...	...
Loch a' Choire Bheithe	NG 18 988037	216	...	133	244	455	119	202	167	113	104	221	162	326	709	2953	...	...
Alltbeithe	NH 28 021072	232	2741	135	230	477	127	211	160	137	71	231	170	406	564	2918	...	...
Glenquoich	NH 28 025032	213	2763	154	248	455	108	189	192	122	87	215	142	356	728	2995	...	...
Lochan na Sgud	NN 27 061984	143	2539	111	245	396	86	157	161	76	76	186	139	317	627	2577	...	...
Kingie Camp	NH 28 102014	146	2167	91	195	327	79	137	141	76	67	164	123	254	505	2156	97	17 Dec
Inchlaggan	NH 28 174016	107	1792	73	157	255	45	98	114	67	51	121	106	199	398	1686	...	...
Invergarry Dam	NH 28 276022	90	...	77	106	182	31	97	115	46	54	93	87	159	308	1354	...	...
Ness																		
Aberchalder Lodge	NH 28 343033	55	...	73	117	156	33	91	125	86	66	87	61	139	254	1289	...	...
Lon Mor	NH 28 336068	160	1373	48	94	263	43	99	113	67	56	109	67	169	296	1425	...	...
Auchterawe	NH 28 368092	41	1232	64	93	159	35	101	121	47	47	53	60	175	306	1260	...	...
Tarff (Ness)																		
Fort Augustus	NH 28 381091	21	1170	49	82	150	35	81	119	53	56	74	59	136	290	1185	48	1 Dec
Moriston (Ness)																		
Strath Cluanie	NH 28 042113	262	...	100	217	561	154	256	179	135	107	178	161	288	509	2846	...	...
Cluanie Dam	NH 28 192100	183	...	101	197	271	69	152	115	58	84	135	107	197	500	1989	...	...
Loyne (Moriston)																		
Loyne Dam	NH 28 204078	229	...	88	176	292	82	125	131	53	89	139	111	207	421	1915		
Moriston (Ness)																		
Corrie Doe	NH 28 183146	285	...	101	197	272	84	122	113	56	66	139	107	230	493	1978	...	...
Ceannacroc Lodge	NH 28 226113	130	...	77	164	240	56	121	109	44	79	116	90	182	398	1675	...	...
Dundreggan Dam	NH 28 358157	104	...	67	86	140	23	91	131	44	56	67	60	129	347	1243	...	...
Loch Ma Stac	NH 28 340210	491	...	90	138	261	49	113	145	72	69	100	70	188	320	1616	...	...
Loch Liath	NH 28 337197	463	...	79	125	188	46	116	138	63	66	100	75	165	282	1442	...	...
Bhlaraidh Headpond	NH 28 355184	375	...	72	81	230	37	103	166	55	54	81	76	166	281	1402	...	...
Loch a'Chrathaich	NH 28 367208	536	...	77	121	164	44	109	136	69	83	101	73	121	351	1449	...	...
Bhlaraidh	NH 28 382165	70	...	62	76	135	24	91	127	38	53	67	61	128	354	1215	...	...
Invermoriston	NH 28 417170	21	1163	52	88	142	28	90	120	46	60	56	52	129	296	1160	...	...
Ness																		
Alltsaigh	NH 28 456192	29	1109	30	62	147	23	85	113	45	67	45	46	125	276	1064	...	...

Station	Grid reference	Altitude m	Av. 1916–1950 mm	Jan	Feb	Mar	Apr	May	Jun	Jul	Aug	Sep	Oct	Nov	Dec	Year	Highest recorded fall in a day mm	date
BEAULY AND NESS—*(contd)*																		
Foyers (Ness)																		
Sronlairig Lodge	NH 28 535070	358	...	94	88	148	37	119	178	56	50	81	107	118	224	1299	...	...
Aberchalder	NH 28 552201	195	1028	27	42	102	28	83	130	61	74	43	73	98	222	983	33	23 Jun
Garthbeg	NH 28 512182	192	998	31	45	116	26	84	122	64	79	50	71	101	210	1000	37	23 Jun
Foyers	NH 28 497211	40	933	29	49	112	24	79	139	57	76	47	69	86	203	971	31	23 Jun
Enrick (Ness)																		
Glenurquhart Forest	NH 28 426284	305	1157	65	113	136	40	89	135	60	88	64	106	99	221	1215	...	...
Balnain	NH 28 447302	122	...	42	82	117	17	80	123	61	65	52	89	119	207	1054	33	1 Dec
Ness																		
Loch Ashie	NH 28 636360	219	845	24	33	66	13	75	171	72	83	40	67	71	138	853	...	...
Inverness	NH 28 668462	4	660	13	38	67	15	58	180	55	65	33	65	62	110	762	58	14 Jun
Coastal																		
Dalcross	NH 28 766520	11	652	16	43	58	11	63	144	52	64	34	65	75	121	746	42	23 Jun
Achareidh	NH 28 868565	18	670	17	51	52	8	71	159	71	70	37	83	63	113	794	48	23 Jun
Balblair	NH 28 873553	29	728	15	59	56	6	66	163	70	72	39	98	68	115	827	53	23 Jun

BANFF, MORAY AND NAIRN RIVER PURIFICATION BOARD AREA, 6358 SQUARE KILOMETRES

HYDROMETRIC AREAS 7 TO 9

Station	Grid reference	Altitude m	Av. 1916–1950 mm	Jan	Feb	Mar	Apr	May	Jun	Jul	Aug	Sep	Oct	Nov	Dec	Year	Highest recorded fall in a day mm	date
Nairn																		
Brin House	NH 28 667299	213	...	15	50	111	20	89	173	81	90	48	90	108	140	1013	...	...
Farr	NH 28 682332	198	...	19	42	79	11	79	167	78	110	71	72	99	150	978	...	...
Cawdor Castle	NH 28 847498	69	806	16	55	65	8	68	226	72	74	40	102	84	128	939	62	23 Jun
Firhall	NH 28 881551	20	...	14	49	53	6	57	169	69	75	38	94	62	105	791	55	23 Jun
Coastal																		
Kintessack	NJ 38 001602	9	...	29	46	62	4	56	161	72	89	46	92	63	103	823	55	23 Jun
Muckle																		
Findhorn																		
Kyllachy	NH 28 790264	328	...	23	46	118	15	107	247	76	67	44	113	143	205	1204	...	...
Shenachie R.F.M. Sta. Area 417 km^2	NH 28 828339	255	1337	...	...	...	...	...	...	...	...	...	...	...	...	...	...	...
Moy	NH 28 766339	305	--	38	59	114	15	107	233	108	82	53	118	131	205	1265	...	...
Drynachan Lodge	NH 28 864397	229	1107	25	70	118	13	92	234	125	88	54	131	148	203	1299	75	23 Jun
Glenferness House	NH 28 937430	213	932	17	47	76	7	67	250	89	96	46	123	94	146	1058	74	23 Jun
Dorback (Findhorn)																		
Duthil Road	NH 28 944332	351	...	19	51	62	15	75	220	76	90	48	126	111	125	1018	...	...
Lochindorb Lodge	NH 28 969354	305	...	29	48	56	6	68	228	74	93	48	126	109	111	997	50	23 Jun
Coastal																		
Forres	NJ 38 047587	50	625	19	44	63	5	44	154	84	98	43	109	67	89	819	59	1 Oct
Findhorn	NJ 38 039644	6	610	25	46	67	5	49	143	63	75	42	91	68	95	769	...	...
Kinloss	NJ 38 067627	5	614	24	46	68	5	49	155	60	82	40	96	77	93	795	60	23 Jun
Covesea Skerries Lighthouse	NJ 38 204713	12	...	24	49	58	11	51	155	60	79	37	79	71	87	760	44	23 Jun
Lossiemouth	NJ 38 213699	6	633	27	47	63	10	56	176	57	87	57	76	93	89	837	40	23 Jun
Lossie																		
Dallas Lodge	NJ 38 108526	168	...	24	53	80	6	53	243	101	142	50	133	104	130	1120	78	23 Jun
Kellas House	NJ 38 165539	136	883	26	61	81	8	54	211	90	125	54	127	115	135	1088	...	...
Allt Creach	NJ 38 183518	238	...	32	51	77	20	51	216	89	126	56	121	135	125	1097	...	...
Glenlatterach Filters	NJ 38 200546	151	789	32	46	86	8	49	217	81	112	43	97	112	111	993	65	23 Jun
Black (Lossie)																		
Loch Romach	NJ 38 073518	155	...	24	57	83	10	55	258	97	126	55	133	95	145	1137	...	...
Lossie																		
Sheriffmills R.F.M. Sta. Area 216 km^2	NJ 38 198626	18	864	...	...	...	...	...	...	...	...	...	...	...	...	...	...	...
Lesmurdie House	NJ 38 226636	35	692	20	36	70	9	53	166	64	97	40	85	93	78	811	47	23 Jun
Elgin	NJ 38 229630	15	...	21	36	79	12	50	188	68	100	40	93	97	85	871	58	23 Jun
Elgin, Kirkhill	NJ 38 249628	11	689	22	38	72	7	47	175	69	100	44	86	107	91	859	41	23 Jun
Blackhills House	NJ 38 271586	94	804	28	52	80	11	59	189	86	113	48	95	109	113	984	57	1 Oct 23 Jun
Spey																		
Loch Spey	NN 27 422938	355	1742	77	136	249	43	108	169	49	69	111	135	229	330	1704	...	...
Mealgarve	NN 27 463959	346	1486	77	118	179	36	114	178	57	69	85	132	223	265	1534	...	...
Glenshero	NN 27 562929	268	...	53	77	118	26	86	147	43	55	85	81	142	295	1209	...	...
Cluny Castle	NN 27 647942	253	1046	41	77	107	17	110	131	40	60	73	71	139	244	1111	35	17 Dec
Truim (Spey)																		
Loch Cuaich	NN 27 702886	408	1200	60	82	129	35	90	119	54	53	54	119	97	213	1104	...	...
Invertruim R.F.M. Sta. Area 401 km^2	NN 27 688964	243	1491	...	...	...	...	...	...	...	...	...	...	...	...	...	...	...
Spey																		
Ruthven Bridge R.F.M. Sta. Area 534 km^2	NN 27 759996	221	1448	...	...	...	...	...	...	...	...	...	...	...	...	...	...	...
Kingussie	NH 28 757005	229	845	27	41	79	13	85	124	38	54	65	58	116	204	903	37	1 Dec
Tromie (Spey)																		
Tromie Dam	NN 27 763883	436	1302	83	99	123	44	137	175	49	79	75	124	153	292	1433	38	17 Dec
Tromie Bridge R.F.M. Sta. Area 130 km^2	NN 27 788995	240	1460	...	...	...	...	...	...	...	...	...	...	...	...	...	...	...
Spey																		
Feshie (Spey)																		
Glenfeshie Lodge	NN 27 846940	357	1093	52	70	80	29	67	157	70	75	48	113	122	177	1060	...	...
Spey																		
Kinrara R.F.M. Sta. Area 1010 km^2	NH 28 883085	210	1364	...	...	...	...	...	...	...	...	...	...	...	...	...	...	...
Achnagoichan	NH 28 913082	305	892	21	47	74	15	71	136	76	63	43	94	167	136	943	44	30 Nov
Druie (Spey)																		
Loch Morlich	NH 28 978097	327	1021	23	46	70	22	69	183	110	84	47	165	189	142	1148	...	...
Glenmore Lodge	NH 28 986095	341	1016	23	55	66	25	77	183	101	96	49	168	189	173	1204	67	23 Jun
Spey																		
Boat of Garten R.F.M. Sta. Area 1270 km^2	NH 28 946191	197	1311	...	...	...	...	...	...	...	...	...	...	...	...	...	...	...
Dulnain (Spey)																		
Dalrachney Beg	NH 28 896229	274	816	18	46	82	12	77	165	61	71	32	93	133	163	954	...	...
Balnaan Bridge R.F.M. Sta. Area 272 km^2	NH 28 976247	224	--	...	...	...	...	...	...	...	...	...	...	...	...	...	...	...

Station	Grid reference	Altitude *m*	Av. 1916–1950 *mm*	Jan	Feb	Mar	Apr	May	Jun	Jul	Aug	Sep	Oct	Nov	Dec	Year	Highest recorded fall in a day *mm*	*date*
BANFF, MORAY AND NAIRN—*(contd)*																		
Spey																		
Grantown R.F.M. Sta. Area 1750 km²	NJ 38 034268	193	1237	...	...	...	...	...	...	...	...	...	...	...	...	...	...	...
Leth-Allt Beag	NJ 38 088229	503	...	56	44	76	27	70	178	119	109	61	157	152	152	1203	...	...
Avon (Spey)																		
Glenlivet	NJ 38 188303	215	...	34	48	81	7	64	185	95	77	58	125	186	124	1085	38	23 Jun
Delnashaugh R.F.M. Sta. Area 544 km²	NJ 38 184352	150	--	...	...	...	...	...	...	...	...	...	...	...	...	...	...	...
Spey																		
Ballindalloch	NJ 38 185369	192	883	28	40	72	8	64	153	65	79	43	88	172	120	933	35	23 Jun
Aberlour R.F.M. Sta. Area 2640 km²	NJ 38 278439	79	1168	...	...	...	...	...	...	...	...	...	...	...	...	...	...	...
Fiddich (Spey)																		
Dufftown, Balvenie Street	NJ 38 326399	168	...	39	47	95	10	55	129	81	104	39	119	154	129	1001	...	...
Spey																		
Boat o' Brig R.F.M. Sta. Area 2850 km²	NJ 38 318518	43	1153	...	...	...	...	...	...	...	...	...	...	...	...	...	...	...
Gordon Castle	NJ 38 350595	32	751	31	39	74	4	39	139	88	115	47	85	94	90	846	50	23 Jun
Coastal																		
Rochomie Res.	NJ 38 441633	94	...	37	46	95	13	66	161	84	133	65	141	122	82	1046	...	...
Buckie	NJ 38 426657	15	...	26	44	73	6	45	127	80	103	47	93	85	88	817	48	23 Jun
Cullen House	NJ 38 503656	46	768	35	52	81	8	49	123	100	120	54	109	79	98	907	...	...
Banff	NJ 38 689647	24	744	45	40	81	11	58	105	49	115	57	106	69	80	816	63	4 Aug
Deveron																		
The Buck	NJ 38 397246	372	...	58	94	102	36	92	139	64	91	45	159	172	146	1198	...	...
Gauch Burn	NJ 38 345248	457	...	70	89	97	31	102	160	71	99	48	155	241	178	1340	...	...
Aldivalloch	NJ 38 359264	358	...	39	102	130	24	101	168	74	94	45	154	227	167	1327	48	23 Jun
Aldunie	NJ 38 358277	450	...	66	89	100	24	98	180	80	111	49	141	229	183	1350	...	...
Cabrach	NJ 38 389271	317	1031	56	88	93	28	93	145	75	97	44	153	185	152	1209	41	23 Jun
Three Burnshead	NJ 38 404294	375	...	56	86	102	32	98	150	76	117	62	136	161	164	1239	...	...
Black Water	NJ 38 326287	457	...	74	89	112	31	102	180	98	114	57	166	241	191	1454	...	...
Bogie (Deveron)																		
Bruntland	NJ 38 449280	329	...	54	93	102	33	95	145	78	117	43	154	187	140	1241	44	23 Jun
Avochie R.F.M. Sta. Area 440 km²	NJ 38 532464	82	991	...	...	...	...	...	...	...	...	...	...	...	...	...	...	...
Isla (Deveron)																		
Drummuir Castle	NJ 38 372441	189	957	38	60	90	15	74	150	79	120	49	127	164	123	1089	43	23 Jun
Keith, Isla Bank Mill	NJ 38 428514	110	943	50	58	95	9	66	165	99	145	45	121	127	95	1075	...	...
Deveron																		
Muiresk R.F.M. Sta. Area 956 km²	NJ 38 705498	25	940	...	...	...	...	...	...	...	...	...	...	...	...	...	...	...
Forglen House	NJ 38 697520	67	834	39	47	72	18	79	128	64	139	61	136	146	94	1023	79	4 Aug

DEE AND DON RIVER PURIFICATION BOARD AREA, 4873 SQUARE KILOMETRES

HYDROMETRIC AREAS 10 TO 12

Station	Grid reference	Altitude *m*	Av. 1916–1950 *mm*	Jan	Feb	Mar	Apr	May	Jun	Jul	Aug	Sep	Oct	Nov	Dec	Year	Highest recorded fall in a day *mm*	*date*
Coastal																		
Kinnaird Head Lighthouse	NJ 38 998675	19	679	28	64	61	13	50	89	48	104	55	107	99	95	813	40	4 Aug
Rattray Head	NK 48 093582	26	720	57	93	79	33	69	89	54	82	63	127	99	105	950	37	4 Aug
Ugie																		
Strichen	NJ 38 946551	53	892	44	75	69	21	75	105	49	113	59	127	132	109	980	61	4 Aug
South Ugie (Ugie)																		
Fedderate Res.	NJ 38 865526	119	930	45	68	85	24	81	109	58	130	56	145	120	99	1021	77	4 Aug
New Deer School	NJ 38 886470	121	889	38	70	73	23	80	114	66	99	54	141	137	97	994	48	23 Jun
Old Deer	NJ 38 978480	46	902	45	85	69	29	76	97	52	111	43	118	129	123	978	...	...
Ugie																		
Forehill W. Wks.	NK 48 097463	40	844	38	75	55	49	57	92	51	80	56	119	95	91	859	38	4 Aug
Ythan																		
Auchterless	NJ 38 714419	85	...	49	53	74	22	79	163	76	115	52	138	143	106	1069	65	23 Jun
Fyvie Castle	NJ 38 764393	61	880	49	76	76	38	80	117	78	122	53	143	132	121	1085	53	23 Jun
Haddo House	NJ 38 868347	57	848	38	81	68	31	74	95	61	81	52	143	116	89	930	43	23 Jun
Esslemont House	NJ 38 934304	30	838	32	77	42	31	61	89	72	69	47	122	66	79	787	...	...
Udny	NJ 38 906232	107	956	52	92	39	44	75	112	59	81	43	136	93	90	918	56	23 Jun
Culterty	NJ 38 999260	3	...	40	71	43	22	47	88	50	87	56	103	74	83	764	42	23 Jun
Don																		
Glenbuchat Lodge	NJ 38 333188	381	...	53	89	128	28	124	162	79	108	54	163	248	169	1405	52	23 Jun
Rinmore	NJ 38 414171	320	...	56	97	99	38	87	139	57	101	43	142	226	152	1236	...	...
Littlewood Park	NJ 38 516185	229	998	58	78	80	44	77	157	54	94	46	139	168	127	1123	83	23 Jun
Tullynessle	NJ 38 560192	191	...	57	76	76	26	58	134	42	74	41	127	135	89	935	...	...
Inverurie	NJ 38 779204	54	...	39	70	44	29	62	132	72	74	44	121	120	84	890	45	23 Jun
Urie (Don)																		
Insch	NJ 38 633280	122	...	58	79	67	26	73	136	49	80	43	145	136	83	977	63	23 Jun
Oldmeldrum	NJ 38 806272	96	...	50	71	45	31	70	109	59	92	55	122	125	98	927	47	23 Jun
Don																		
Dyce	NJ 38 883125	58	...	54	87	46	42	64	115	45	84	42	143	89	83	894	43	23 Jun
Craibstone	NJ 38 872107	91	861	58	82	47	43	70	123	51	95	49	131	81	87	916	46	23 Jun
Aberdeen, Grandholm Wks.	NJ 38 928096	9	868	59	87	45	45	65	125	51	95	62	155	89	79	960	...	...
Bridge of Don	NJ 38 938097	35	...	40	74	39	38	54	123	58	78	48	119	81	75	828	...	...
Dee																		
Derry Lodge	NO 37 036932	427	1115	32	95	56	29	109	122	19	66	47	126	165	168	1033	39	2 Dec
Clunie (Dee)																		
Braemar	NO 37 152914	339	927	49	81	51	49	94	108	26	74	50	122	122	118	943	32	19 Oct
Braemar, Irrigation Farm	NO 37 148921	323	979	52	83	71	45	96	109	29	75	50	125	153	156	1043	...	...
Dee																		
Balmoral	NO 37 260946	283	845	39	74	53	48	80	115	44	81	40	126	147	113	960	30	5 Oct
Gairn (Dee)																		
Corndavon Lodge	NJ 38 228021	427	931	51	63	56	25	99	117	53	107	61	125	168	145	1069	...	...
Gairnshiel Lodge	NJ 38 294007	335	900	41	52	51	33	103	141	50	104	33	128	176	123	1037	36	16 Nov
Muick (Dee)																		
Alltnaguibhsaich	NO 37 298857	418	1183	56	95	45	57	78	93	37	113	47	146	109	140	1017	...	...
Inchnabobart	NO 37 310875	387	1125	61	91	46	54	74	97	49	114	43	143	108	138	1019	37	3 Aug

Station	Grid reference	Altitude *m*	Av. 1916–1950 *mm*	Jan	Feb	Mar	Apr	May	Jun	Jul	Aug	Sep	Oct	Nov	Dec	Year	Highest recorded fall in a day *mm*	*date*
DEE AND DON—*(contd)*																		
Birkhall Sawmill	NO 37 348937	229	880	46	78	32	45	59	110	40	99	29	112	117	100	866	...	...
Dee																		
Ballater, Irrigation Farm	NO 37 380965	193	872	47	70	42	48	59	117	31	110	29	122	141	119	935	41	3 Aug
Dinnet	NJ 38 446025	177	...	48	65	37	40	59	132	38	81	34	116	116	93	860	30	23 Jun
Tanar (Dee)																		
Glen Tanar House	NO 37 474958	169	949	58	87	30	64	49	123	46	86	28	130	93	107	900	40	23 Jun
Tarland (Dee)																		
Tarland	NJ 38 488040	142	...	57	64	40	33	59	136	38	71	33	115	121	93	858	...	...
Aboyne	NO 37 542982	116	764	52	71	40	36	47	103	42	76	27	131	107	93	825	...	...
Dee																		
Lumphanan	NJ 38 575036	145	807	69	68	49	43	52	129	44	82	43	129	121	93	924	...	...
Woodend R.F.M. Sta. Area 1370 km²	NO 37 632960	67	1146	...	...	...	...	...	...	...	...	...	...	...	...	...	...	...
Invercannie W.Wks	NO 37 666964	64	839	47	96	37	47	53	98	51	86	30	142	79	87	853	36	5 Oct
Feugh (Dee)																		
Bridge of Dye	NO 37 652862	169	1141	86	147	29	102	102	113	59	129	49	176	92	148	1232	53	5 Oct
Strachan	NO 37 696917	120	...	55	116	34	73	75	143	55	107	33	192	69	113	1065	81	5 Oct
Leuchar (Dee)																		
Drum Castle	NJ 38 796005	98	840	64	105	37	48	64	115	54	91	54	149	70	93	942	40	5 Oct
Dee																		
Bieldside, Kerrera	NJ 38 872036	122	902	58	116	34	52	72	134	49	115	38	132	88	88	976	...	...
Bieldside, Cairnlee	NJ 38 882028	73	...	53	110	33	50	79	139	53	103	39	137	82	86	965	...	...
Aberdeen	NJ 38 915042	52	836	47	91	27	44	63	126	52	107	36	125	85	81	882	51	23 Jun
Aberdeen, Rubislaw	NJ 38 916058	61	855	53	97	35	47	65	133	54	101	42	136	88	84	933	...	...
Craigiebuckler	NJ 38 903045	67	890	52	104	37	51	67	134	60	118	41	136	96	86	981	54	23 Jun

NORTH AND SOUTH ESK AREA, 2025 SQUARE KILOMETRES

HYDROMETRIC AREA 13

Station	Grid reference	Altitude *m*	Av. 1916–1950 *mm*	Jan	Feb	Mar	Apr	May	Jun	Jul	Aug	Sep	Oct	Nov	Dec	Year	Highest recorded fall in a day *mm*	*date*
Coastal																		
Girdleness Lighthouse	NJ 38 972054	12	725	32	75	29	29	47	108	53	80	36	129	86	90	796	...	...
Cowie																		
Bervie																		
North Esk																		
Glen Lee	NO 37 374815	471	...	128	166	51	113	111	140	47	146	57	201	171	145	1476	...	...
Bridge of Lee	NO 37 395803	280	...	102	187	57	114	109	127	35	135	53	209	161	128	1417	...	...
Inchgrundle	NO 37 410793	270	1256	112	181	46	113	106	119	32	132	55	213	155	135	1400	66	5 Oct
Invermark Lodge	NO 37 431803	273	1172	105	179	44	111	101	117	25	129	52	203	150	138	1352	62	5 Oct
Whitehillocks	NO 37 448800	258	1132	97	181	41	111	109	112	22	120	52	197	147	139	1327	61	5 Oct
Cruick (North Esk)																		
Stracathro Hospital	NO 37 626655	37	829	76	106	27	90	81	84	43	105	31	136	37	82	899	...	...
North Esk																		
Edzell	NO 37 623690	49	...	60	109	43	84	76	101	38	91	65	137	41	93	938	...	...
Luther (North Esk)																		
Auchenblae	NO 37 726787	107	860	57	109	34	84	83	109	45	117	43	147	77	88	994	47	5 Oct
Fettercairn	NO 37 669782	171	990	78	118	43	104	95	97	47	131	47	167	69	113	1110	50	5 Oct
Fasque	NO 37 649757	98	944	71	114	51	87	79	105	51	123	56	164	81	112	1094	47	5 Oct
North Esk																		
Kinnaber W. Wks.	NO 37 713621	9	719	43	88	23	76	70	90	49	108	29	97	31	56	761	36	5 Oct
Coastal																		
Montrose, Wellington Park	NO 37 722582	5	731	45	100	23	73	61	92	39	116	31	113	31	62	787	41	5 Oct
Prosen (South Esk)																		
Cormuir	NO 37 306667	265	1183	102	168	51	109	127	135	58	76	63	185	109	117	1300	...	...
Noran (South Esk)																		
Den of Ogil	NO 37 439620	202	1018	79	122	29	118	85	114	42	116	62	127	66	115	1076	46	9 Apr
Glenogil W. Wks.	NO 37 445625	210	1036	87	110	29	122	85	113	37	120	95	129	59	124	1110	48	9 Apr
South Esk																		
Montrose	NO 37 707617	57	728	47	102	22	84	74	93	47	115	29	113	39	64	829	43	5 Oct
Coastal																		
Usan	NO 37 724544	30	...	39	84	17	54	53	84	32	102	29	95	26	47	661	30	3 Aug 5 Oct
Arbroath	NO 37 649411	29	667	42	64	17	61	48	102	33	144	41	83	35	54	725	56	3 Aug
Muirheads Res.	NO 37 607443	77	...	65	104	19	101	67	107	29	160	37	106	25	68	889	...	...

TAY RIVER PURIFICATION BOARD AREA, 7130 SQUARE KILOMETRES

HYDROMETRIC AREAS 14 TO16

Station	Grid reference	Altitude *m*	Av. 1916–1950 *mm*	Jan	Feb	Mar	Apr	May	Jun	Jul	Aug	Sep	Oct	Nov	Dec	Year	Highest recorded fall in a day *mm*	*date*
Coastal																		
Fallaws	NO 37 503405	177	892	92	127	22	123	81	117	65	187	62	126	39	88	1126	...	...
Crombie Res.	NO 37 522404	161	908	92	129	24	126	76	118	77	173	56	118	41	91	1120	61	4 Aug
Monikie Res.	NO 37 502383	151	863	84	135	21	128	75	115	74	155	56	105	47	84	1080	53	4 Aug
Carnoustie	NO 37 558344	9	740	52	110	17	93	56	111	75	141	51	95	43	70	914	46	3 Aug
Buddon Ness	NO 37 539309	5	714	55	105	16	100	55	97	70	130	51	80	42	68	868	44	3 Aug
Dighty																		
Clatto Res.	NO 37 368344	152	824	73	102	22	106	81	107	46	164	55	99	45	87	986	67	4 Aug
Coastal																		
Dundee	NO 37 422313	45	789	63	88	17	114	59	101	46	131	61	93	49	68	889	50	9 Apr
Stobsmuir Res.	NO 37 414321	87	767	77	105	18	121	65	106	47	130	54	92	47	72	935	53	9 Apr
Balruddery	NO 37 303329	131	...	56	85	21	89	77	124	54	143	49	79	30	78	884	...	...
Dundee, Camperdown Wks.	NO 37 382318	87	791	72	102	18	99	70	123	40	137	53	95	45	73	929	44	4 Aug
Mylnefield	NO 37 341299	30	714	52	89	19	104	54	106	41	109	53	74	46	70	817	42	9 Apr
Dundriven	NO 37 251343	213	...	59	82	20	90	81	153	36	136	55	113	29	86	942	49	5 Oct
Fingask	NO 37 223294	166	784	58	119	28	122	84	159	53	90	30	117	41	84	984	...	...
Errol Park	NO 37 247226	41	645	46	86	22	85	63	114	28	70	39	95	38	56	743	...	...
Tay																		
Cononish	NN 27 301285	288	...	190	282	300	133	230	159	84	84	269	165	246	410	2552	63	17 Dec
Glendochart	NN 27 453274	168	2234	187	281	189	127	185	203	52	100	164	151	152	401	2193	...	...
Auchlyne West	NN 27 485306	427	2403	167	274	223	147	151	201	61	99	173	153	212	373	2234	...	...

Station	Grid reference	Altitude	Av. 1916–1950	Rainfall in millimetres													Highest recorded fall in a day	
		m	*mm*	Jan	Feb	Mar	Apr	May	Jun	Jul	Aug	Sep	Oct	Nov	Dec	Year	*mm*	*date*
TAY—*(contd)*																		
Glenoglehead	NN 27 558281	287	1860	197	278	134	119	156	139	28	81	138	132	123	274	1799	58	17 Dec
Kinnell	NN 27 577329	116	1581	101	153	119	83	113	117	29	53	125	106	122	268	1390	...	...
Lochay (Tay)																		
Ben Challum	NN 27 396334	457	2963	209	291	319	184	244	208	103	125	286	160	318	499	2946	...	...
Badvaim	NN 27 412338	274	2771	179	256	353	132	189	218	79	94	234	145	299	466	2645	...	...
Innisdaimh	NN 27 468346	527	2566	212	286	245	134	193	182	42	90	177	147	196	347	2251	...	...
Lairig nan Lunn	NN 27 447390	524	2370	182	258	312	127	174	196	61	98	201	141	235	447	2431	...	...
Murlaganmore	NN 27 538342	229	1871	142	197	177	108	148	166	35	79	174	126	173	359	1883	...	...
Lochay Power Sta.	NN 27 546350	114	...	108	151	140	82	129	136	25	67	127	119	144	254	1481	56	17 Dec
Tay																		
Achmore Burn	NN 27 587296	430	...	205	266	123	117	140	150	41	69	137	148	128	285	1809	...	...
Meall Liath	NN 27 583356	539	1602	138	169	132	90	139	172	38	99	152	117	109	265	1621	...	...
Finlarig Power Sta., North	NN 27 580348	183	1538	123	161	129	80	125	148	30	63	119	117	84	292	1472	...	...
Lawers Dam	NN 27 597391	515	1913	135	187	156	120	158	147	43	100	164	135	118	281	1743	...	...
Allt a'Mhoirneas	NN 27 604383	457	1832	135	169	161	107	151	154	47	101	146	141	130	292	1734	...	...
Allt meall nan Damh	NN 27 663317	457	1862	188	255	103	102	153	140	38	77	128	127	103	261	1674	...	...
Braentrian	NN 27 662342	259	1685	137	204	104	88	143	130	31	73	117	122	102	235	1487	46	17 Dec
Lochan nan Cat	NN 27 648424	719	...	203	254	231	129	175	183	46	142	183	161	140	317	2165	...	...
Ardtalnaig	NN 27 702394	130	1318	107	137	90	91	103	126	28	69	95	99	74	158	1177	28	5 Oct 17 Dec
Beinn Bhreac	NN 27 716398	518	...	132	218	101	116	122	130	34	88	110	111	89	185	1435	...	...
Kenmore	NN 27 761452	116	1088	101	119	64	66	96	159	32	70	77	100	61	141	1085	35	25 Jan
Lyon (Tay)																		
Allt Tarabhan	NN 27 371396	366	...	164	260	306	156	195	260	99	66	230	188	312	504	2740	...	...
Invermearan	NN 27 402419	366	2628	164	267	354	124	189	203	101	95	206	153	270	480	2605	...	...
Lubreoch Dam	NN 27 457417	312	2065	164	274	212	90	159	150	52	77	166	114	185	395	2037	...	...
Phubuill	NN 27 455435	546	2117	165	279	238	90	173	153	79	84	181	116	226	415	2199	...	...
Giorra Dam	NN 27 510465	419	1812	145	225	185	97	133	144	43	77	148	123	178	331	1830	...	...
Meggernie Castle	NN 27 552462	229	1630	137	197	136	97	131	126	43	69	109	129	140	287	1601	58	17 Dec
Coire Riadhailt	NN 27 576399	610	...	203	235	151	145	187	177	55	95	180	165	162	364	2118	...	...
Finglen No.2	NN 27 625438	472	2065	197	269	149	150	206	159	36	104	151	143	122	275	1959	...	...
Glenlyon House	NN 27 734472	130	1231	129	159	89	68	94	141	35	76	103	107	85	173	1260	37	25 Jan
Tay																		
Aberfeldy	NN 27 854489	103	985	85	115	42	74	85	155	32	85	56	116	49	131	1025	...	...
Derculich	NN 27 891526	107	923	91	120	46	52	83	182	30	93	63	113	57	122	1054	...	...
Pitnacree R.F.M.Sta. Area 1150 km²	NN 27 924534	61	1946	...	...	...	...	...	...	...	...	...	...	...	...	...	...	...
Tummel (Tay)																		
Gaur Power Sta.	NN 27 464572	242	1388	93	140	137	54	83	129	33	52	98	84	136	253	1291	...	...
Ericht (Tummel)																		
Ericht Dam	NN 27 511629	365	1457	103	154	148	57	103	132	42	54	116	89	172	270	1441	...	...
Tummel (Tay)																		
Loch Rannoch Power House	NN 27 531582	206	1149	80	121	108	56	82	125	41	47	90	80	109	223	1162	39	17 Dec
Dall	NN 27 593562	232	...	83	154	90	56	79	123	23	57	91	118	76	199	1148	...	...
Errochty Switching Sta.	NN 27 770590	148	...	89	132	54	45	84	133	27	77	63	124	65	162	1055	34	5 Oct
Domnaheich	NN 27 792579	195	1052	112	137	52	49	95	137	28	82	66	116	57	157	1089	...	...
Fincastle House	NN 27 871622	299	933	78	111	57	45	101	201	24	85	64	111	60	132	1069	(36	11 May)
Garry (Tummel)																		
Edendon Bridge	NN 27 715706	317	1233	97	155	95	46	135	143	27	70	77	121	106	241	1312	...	...
Sronphadruig Lodge	NN 27 714791	497	1657	178	192	171	74	161	201	53	96	112	152	136	323	1849	...	...
Dalnacardoch	NN 27 727672	434	...	104	135	109	57	107	150	42	64	84	147	124	214	1337	...	...
Errochty (Garry)																		
Trinafour	NN 27 725649	262	...	99	123	104	49	103	149	39	64	76	129	120	170	1227	43	17 Dec
Bruar (Garry)																		
Bruar Intake	NN 27 823726	381	...	89	94	89	32	140	126	38	62	90	103	100	174	1137	...	...
Garry (Tummel)																		
Blair Castle	NN 27 864658	137	910	54	80	55	33	93	160	34	55	66	98	67	141	935	31	11 May
Tilt (Garry)																		
Garry (Tummel)																		
Tummel (Tay)																		
Faskally	NN 27 918599	94	900	59	92	51	34	97	213	25	75	67	98	52	117	979	40	20 Jun
Pitlochry	NN 27 945588	144	854	80	93	54	47	97	212	27	74	66	89	50	121	1009	49	20 Jun
Tay																		
Lochan Oisinneach Mor	NO 37 026543	427	...	99	145	37	85	118	206	37	107	57	131	43	127	1191	...	...
Capel Hill West	NO 37 029519	354	...	93	113	42	93	108	207	36	112	57	135	51	142	1187	...	...
Loch Ordie	NO 37 029501	299	...	95	123	36	94	107	203	31	111	57	129	32	122	1139	...	...
Braan (Tay)																		
Glenquaich Lodge	NN 27 866368	274	...	140	232	63	138	141	163	49	108	80	166	73	162	1513	55	5 Oct
Tay																		
Caputh R.F.M.Sta. Area 3210 km²	NO 37 088395	35	1648	...	...	...	...	...	...	...	...	...	...	...	...	...	...	...
Isla (Tay)																		
Sheilin'	NO 37 196756	466	...	140	168	74	127	151	159	24	137	126	138	156	206	1605	...	...
Tulchan Lodge	NO 37 185723	442	...	121	155	66	114	142	156	23	136	113	133	150	201	1510	...	...
The Linns	NO 37 199704	396	...	114	140	58	104	142	142	22	135	109	140	117	183	1406	...	...
Auchavan	NO 37 191697	366	...	114	117	66	99	132	137	20	97	104	129	94	201	1311	...	...
Dalvanie	NO 37 186657	311	1161	114	140	56	112	140	152	51	94	76	137	102	137	1311	...	...
Forter R.F.M.Sta. Area 70.7 km²	NO 37 187647	285	1427	...	...	...	...	...	...	...	...	...	...	...	...	...	...	...
Newton (Isla)																		
Newton R.F.M.Sta. Area 15.4 km²	NO 37 230605	256	1288	...	...	...	...	...	...	...	...	...	...	...	...	...	...	...
Melgam (Isla)																		
Barney	NO 37 251634	341	1144	109	150	53	119	104	129	33	81	61	165	67	136	1209	...	...
Glenhead	NO 37 273648	366	1191	122	152	56	111	104	132	41	107	66	180	75	133	1279	...	...
Clintlaw	NO 37 258616	317	...	109	163	34	122	112	125	28	98	67	157	52	152	1219	...	...
Backwater	NO 37 255589	259	1068	94	126	36	109	102	109	29	102	61	147	51	127	1092	...	...
Dykends	NO 37 253569	229	979	99	155	36	123	105	129	36	95	66	142	44	132	1163	...	...
Loch of Lintrathen R.F.M.Sta. Area 40.9 km²	NO 37 275558	212	1166	...	...	...	...	...	...	...	...	...	...	...	...	...	...	...
Quharily (Melgam)																		
Glendye	NO 37 296621	320	1206	121	178	39	136	113	137	33	123	57	189	69	147	1342	...	...
Longdrum	NO 37 278619	319	1132	105	147	43	133	105	127	28	114	63	169	69	135	1239	...	...
Westerton	NO 37 282598	274	1081	103	150	38	127	98	113	25	113	62	160	56	152	1198	...	...
Newton	NO 37 285573	224	1012	95	119	33	121	94	126	31	105	62	142	39	121	1088	...	...

Station	Grid reference	Altitude m	Av. 1916–1950 mm	Jan	Feb	Mar	Apr	May	Jun	Jul	Aug	Sep	Oct	Nov	Dec	Year	Highest recorded fall in a day mm	date
TAY—*(contd)*																		
Loch of Lintrathen R.F.M.Sta. Area 24.7 km^2	NO 37 280559	199	1107	...	...	...	...	...	...	...	...	...	...	...	...	...	...	...
Melgam (Isla)																		
Balnakeilly	NO 37 281554	209	933	76	95	33	89	97	132	36	102	67	120	39	107	995	...	...
Lintrathen	NO 37 274539	210	897	74	100	31	85	91	116	36	97	58	111	37	97	933	44	5 Oct
Dean (Isla)																		
Glamis Castle	NO 37 388486	61	859	74	122	19	102	67	159	43	111	47	126	39	88	999	45	5 Oct
Ericht (Isla)																		
Glenshee Lodge	NO 37 133684	335	1153	97	134	59	70	141	134	25	80	89	127	96	159	1211	42	5 Oct
Ardle (Ericht)																		
Straloch	NO 37 036642	290	...	70	114	63	58	114	161	25	76	79	116	69	142	1087	...	...
Pitcarmick	NO 37 087568	198	...	89	133	36	84	129	168	28	86	67	119	54	134	1126	50	5 Oct
Lornty (Ericht)																		
Craigsheal Burn	NO 37 058532	451	...	102	147	39	83	122	216	41	103	65	140	53	137	1247	...	...
Craig Wood South	NO 37 060518	344	...	80	129	39	91	121	209	36	106	60	138	53	136	1199	...	...
Loch Benachally	NO 37 078497	308	...	103	141	23	109	123	215	38	125	62	144	37	118	1236	...	...
Loch Benachally South	NO 37 078491	323	...	69	114	29	89	117	197	32	124	60	131	33	125	1120	...	...
Meall Dubh	NO 37 074533	381	...	104	129	31	82	133	190	37	106	65	141	63	146	1228	...	...
Ericht (Isla)																		
Blairgowrie, Burnhead	NO 37 174459	134	...	90	131	21	77	88	174	37	101	56	99	28	82	983	...	...
Lunan (Isla)																		
Cardney House	NO 37 051452	107	980	72	122	29	93	115	179	38	138	60	130	35	107	1116	51	5 Oct
Capel Hill North	NO 37 038532	411	...	105	141	38	89	117	206	38	108	62	137	51	132	1224	...	...
Blairgowrie	NO 37 175438	70	759	61	89	18	65	86	150	39	114	50	91	28	78	869	37	3 Aug
Isla (Tay)																		
Burrelton	NO 37 207378	52	...	65	86	23	67	72	133	41	119	48	107	28	79	868	...	...
Tay																		
Ballathie R.F.M.Sta. Area 4580 km^2	NO 37 147367	26	1471	...	...	...	...	...	...	...	...	...	...	...	...	...	...	...
Almond (Tay)																		
Almond Intake	NN 27 758332	375	...	198	241	104	139	153	198	41	102	119	157	109	207	1769	...	...
Fendoch Burn	NN 27 885297	451	...	178	211	61	168	132	178	66	119	132	114	95	192	1646	...	...
Tulchan House	NN 27 950285	152	...	94	126	58	118	107	185	56	119	91	129	70	152	1304	...	...
Methven W.Wks.	NN 27 992267	137	940	69	135	43	89	85	215	55	114	83	100	51	117	1154	...	...
Tay																		
Inveralmond	NO 37 099258	8	...	47	83	24	79	74	143	38	109	48	90	33	100	869	33	23 Jun
Montague Res.	NO 37 191281	162	823	63	122	28	108	77	159	31	158	60	115	39	79	1039	...	...
Perth	NO 37 101239	23	780	57	97	31	91	70	144	30	107	57	97	36	84	900	31	3 Aug
Earn (Tay)																		
Lochearnhead	NN 27 592239	113	1599	123	170	113	107	116	125	29	74	127	110	113	211	1419	...	...
Edinchip	NN 27 575224	171	1746	152	209	145	136	146	151	30	91	131	134	121	212	1657	...	...
Glen Tarken	NN 27 655283	357	...	173	198	86	131	134	139	30	80	115	136	115	232	1569	...	...
Aberuchill R.F.M.Sta. Area 177 km^2	NN 27 754216	62	1859	...	...	...	...	...	...	...	...	...	...	...	...	...	...	...
Ruchill (Earn)																		
Ben Vorlich	NN 27 648166	433	...	255	286	152	180	191	185	44	143	185	184	149	291	2246	...	...
Auchinner	NN 27 696157	198	...	188	251	114	166	174	155	47	125	135	149	127	238	1869	...	...
Findhu Glen	NN 27 735141	274	...	191	258	131	161	164	175	45	143	155	129	137	247	1935	...	...
Dalchruin	NN 27 716168	174	...	186	249	107	163	162	143	40	124	123	143	105	228	1773	...	...
Earn (Tay)																		
Allt an Druichd	NN 27 679307	533	...	149	154	79	117	151	152	31	90	125	137	116	252	1553	...	...
Lednock Dam	NN 27 729287	347	1547	162	184	80	117	127	154	37	91	117	145	90	206	1511	...	...
Meall nan Fiadh	NN 27 712275	564	...	159	150	89	154	145	159	37	94	125	144	105	256	1616	...	...
Turret (Earn)																		
Glenturret Lodge	NN 27 802293	381	1650	155	231	102	147	170	191	46	152	119	183	104	236	1836	...	...
Turret Dam	NN 27 822263	341	1451	140	170	74	139	112	159	59	121	106	159	61	193	1494	...	...
Lower Barvick	NN 27 843263	335	...	109	165	66	127	117	160	41	114	107	137	107	244	1493	...	...
Earn (Tay)																		
Strathearn Hydro	NN 27 867223	122	1041	80	122	33	97	95	128	54	76	78	127	51	119	1059	51	5 Oct
Kinkell Bridge R.F.M.Sta. Area 590 km^2	NN 27 933167	15	1608	...	...	...	...	...	...	...	...	...	...	...	...	...	...	...
Gleneagles	NN 27 918112	152	...	79	154	50	101	94	166	56	125	81	105	62	107	1179	33	3 Sep 5 Oct
Kippen House	NO 37 018130	137	...	83	114	35	101	92	128	64	129	56	105	49	99	1055	...	...
Water of May (Earn)																		
Knowehead	NO 37 036096	299	...	108	150	53	103	94	170	60	166	88	105	57	119	1274	...	...
Earn (Tay)																		
Chapel Burn	NO 37 058094	308	1227	99	150	46	104	98	171	58	170	86	105	56	113	1257	...	...
Slateford Burn	NO 37 087097	236	1197	86	156	47	118	102	165	57	173	86	103	61	109	1263	...	...
Strathallan School	NO 37 093188	18	...	50	106	27	96	80	144	30	115	53	99	40	73	913	38	3 Aug
Upper Farg	NO 37 111108	166	1104	77	134	38	113	94	155	50	148	76	101	55	110	1153	33	5 Oct
Glenfarg W. Wks.	NO 37 127127	206	1017	81	151	32	105	80	138	43	138	72	109	41	89	1081	...	...
Ramsheugh	NO 37 136151	94	934	72	127	27	101	71	131	44	129	64	105	40	85	996	...	...
Coastal																		
Balmerino Manse	NO 37 367245	61	722	59	96	21	131	63	110	30	128	60	84	36	77	896	49	9 Apr
Newport, Tayfield	NO 37 422274	30	773	67	102	16	123	60	107	39	117	52	91	39	72	885	55	9 Apr
Eden																		
Over Rankeillour	NO 37 327134	52	...	54	108	18	85	65	128	42	122	51	96	46	69	884	38	3 Aug
Letham	NO 37 306144	64	...	59	103	31	106	70	135	43	131	54	96	41	73	943	...	...
Pitlair	NO 37 317122	35	773	54	104	20	78	67	125	40	131	45	101	48	73	887	38	5 Oct
Cupar	NO 37 388152	25	736	47	101	15	85	59	111	42	125	42	92	54	73	847	57	3 Aug
Clatto Res.	NO 37 366078	154	870	52	101	25	97	81	107	43	127	47	107	54	81	920	47	3 Aug
Leuchars	NO 37 468208	10	678	57	102	11	100	51	90	44	140	53	80	42	69	839	63	3 Aug
Coastal																		
St.Andrews	NO 37 503167	18	713	55	86	12	86	53	87	61	147	45	66	37	61	795	65	3 Aug
St.Andrews W.Wks.	NO 37 507148	66	735	56	109	14	107	65	98	54	157	39	88	34	82	904	60	3 Aug
Cameron W.Wks.	NO 37 478114	146	823	66	121	25	119	88	115	63	182	54	98	42	106	1080	60	3 Aug

FORTH RIVER PURIFICATION BOARD AREA, 3078 SQUARE KILOMETRES
HYDROMETRIC AREAS 17 AND 18

Station	Grid reference	Altitude m	Av. 1916–1950 mm	Jan	Feb	Mar	Apr	May	Jun	Jul	Aug	Sep	Oct	Nov	Dec	Year	Highest recorded fall in a day mm	date
Coastal																		
Isle of May Lighthouse	NT 36 654994	55	521	27	47	24	77	69	82	42	144	30	60	61	72	735	51	3 Aug
Belliston	NO 37 500055	82	775	45	87	23	103	77	102	43	156	46	75	47	77	883	63	3 Aug
Ovenstone W. Wks.	NO 37 538051	51	707	44	82	21	84	70	101	47	187	45	70	48	75	876	65	4 Aug
Leven, Silverburn	NO 37 393019	9	716	41	111	26	93	68	127	37	124	42	81	70	75	895	42	3 Aug
Leven																		
Ochil Hills Hospital	NO 37 098074	235	1206	67	153	67	89	95	153	49	165	93	95	74	109	1212	...	...
Milnathort	NO 37 124047	145	923	63	117	43	81	69	128	38	122	79	77	67	93	977	...	...
Lochglow Res.	NT 36 094961	277	1101	72	117	77	107	88	145	53	133	93	104	98	141	1228	...	...
Loch Leven Sluices	NT 36 171994	110	942	54	135	47	109	75	165	51	120	63	89	88	92	1086	(40	5 Oct)
Lothrie Res.	NO 37 222038	206	947	66	153	41	110	78	134	41	122	63	85	70	102	1063	39	5 Oct
Markinch, Rothes	NO 37 283015	60	831	52	136	33	111	79	129	41	131	53	89	64	93	1012	36	5 Oct
Ore (Leven)																		
Craigaveral	NT 36 087940	232	...	63	141	62	114	83	133	49	129	79	100	88	125	1167	...	...
Roscobie Res.	NT 36 090934	201	1013	65	126	61	111	81	132	52	138	77	98	88	121	1149	...	...
Cowdenbeath S. Wks.	NT 36 170917	119	...	57	128	63	115	88	147	51	121	69	73	83	98	1090	33	5 Oct
Ballingry	NT 36 180972	130	...	52	106	44	94	74	162	50	119	66	81	84	93	1024	39	5 Oct
Leven																		
Carriston Res.	NO 37 327035	88	869	48	141	30	107	85	105	41	125	54	84	72	88	979	38	5 Oct
Coul Res.	NO 37 271037	131	898	55	130	35	97	79	132	39	120	57	89	60	96	989	41	5 Oct
Coastal																		
Kirkcaldy, Carberry	NT 36 284948	99	785	47	115	40	107	75	119	56	129	50	72	68	76	954	33	3 Aug
Kirkcaldy	NT 36 268912	32	755	39	103	46	84	69	120	45	119	47	61	65	69	867	37	3 Aug
Cullaloe Res.	NT 36 186872	90	...	49	106	41	107	65	116	55	131	61	70	67	75	945	32	3 Aug
Pitreavie	NT 36 117848	40	...	39	84	28	82	55	122	51	121	76	72	45	69	843	31	3 Aug
Tulliallan	NS 26 936881	27	860	45	101	36	54	63	118	50	116	76	79	61	99	898	34	13 Aug
Forth																		
Achaidh Burn	NN 27 369043	366	2533	265	268	207	146	209	171	70	136	235	160	213	380	2461	...	...
Comer	NN 27 385041	183	...	237	257	215	145	197	164	56	117	222	152	208	383	2353	...	...
Beinn Uamha	NN 27 388062	274	2355	203	265	211	152	196	155	57	117	206	157	183	376	2278	...	...
Glen Dhu	NN 27 400035	152	2205	212	235	174	133	193	157	48	116	213	146	191	324	2142	...	...
Beinn a' Bhan	NN 27 400017	457	...	236	295	235	153	209	191	76	138	230	165	209	383	2521	...	...
Loch Dhu	NN 27 429037	99	2059	177	256	186	143	173	144	56	108	199	143	177	305	2066	57	5 Oct
Gribloch Gardens	NS 26 639935	162	1286	91	140	79	108	105	162	40	97	119	101	90	142	1275	31	5 Oct
Goodie (Forth)																		
Thornhill	NN 27 664002	38	...	84	130	81	77	90	153	36	82	86	81	75	149	1125	...	...
Forth																		
Meiklewood	NS 26 727957	12	1119	78	145	64	74	99	161	37	109	89	86	71	126	1139	...	...
Touch Res. No.3	NS 26 732917	231	1489	101	196	108	117	123	175	52	151	123	114	115	169	1544	45	5 Oct
Touch Res. No.2	NS 26 735918	215	1504	97	191	96	120	116	175	54	151	119	114	102	148	1482	46	5 Oct
Touch Res. No.1	NS 26 746917	139	1413	100	183	97	111	105	166	43	138	106	107	105	156	1417	44	5 Oct
Teith (Forth)																		
Blaircreach	NN 27 441178	140	2578	203	247	219	127	197	217	47	93	211	143	173	326	2203	...	...
Tulloch Lodge	NN 27 530208	137	...	162	194	143	120	149	148	36	86	151	124	143	283	1740	...	...
Meall Reamhar	NN 27 517233	564	...	205	208	171	168	184	184	46	102	196	141	196	323	2121	...	...
Strathyre	NN 27 561167	130	2053	195	249	137	141	175	185	51	113	152	140	125	280	1943	55	29 Jan
Vennachar (Teith)																		
Upper Glengyle	NN 27 360147	213	2460	243	243	229	141	178	204	63	77	216	141	193	368	2297	...	...
Glengyle	NN 27 388133	122	2331	231	258	248	135	191	173	61	97	229	153	197	387	2360	72	17 Dec
Allt-a-Choin	NN 27 414143	381	2349	241	218	242	140	166	165	53	65	205	165	191	359	2212	...	...
Stronachlachar	NN 27 401103	117	...	193	232	200	130	162	143	52	81	181	158	180	333	2046	79	17 Dec
Loch Katrine, Tunnel Hill Top	NN 27 417088	253	2115	230	281	192	128	169	142	46	86	194	146	159	343	2116	...	...
Strone Bridge	NN 27 454106	128	1900	167	215	159	97	151	150	49	92	155	129	143	277	1785	...	...
Ben Venue	NN 27 473077	274	...	208	223	154	104	155	154	28	102	156	105	128	305	1824	...	...
Glen Finglas	NN 27 491126	335	2051	187	212	197	127	161	166	67	95	192	156	174	329	2064	...	...
Glen Meann	NN 27 522125	244	2025	222	207	182	127	146	182	55	85	177	141	164	309	1995	...	...
Achnahard	NN 27 529094	198	1751	220	258	136	128	146	174	39	85	138	136	146	260	1867	...	...
Brig o' Turk	NN 27 537063	84	1646	156	211	133	104	132	148	47	97	152	133	142	232	1687	51	5 Oct
Loch Drunkie	NN 27 549041	122	1674	135	192	135	99	125	155	56	91	157	121	125	226	1615	...	...
Loch Vennachar	NN 27 598063	84	1500	114	187	125	82	136	149	40	106	143	121	120	198	1520	53	5 Oct
Teith (Forth)																		
Callander	NN 27 634080	107	...	117	197	123	89	149	153	33	103	137	109	116	(203)	1529	28	22 Apr
Bridge of Teith R.F.M. Sta. Area 518 km²	NN 27 725011	15	1996	...	...	...	...	...	...	...	...	...	...	...	...	...	...	...
Allan Water (Forth)																		
Blackford	NN 27 898080	158	1215	99	165	61	104	102	138	55	136	97	114	81	114	1266	...	...
Kinbuck R.F.M. Sta. Area 161 km²	NN 27 792053	93	--	...	...	...	...	...	...	...	...	...	...	...	...	...	...	...
Ledcameroch	NN 27 788019	98	...	68	103	66	61	86	145	45	111	85	93	69	114	1047	31	5 Oct
Bridge of Allan, Park Cottage	NS 26 792974	12	966	56	127	49	69	77	132	35	121	85	101	64	108	1025	...	...
Forth																		
Stirling, Batterflats	NS 26 786925	38	...	73	129	62	75	91	122	33	117	86	101	68	114	1071	45	5 Oct
Stirling	NS 26 793936	46	978	63	115	54	63	83	109	34	111	81	97	63	98	970	43	5 Oct
North Third Res. No.3	NS 26 733886	301	1458	104	185	98	136	104	166	53	174	132	112	111	175	1551	...	...
Earl's Hill	NS 26 725882	335	...	113	171	101	117	101	150	47	177	119	105	108	183	1491	52	13 Aug
North Third Res. No.2	NS 26 741890	256	1410	102	147	89	113	97	157	51	165	119	112	100	161	1413	...	...
North Third Res. No.1	NS 26 756895	175	1374	84	164	85	108	97	178	40	156	114	111	93	186	1418	...	...
Devon (Forth)																		
Frandy Burn	NN 27 933032	319	1788	98	164	113	127	153	201	57	140	103	121	133	198	1608	...	...
Frandy	NN 27 942043	290	1683	93	178	106	109	137	173	47	151	111	110	109	183	1507	44	5 Oct
Glensherup Res. No.2	NN 27 959038	287	1643	126	207	104	128	128	187	56	197	125	137	111	164	1671	...	...
Glensherup Res. No.1	NN 27 964048	257	1606	127	179	105	134	131	197	45	174	115	129	114	169	1618	50	5 Oct
Glenquey Res.	NN 27 982029	277	1578	100	174	122	117	126	208	50	148	133	124	141	172	1616	51	5 Oct
Tillicoultry Cemetery	NS 26 925970	24	945	49	115	57	82	87	160	47	133	94	115	78	110	1128	38	5 Oct
Glenochil R.F.M. Sta. Area 181 km²	NS 26 858960	5	1273	...	...	...	...	...	...	...	...	...	...	...	...	...	...	...
Forth																		
Gartmorn Res.	NS 26 915937	52	925	40	106	50	69	75	145	28	132	85	101	60	91	983	36	5 Oct
Coastal																		
Carron																		
Waterhead	NS 26 651836	259	1715	106	215	124	108	127	201	59	144	180	93	168	209	1732	...	...

Station	Grid reference	Altitude m	Av. 1916–1950 mm	Jan	Feb	Mar	Apr	May	Jun	Jul	Aug	Sep	Oct	Nov	Dec	Year	Highest recorded fall in a day mm	date
FORTH—*(contd)*																		
Kirkburn	NS 26 701839	226	1581	122	164	127	117	97	150	43	167	137	108	133	198	1562	...	...
Craigannet	NS 26 717839	226	...	125	180	136	125	125	138	47	187	131	114	122	199	1630	65	13 Aug
Earlsburn Res.	NS 26 702893	366	1611	108	187	102	143	112	187	57	167	167	105	99	214	1648	...	...
Buckieburn Res.	NS 26 742856	251	1539	111	205	127	133	113	175	41	180	136	106	125	222	1673	...	...
Buckieburn Filters	NS 26 751851	232	1546	109	189	125	134	122	172	44	182	140	105	120	224	1666	...	...
Loch Coulter	NS 26 757858	217	...	115	165	98	145	112	163	40	167	117	105	92	184	1503	...	...
Little Denny Res.	NS 26 799816	99	1185	78	134	111	95	85	165	44	133	117	95	133	174	1365	...	...
Little Denny Filters	NS 26 805820	74	1187	63	131	115	95	80	156	39	136	107	91	116	180	1308	...	...
Cumbernauld	NS 26 772760	94	1147	62	110	108	85	83	136	38	148	126	95	113	176	1279	58	13 Aug
Falkirk, Windsor Road	NS 26 872795	49	...	45	122	79	73	73	148	39	118	79	79	91	134	1081	...	...
Falkirk	NS 26 879803	32	945	47	114	71	62	68	151	33	109	68	87	88	102	1001	...	...
Coastal																		
Grangemouth	NS 26 929819	4	–	37	105	52	58	66	132	28	104	57	81	61	93	874	33	5 Oct
Gardrum Moss	NS 26 885757	201	...	56	149	101	78	79	163	45	134	83	90	105	152	1235	...	...
Polmont, Parkhill	NS 26 929788	58	...	45	115	72	65	74	146	35	109	63	81	75	114	994	36	17 Dec
Laurieston, Inglewood	NS 26 914794	41	943	49	111	68	62	76	130	33	112	72	71	77	108	969	...	
Avon																		
Forrestburn Res.	NS 26 874651	213	...	52	105	91	76	84	175	43	137	86	81	114	144	1189	55	14 Jun
Armadale S. Wks.	NS 26 938696	127	966	54	95	69	64	75	149	40	129	81	77	104	121	1058	...	...
Lochcote	NS 26 974737	168	917	56	99	52	77	67	168	43	121	68	76	74	88	989	31	13 Aug
Carribber Res.	NS 26 972749	107	892	57	100	52	80	63	175	40	127	60	77	73	102	1007	...	...
Cockleroi Res.	NS 26 994749	159	900	47	97	53	76	81	165	51	131	66	73	68	101	1009	38	13 Aug
Grange	NT 36 000787	131	824	41	97	40	72	68	149	33	113	53	72	55	84	878	...	...
Linlithgow, Craigenroan	NT 36 001766	88	849	41	111	55	76	76	155	48	133	65	82	73	97	1013	38	13 Aug
Coastal																		
Beecraigs	NT 36 013743	145	997	56	97	52	86	75	153	58	133	72	78	84	105	1049	41	13 Aug
Dalmeny House	NT 36 163779	6	716	37	79	12	61	53	120	34	134	39	50	39	64	721	31	3 Aug 13 Aug

LOTHIANS RIVER PURIFICATION BOARD AREA, 1545 SQUARE KILOMETRES

HYDROMETRIC AREAS 19 AND 20

Station	Grid reference	Altitude m	Av. 1916–1950 mm	Jan	Feb	Mar	Apr	May	Jun	Jul	Aug	Sep	Oct	Nov	Dec	Year	Highest recorded fall in a day mm	date
Almond																		
Calderhead	NS 26 878613	241	...	55	96	66	69	77	156	52	120	99	94	100	134	1117	...	...
Whitburn S.Wks.	NS 26 950655	160	...	54	84	71	60	70	138	38	128	83	75	100	114	1015	44	13 Aug
Almond Weir R.F.M.Sta. Area 43.8 km²	NT 36 004652	128	1062	...	...	...	...	...	...	...	...	...	...	...	...	...	...	...
Breich (Almond)																		
Fauldhouse S.Wks.	NS 26 940603	201	...	49	98	74	64	79	143	47	131	110	92	106	137	1130	41	13 Aug
Stoneyburn S.Wks.	NS 26 983626	180	...	44	91	54	49	63	106	42	125	86	73	82	103	917	40	13 Aug
Breich Weir R.F.M.Sta. Area 51.8 km²	NT 36 014639	136	980	...	...	...	...	...	...	...	...	...	...	...	...	...	...	...
Almond																		
Pateshill Res.	NS 26 984595	280	1074	54	111	89	55	81	126	58	144	117	85	113	126	1161	44	13 Aug
Addiewell	NT 36 000625	191	957	47	97	55	25	71	114	51	131	92	77	101	119	978	...	...
Calder House	NT 36 069671	116	...	56	95	44	59	62	143	47	134	62	58	66	85	910	42	13 Aug
Linhouse (Almond)																		
Crosswood	NT 36 056575	290	1032	49	88	56	55	67	132	56	148	93	63	90	111	1007	49	13 Aug
Harburn	NT 36 045607	244	...	53	108	60	57	67	131	61	152	83	71	85	105	1033	...	...
Morton	NT 36 076631	229	977	52	107	50	70	59	162	58	153	76	59	80	104	1028	48	13 Aug
Bog (Linhouse)																		
Almond																		
Pumpherston	NT 36 072695	131	...	49	69	33	50	61	136	43	127	59	53	61	79	820	...	...
Almondell R.F.M.Sta. Area 229 km²	NT 36 086686	73	968	...	...	...	...	...	...	...	...	...	...	...	...	...	...	...
Uphall No.7	NT 36 018712	195	899	44	79	42	60	65	135	49	126	66	70	77	85	898	...	...
Uphall No.8	NT 36 024708	176	883	42	86	38	62	60	133	47	121	65	72	72	83	881	...	...
Newliston	NT 36 110735	46	...	47	91	29	54	55	116	38	136	55	50	59	77	807	36	13 Aug
Kirknewton	NT 36 116653	207	...	58	86	33	65	55	144	44	154	65	59	62	86	913	...	...
Ratho Park	NT 36 152708	70	...	43	70	26	58	44	131	46	144	44	46	52	70	774	46	13 Aug
Turnhouse	NT 36 159739	35	685	43	73	19	56	45	117	58	147	42	47	42	69	760	42	3 Aug
Craigiehall R.F.M.Sta. Area 368 km²	NT 36 165752	23	914	...	...	...	...	...	...	...	...	...	...	...	...	...	...	...
Edinburgh, East Craigs	NT 36 183735	61	699	35	69	17	58	33	100	39	120	35	38	35	58	636	44	13 Aug
Edinburgh, Essex Road	NT 36 180758	26	...	43	79	18	63	45	109	47	147	47	47	43	75	764	...	...
Davidsons Mains	NT 36 199757	61	715	42	74	15	65	43	111	41	155	45	47	44	69	751	47	3 Aug
Coastal																		
Edinburgh, Silverknowes	NT 36 211760	40	...	41	70	15	63	39	105	38	151	40	48	39	66	717	45	3 Aug
Leith																		
Harperrig	NT 36 101616	277	...	44	80	43	61	65	152	59	158	60	67	66	95	949	51	13 Aug
Riccarton	NT 36 176692	94	...	44	81	26	70	42	127	45	160	50	51	61	72	829	52	13 Aug
Edinburgh, Hillview Road	NT 36 194734	73	...	45	73	17	68	42	112	39	161	40	47	42	72	760	54	13 Aug
Murrayfield R.F.M.Sta. Area 107 km²	NT 36 228732	37	935	...	...	...	...	...	...	...	...	...	...	...	...	...	...	...
Edinburgh, Royal Botanic Garden	NT 36 245755	26	678	45	81	17	64	38	111	39	154	40	41	52	61	742	45	13 Aug
Figgate																		
Clubbiedean Res.	NT 36 203670	236	817	59	97	27	76	50	134	49	200	50	63	67	85	956	66	13 Aug
Edinburgh, Blackford Hill	NT 36 258706	134	699	40	65	14	49	37	108	40	188	38	42	53	59	731	55	13 Aug
Liberton, The Inch	NT 36 277708	46	750	40	65	21	45	37	97	39	188	36	46	54	67	736	52	13 Aug
Edinburgh, Astley Ainslie Hospital	NT 36 251713	82	725	42	74	17	54	41	114	43	171	38	45	57	68	765	46	13 Aug
Brunstane																		
Fairmilehead W.Wks.	NT 36 248683	180	792	47	69	24	45	44	119	42	168	43	49	51	65	767	45	13 Aug
Liberton, Alnwick Hill	NT 36 273690	124	730	41	67	22	41	36	107	45	205	37	49	56	61	766	59	13 Aug
Esk																		
North Esk Res.	NT 36 154578	351	...	61	109	64	55	73	171	51	179	91	76	114	117	1160	57	13 Aug
Newhall House	NT 36 175566	268	...	43	88	54	40	58	166	33	181	81	79	86	113	1023	70	13 Aug
Penicuik	NT 36 233599	189	948	47	83	39	45	45	113	56	214	63	70	87	93	956	82	13 Aug
Dalmore Weir R.F.M.Sta. Area 81.6 km²	NT 36 252616	132	1003	...	...	...	...	...	...	...	...	...	...	...	...	...	...	...
Glencorse (Esk)																		
Harehill	NT 36 184615	373	...	53	103	52	52	76	154	51	223	81	73	83	97	1097	...	...
Loganlea	NT 36 199626	247	980	51	93	40	63	61	167	53	210	68	78	87	79	1049	70	13 Aug
Glen Cottage	NT 36 223635	225	968	54	93	41	59	54	143	59	211	62	71	91	87	1024	69	13 Aug
Glencorse Filters	NT 36 225631	194	915	51	89	39	54	50	130	56	206	59	63	86	85	968	67	13 Aug
Bush House	NT 36 244636	184	906	48	83	37	49	48	118	55	198	61	58	82	82	919	57	13 Aug

Station	Grid reference	Altitude m	Av. 1916–1950 mm	Jan	Feb	Mar	Apr	May	Jun	Jul	Aug	Sep	Oct	Nov	Dec	Year	Highest recorded fall in a day mm	date
LOTHIANS—*(contd)*																		
Martyr's Cross	NT 36 229623	229	946	45	80	33	49	49	120	56	203	54	69	77	73	906	...	...
North Esk																		
Dalkeith House	NT 36 334679	46	...	33	51	13	34	35	85	45	210	42	62	58	66	733	...	...
South Esk (North Esk)																		
Gladhouse Res.	NT 36 299544	279	879	47	72	28	38	49	100	63	172	70	78	88	81	886	57	13 Aug
Rosebery	NT 36 308570	229	855	50	73	31	42	41	101	53	201	63	78	94	89	915	67	13 Aug
Portmore Res.	NT 36 260507	305	956	49	80	31	42	57	125	63	178	79	82	105	96	987	59	13 Aug
Middleton House	NT 36 367582	229	...	55	75	19	34	44	99	46	201	56	61	96	81	865	...	...
Gorebridge	NT 36 342620	171	765	40	59	21	38	32	97	50	221	55	75	87	74	847	...	...
Prestonholm R.F.M.Sta. Area 112 km	NT 36 325623	77	921	...	...	...	...	...	...	...	...	...	...	...	...	...	...	...
Greenhall School	NT 36 342629	158	...	38	51	10	33	27	90	47	199	47	66	84	69	760	56	3 Aug
Gourlaw Filters	NT 36 283612	181	...	39	65	24	34	33	112	55	205	51	61	73	72	825	63	13 Aug
St.Joseph's	NT 36 295620	146	...	36	73	29	33	37	102	44	207	56	61	75	77	830	...	...
Esk																		
Carberry Tower	NT 36 364696	70	...	35	44	10	42	35	97	45	213	36	63	59	66	746	...	...
Edinburgh, The Drum	NT 36 302688	88	...	39	56	19	39	35	101	47	198	48	47	55	66	750	56	13 Aug
Inveresk, Wedderburn House	NT 36 349715	23	...	35	47	11	45	33	100	40	197	43	52	54	58	715	61	3 Aug
Musselburgh R.F.M.Sta. Area 326 km	NT 36 339723	3	892	...	...	...	...	...	...	...	...	...	...	...	...	...	...	...
Coastal																		
Prestonpans, Burnside	NT 36 390744	17	...	36	39	10	52	41	95	36	176	38	57	54	56	689	...	...
St. Germains	NT 36 426748	46	...	27	38	11	51	44	93	35	190	42	59	55	55	699	65	3 Aug
Greendykes	NT 36 436736	76	...	30	41	9	47	36	100	33	205	43	66	62	61	733	...	...
Longniddry	NT 36 437763	15	...	34	42	12	43	46	88	29	191	40	58	56	57	697	63	3 Aug
Elvingston	NT 36 461742	85	...	27	40	12	45	42	100	34	223	37	73	73	59	765	71	3 Aug
Gosford House	NT 36 454786	8	...	30	50	14	49	45	90	31	173	38	54	54	59	685	57	13 Aug
East Fortune Hospital	NT 36 553793	31	633	29	48	14	43	53	86	37	197	38	76	76	58	754	60	3 Aug
Peffer																		
Drem, New Mains	NT 36 505813	18	...	31	49	19	53	51	80	44	185	37	61	67	73	750	...	...
Gullane	NT 36 480827	27	...	29	57	27	62	53	106	38	174	52	54	64	69	783	...	...
Coastal																		
Ferrygate	NT 36 526844	24	...	31	60	15	59	56	85	33	166	36	54	65	68	728	...	...
La Sagesse Convent	NT 36 572832	46	...	36	61	18	61	50	98	27	180	34	56	63	63	748	...	...
Tantallon Castle	NT 36 594850	30	...	25	53	20	56	57	95	30	163	35	65	65	66	731	...	...
Scoughall	NT 36 615832	8	...	33	57	18	44	55	110	29	173	32	69	81	63	763	...	...
Tyne																		
Hagbrae	NT 36 375618	198	...	38	58	17	32	35	96	54	235	55	71	93	74	857	79	3 Aug
Ford	NT 36 387644	128	700	36	50	15	33	30	94	51	231	51	65	76	70	801	79	3 Aug
Winton Castle	NT 36 438698	90	...	29	43	10	44	31	118	44	219	42	75	74	59	788	...	...
Humbie (Tyne)																		
Woodcote Park	NT 36 458602	244	...	51	73	18	57	59	96	76	189	57	83	113	92	963	...	...
Humbie House	NT 36 469640	165	...	39	60	11	40	39	101	71	203	45	83	98	74	865	53	3 Aug
Stobshiel Res.	NT 36 501622	274	...	53	57	18	45	56	97	69	230	56	94	99	88	963	...	...
Stobshiel Filters	NT 36 498638	198	...	47	53	15	38	48	92	67	207	51	86	100	80	883	60	3 Aug
Tyne																		
Samuelston	NT 36 486711	64	...	28	41	10	38	33	95	42	204	39	71	79	54	737	78	3 Aug
Colstoun (Tyne)																		
West Hopes	NT 36 551622	247	931	55	91	18	60	69	114	59	232	58	104	116	99	1075	76	3 Aug
Yester House	NT 36 541673	137	...	39	65	12	47	39	98	46	194	35	80	95	76	826	...	...
Lammerloch Res.	NT 36 517634	274	858	54	71	20	52	65	103	72	233	58	98	102	92	1019	...	...
Skedsbush	NT 36 512657	183	...	37	57	8	43	35	79	39	203	39	76	83	74	773	63	3 Aug
Lennoxlove	NT 36 510713	64	...	28	44	10	37	41	96	41	209	38	75	84	61	764	...	...
Tyne																		
Haddington	NT 36 513736	49	650	25	38	12	34	38	92	36	202	37	77	79	61	732	70	3 Aug
Morham Church	NT 36 557726	70	...	27	44	9	35	38	93	36	181	32	73	90	65	723	...	...
Hailes Castle	NT 36 575758	30	...	31	59	8	43	38	90	39	203	40	74	81	63	770	67	3 Aug
East Linton R.F.M.Sta. Area 308 km²	NT 36 591768	17	759	...	...	...	...	...	...	...	...	...	...	...	...	...	...	...
East Linton, Smeaton	NT 36 594786	30	660	29	41	12	36	41	98	29	193	33	66	73	59	711	62	3 Aug
Biel																		
Donolly Res.	NT 36 580690	162	778	42	62	11	52	44	92	40	209	42	100	93	72	860	...	...
White Castle	NT 36 619684	301	...	45	73	16	58	58	100	35	229	52	113	129	90	999	...	...
Thorters Res.	NT 36 607695	221	769	44	58	12	55	52	101	45	207	42	102	109	84	912	...	...
Whittingehame, West Lodge	NT 36 592724	107	...	39	74	11	50	41	96	43	202	40	77	79	77	831	81	3 Aug
Belton House	NT 36 645766	30	...	32	73	14	59	45	108	36	200	34	75	78	67	820	56	3 Aug
Coastal																		
Dunbar	NT 36 672791	23	596	29	47	12	37	43	97	31	175	30	60	68	56	686	71	3 Aug
Broxmouth	NT 36 699774	15	...	35	66	14	44	39	98	33	189	32	64	65	64	745	57	3 Aug
Barns Ness Lighthouse	NT 36 723772	8	...	31	70	13	53	34	75	28	167	29	75	65	57	696	52	3 Aug

TWEED RIVER PURIFICATION BOARD AREA, 4574 SQUARE KILOMETRES

HYDROMETRIC AREA 21

Station	Grid reference	Altitude m	Av. 1916–1950 mm	Jan	Feb	Mar	Apr	May	Jun	Jul	Aug	Sep	Oct	Nov	Dec	Year	Highest recorded fall in a day mm	date
Coastal																		
Blackcastle Hill	NT 36 717719	308	...	52	93	12	41	52	104	46	236	57	76	77	72	918	...	...
Dunglass	NT 36 763720	67	...	39	91	15	45	35	101	34	215	47	68	65	67	821	(65	3 Aug)
Redheugh	NT 36 822699	88	711	40	87	12	85	43	67	42	210	40	74	61	73	835	...	...
Tweed																		
Hawkshaw	NT 36 083204	335	1369	97	160	76	66	94	114	51	114	137	125	140	165	1339	...	...
Fruid (Tweed)																		
Garelot Dod	NT 36 126173	671	...	114	165	86	241	216	152	74	196	223	175	188	356	2187	...	...
Brawn's Dod	NT 36 103195	472	...	147	178	69	109	117	122	51	145	178	135	147	216	1613	...	...
Shiel Hope	NT 36 095187	381	...	150	223	94	112	122	107	43	125	163	122	133	171	1565	...	...
Fruid R.F.M. Sta. Area 23.7 km²	NT 36 088205	277	1755	...	...	...	...	...	...	...	...	...	...	...	...	...	...	...
Fruid Dam	NT 36 089205	312	...	112	166	93	95	104	112	48	138	151	119	157	190	1486	53	13 Aug
Tweed																		
Upper Menzion	NT 36 102207	311	...	112	203	86	91	91	99	48	147	165	132	140	165	1481	...	...
The Rig	NT 36 090223	344	...	109	157	84	79	84	112	46	142	165	135	140	165	1417	...	...
Quarter Hill	NT 36 097235	365	1302	99	196	71	66	69	112	46	127	122	122	145	178	1351	...	...
Talla (Tweed)																		
Loch Craig Head	NT 36 162173	709	...	203	305	168	188	145	114	69	180	201	150	178	320	2220	...	...
Gameshope Loch	NT 36 132162	579	2114	216	286	152	178	211	142	63	183	244	178	203	302	2358	...	...

Station	Grid reference	Altitude m	Av. 1916–1950 mm	Jan	Feb	Mar	Apr	May	Jun	Jul	Aug	Sep	Oct	Nov	Dec	Year	Highest recorded fall in a day mm	date
TWEED—*(contd)*																		
Gameshope Farm	NT 36 139176	466	2009	191	259	119	135	196	132	63	178	218	178	170	249	2088	...	...
Talla Linns Foot	NT 36 134204	294	1550	140	203	94	125	122	102	61	168	170	142	163	223	1712	...	...
Victoria Lodge	NT 36 106231	274	1274	90	139	71	79	75	99	43	129	119	120	136	175	1274	54	13 Aug
Tweed																		
Stanhope No.2	NT 36 155273	381	...	118	129	95	93	102	112	63	166	115	147	161	161	1463	...	...
Stanhope Farm	NT 36 123296	226	1139	70	108	65	56	57	106	47	130	116	113	131	143	1143	47	13 Aug
Stobo Castle	NT 36 179368	181	978	53	113	38	54	52	101	54	150	97	84	106	118	1021	59	13 Aug
Lyne (Tweed)																		
Lyne Ford R.F.M. Sta. Area 378 km²	NT 36 206397	171	1364	...	...	...	...	...	...	...	...	...	...	...	...	...	...	...
Baddinsgill Res.	NT 36 126554	335	1206	54	113	71	53	84	160	71	165	107	76	128	134	1214	56	13 Aug
West Linton	NT 36 150520	244	1001	41	78	44	34	56	144	54	157	82	67	98	106	960	43	13 Aug
West Linton, Woodlea Cottage	NT 36 159507	244	940	43	79	37	43	59	137	49	144	85	67	76	97	915	47	13 Aug
Craigengar	NT 36 095554	457	...	61	152	66	64	78	152	63	178	106	81	104	135	1240	...	...
Upper West Water	NT 36 101534	373	...	48	100	58	53	72	146	57	155	107	74	107	127	1104	...	...
West Water Res.	NT 36 123519	290	...	46	84	49	36	64	156	58	131	92	76	88	121	1002	...	...
Blyth Bridge	NT 36 141463	253	918	39	81	37	34	51	120	50	147	92	71	82	105	907	51	13 Aug
Tweed																		
Peebles R.F.M. Sta. Area 694 km²	NT 36 257400	155	1252	...	...	...	...	...	...	...	...	...	...	...	...	...	...	...
Glentress	NT 36 283397	165	773	38	93	25	46	47	111	47	123	77	85	107	106	901	(35	1 Nov)
Glen House	NT 36 296330	229	...	65	159	(49)	77	70	95	55	145	114	85	91	146	1152	...	...
Traquair School	NT 36 331344	168	827	49	116	25	76	45	51	43	131	73	79	79	102	867	...	...
Elibank	NT 36 396367	134	876	52	109	23	65	52	97	42	122	78	87	80	103	912	...	...
Peel Hospital	NT 36 431351	137	...	52	113	27	67	69	76	44	129	79	89	79	109	930	31	13 Aug
Caddonhead	NT 36 401411	320	...	66	122	41	84	86	102	61	175	91	89	114	122	1153	...	...
Craig Head	NT 36 417403	274	988	61	117	31	71	71	109	51	155	79	107	89	122	1062	...	...
Fairnilee	NT 36 456332	137	856	61	122	29	80	67	75	47	133	88	101	88	107	999	35	13 Aug
Ettrick (Tweed)																		
Crook Cottage	NT 36 218123	283	...	119	284	138	109	157	108	55	164	184	127	147	249	1838	52	3 Sep
Ettrick School	NT 36 266145	232	1505	110	186	105	98	109	118	45	161	154	124	132	199	1542	...	...
Ettrickbridge	NT 36 390243	168	...	69	136	33	77	63	82	43	143	88	109	88	108	1039	37	13 Aug
Bowhill	NT 36 427273	183	871	52	131	28	69	54	85	42	150	80	109	85	95	979	32	13 Aug
Yarrow (Ettrick)																		
Quaveburn Cottage	NT 36 228181	258	1722	139	321	107	135	107	115	60	175	190	144	158	278	1929	...	...
Megget-Head	NT 36 152200	451	1735	84	254	94	102	112	127	66	165	185	132	147	221	1689	...	...
Gordon Arms	NT 36 308249	229	...	67	140	45	87	65	87	40	153	85	118	99	134	1119	44	13 Aug
Philiphaugh R.F.M. Sta. Area 233 km²	NT 36 439277	127	1488	...	...	...	...	...	...	...	...	...	...	...	...	...	...	...
Ettrick (Tweed)																		
Selkirk, Byethorn	NT 36 475286	183	782	48	124	17	77	51	86	43	146	63	99	66	87	908	...	...
Lindean R.F.M. Sta. Area 502 km²	NT 36 486315	98	1455	...	...	...	...	...	...	...	...	...	...	...	...	...	...	...
Tweed																		
Boleside R.F.M. Sta. Area 1510 km²	NT 36 498334	91	1270	...	...	...	...	...	...	...	...	...	...	...	...	...	...	...
Gala (Tweed)																		
Stow	NT 36 458453	196	...	66	102	32	79	62	79	37	154	78	77	100	106	972	43	13 Aug
Knowsdean Res.	NT 36 436393	272	934	62	128	30	67	74	111	43	148	83	103	86	118	1053	39	13 Aug
Galashiels R.F.M. Sta. Area 207 km²	NT 36 479374	120	975	...	...	...	...	...	...	...	...	...	...	...	...	...	...	...
Tweed																		
Leader (Tweed)																		
Oxton	NT 36 496536	221	820	49	85	16	50	61	91	52	195	58	81	105	103	945	...	...
Lauder	NT 36 521488	183	816	50	115	21	63	64	95	49	129	72	80	96	96	929	38	3 Aug
Scoured Rig	NT 36 584519	354	...	61	145	14	91	74	114	48	170	79	107	94	86	1083	...	...
Blythe	NT 36 585495	262	...	50	93	18	64	77	112	49	155	87	104	87	97	995	36	3 Aug 13 Aug
Carolside	NT 36 561399	107	753	47	102	17	74	60	88	40	125	79	82	74	96	885	37	3 Aug
Cowdenknowes	NT 36 575372	91	750	46	120	17	77	60	89	39	121	71	80	65	87	872	...	...
Tweed																		
Newtown St. Boswell's S. Wks.	NT 36 581317	88	...	41	95	13	63	44	74	36	126	68	77	69	77	781	...	...
Dryburgh R.F.M. Sta. Area 2080 km²	NT 36 588320	67	1166	...	...	...	...	...	...	...	...	...	...	...	...	...	...	...
Mertoun	NT 36 647321	69	...	31	61	12	63	51	78	41	125	63	78	79	81	764	43	3 Aug
Floors Castle	NT 36 707345	59	659	24	69	11	61	51	106	42	125	49	82	78	85	783	35	3 Aug
Teviot (Tweed)																		
Braehead	NT 36 406007	267	...	114	174	84	122	105	103	67	166	136	102	112	151	1436	...	...
Borthwick (Teviot)																		
Craik	NT 36 349082	212	1426	91	158	86	92	94	94	52	151	145	105	117	184	1369	...	...
Roberton Filters	NT 36 428151	259	...	74	113	44	73	60	86	32	145	94	100	77	109	1006	52	13 Aug
Teviot (Tweed)																		
Hawick S. Wks.	NT 36 512156	96	822	48	74	24	66	48	118	46	133	70	90	76	95	886	36	21 Jun
Hawick R.F.M. Sta. Area 326 km²	NT 36 522159	88	–	...	...	...	...	...	...	...	...	...	...	...	...	...	...	...
Rule (Teviot)																		
Wauchope	NT 36 587097	155	905	52	93	31	60	59	89	50	124	118	90	79	92	937	35	3 Aug
Spital Tower	NT 36 585177	134	790	33	70	19	55	49	104	53	129	61	95	73	87	828	37	3 Aug
Ale (Teviot)																		
Shiringscleugh	NT 36 359153	308	...	(113)	155	(80)	89	77	84	38	144	118	86	104	125	1214	...	...
Alemoor Loch	NT 36 406156	274	1139	99	150	64	83	74	96	34	125	114	107	89	137	1171	...	...
Teviot (Tweed)																		
Monteviot House	NT 36 643250	64	703	33	69	16	52	48	59	48	127	59	85	79	76	750	...	...
Jed (Teviot)																		
Edgerston Tofts	NT 36 700109	241	...	42	85	17	53	49	126	47	149	57	100	86	81	892	43	3 Aug
Jedneuk	NT 36 656234	82	679	25	62	12	42	47	97	56	123	61	75	72	62	735	...	...
Teviot (Tweed)																		
Ormiston Mill R.F.M. Sta. Area 1110 km²	NT 36 702280	46	1046	...	...	...	...	...	...	...	...	...	...	...	...	...	...	...
Eden (Tweed)																		
East Gordon Hospital	NT 36 658434	157	706	43	86	17	80	79	66	46	145	68	85	83	88	886	...	...
Tweed																		
Lochton	NT 36 776389	37	654	27	67	11	66	45	79	37	150	42	91	76	78	770	51	3 Aug

Station	Grid reference	Altitude m	Av. 1916–1950 mm	Jan	Feb	Mar	Apr	May	Jun	Jul	Aug	Sep	Oct	Nov	Dec	Year	Highest recorded fall in a day mm	date
TWEED—*(contd)*																		
Till (Tweed)																		
Linhope	NT 36 963160	226	1033	64	149	25	126	72	79	49	181	50	153	119	127	1192	57	13 Aug
Lilburn Tower	NU 46 026243	70	787	43	81	19	80	43	88	47	152	31	109	70	98	861	44	13 Aug
Middleton Hall	NT 36 988254	131	803	50	91	19	89	47	83	49	185	32	114	81	96	936	57	13 Aug
Glen (Till)																		
Old Graden	NT 36 798299	152	761	35	75	17	62	43	84	54	161	44	103	77	93	846	47	3 Aug
Mindrum	NT 36 842327	76	719	41	81	15	77	37	82	43	165	39	101	88	90	858	51	13 Aug
Goldscleugh	NT 36 913233	305	...	76	184	25	138	69	90	70	207	62	137	154	104	1318	...	...
Hethpool	NT 36 895282	128	874	48	102	18	108	45	84	57	208	37	112	93	107	1019	82	13 Aug
Till (Tweed)																		
Pallinsburn	NT 36 911382	41	625	33	81	11	57	54	62	27	141	25	114	85	93	782	56	13 Aug
Etal R.F.M. Sta. Area 648 km²	NT 36 927396	25	838	...	...	...	...	...	...	...	...	...	...	...	...	...	...	...
Tweed																		
Norham R.F.M. Sta. Area 4380 km²	NT 36 898477	4	1024	...	...	...	...	...	...	...	...	...	...	...	...	...	...	...
Tweedhill	NT 36 929513	15	684	41	83	21	68	45	66	28	167	44	115	81	79	838	63	13 Aug
Whiteadder (Tweed)																		
Hungry Snout R.F.M. Sta. Area 45.6 km²	NT 36 663633	215	884	...	...	...	...	...	...	...	...	...	...	...	...	...	...	...
Dye (Whiteadder)																		
Hungry Snout	NT 36 665633	218	...	57	86	18	62	73	136	40	231	53	115	117	94	1083	77	3 Aug
Watchwater (Dye)																		
Watchwater Res. R.F.M. Sta. Area 10.7 km²	NT 36 664566	252	...	...	...	...	...	...	...	...	...	...	...	...	...	...	...	...
Rawburn Filters	NT 36 677563	248	...	64	109	26	87	88	86	40	185	73	114	102	109	1083	...	...
Whiteadder (Tweed)																		
Whitchester	NT 36 721589	255	856	64	145	26	99	70	95	37	202	56	120	105	113	1132	61	3 Aug
Manderston House	NT 36 809547	108	791	42	95	19	85	56	73	31	187	53	96	77	83	899	61	13 Aug
Blackadder (Whiteadder)																		
Marchmont House	NT 36 743484	152	779	45	101	20	78	66	95	39	154	65	88	91	92	934	42	13 Aug
Kimmerghame House	NT 36 815513	76	747	40	100	24	81	53	78	42	156	57	85	80	86	880	...	...
Duns Castle	NT 36 775538	137	791	45	88	19	75	67	83	35	182	52	82	81	85	896	59	3 Aug
Tweed																		
Berwick-on-Tweed	NU 46 002528	23	592	31	55	16	55	44	67	20	149	34	82	52	65	669	49	13 Aug

Northern Ireland

FOYLE AND NORTH-WESTERN STREAMS AREA, 2047 SQUARE KILOMETRES (within Northern Ireland)

Station	Grid reference	Altitude m	Av. 1916–1950 mm	Jan	Feb	Mar	Apr	May	Jun	Jul	Aug	Sep	Oct	Nov	Dec	Year	Highest recorded fall in a day mm	date
Coastal																		
Creggan Res. No. 2	IC 24 417172	88	...	32	130	125	59	84	136	83	65	94	84	132	220	1244	(26	19 Jun)
Creggan Res. No. 1	IC 24 415171	96	1172	31	127	125	57	83	134	81	64	94	83	131	219	1229	25	19 Jun
Foyle																		
Milltown	IH 23 576755	168	...	88	154	102	84	111	131	51	63	(83)	91	115	(178)	1249	18	27 Jan
Beragh Police Sta.	IH 23 546672	102	...	79	140	116	100	95	141	45	73	65	112	110	187	1262	32	9 Jun
Killyclogher Filter House	IH 23 473746	113	...	71	127	107	78	100	151	58	54	85	97	105	165	1200	27	3 Sep
Drumlish Forest	IH 23 353665	110	...	75	155	109	81	120	131	58	61	110	96	118	201	1315	37	3 Sep
Drumragh (Foyle)																		
Seskinore Forest	IH 23 490640	110	...	81	128	93	89	95	142	46	66	65	91	109	180	1186	(25	4 Sep)
Edenfel	IH 23 466719	88	1087	66	127	96	68	97	133	52	56	70	93	102	166	1124	20	9 Dec
Fairy Water (Foyle)																		
Lough Bradan No. 1	IH 23 261712	186	...	70	183	105	95	127	135	67	61	105	91	116	189	1345	...	...
Lough Bradan No. 2	IH 23 262715	210	...	77	184	116	98	126	138	72	63	100	95	127	203	1401	...	...
Lough Bradan No. 3	IH 23 253713	195	...	79	171	138	93	136	150	81	63	118	93	151	205	1478	...	...
Lough Bradan	IH 23 259725	169	...	90	180	145	105	130	162	87	68	119	94	146	250	1575	38	22 Jun
Foyle																		
Lislap Forest	IH 23 485823	209	1319	71	175	106	104	112	159	71	75	99	111	137	188	1409	29	9 Sep 30 Nov
Carrigans	IH 23 435810	113	...	68	140	105	81	108	143	69	73	102	88	119	186	1282	26	3 Sep
Owenkillew (Foyle)																		
Davagh Forest	IH 23 695862	183	...	100	205	132	114	150	153	48	72	107	130	144	222	1578	...	...
Owenreagh (Owenkillew)																		
Lough Carn	IH 23 575793	180	...	80	180	145	97	(112)	97	(63)	63	104	(114)	147	251	1454	...	...
Bauck Hill, Old site	IH 23 593783	202	...	112	187	127	106	(114)	(146)	68	65	96	111	147	230	1509	...	...
Owenkillew (Foyle)																		
Gortin	IH 23 491858	117	1262	65	161	106	86	123	147	53	74	93	93	115	199	1314	31	30 Nov
Glenelly (Owenkillew)																		
Goles Forest	IH 23 682941	241	...	108	257	153	141	135	143	100	75	97	182	171	274	1835	...	...
Plumbridge Police Sta.	IH 23 485912	91	...	52	149	115	67	98	163	61	60	90	87	117	205	1265	36	22 Jun
Derg (Foyle)																		
Glenderg Forest	IH 23 090819	180	...	94	206	204	128	177	196	99	106	155	105	245	370	2085	...	...
Killeter	IH 23 176796	119	...	74	154	138	94	127	148	65	67	128	97	142	227	1462	...	...
Cross Hill	IH 23 090845	305	...	69	198	253	112	182	282	108	100	177	99	296	495	2370	...	...
Castlederg S. Wks.	IH 23 272846	49	...	54	137	91	81	80	135	51	72	86	68	106	175	1137	37	3 Sep
Baron's Court	IH 23 364825	90	1031	56	134	84	80	94	115	61	68	83	88	100	178	1139	29	3 Sep
Baronscourt Forest	IH 23 361843	88	...	54	146	97	77	102	136	60	72	95	81	100	171	1192	30	3 Sep
Foyle																		
Ligfordrum Forest	IH 23 418935	161	...	74	133	102	74	120	169	69	59	100	76	114	208	1298	(41	20 Feb)
Sion Mills	IH 23 339932	30	1081	35	117	88	63	99	129	50	59	79	62	104	177	1063	(28	3 Sep)
Strabane Convent	IH 23 352974	44	...	49	146	96	53	91	144	59	52	88	80	103	162	1123	27	3 Sep
Strabane County Secondary School	IH 23 349989	15	...	(46)	140	82	41	86	121	49	40	78	59	97	155	994	...	...
Dunnamanagh Police Sta.	IC 24 442032	62	...	47	135	101	59	91	152	63	39	82	74	111	173	1127	(26	29 Jul)

LOUGH FOYLE, EAST AREA, 895 SQUARE KILOMETRES

Station	Grid reference	Altitude m	Av. 1916–1950 mm	Jan	Feb	Mar	Apr	May	Jun	Jul	Aug	Sep	Oct	Nov	Dec	Year	Highest recorded fall in a day mm	date
Faughan																		
Learmount Forest	IC 24 578022	140	...	71	175	136	77	106	119	81	57	94	122	144	202	1385	(39	23 Apr)
Loughermore Forest	IC 24 584128	174	...	52	125	120	71	93	124	89	60	95	100	122	221	1273	(28	23 Jun)
Slievekirk	IC 24 455088	296	...	36	134	121	60	106	174	101	46	90	89	117	225	1299	...	...
Drumahoe	IC 24 465146	25	...	32	108	111	52	82	125	64	43	81	79	117	187	1081	28	30 Nov
Altnagelvin Cemetery	IC 24 453151	55	...	26	105	106	51	82	124	67	46	81	70	118	194	1070	22	30 Nov
Cloghole P. Sta.	IC 24 489200	9	...	33	101	107	53	70	127	59	64	81	79	121	191	1087	23	8 Aug
Coastal																		
Ballykelly Forest	IC 24 612220	27	...	38	93	96	52	62	122	44	57	64	93	103	169	993	32	23 Jun
Ballykelly, No.3	IC 24 624234	2	909	33	82	96	48	61	116	43	56	60	91	93	159	938	38	23 Jun
Ballykelly, No.1	IC 24 624234	2	877	30	78	89	45	57	116	38	53	60	87	88	154	895	39	23 Jun
Roe																		
Curradda Bridge	IH 23 776054	233	...	102	334	137	121	109	167	74	81	90	194	145	222	1777	...	...
Altnaheglish (Owenrigh)																		
Labbyheige	IC 24 723021	347	1508	113	271	152	134	118	138	100	90	86	168	127	241	1738	...	...
Glack Burn	IC 24 707040	292	1406	109	247	132	147	101	127	77	86	78	162	121	215	1603	...	...
Altabrogy Burn	IC 24 701033	292	1481	111	269	137	149	115	155	84	89	90	170	133	214	1717	...	...
Glenedra (Altnaheglish)																		
Spelhoagh Mountain	IH 23 704985	404	1686	124	318	142	171	115	141	84	108	97	195	141	273	1909	...	...
Glenedra Hill	IH 23 696992	404	1707	112	296	153	145	128	162	94	98	91	185	177	298	1939	...	...
Glenedra Lodge	IC 24 705005	287	1553	108	289	144	150	127	146	89	100	95	185	157	252	1842	...	...
Altagarran Mountain	IC 24 711012	372	1499	110	276	134	149	118	139	92	101	86	181	129	232	1747	...	...
Glenedra Intake	IC 24 682024	223	1466	109	270	137	145	109	137	75	94	85	172	140	213	1686	...	...
Altnaheglish (Owenrigh)																		
Banagher Filters	IC 24 670046	111	1328	93	214	127	111	92	111	59	83	77	145	122	199	1434	32	12 Oct
Roe																		
Gortnamoyagh Forest	IC 24 740153	137	...	74	181	108	83	79	137	91	71	59	134	106	177	1300	38	23 Jun
Dogleap House	IC 24 679205	40	...	48	126	100	67	59	136	54	55	67	102	91	174	1078	47	23 Jun
Ballyhanna Forest	IC 24 707283	223	...	45	89	112	59	99	197	107	62	88	118	105	191	1271	87	23 Jun
Drenagh	IC 24 689236	24	954	43	113	93	62	71	138	71	59	65	105	92	175	1087	47	23 Jun
Ballycarton Forest	IC 24 677293	9	...	37	92	85	67	67	134	68	47	76	88	92	157	1009	43	23 Jun
Coastal																		
Magilligan Police Sta.	IC 24 676335	3	...	45	100	102	56	68	106	65	42	74	99	120	152	1030	25	18 Feb

GENERAL TABLE, NORTHERN IRELAND

BANN AND LOUGH NEAGH AREA, 5404 SQUARE KILOMETRES (within Northern Ireland)

Station	Grid reference	Altitude m	Av. 1916–1950 mm	Jan	Feb	Mar	Apr	May	Jun	Jul	Aug	Sep	Oct	Nov	Dec	Year	Highest recorded fall in a day mm	date
Bann																		
Deer's Meadow	IJ 33 260257	378	1788	165	325	84	333	180	165	53	102	102	179	129	205	2022	...	...
Ott Mountain	IJ 33 280272	454	1831	183	335	75	325	175	165	53	102	102	206	129	205	2054	...	...
Rathfriland Police Sta.	IJ 33 202335	149	...	86	134	49	141	82	101	40	69	56	142	90	115	1103	56	9 Jun
Huntly S. Wks.	IJ 33 118467	55	--	48	104	56	72	62	104	47	52	104	117	71	95	932	49	9 Jun
Cusher (Bann)																		
Whitecross	IH 23 987314	125	1046	104	159	75	126	101	120	53	76	96	180	109	152	1352	55	9 Jun
Glenanne	IH 23 978354	134	...	81	148	60	108	90	128	36	66	93	150	97	134	1191	57	9 Jun
Ballylane W. Wks.	IH 23 965352	143	...	87	151	71	110	99	116	55	71	93	146	105	151	1256	55	9 Jun
Tandragee S. Wks.	IJ 33 038457	18	...	63	138	61	81	70	132	43	59	68	129	84	119	1049	52	9 Jun
Ballybay (Bann)																		
Gosford Castle Forest, Old site	IH 23 963413	87	...	75	145	69	91	77	122	51	68	70	133	86	140	1129	53	9 Jun
Gosford Castle Forest, New site	IH 23 965412	88	...	75	145	69	91	77	122	51	68	70	133	86	140	1129	53	9 Jun
Bann																		
Portadown S. Wks. No. 2	IJ 33 019555	16	...	56	121	64	77	84	109	40	69	55	123	80	109	987	46	9 Jun
Portadown S. Wks. No. 1	IJ 33 019556	18	...	54	111	47	73	72	103	34	66	49	120	68	93	892	44	9 Jun
Derrykerran	IJ 33 007582	14	...	(51)	(114)	61	80	69	128	44	78	53	127	77	112	994	...	...
The Birches Police Sta.	IH 23 942603	20	...	59	118	68	85	69	118	35	72	61	133	73	123	1014	45	9 Jun
Blackwater North Branch (Blackwater)																		
Aghintaine Castle	IH 23 459511	110	...	75	121	106	81	100	118	55	73	74	91	127	193	1215	(26	3 Sep)
Fardross Forest	IH 23 522484	125	...	93	157	102	109	121	137	47	109	74	78	139	199	1366	(40	27 Jan)
Knockmany Forest	IH 23 537546	73	--	85	157	83	100	90	136	52	84	64	89	128	176	1243	(26	4 Nov)
Favor Royal Forest	IH 23 615532	50	...	75	163	86	97	72	132	47	103	51	114	110	170	1222	(31	9 Jun)
Ballygawley Res.	IH 23 631569	100	...	69	171	76	94	80	126	31	86	57	105	105	168	1168	...	...
Auchnacloy S. Wks.	IH 23 664519	45	938	80	157	81	93	61	106	30	75	51	128	103	143	1108	(31	20 Feb)
Tynan (Blackwater)																		
Tynan Abbey	IH 23 759422	48	870	62	106	73	89	77	146	32	76	63	172	88	125	1109	(33	7 Oct)
Blackwater (Lough Neagh)																		
Benburb Priory	IH 23 816522	53	...	60	124	68	94	77	114	31	81	55	135	88	127	1055	35	9 Jun
Moy Res.	IH 23 836564	59	...	73	103	79	88	76	110	28	89	65	142	89	150	1092	...	...
Callan (Blackwater)																		
Clay Lake	IH 23 836326	184	...	106	142	74	144	110	134	48	86	89	191	111	153	1388	(44	4 Nov)
Seagahan Filters	IH 23 897382	122	...	81	147	63	113	87	113	51	73	72	156	99	123	1180	46	9 Jun
Armagh	IH 23 878458	62	845	69	117	61	101	69	103	36	67	69	144	85	111	1031	48	9 Jun
Blackwater (Lough Neagh)																		
Dungannon Park	IH 23 802606	73	1020	84	152	91	111	91	118	39	83	65	151	105	170	1260	35	12 Oct
Tall (Blackwater)																		
Loughgall	IH 23 907519	25	782	64	119	59	78	61	105	39	69	54	145	85	108	985	45	9 Jun
Torrent (Blackwater)																		
Cappagh Res.	IH 23 684675	173	...	105	197	95	131	109	103	43	84	81	137	121	194	1400	27	8 Feb
Lough Neagh (Bann)																		
Washing Bay P. Sta.	IH 23 902668	18	...	67	136	80	84	81	98	43	88	64	142	88	151	1122	36	9 Jun
Lurgan Cemetery	IJ 33 073569	55	...	55	118	67	65	77	118	40	67	58	124	72	108	971	49	9 Jun
Crumlin (Lough Neagh)																		
Stoneyford Res. No.1	IJ 33 221701	135	1047	61	172	87	86	81	101	57	72	74	124	83	141	1137	37	9 Jun
Stoneyford Res. No.2	IJ 33 218701	135	...	63	176	83	89	81	98	55	71	73	124	81	137	1132	37	9 Jun
Leathemstown Res.	IJ 33 214723	164	1050	72	188	83	104	93	112	58	88	75	130	95	132	1229	41	9 Jun
Aghalee Police Sta.	IJ 33 120649	35	...	56	119	60	70	81	102	40	74	71	116	79	125	995	29	9 Jun
Main (Lough Neagh)																		
Cloughmills Police Sta.	ID 34 068179	98	...	58	105	130	56	106	111	70	60	81	130	129	197	1233	31	9 Jun
Clogh (Main)																		
Collin Top	ID 34 207162	381	...	79	188	183	122	155	142	84	86	109	170	193	272	1783	...	...
Dungonnell Filters	ID 34 183174	244	...	64	189	172	119	143	122	79	78	104	171	201	273	1715	38	9 Apr
Carnlea	ID 34 074102	110	...	54	121	112	56	102	128	71	66	71	113	100	180	1174	...	...
Main (Lough Neagh)																		
Lisnafillan	ID 34 075023	38	...	55	112	105	57	93	116	60	79	71	103	94	169	1114	34	9 Jun
Braid (Main)																		
Quolie Res.	ID 34 178128	252	1417	64	159	144	102	123	133	73	73	91	167	164	239	1532	36	9 Jun
Ballymena, Harryville	ID 34 107028	43	1103	54	113	110	60	85	125	64	76	73	107	103	177	1146	37	9 Jun
Kells (Main)																		
Carninard Mountain	IJ 33 313983	347	...	97	267	141	117	130	134	37	117	99	193	146	235	1713	...	...
Brunt Hill No.1	ID 34 318013	351	...	88	273	123	128	119	125	50	117	90	196	126	217	1652	...	...
Brunt Hill No.2	ID 34 309006	317	...	76	198	131	123	108	128	38	107	76	180	114	192	1472	...	...
Braetown	ID 34 288018	306	...	77	279	141	119	103	140	57	95	92	180	150	248	1681	...	...
Lowtown	ID 34 276005	271	...	83	232	159	142	115	121	46	89	97	180	166	268	1699	53	3 Oct
Killylane Res.	IJ 33 279986	223	1419	85	218	178	124	111	120	40	93	105	180	152	245	1650	59	30 Jan
Glen Head	ID 34 253019	299	...	83	191	116	112	106	128	48	85	90	178	141	248	1525	...	...
Tardree Forest, New site	IJ 33 188932	209	...	72	144	99	91	110	145	59	88	101	118	104	190	1321	38	9 Jun
Kells Police Sta.	IJ 33 144968	47	...	68	121	91	68	96	143	62	83	79	114	104	184	1213	38	9 Jun
Main (Lough Neagh)																		
Randalstown	IJ 33 083907	27	1046	56	110	78	69	95	115	58	79	75	111	83	143	1070	37	9 Jun
Sixmile (Lough Neagh)																		
Ballyboley Forest, New site	IJ 33 315970	271	...	88	235	123	102	118	132	34	108	91	172	140	218	1560	(32	26 Mar)
Ballyclare S. Wks.	IJ 33 282901	61	...	78	194	104	89	99	122	44	91	74	128	113	172	1308	(43	9 Jun)
Castle Upton	IJ 33 228857	49	...	81	157	88	86	96	106	52	84	65	109	97	164	1185	43	9 Jun
Islandreagh	IJ 33 187857	53	...	65	137	94	85	91	123	59	71	77	112	91	158	1162	40	9 Jun
Muckamore Forest	IJ 33 162856	35	...	55	123	81	79	86	106	51	61	72	101	77	135	1026	39	9 Jun
Greenmount	IJ 33 152847	32	...	51	116	79	71	80	109	53	66	67	99	71	131	992	41	9 Jun
Aldergrove	IJ 33 147798	69	...	59	121	85	68	84	120	53	80	74	109	78	131	1062	40	9 Jun
Killymoon (Ballinderry)																		
Pomeroy Forest	IH 23 705724	165	...	95	189	108	109	99	140	42	81	75	144	110	191	1384	33	12 Oct
Cookstown No.1	IH 23 813743	63	1004	77	179	84	87	93	124	37	77	73	129	98	169	1226	27	9 Jun
Cookstown No.2	IH 23 816748	77	990	72	172	84	78	94	121	35	78	70	134	92	163	1191	26	9 Jun
Lissan (Ballinderry)																		
Lough Fea W. Wks.	IH 23 767863	218	...	111	215	115	122	135	143	57	80	94	183	139	247	1640	33	3 Oct
Lough Neagh (Bann)																		
Ballyronan Police Sta.	IH 23 946853	18	...	50	136	71	68	81	90	38	74	57	120	70	152	1008	36	9 Jun

Station	Grid reference	Altitude m	Av. 1916–1950 mm	Jan	Feb	Mar	Apr	May	Jun	Jul	Aug	Sep	Oct	Nov	Dec	Year	Highest recorded fall in a day mm	date
BANN AND LOUGH NEAGH—*(contd)*																		
Moyola (Lough Neagh)																		
Derrynoyd Forest ...	IH 23 770956	88	...	90	245	120	102	97	106	47	75	80	157	133	218	1470	36	12 Oct
Moydamlaght P. Sta.	IH 23 741990	178	...	110	274	166	137	107	149	92	83	98	178	150	282	1825	32	12 Oct
Iniscarn Forest	IH 23 828904	162	...	105	221	120	101	139	128	46	91	85	156	110	234	1537	(36	27 Mar)
Bann																		
Toomebridge	IH 23 988905	15	920	61	143	77	61	76	101	51	78	72	125	71	139	1056	36	9 Jun
Bellaghy Police Sta.	IH 23 952966	25	...	60	130	83	56	95	109	62	91	77	130	85	154	1131	28	9 Jun
Portglenone Forest ...	IC 24 988027	58	...	57	154	94	58	98	145	67	63	70	114	89	159	1167	...	...
Portglenone Monastery ...	IC 24 977037	20	...	51	151	88	54	87	138	68	68	71	119	79	146	1120	(42	25 Jun)
Claudy (Bann)																		
Upperlands	IC 24 874047	73	1106	71	154	113	53	135	126	60	61	86	111	100	185	1254	30	20 Jun
Bann																		
Kilrea Police Sta.	IC 24 922125	43	...	60	137	94	48	106	174	55	63	87	124	81	159	1188	34	23 Jun
Agivey (Bann)																		
Garvagh Forest No. 1 ...	IC 24 836159	88	...	83	164	123	65	92	123	72	65	68	111	103	215	1285	31	23 Jun
Moneydig ...	IC 24 890170	34	1078	61	98	107	44	94	114	65	58	83	113	89	179	1105	26	23 Jun
Cam Forest	IC 24 765221	226	...	63	152	125	68	93	159	113	85	90	128	103	227	1407	52	23 Jun
Macosquin (Bann)																		
Springwell Forest	IC 24 771275	177	...	62	152	125	71	112	146	87	58	93	120	123	224	1373	(44	20 Feb)
Coleraine, Carthall Filters	IC 24 829324	51	...	52	107	123	45	77	144	74	48	69	90	107	165	1101	27	9 Jun
Bann																		
Ballymoney	IC 24 943259	55	...	44	68	112	37	102	124	73	44	82	116	83	146	1033	36	20 Jun
Coleraine, The Cutts	IC 24 854302	4	1039	48	84	123	39	73	124	64	37	66	93	101	166	1018	23	9 Jun
Somerset Forest	IC 24 840301	43	...	50	97	123	41	77	126	63	44	63	94	99	162	1039	25	9 Jun

BUSH AND NORTH-EASTERN STREAMS AREA, 921 SQUARE KILOMETRES

Coastal																		
Portrush	IC 24 854411	9	926	42	92	103	46	92	138	71	41	84	98	111	152	1070	27	18 Jun 22 Jun
Bush																		
Slievenamaddy ...	ID 34 145213	378	...	95	205	185	128	157	143	109	84	133	203	178	255	1877	...	...
Skerry Hill, North ...	ID 34 133216	329	...	94	209	184	107	168	135	105	90	157	195	170	255	1871	(35	9 Sep)
Altnahinch Dam	ID 34 123230	249	1501	81	203	185	133	142	129	111	83	122	179	155	227	1751	...	...
Altnahinch Filters ...	ID 34 117236	226	...	79	192	176	102	140	119	91	72	122	185	160	230	1668	(33	4 Nov)
Ballybraddin Forest	ID 34 109247	192	...	69	176	184	120	130	113	82	65	107	178	156	206	1586	33	3 Oct
Orra More, Ballybraddin	ID 34 125261	396	...	80	209	180	137	125	136	108	80	104	191	163	237	1750	...	...
Breen Forest	ID 34 122338	131	...	61	151	125	121	120	85	71	75	111	177	134	206	1437	(43	20 Feb)
Armoy Police Sta. ...	ID 34 066329	100	...	55	104	108	72	108	92	70	42	91	145	113	189	1187	29	3 Oct
Lisnisk	IC 24 916335	53	...	(38)	(76)	(102)	35	89	127	67	35	77	117	87	159	1009	40	20 Jun
Bushmills Police Sta.	IC 24 942407	12	...	36	89	90	34	79	140	48	34	83	118	101	150	1001	40	20 Jun
Coastal																		
Runkerry ...	IC 24 936434	8	...	33	69	90	37	77	144	47	35	70	113	97	159	972	51	20 Jun
Ballycastle, Glenowen	ID 34 125405	30	...	57	111	99	54	93	82	48	53	81	154	123	168	1123	29	9 Jun
Ballypatrick Forest ...	ID 34 176387	152	--	80	203	107	118	93	78	45	79	93	212	145	192	1446	36	19 Feb
Ballycastle Forest ...	ID 34 117400	73	...	68	135	105	70	104	85	49	58	89	161	128	178	1231	(32	20 Feb)
Cushendun	ID 34 248329	9	...	59	157	133	121	91	73	41	79	79	147	150	226	1357	31	29 Nov
Bush Head	ID 34 159229	363	...	89	229	199	140	169	143	111	105	132	206	191	306	2019	...	...
Orra More, South	ID 34 142252	375	...	(81)	202	208	108	150	147	121	107	129	184	185	281	1904	...	...
Beaghs Forest ...	ID 34 162273	248	...	87	189	202	134	146	116	94	91	125	202	187	312	1885	(45	9 Apr)
Orra Beg ...	ID 34 143277	335	...	84	217	197	129	149	136	112	88	128	196	178	254	1868	...	...
Parkmore Forest	ID 34 203207	235	--	83	228	198	129	150	129	90	81	115	169	216	350	1938	42	9 Apr
Crockravar No. 2	ID 34 240199	347	...	81	185	180	135	132	122	66	81	84	196	168	259	1689	...	...
Loughgarve	ID 34 214184	305	...	76	201	160	135	150	132	74	79	94	191	61	264	1615	...	...
Glenarm Forest	ID 34 301123	27	...	73	225	130	122	98	106	35	82	90	155	155	212	1482	43	9 Jun

LAGAN AND EASTERN STREAMS AREA, 1979 SQUARE KILOMETRES

Coastal																		
Kilwaughter	ID 34 354007	143	...	84	222	142	116	132	117	46	113	86	160	133	217	1567	40	9 Jun
Larne ...	ID 34 393026	43	1086	84	186	119	91	109	102	46	99	79	136	108	175	1333	47	9 Jun
Larne, Greenland Cemetery ...	ID 34 392036	98	...	78	186	116	91	157	96	50	97	79	152	113	182	1397	(48	20 Feb)
Ballycarry ...	IJ 33 459949	61	1057	54	144	103	83	90	98	31	97	75	136	97	159	1166	(46	9 Jun)
Whitehead Pavilion ...	IJ 33 476921	12	...	53	150	81	84	84	87	32	99	65	117	90	130	1072	...	...
Lough Mourne Res.	IJ 33 418923	186	1132	72	208	118	112	115	126	54	109	87	163	116	200	1480	46	9 Jun
Copeland Res. ...	IJ 33 429916	128	1056	64	175	102	103	105	115	47	110	77	147	115	178	1338	44	9 Jun
Belmont	IJ 33 432908	82	...	54	156	93	84	99	96	40	98	67	126	97	154	1165	40	9 Jun
Kilroot	IJ 33 441894	22	...	50	136	83	72	79	93	41	93	58	111	88	145	1049	35	9 Jun
Carrickfergus ...	IJ 33 414876	7	...	56	161	94	83	80	97	40	88	68	125	97	155	1142	37	3 Oct
Woodburn																		
North Carn Forest ...	IJ 33 366922	247	...	69	210	107	116	103	129	47	132	81	179	110	206	1488	(45	29 Jan)
Woodburn, North	IJ 33 367905	217	1234	72	170	109	100	105	124	54	104	87	171	107	207	1411	44	9 Jun
Woodburn No.2	IJ 33 362900	255	...	69	180	103	98	104	127	52	112	81	179	112	211	1428	...	...
Boydstown (Woodburn)																		
Woodburn No.4	IJ 33 354883	223	...	69	184	89	103	88	114	48	99	81	164	98	191	1327	...	...
Woodburn No.1	IJ 33 353896	259	...	67	223	84	107	90	122	49	107	81	174	98	178	1381	...	...
Knockagh ...	IJ 33 365881	218	1118	66	170	80	91	86	114	43	98	77	151	86	171	1235	...	...
Woodburn No.3	IJ 33 361896	233	...	69	187	99	99	97	127	52	105	83	175	108	182	1382	...	...
Woodburn, South	IJ 33 370893	186	1184	66	184	104	98	93	117	50	97	80	170	107	195	1362	46	3 Oct
Coastal																		
Dorisland Res. ...	IJ 33 386879	77	1021	64	173	113	85	90	117	51	92	81	154	119	196	1338	44	3 Oct
Monkstown, Upper ...	IJ 33 344872	239	...	69	178	95	108	80	119	50	84	80	157	96	179	1296	46	9 Jun
Monkstown	IJ 33 341838	52	...	64	159	100	102	84	126	33	72	77	127	84	168	1197	41	9 Jun
Newtownabbey	IJ 33 340802	32	...	72	158	106	94	83	108	33	80	89	129	95	168	1214	58	9 Jun
Belfast P. Sta. ...	IJ 33 352769	5	961	69	160	90	91	72	112	26	69	95	122	91	127	1126	56	9 Jun
Lagan																		
Lisadian, Laurelgrove	IJ 33 219591	63	902	61	136	68	99	84	106	61	81	90	117	84	123	1110	43	9 Jun
Halliday's Bridge	IJ 33 212597	37	...	56	129	64	80	81	97	54	76	92	109	79	122	1041	45	9 Jun
Hillsborough	IJ 33 251577	116	900	67	150	67	106	77	103	47	67	85	118	82	120	1087	49	9 Jun
Blaris Cemetery	IJ 33 255626	42	...	58	134	67	78	78	110	44	73	79	102	81	119	1022	46	9 Jun
Lisburn W. Wks., Old gauge	IJ 33 254663	88	...	67	164	85	81	88	119	52	97	99	126	91	142	1210	...	...
Lisburn W. Wks., New gauge	IJ 33 254663	88	...	59	150	79	81	88	112	51	84	97	117	87	146	1151	41	9 Jun
Lisburn, Seymour Park ...	IJ 33 276639	40	...	64	154	71	100	83	100	41	72	91	118	102	135	1131	...	...
Tullynacross S.Wks., New site	IJ 33 282655	23	...	58	147	72	89	87	99	39	76	89	107	88	127	1078	43	9 Jun
Lagmore Res. ...	IJ 33 265690	115	1133	66	176	92	100	107	116	55	87	96	114	90	160	1261	(43	1 Sep)
Dunmurry, Seymour House ...	IJ 33 288681	32	...	65	167	81	(97)	96	90	36	80	90	106	93	135	1135	...	...

GENERAL TABLE, NORTHERN IRELAND

Station	Grid reference	Altitude m	Av. 1916–1950 mm	Jan	Feb	Mar	Apr	May	Jun	Jul	Aug	Sep	Oct	Nov	Dec	Year	Highest recorded fall in a day mm	date
LAGAN—*(contd)*																		
Black Mountain	IJ 33 278727	295	...	82	179	97	107	111	110	67	86	101	132	79	164	1315	38	9 Jun
Dunmurry S. Wks.	IJ 33 298679	12	...	65	161	78	95	95	91	35	85	93	125	95	132	1151	50	1 Sep
Knockbreckan Res.	IJ 33 364661	116	...	72	196	83	115	90	104	34	79	78	133	89	132	1205	47	9 Jun
Newtownbreda	IJ 33 345691	48	...	72	173	83	107	94	109	32	75	75	130	101	128	1179	49	9 Jun
Rosetta	IJ 33 346712	24	...	65	161	79	103	92	97	35	79	65	122	96	125	1119	47	9 Jun
Finaghy	IJ 33 309700	15	...	70	165	102	103	77	97	40	81	95	115	104	143	1192	49	1 Sep
Oldpark Filters	IJ 33 322772	58	1163	85	178	121	114	88	114	49	88	95	139	104	192	1367	47	9 Jun
Divis Mountain	IJ 33 287751	363	...	90	179	97	102	114	121	79	99	99	142	95	199	1415	47	9 Jun
Coastal																		
Knockbreda	IJ 33 359708	50	...	71	185	82	116	97	103	36	77	69	137	95	130	1197	47	9 Jun
Belmont	IJ 33 389748	37	...	56	134	72	102	87	107	31	71	70	134	85	106	1055	53	9 Jun
Stormont Castle	IJ 33 402749	56	–	66	177	75	106	99	107	33	72	74	145	83	117	1155	47	9 Jun
Ballycultra	IJ 33 414800	37	...	71	190	73	116	80	99	32	93	93	133	83	121	1183	(33	1 Sep)
Creighton's Green	IJ 33 431789	119	...	65	184	76	116	102	106	25	86	82	136	85	123	1186	47	9 Jun
Ballysallagh Res.	IJ 33 455785	72	1019	59	174	73	89	92	81	31	84	80	141	93	133	1130	34	3 Oct
Helen's Bay	IJ 33 453825	43	...	58	159	66	82	86	99	35	83	75	127	84	122	1076	43	9 Jun
Conlig Res.	IJ 33 499786	61	944	55	158	77	76	91	86	39	83	79	129	99	127	1100	44	9 Jun
Orlock Res.	IJ 33 561827	4	...	44	131	71	73	89	74	31	84	76	125	88	100	983	39	3 Oct
Lough Cowey Res.	IJ 33 597546	10	837	62	172	43	101	68	69	25	49	55	121	88	97	951	(32	4 Nov)
Portavogie	IJ 33 660596	5	...	51	146	66	97	69	59	32	52	61	111	88	104	936	41	4 Nov
Ballywalter Park	IJ 33 623681	11	867	50	147	71	97	81	59	35	89	68	127	96	107	1025	44	4 Nov
Woburn House	IJ 33 604746	5	...	45	143	47	63	71	63	33	94	69	115	88	113	944	(18	25 Jan)
Donaghadee, New Road	IJ 33 585803	18	...	49	141	68	86	81	71	46	88	80	125	86	112	1031	41	3 Oct
Ballywatticock	IJ 33 516721	17	...	50	164	62	95	68	77	27	74	69	132	84	96	997	40	4 Nov
Whitespots Res.	IJ 33 487753	55	...	51	171	77	94	80	99	36	83	81	134	87	119	1113	45	9 Jun
Dundonald P. Sta.	IJ 33 422732	20	...	61	188	68	102	83	86	27	68	74	146	74	109	1087	38	3 Oct
Tullyhubbert P. Sta.	IJ 33 417662	85	...	80	209	79	123	77	92	40	74	73	130	84	122	1183	39	9 Jun
Comber Cemetery	IJ 33 467694	11	...	74	212	72	103	79	91	37	81	75	177	86	108	1197	42	16 Oct
Reagh Island	IJ 33 524641	12	...	47	147	53	94	55	65	25	70	69	111	82	86	906	38	4 Nov
Killyleagh Primary School	IJ 33 522531	9	...	52	129	44	80	72	72	41	58	55	103	79	87	872	41	4 Nov
Moore's Point	IJ 33 524507	12	...	52	152	48	85	79	91	42	56	56	113	72	97	944	42	9 Jun
Hare Island	IJ 33 506494	3	...	60	167	52	101	84	100	38	60	72	105	75	111	1025	44	9 Jun
Quoile																		
Drumkeeragh Forest	IJ 33 338463	256	...	89	271	101	161	99	155	67	62	78	161	92	159	1497	63	9 Jun
Ballycloghan	IJ 33 413607	61	...	73	208	68	134	77	92	40	74	65	167	83	116	1196	43	9 Jun
Crossgar	IJ 33 460524	36	...	70	195	46	119	82	93	29	56	57	139	85	115	1088	40	9 Jun
Downpatrick, la Salle	IJ 33 489443	26	...	70	196	63	120	92	120	31	57	61	122	79	104	1115	62	9 Jun
Downpatrick, Technical School	IJ 33 485442	6	...	66	186	59	114	85	103	36	54	55	109	71	95	1032	51	9 Jun
Coastal																		
Killough W. Wks.	IJ 33 526378	2	...	50	170	42	104	65	90	28	46	59	107	67	98	927	38	9 Jun
Killough	IJ 33 540353	15	...	61	180	47	119	81	106	38	55	71	105	90	141	1092	38	9 Jun
Seaforde Police Sta.	IJ 33 405418	56	1051	72	219	70	152	94	109	40	63	49	139	89	114	1208	52	9 Jun
Castlewellan Forest	IJ 33 350366	67	...	90	309	45	175	101	149	52	93	49	141	91	132	1427	69	9 Jun
Newcastle, Murlough Upper	IJ 33 379323	6	...	95	309	50	170	103	137	29	63	59	139	79	161	1393	83	9 Jun

NEWRY AND SOUTH-EASTERN STREAMS AREA, 929 SQUARE KILOMETRES (within Northern Ireland)

Station	Grid reference	Altitude m	Av. 1916–1950 mm	Jan	Feb	Mar	Apr	May	Jun	Jul	Aug	Sep	Oct	Nov	Dec	Year	Highest recorded fall in a day mm	date
Coastal																		
Foffanny Res. No.1	IJ 33 285292	283	–	158	309	70	309	181	184	60	97	98	230	139	182	2017	94	9 Jun
Tollymore Park, New site	IJ 33 344328	93	–	84	361	56	191	137	162	36	97	56	168	82	156	1586	95	9 Jun
Newcastle, Slievenamaddy Avenue	IJ 33 368309	23	...	(102)	337	64	196	128	123	41	77	69	142	104	183	1564	56	9 Jun
Glasdrumman	IJ 33 371221	46	...	91	214	83	133	101	(114)	44	61	72	135	96	159	1305	42	18 Feb
Slieve Lamagan	IJ 33 322255	399	1832	138	348	98	245	164	182	81	118	113	196	140	231	2053	...	...
Annalong Valley	IJ 33 355225	130	1228	106	272	93	195	120	125	75	75	76	167	104	194	1602	50	3 Oct
Slieve Bearnagh	IJ 33 314279	593	1884	135	340	114	234	168	181	110	119	122	189	139	249	2099	...	...
Lough Shannagh	IJ 33 298259	387	1833	142	347	111	237	165	176	107	122	117	193	137	245	2099	...	...
Silent Valley W. Wks.	IJ 33 305216	129	1341	126	318	92	242	128	141	65	86	88	166	118	218	1787	52	7 Feb
Mourne Park Forest	IJ 33 282166	76	...	98	239	69	171	97	123	31	60	62	81	98	157	1287	43	9 Jun
Kilkeel Police Sta.	IJ 33 304146	27	...	87	192	58	147	87	113	29	47	61	74	84	127	1107	42	9 Jun
Kilkeel	IJ 33 315140	18	...	69	152	59	134	85	113	29	47	63	77	77	115	1019	43	9 Jun
Greencastle	IJ 33 245117	4	...	(83)	177	47	143	71	134	39	44	49	71	87	129	1075	45	9 Jun
Rostrevor Forest	IJ 33 185179	43	...	139	257	63	240	102	119	37	72	62	122	110	136	1460	(70	31 Jan)
Crotlieve, South	IJ 33 205223	192	...	157	270	61	(241)	78	133	64	99	70	157	155	124	1609	...	...
Newry																		
Poyntzpass Police Sta.	IJ 33 059396	27	...	78	132	56	92	73	119	51	46	87	110	79	98	1020	51	9 Jun
Bessbrook No.1	IJ 33 051284	76	1019	93	156	68	127	90	114	53	70	77	156	93	129	1223	57	9 Jun
Bessbrook No.2	IJ 33 051284	76	...	91	154	70	121	87	112	51	66	72	156	92	134	1207	57	9 Jun
Newry Abbey	IJ 33 088272	10	...	103	153	67	131	90	113	44	64	60	157	100	128	1210	61	9 Jun
Newry, Bridge Street House	IJ 33 082258	4	983	104	159	68	127	90	112	47	66	61	155	96	129	1214	60	9 Jun
Newry Urban S. Wks.	IJ 33 091247	1	...	103	148	66	124	91	109	40	62	56	138	97	121	1155	60	9 Jun
Narrow Water Forest	IJ 33 113208	9	...	152	179	70	167	113	127	47	71	75	153	115	144	1411	69	9 Jun
Castletown																		
Bonds Mountain Forest	IH 23 925291	224	...	117	175	86	167	115	152	68	92	104	213	115	185	1590	52	12 Oct
Kilcurry (Castletown)																		

ERNE AND WESTERN STREAMS AREA, 2015 SQUARE KILOMETRES (within Northern Ireland)

Station	Grid reference	Altitude m	Av. 1916–1950 mm	Jan	Feb	Mar	Apr	May	Jun	Jul	Aug	Sep	Oct	Nov	Dec	Year	Highest recorded fall in a day mm	date
Drowes																		
Big Dog Forest	IH 23 039493	204	...	103	154	111	125	114	137	96	78	103	85	142	154	1402	(27	5 May)
Garrison	IG 13 941517	34	...	65	131	91	99	75	121	59	64	88	70	124	233	1219	(36	9 Dec)
Finn (Erne)																		
Rosslea Forest	IH 23 541313	70	...	68	140	87	118	87	149	42	86	85	110	125	177	1275	30	9 Jun
Erne																		
Crom Castle	IH 23 356240	58	999	81	141	74	109	93	149	53	81	73	113	102	166	1235	(26	3 Sep)
Newtownbutler Police Sta.	IH 23 417262	66	...	73	134	69	108	84	131	47	83	70	119	106	167	1192	33	9 Dec
Colebrooke (Erne)																		
Mullaghfad Forest	IH 23 497423	162	...	109	196	113	149	107	123	42	103	79	103	150	203	1477	(62	31 Dec)
Doon Forest, New site	IH 23 447370	189	...	90	159	113	125	114	138	61	113	100	117	147	220	1497	(40	13 Nov)
Alderwood W. Wks.	IH 23 487462	150	...	80	133	99	94	109	121	41	103	84	97	142	175	1277	...	...
Colebrooke	IH 23 409446	76	1048	81	140	103	101	107	120	42	79	71	109	132	195	1282	25	9 Jun
Clabby W. Wks.	IH 23 408508	138	...	79	139	91	80	106	143	47	68	75	105	123	177	1234	27	3 Sep

Station	Grid reference	Altitude *m*	Av. 1916–1950 *mm*	Jan	Feb	Mar	Apr	May	Jun	Jul	Aug	Sep	Oct	Nov	Dec	Year	Highest recorded fall in a day *mm*	*date*
ERNE—*(contd)*																		
Tempo (Colebrooke)																		
Pubble Forest, New site	IH 23 335465	91	...	71	138	93	97	122	113	70	65	73	102	147	183	1274	25	3 Sep
Colebrooke (Erne)																		
Lisnaskea Creamery	IH 23 341351	49	...	77	113	68	85	85	98	51	78	68	109	97	171	1100	31	9 Dec
Bunlougher Forest	IH 23 405350	157	...	80	141	75	120	94	117	53	100	85	126	125	181	1298	(36	4 Sep)
Glasdrumman Res....	IH 23 390340	107	...	82	132	66	115	93	120	43	93	89	128	116	177	1254	31	3 Sep
Claddagh (Erne)																		
Moher	IH 23 217302	59	...	103	183	79	113	104	128	46	83	57	92	131	183	1302	31	30 Nov
Arney (Erne)																		
Ballintempo Forest	IH 23 054413	119	...	105	201	109	107	104	120	60	83	93	98	157	233	1468	(36	4 Sep)
Marble Arch P. Sta.	IH 23 124356	87	...	90	193	(108)	96	100	111	50	89	65	93	131	191	1316	28	9 Aug
Florence Court, New site	IH 23 175344	84	1235	121	216	89	127	103	117	38	68	64	101	120	209	1371	28	3 Sep
Florencecourt Police Sta.	IH 23 192354	53	...	93	207	92	109	95	125	45	70	58	92	98	194	1278	25	3 Sep 9 Dec
Florencecourt Primary School	IH 23 186352	62	...	(102)	(209)	81	111	92	125	43	74	65	90	105	191	1288	...	...
Erne																		
Castle Coole	IH 23 259430	62	...	82	125	76	79	95	119	33	74	78	91	110	182	1143	...	...
Bellanaleck W. Wks.	IH 23 244390	28	...	77	145	86	84	86	129	44	72	72	80	111	193	1179	36	9 Dec
Bellevue	IH 23 253410	61	1007	61	115	70	66	84	112	42	71	66	84	105	170	1045	30	9 Dec
Sillees (Erne)																		
Lough Navar Forest	IH 23 062546	148	...	100	190	118	118	119	134	80	73	122	72	157	276	1557	39	9 Dec
Belmore Forest	IH 23 121438	114	...	103	219	128	113	123	156	63	105	71	95	161	284	1621	(39	27 Mar)
Erne																		
Portora Royal School	IH 23 224448	75	...	77	120	87	75	105	123	50	69	65	78	118	188	1153	(44	10 Jan)
Portora Sluice	IH 23 222453	46	...	93	148	105	95	113	135	47	73	78	94	112	181	1273	27	9 Dec
Breagho Bridge	IH 23 265475	103	...	75	126	77	87	104	120	50	60	87	88	110	172	1155	25	9 Dec
Ballinamallard (Erne)																		
Ballinamallard S. Wks.	IH 23 261528	53	...	73	126	95	73	95	135	44	59	82	90	120	171	1163	27	22 Jun
Irvinestown Police Sta.	IH 23 236582	90	1037	68	113	82	74	97	124	47	51	97	89	106	175	1123	27	9 Dec
Irvinestown P. Sta....	IH 23 230583	85	...	71	125	86	77	102	123	50	57	95	88	113	184	1173	26	9 Dec
Erne																		
Castle Archdale Forest No. 1	IH 23 189593	66	1040	71	134	88	77	101	118	54	51	92	83	114	193	1175	27	3 Sep
Kesh (Erne)																		
Kesh Forest	IH 23 181664	82	....	78	129	110	81	106	114	55	53	106	74	128	213	1247	24	26 Jun
Erne																		
Belleek	IG 13 940590	47	1229	55	109	81	74	94	134	72	59	106	74	111	203	1173	46	9 Dec
Belleek Police Sta.	IG 13 939591	50	...	62	120	83	89	101	135	68	63	109	84	119	216	1249	46	9 Dec

Station	Grid reference	Altitude m	Av. 1916–1950 mm	Rainfall in millimetres Jan	Feb	Mar	Apr	May	Jun	Jul	Aug	Sep	Oct	Nov	Dec	Year	Highest recorded fall in a day mm	date
The Channel Islands																		
ALDERNEY AREA. 8.03 SQUARE KILOMETRES																		
Coastal																		
Alderney Airport	WA555 564065	88	...	96	93	20	127	41	75	26	79	32	140	72	105	906	22	24 Oct 11 Dec
GUERNSEY AREA. 63.7 SQUARE KILOMETRES																		
Coastal																		
Les Quartiers	WV554 323804	12	802	114	106	20	136	44	79	36	84	26	181	85	110	1022	38	14 Oct
Haut Nez	WV554 311762	107	878	109	96	18	126	41	75	40	89	25	202	101	120	1042	36	14 Oct
Guernsey Airport	WV554 294759	104	862	119	99	19	130	42	88	41	91	25	180	96	117	1049	36	14 Oct
Kings Mills	WV554 292787	14	...	114	100	20	129	42	80	37	92	24	189	85	111	1022	33	14 Oct
SARK AREA. 5.44 SQUARE KILOMETRES																		
Coastal																		
Sark, La Cloture	WV554 456736	84	...	86	99	18	115	43	53	31	69	29	193	70	115	920	44	27 Oct
JERSEY AREA. 116 SQUARE KILOMETRES																		
Coastal																		
St. Ouen, La Brecquette	WV554 558542	17	823	105	78	21	109	45	64	28	56	38	152	92	121	910	41	27 Oct
St. Brelade, La Sergente	WV554 566486	58	818	102	86	25	115	55	62	41	67	35	158	100	125	971	37	27 Oct
St. Ouen's Manor	WV554 583530	79	805	112	88	22	111	54	68	34	58	39	149	88	123	947	36	27 Oct
Jersey Airport	WV554 587508	83	833	135	87	21	117	58	64	39	74	39	156	100	134	1023	38	27 Oct
Beaumont Marsh	WV554 616499	5	...	134	101	26	128	76	85	49	81	40	157	110	145	1133	32	22 Jan
Handois Res.	WV554 631536	76	959	121	93	22	128	81	81	47	89	45	159	95	143	1105	29	14 Oct
Millbrook Res....	WV554 631507	16	871	122	88	25	125	69	99	43	76	40	143	95	154	1079	28	22 Jan
Trinity	WV554 659539	104	912	120	98	28	137	67	86	47	92	48	145	94	147	1109	29	14 Oct
St. Helier	WV554 651483	9	765	113	89	23	115	56	85	31	74	37	150	87	133	993	28	27 Oct
St. Louis Observatory	WV554 660493	55	847	123	90	24	119	58	91	35	64	43	163	92	152	1055	29	27 Oct
Grands Vaux Res.	WV554 664507	18	...	116	91	24	118	58	88	40	81	42	163	94	148	1062	29	14 Oct
St. Saviour, Petit Menage	WV554 662481	9	844	111	87	21	120	57	85	35	73	42	157	88	140	1015	29	27 Oct

BRITISH RAINFALL 1966

Part II Summary tables, maps and graphs with discussion

1 MAIN CHARACTERISTICS OF THE YEAR

General summary

The year 1966 was, for the most part, cold and dull with rainfall above average over all countries of the United Kingdom; indeed sunshine was below average during nine months of the year.

The year commenced mild and wet; but in mid-January easterly winds prevailed with considerable snowfall in south Wales and the West Country. The first four months were characterized by changes from mild weather to sharp wintry spells. March was dry in eastern and southern England where no measurable rainfall occurred for about three weeks, but in north-west Scotland rainfall was about twice the average. In early April there were some scattered falls of snow in the north and in mid-month heavy falls over south-west Wales and southern England. May was mostly unsettled and wet, and June had frequent thundery rain.

At the commencement of the second half of the year, a feature of July was the cold dull weather over south-east England whilst in south-west Scotland it was warm and sunny. Some of the heaviest storms of the year were experienced over southern and eastern Scotland in August and also over areas of the Lake District and northern Pennines. The period April to August was the wettest over England and Wales since 1931. The Lake District suffered further floods and heavy rain early in September, but another significant weather feature was the dry period during the second half of the month when most counties in England and Wales experienced a fortnight without rain.

The remaining months October to December were generally unsettled with heavy rain in the extreme south-east of England during October. November was mostly cold with northerly winds whilst the first day of December heralded one of the deepest depressions experienced by Britain in recent years. It produced gales and rain in most areas. Overall, December was the wettest month in Scotland and Northern Ireland since 1929, and owing to heavy rain on the 16th and 17th there was flooding in parts of Scotland.

Notes on conditions month by month

January. Mild and wet early in the month and again during the last 10 days, otherwise wintry showers associated with anticyclones to east and north of the British Isles.

February. Mild cyclonic weather generally, interrupted by a brief cold spell when easterly winds brought snow to most areas between about the 11th and 17th.

March. Predominantly anticyclonic and dry, especially over England. North-westerly winds with wintry showers of a thundery nature in most districts towards mid-month and again towards the end of the month.

April. Mainly unsettled and wet. Snow occurred from about the 11th to the 21st when easterly winds prevailed.

May. Wet, cyclonic weather generally, but warm and sunny during the last few days.

June. Mainly wet and unsettled with frequent thunderstorms at times.

July. Cool with sunny spells, but unsettled cyclonic weather with rain or showers, heavy at times, from the 5th to 16th, and again from the 24th onwards.

August. Generally cyclonic with heavy, thundery rain at times, especially in the north.

September. Mild, cyclonic weather with rain and thunderstorms, heavy at times, for much of the first half of the month, followed by anticyclonic, mainly dry weather, apart from heavy rain in the south-east late in the month.

October. Mainly cyclonic. Cold northerly winds with wintry showers from about the 23rd.

November. Cold, mainly cyclonic weather with winds frequently from a northerly quarter. A spell of wet, very stormy weather in most areas around mid-month. Snow showers and strong to gale north-westerly winds in most areas late in the month.

December. Cyclonic, wet weather generally. Mild in the south, colder farther north. Gales and heavy rain in Scotland and northern England early in the month and again around mid-month. Snow showers over northern districts on the 23rd and 24th in a north-westerly airstream.

Note: Details of heavy falls on individual occasions are given in Sections 5 and 6.

TABLE 1

Estimated areal rainfall, monthly, annual and seasonal, for parts of the United Kingdom

A Averages for 1916–50 in millimetres

	Jan.	Feb.	Mar.	Apr.	May	June	July	Aug.	Sept.	Oct.	Nov.	Dec.	Year	Winter half-year Oct.–Mar.	Summer half-year Apr.–Sept.	Area* *km*²
England	83	60	52	57	59	52	75	75	69	83	88	79	832	444	388	130 297
Wales	149	105	85	82	86	79	106	116	114	148	145	142	1357	774	583	20 756
England and Wales . .	92	66	57	60	63	55	79	81	76	92	95	88	904	490	414	151 053
Scotland	154	106	89	88	87	87	114	122	128	158	143	143	1419	793	626	78 734
Northern Ireland . . .	109	76	66	67	72	71	96	102	96	111	104	111	1081	577	504	14 133
Isle of Man	140	87	79	73	79	77	86	105	115	141	132	136	1250	715	535	572
Great Britain	112	80	68	70	72	66	92	95	93	114	113	106	1081	592	489	229 787

* Values derived from: London, Central Office of Information. *Britain—an official handbook*, 1964. London, HMSO, 1964.

B Values for 1966 in millimetres

	Jan.	Feb.	Mar.	Apr.	May	June	July	Aug.	Sept.	Oct.	Nov.	Dec.	Year	Winter half-year Oct. 1965–Mar. 1966	Summer half-year Apr.–Sept. 1966	12 Months Oct. 1965–Sept. 1966
England	51	116	32	98	65	79	72	101	47	117	74	106	959	475	462	937
Wales	81	171	70	143	104	111	91	103	79	176	125	210	1464	841	631	1472
England and Wales . .	55	123	37	104	71	84	75	101	51	125	81	120	1029	524	486	1010
Scotland	80	124	139	68	111	147	71	111	116	131	153	241	1491	745	624	1369
Northern Ireland . . .	70	151	92	93	93	119	51	70	79	123	105	167	1213	660	505	1165
Isle of Man	91	228	85	143	129	106	29	115	120	114	110	181	1450	880	642	1522
Great Britain	64	123	72	92	85	105	73	105	73	127	105	162	1188	601	533	1134

C Values for 1966, percentage of annual average (1916–50)

	Jan.	Feb.	Mar.	Apr.	May	June	July	Aug.	Sept.	Oct.	Nov.	Dec.	Year	Winter half-year Oct. 1965–Mar. 1966	Summer half-year Apr.–Sept. 1966	12 Months Oct. 1965–Sept. 1966
England	6·2	13·9	3·9	11·8	7·9	9·6	8·7	12·1	5·6	14·1	8·9	12·8	115	57	56	113
Wales	5·9	12·6	5·1	10·6	7·7	8·2	6·7	7·6	5·8	13·0	9·2	15·5	108	62	47	109
England and Wales. . .	6·1	13·7	4·1	11·5	7·8	9·3	8·3	11·2	5·7	13·9	8·9	13·3	114	58	54	112
Scotland	5·6	8·7	9·8	4·8	7·9	10·3	5·0	7·8	8·2	9·3	10·7	17·0	105	53	44	97
Northern Ireland . . .	6·5	14·1	8·5	8·7	8·6	10·9	4·6	6·5	7·3	11·4	9·7	15·4	112	61	47	108
Isle of Man	7·3	18·2	6·8	11·4	10·3	8·5	2·3	9·2	9·6	9·1	8·8	14·5	116	70	51	121
Great Britain	5·9	11·4	6·7	8·5	7·8	9·7	6·8	9·7	6·8	11·8	9·8	15·0	110	56	49	105

2 MONTHLY, ANNUAL AND SEASONAL RAINFALL

This section presents, for a selection of stations, tabulated data expressed as a percentage of average annual rainfall for the standard period 1916–50, a series of monthly, seasonal and annual equipercental maps and a map of total rainfall in 1966 expressed in absolute measure.

Equipercental maps in their present form were first introduced in *British Rainfall* 1961. The monthly and seasonal maps are based on the records of a rather limited number of stations (about 550) and consequently give a broad and somewhat simplified picture of the equipercental distribution of rainfall. The annual map, however, is based on all the reliable data available and therefore is much more detailed, so much so that a few minor features too small to reproduce clearly in this publication have had to be suppressed. This map supersedes earlier versions based on less abundant data.

For the purpose of comparisons, maps of monthly and seasonal average rainfall expressed as a percentage of standard period annual average are published in the *British Rainfall Supplement* 1961–1965. The monthly equipercental maps are also available in booklet No. 2A issued in conjunction with the map of average annual rainfall for 1916–50 published by the Ordnance Survey in the Ministry of Housing and Local Government planning series.

Monthly maps in absolute measure are published in the *Monthly Weather Report*. Isohyets are given in inches with millimetre equivalents until December 1966 and in millimetres with inch equivalents thereafter.

TABLE 2

Monthly, annual and seasonal rainfall for individual stations, percentage of annual average, 1916–50

Area and station	Jan.	Feb.	Mar.	Apr.	May	June	July	Aug.	Sept.	Oct.	Nov.	Dec.	Year	Winter half-year Oct. 1965–Mar. 1966	Summer half-year Apr.–Sept. 1966	Seasonal year Oct. 1965–Sept. 1966
Northumbrian																
Acklington	4·8	10·5	1·4	12·7	7·9	9·7	5·7	25·1	5·1	10·3	7·8	10·5	111	50	66	117
Whickhope	5·9	12·5	5·8	9·1	7·8	9·7	4·6	10·8	8·4	8·6	9·5	11·5	104	57	50	107
Hexham Res.	5·2	12·4	4·9	10·3	6·6	10·0	6·4	17·0	9·0	13·6	8·7	10·9	115	57	59	116
Durham	7·6	14·7	1·4	11·9	8·5	9·3	10·1	16·8	4·1	15·1	7·3	7·4	114	60	61	121
Hury Res.	6·2	13·8	4·8	9·7	7·1	8·3	5·1	11·5	6·3	11·4	8·7	12·6	105	52	48	100
Yorkshire Ouse and Hull																
Mulgrave Castle	5·6	13·3	2·7	13·9	4·8	10·2	8·1	19·3	3·1	17·1	10·8	10·2	119	57	59	116
Scarborough	5·3	13·7	3·9	15·5	9·3	10·1	12·1	17·6	3·6	10·4	10·6	7·2	119	63	68	132
Hull	6·2	18·9	4·1	11·5	6·6	10·5	11·9	15·9	8·0	11·2	9·6	7·6	122	68	64	132
Richmond, Green Howard Rd	7·2	18·5	2·4	15·1	11·1	7·2	4·9	15·4	3·1	14·5	9·0	7·1	115	58	57	115
Ramsgill	6·2	21·5	7·3	12·3	11·0	8·1	5·0	9·7	6·2	12·2	13·1	12·4	125	77	52	130
York, Acomb Landing	6·0	20·1	2·3	13·0	8·7	7·9	9·3	20·1	3·0	9·7	6·3	8·1	115	65	62	127
Ampleforth	6·7	18·5	3·4	12·7	8·7	10·4	9·0	16·4	5·5	7·8	6·9	6·4	112	58	63	121
Malham Tarn	4·8	16·0	7·3	7·3	10·1	9·3	7·9	9·0	7·8	7·7	9·0	15·9	112	62	51	113
Bradford	5·5	19·5	4·0	14·2	12·4	11·0	6·4	18·9	5·6	10·4	10·1	12·3	130	70	69	139
Goole S. Wks	5·0	16·7	4·1	9·4	6·7	10·7	13·6	21·7	5·8	10·6	7·7	6·5	119	60	68	128
Thorne	5·2	18·1	3·0	13·3	6·9	13·1	12·5	17·7	5·8	11·9	7·2	7·7	122	60	69	129
Trent																
Keele	6·1	10·8	5·7	9·3	9·1	11·3	12·1	12·1	6·3	12·4	7·0	15·7	118	55	60	115
Edgbaston	4·5	14·5	4·5	10·4	10·2	11·0	9·2	12·2	3·5	14·0	7·5	12·7	114	57	57	114
Nanpantan Res.	5·5	15·0	4·0	15·7	7·7	10·7	10·7	12·8	4·5	14·8	6·6	10·9	119	61	62	123
Mansfield	7·5	20·6	4·5	15·6	8·4	10·2	7·4	11·0	5·9	13·2	9·8	9·2	123	75	59	133
Lincolnshire																
Manby	6·0	17·0	3·5	10·0	7·5	16·1	19·1	10·1	4·8	11·7	11·8	9·4	127	74	68	142
Revesby Res.	6·6	15·0	3·4	9·6	7·3	12·2	10·2	11·6	4·1	10·1	10·6	8·8	109	61	55	116
Welland and Nene																
Eye Brook Res.	5·1	13·0	1·3	14·9	6·3	8·6	10·4	15·3	5·8	10·9	7·9	12·3	112	50	61	111
Ravensthorpe	4·6	11·4	1·3	12·8	7·5	9·7	9·9	18·3	6·3	14·5	7·0	14·1	117	48	65	112
Great Ouse																
Foxcote P. Sta.	4·0	14·2	3·7	12·8	10·0	11·3	12·0	11·3	7·2	17·2	6·9	10·4	121	51	65	115
Cardington	4·7	11·9	2·0	14·4	7·1	14·1	9·9	15·8	8·9	17·2	9·5	13·2	129	49	70	119
Odsey	4·7	10·9	1·5	10·0	7·3	13·8	11·9	17·8	5·6	13·1	8·2	13·9	119	47	66	113
Cambridge, Botanic Garden	5·1	8·1	1·9	9·7	5·6	9·8	16·1	14·3	4·4	12·0	7·3	13·1	107	40	60	100
Mildenhall	3·3	9·1	1·7	7·7	5·0	12·2	14·1	13·7	4·6	15·6	9·4	12·2	109	40	57	97
Sandringham House	5·9	13·3	4·1	8·0	8·7	12·3	11·1	17·1	4·8	9·1	15·9	12·3	123	57	62	119
East Suffolk and Norfolk																
Bressingham Hall	4·1	7·7	3·2	7·6	5·9	6·7	13·5	9·1	3·0	11·4	12·6	11·9	97	43	46	89
Sprowston	4·8	11·2	3·7	7·4	4·3	10·9	13·5	9·6	2·3	10·6	15·1	9·5	103	51	48	99
Lowestoft	5·7	9·1	2·3	8·1	4·7	7·9	15·2	9·9	2·7	21·4	19·3	11·9	118	43	49	91
Belstead Hall	5·2	8·1	2·5	10·5	8·3	9·3	15·3	9·5	3·9	15·7	12·8	12·5	114	44	57	101

TABLE 2 (*continued*)

Monthly, annual and seasonal rainfall for individual stations, percentage of annual average, 1916–50

Area and station	Jan.	Feb.	Mar.	Apr.	May	June	July	Aug.	Sept.	Oct.	Nov.	Dec.	Year	Winter half-year Oct. 1965–Mar. 1966	Summer half-year Apr.–Sept. 1966	Seasonal year Oct. 1965–Sept. 1966
Essex																
Felsted School	4·8	10·2	1·7	11·6	7·7	9·7	13·0	10·1	6·2	11·3	7·7	15·1	109	50	58	108
Shoeburyness	6·5	9·4	1·6	12·7	3·8	10·3	20·0	13·0	2·1	18·2	12·4	13·6	124	44	62	106
Lee																
Meesden	5·0	11·6	1·3	9·6	7·3	9·7	14·5	11·9	3·7	14·0	7·3	12·9	109	47	57	104
Darnicle Hill P. Sta.	4·7	12·5	1·8	13·9	7·8	15·4	11·5	13·2	5·3	14·5	5·6	11·5	118	51	67	118
London, left bank																
St James's Park	7·0	11·0	2·3	16·4	6·7	12·9	13·7	15·2	5·5	15·2	6·1	13·5	125	52	70	123
Thames																
Cirencester, Chesterton Grove	6·8	13·3	2·9	12·3	7·8	3·9	7·5	12·6	4·9	14·2	6·9	10·1	103	56	49	105
Oxford	4·7	11·5	1·5	12·5	8·3	8·7	12·4	12·1	8·0	18·8	4·3	12·3	115	42	62	104
Maidenhead S. Wks	5·1	12·0	2·1	15·3	7·2	13·0	12·1	10·8	5·0	19·3	7·5	10·8	120	48	63	111
Hindhead W. Wks	6·0	17·8	2·2	12·3	8·3	7·3	8·5	8·2	3·8	18·7	6·7	9·5	109	58	48	106
Mickleham	4·9	15·7	1·6	11·5	8·5	6·7	12·1	7·3	4·5	15·8	6·9	11·0	107	51	51	101
London, right bank																
Kew	5·2	11·3	1·8	15·1	8·0	10·9	11·4	12·7	4·5	16·4	5·5	12·2	115	49	63	112
Bromley	6·0	13·1	3·1	12·8	7·7	11·1	13·3	13·5	5·0	19·2	7·3	14·3	126	53	63	116
Kent																
Falconhurst	4·0	16·2	1·4	13·7	6·1	9·8	11·3	8·1	5·2	18·3	8·4	11·3	114	51	54	105
Goudhurst	6·4	17·7	2·3	12·5	5·1	11·1	9·8	11·6	2·9	22·4	9·7	11·9	123	56	53	109
East Malling	5·4	12·9	1·7	12·0	5·1	8·9	17·8	7·7	4·7	22·7	9·4	13·9	122	47	56	103
Herne Bay	9·6	9·1	1·3	11·3	3·2	11·2	16·2	9·7	1·4	29·7	16·7	13·0	132	46	53	99
Ashford, Belmore	6·2	12·9	2·4	10·3	4·8	11·3	14·8	10·4	2·3	33·3	14·3	16·1	139	51	54	105
Dover	9·2	13·4	3·1	10·0	5·5	9·9	20·0	12·3	1·7	32·5	22·0	15·1	155	56	59	115
Sussex																
Seaford, Newlands	8·0	13·8	2·5	11·1	5·1	9·4	8·1	8·2	3·5	25·9	8·5	11·6	116	57	45	102
Knepp Castle	5·8	15·1	2·4	14·6	8·5	9·4	9·4	10·0	4·3	17·6	7·5	9·7	114	58	56	114
Worthing	7·1	16·7	2·7	13·9	4·9	10·4	7·9	9·5	3·3	20·6	6·9	9·7	114	60	50	110
Hampshire																
Farlington Res.	6·3	14·7	2·9	13·6	7·0	11·5	9·7	11·1	2·8	20·6	8·1	9·8	118	58	56	114
Southampton	6·5	17·7	3·2	12·1	7·5	8·6	7·0	9·4	6·2	18·6	6·2	8·3	111	59	51	110
Wootton House	5·8	15·2	2·5	13·8	8·5	10·1	12·7	10·5	4·2	14·5	6·4	10·3	115	53	60	113
Andover S. Wks	6·8	16·5	2·1	15·1	9·5	9·1	8·1	15·3	8·5	21·2	6·1	10·0	128	56	66	122
Isle of Wight																
Ryde	9·1	15·9	2·2	13·2	7·1	8·7	8·1	13·8	3·5	20·2	8·3	8·6	119	68	54	122
Avon and Dorset																
Stourhead	6·1	14·7	2·9	13·7	6·4	4·9	8·5	12·8	2·5	15·9	10·1	13·4	112	57	49	106
Blandford, Kings Road	7·7	17·1	2·1	12·2	6·7	4·5	4·3	10·2	3·0	22·0	9·7	10·5	110	61	41	102
Weymouth	12·3	15·1	4·0	14·6	5·2	4·8	4·3	12·6	3·0	24·9	11·7	9·5	122	68	45	113
Beaminster	7·9	15·1	3·7	12·9	8·0	5·3	5·6	11·3	2·3	17·3	9·4	9·7	109	63	45	108
Devon, south																
Exeter	15·2	14·6	2·8	15·4	6·5	5·8	4·3	7·9	2·4	18·7	6·0	8·0	108	66	42	108
Totnes	12·8	18·0	4·2	18·6	7·7	4·5	2·9	10·6	2·8	15·5	6·0	9·3	113	76	47	123
Cornwall																
Mount Batten	12·2	16·6	4·0	15·6	8·0	4·5	4·6	12·8	2·7	20·8	6·3	11·0	119	77	48	125
St Austell	11·0	15·6	3·4	12·8	6·6	4·8	3·4	11·5	3·3	17·1	6·1	14·1	110	67	42	110
Tresco Abbey	9·0	13·4	3·6	15·1	6·6	12·4	3·2	8·9	5·5	22·2	5·0	10·5	115	58	52	110
Rosewarne	10·1	22·0	4·3	17·1	10·3	8·2	2·8	11·7	5·8	28·0	5·7	12·5	139	80	56	136
Newquay	9·5	14·8	3·6	12·4	5·9	7·2	4·1	12·3	4·9	17·1	6·3	11·9	110	62	47	109
Bude	8·3	9·3	3·6	13·1	6·8	9·5	6·9	11·1	5·4	12·7	7·2	10·9	105	52	53	105
Devon, north																
Okehampton, Pleasure Gardens	8·7	14·2	4·4	12·1	7·9	4·5	5·2	11·4	3·9	12·5	8·1	15·1	108	67	45	113
Chivenor	7·0	8·9	3·3	13·9	6·6	8·7	4·4	11·7	3·9	13·2	7·3	9·8	99	49	49	98
Somerset																
Taunton, Vivary Park	8·5	15·3	3·1	14·9	7·0	9·1	5·5	9·0	2·6	16·6	8·2	9·9	110	56	48	104
Street S. Wks	7·2	12·0	3·6	12·9	10·9	8·4	8·9	12·9	2·1	15·4	11·0	9·9	115	51	56	107
Bristol Avon																
West Lavington	6·8	12·9	1·9	11·4	7·1	5·9	9·1	11·7	3·4	20·3	7·5	9·5	107	54	49	102
Long Ashton	4·8	11·1	3·2	11·4	7·5	6·7	5·8	8·3	6·0	11·8	9·5	10·5	97	54	46	100
Severn																
Lower Hazel	6·7	11·8	3·6	12·6	7·5	6·9	5·1	8·7	5·0	15·1	10·3	10·5	104	61	46	106
Lake Vyrnwy	6·4	15·8	6·9	8·9	8·1	6·3	6·1	6·0	5·9	8·6	11·2	17·8	108	72	41	113
Shawbury	5·9	12·7	4·7	11·8	8·7	14·4	10·7	12·7	4·4	14·5	10·4	15·1	126	58	63	120
Anchor	5·0	13·4	4·7	8·1	8·1	4·8	4·9	6·3	4·7	10·5	11·8	11·1	93	54	37	91
Oakly Park	5·4	13·0	5·1	10·5	10·2	9·8	5·5	12·5	4·8	12·7	10·0	9·2	109	53	53	106
Dingle Leys	6·3	15·3	5·6	12·9	9·3	9·9	8·3	10·1	4·1	14·8	11·3	12·9	121	59	55	114
Finham S. Wks	4·1	14·5	4·8	12·6	10·5	11·3	8·5	15·3	4·8	15·6	5·8	13·0	121	53	63	116
Wellesbourne	3·0	10·8	3·6	11·5	7·4	8·6	10·7	10·5	5·0	16·5	5·6	10·6	104	45	54	98
Cheltenham	4·8	11·9	4·4	10·2	9·9	6·8	10·1	8·9	4·0	15·5	8·5	12·9	108	48	50	98
Wye																
Llandrindod Wells	5·7	12·4	5·8	7·5	8·9	7·7	6·4	8·7	6·2	13·5	11·4	13·0	107	61	45	106
Glasbury	7·3	10·5	5·2	12·5	9·1	6·6	4·9	7·0	4·0	14·1	9·8	12·7	104	59	44	103
Ross-on-Wye	6·2	15·0	3·4	13·2	7·6	4·3	4·9	11·3	4·3	15·1	11·2	7·6	104	55	46	100

TABLE 2 (*continued*)

Monthly, annual and seasonal rainfall for individual stations, percentage of annual average, 1916–50

Area and station	Jan.	Feb.	Mar.	Apr.	May	June	July	Aug.	Sept.	Oct.	Nov.	Dec.	Year	Winter half-year Oct. 1965– Mar. 1966	Summer half-year Apr.– Sept. 1966	Seasonal year Oct. 1965– Sept. 1966
Usk																
Blaenau-isaf	5·5	19·2	5·3	11·4	9·4	7·7	7·1	8·8	6·9	11·3	9·5	19·6	122	72	51	123
Gilwern	10·0	15·4	4·7	13·8	6·2	6·0	3·7	7·2	4·9	14·4	10·0	13·7	110	65	42	107
Glamorgan																
Treherbert, Tyn-y-Waun, No. 1	6·9	14·6	4·5	10·1	8·3	7·7	4·9	8·7	5·1	10·0	6·9	17·8	105	67	45	112
South West Wales																
Golden Grove Farm	6·9	15·2	4·4	13·2	9·3	6·5	5·2	9·2	5·7	17·4	9·3	15·9	118	66	49	115
Cwmoernant Res.	5·6	13·9	4·5	12·8	9·1	5·9	6·2	10·5	6·0	16·6	7·2	14·0	112	65	51	115
Brawdy	8·1	12·4	3·7	14·3	7·0	8·2	11·4	6·8	4·8	17·8	5·5	7·9	108	62	53	115
Gwynedd																
Aberdovey	3·2	10·1	5·2	8·1	6·7	11·0	6·9	4·5	4·4	12·3	7·2	14·6	94	53	42	95
Llwydiarth Esgob	5·8	11·4	6·7	9·0	7·1	9·5	6·9	7·8	8·3	8·8	7·7	14·2	103	63	49	112
Bryn Adda	3·3	8·0	6·7	8·5	6·3	12·4	9·1	6·8	8·6	11·3	9·1	12·9	103	48	52	99
Bodnant	4·2	13·4	4·8	10·7	5·1	8·1	7·0	6·1	6·8	16·0	11·9	14·4	109	61	44	105
Dee and Clwyd																
Bala Sluices	5·9	11·1	6·3	6·5	6·3	7·1	6·3	4·7	7·0	12·3	11·2	16·3	101	62	38	100
Alwen	4·8	11·1	5·9	8·0	5·8	6·1	9·2	6·3	5·7	11·4	12·5	18·3	105	58	41	99
Pack Saddle Res.	7·1	10·3	5·2	13·2	8·9	4·9	7·8	8·2	5·3	11·5	10·5	20·0	113	58	48	106
Mersey and Weaver																
Heaton Res.	4·4	13·7	5·7	8·5	8·7	9·6	8·5	7·8	7·0	10·5	8·5	16·1	109	60	50	110
Knott Hill Res.	4·1	14·2	5·5	8·3	8·9	9·7	10·3	13·0	8·4	8·9	11·2	13·3	116	61	59	119
Ringway	3·6	12·5	4·8	10·0	9·5	15·3	15·2	11·2	7·5	9·4	9·2	15·4	124	55	69	124
Lancashire																
Stonyhurst	4·2	11·7	6·9	6·2	9·2	11·0	10·2	10·3	9·2	9·4	9·1	16·2	114	56	56	112
Squires Gate	5·1	11·8	4·9	8·6	6·9	7·9	5·9	10·0	9·5	13·5	9·1	14·2	107	51	49	100
Ambleside	7·9	14·0	8·7	6·1	10·0	9·6	3·6	8·3	11·9	8·1	9·5	15·3	113	60	49	110
Poaka Beck	6·2	14·1	6·7	7·0	8·1	11·1	7·2	7·9	11·9	9·2	7·5	14·6	111	55	53	108
Cumberland																
Dale Head Hall	7·3	15·1	7·9	6·4	8·5	8·1	3·7	7·9	9·7	7·1	11·1	12·8	106	56	44	100
Newton Rigg	5·0	10·4	5·5	7·2	5·2	9·9	5·1	13·1	12·7	12·2	10·1	12·3	109	49	53	102
Carlisle S. Wks	3·2	7·1	5·7	5·8	7·1	15·5	7·0	12·0	15·1	8·9	8·7	11·9	108	43	63	106
Isle of Man																
Douglas	7·2	15·9	7·1	10·4	10·3	8·6	2·1	7·8	10·3	7·8	9·0	13·8	110	68	49	117
Solway																
Eskdalemuir	6·9	12·7	7·7	5·3	7·6	7·1	3·5	7·7	11·5	7·4	6·9	13·5	98	54	43	97
Dumfries	5·8	10·9	4·9	7·9	7·8	5·0	3·3	9·6	12·0	8·5	8·0	10·0	94	48	46	94
Glenlee	7·4	11·8	8·0	6·6	8·3	5·2	2·2	7·8	9·8	9·6	9·5	17·1	103	59	40	98
Stranraer Filters	4·9	9·7	8·3	6·1	9·9	4·8	3·3	8·5	9·5	9·7	9·8	20·1	105	51	42	93
Ayrshire																
Prestwick	4·9	8·1	7·7	3·5	6·8	13·3	3·6	12·5	11·0	8·7	11·3	15·9	107	46	51	97
Clyde																
Camps Res.	6·1	8·1	5·8	4·0	6·3	8·2	3·8	11·0	9·9	8·5	9·1	12·6	93	47	43	90
Sloy Power Sta.	8·4	10·6	10·7	5·2	6·5	6·8	2·4	3·4	9·8	5·3	9·5	15·3	94	59	34	93
Lephinmore	6·3	7·1	10·0	4·6	6·9	9·9	3·3	2·3	9·6	7·1	10·8	16·7	95	51	37	88
Kintyre and south-western islands																
Largie Castle	6·5	10·4	10·7	6·7	7·9	7·5	3·4	3·7	8·9	9·5	11·2	12·4	99	55	38	93
Lochy and Linnhe																
Fersit	6·7	10·5	13·3	3·3	7·3	8·5	3·0	3·6	7·2	6·0	11·5	26·0	107	60	33	93
Ardgour House	5·4	6·6	14·3	3·8	7·7	8·3	4·0	3·7	9·9	4·8	10·3	18·5	97	54	37	92
Shiel, Alsh and Maree																
Polloch House	6·3	7·9	10·6	4·3	9·1	6·5	5·3	4·9	8·4	6·4	10·9	17·9	99	53	39	92
Rhubana	4·2	5·4	10·5	3·1	8·1	7·8	6·9	4·5	9·1	5·6	8·7	17·2	91	47	39	87
Ratagan	3·4	8·1	20·0	4·2	7·8	7·4	5·0	3·4	7·3	6·4	12·6	23·5	109	62	35	97
Poolewe	2·4	3·9	17·4	2·1	8·1	10·0	6·8	3·7	8·2	8·6	12·2	22·2	106	52	39	91
Inner and Outer Hebrides																
Tiree	5·3	8·2	8·3	4·3	7·6	5·6	5·9	2·9	10·1	6·4	9·5	14·9	89	53	36	89
Portree	5·5	5·7	14·6	4·1	8·4	7·0	5·1	3·9	7·9	5·1	11·5	18·8	98	56	36	92
Benbecula	6·8	5·6	13·0	3·8	8·7	6·1	5·7	2·8	10·8	8·4	10·8	16·9	99	55	38	93
Laxford																
Lochmore Lodge	3·4	3·9	17·3	2·3	8·0	11·3	9·0	4·9	9·3	10·9	16·1	20·7	117	51	45	96
Naver and Thurso																
Strathy	4·7	5·5	12·3	1·1	6·7	14·6	7·6	5·8	9·3	16·2	11·6	17·1	113	48	45	93
Orkney and Shetland Islands																
Lerwick	9·4	8·8	12·4	3·0	11·1	6·5	5·4	3·1	7·7	12·4	9·6	16·7	106	67	37	104
Kirkwall	10·8	8·9	11·1	1·8	8·0	6·5	8·6	8·3	7·2	14·4	11·7	22·2	119	61	40	101
Wick Water to Conon																
Wick	6·0	8·8	10·6	1·7	8·9	14·0	11·4	11·3	10·0	17·9	13·4	18·1	132	57	57	114
Achany	3·9	6·1	15·9	2·6	8·2	12·1	4·6	6·1	7·8	9·4	14·2	24·6	115	53	41	94
Geanies House	3·1	7·9	8·0	3·1	10·2	23·5	12·7	10·1	6·1	13·7	6·6	16·3	121	47	66	113
Fairburn House	3·8	8·2	12·5	2·7	8·2	12·9	5·2	7·5	6·2	9·5	13·7	24·8	115	53	43	96

TABLE 2 (*continued*)

Monthly, annual and seasonal rainfall for individual stations, percentage of annual average, 1916–50

Area and station	Jan.	Feb.	Mar.	Apr.	May	June	July	Aug.	Sept.	Oct.	Nov.	Dec.	Year	Winter half-year Oct. 1965–Mar. 1966	Summer half-year Apr.–Sept. 1966	Seasonal year Oct. 1965–Sept. 1966
Beauly and Ness																
Kingie Camp	4·2	9·0	15·1	3·6	6·3	6·5	3·5	3·1	7·6	5·7	11·7	23·3	100	57	31	87
Inverness	2·1	5·8	10·1	2·2	8·7	27·3	8·3	9·9	5·0	9·8	9·4	16·7	115	42	61	103
Banff, Moray and Nairn																
Cluny Castle	3·9	7·3	10·2	1·6	10·5	12·5	3·8	5·7	7·0	6·8	13·3	23·4	106	48	41	89
Achnagoichan	2·4	5·3	8·3	1·7	8·0	15·3	8·5	7·0	4·8	10·6	18·7	15·3	106	38	45	83
Gordon Castle	4·1	5·2	9·8	0·5	5·1	18·6	11·8	15·3	6·3	11·4	12·5	12·0	113	45	58	103
Dee and Don																
New Deer School	4·3	7·9	8·3	2·6	9·0	12·9	7·5	11·1	6·1	15·9	15·4	10·9	112	51	49	100
Craibstone	6·8	9·5	5·5	5·0	8·1	14·3	5·9	11·0	5·7	15·2	9·3	10·1	106	46	50	96
Braemar	5·2	8·7	5·5	5·3	10·1	11·7	2·8	7·9	5·4	13·1	13·2	12·7	102	41	43	85
North and South Esk																
Inchgrundle	8·9	14·4	3·7	9·0	8·5	9·5	2·6	10·5	4·4	17·0	12·3	10·8	112	53	45	97
Tay																
Dundee	8·1	11·2	2·1	14·4	7·5	12·8	5·8	16·5	7·7	11·7	6·2	8·7	113	47	65	112
Ardtalnaig	8·1	10·4	6·9	6·9	7·8	9·5	2·1	5·2	7·2	7·5	5·6	12·0	89	52	39	91
Faskally	6·5	10·2	5·7	3·8	10·8	23·6	2·8	8·4	7·4	10·9	5·7	13·0	109	47	57	104
Perth	7·3	12·4	3·9	11·6	9·0	18·5	3·9	13·7	7·4	12·4	4·7	10·7	115	52	64	117
Forth																
Stirling	6·4	11·8	5·5	6·5	8·5	11·1	3·5	11·3	8·2	9·9	6·5	10·1	99	52	49	101
Lothians																
Edinburgh, Blackford Hill	5·7	9·3	2·0	6·9	5·2	15·4	5·7	26·9	5·4	5·9	7·5	8·5	104	41	65	106
Tweed																
Lauder	6·2	14·1	2·5	7·7	7·8	11·6	5·9	15·8	8·8	9·8	11·8	11·8	114	52	58	109
Wauchope	5·7	10·3	3·5	6·6	6·5	9·8	5·6	13·7	13·0	10·0	8·7	10·1	103	41	55	97
Pallinsburn	5·3	12·9	1·8	9·1	8·7	9·9	4·3	22·6	4·0	18·2	13·6	14·8	125	53	59	112
Foyle and north-western streams																
Edenfel	6·1	11·6	8·8	6·2	8·9	12·2	4·7	5·1	6·4	8·5	9·4	15·3	103	57	43	101
Lough Foyle, east																
Banagher Filters	7·0	16·1	9·6	8·3	6·9	8·4	4·5	6·2	5·8	10·9	9·2	15·0	108	61	40	101
Bann and Lough Neagh																
Armagh	8·2	13·8	7·2	12·0	8·2	12·2	4·2	8·0	8·1	17·0	10·0	13·1	122	62	53	114
Stoneyford Res. No. 1	5·8	16·4	8·3	8·2	7·7	9·6	5·4	6·8	7·1	11·8	7·9	13·5	109	65	45	110
Coleraine, The Cutts	4·6	8·1	11·8	3·7	7·1	11·9	6·2	3·6	6·4	9·0	9·7	15·9	98	55	39	93
Lagan and eastern streams																
Belfast P. Sta.	7·2	16·6	9·4	9·5	7·5	11·7	2·7	7·2	9·9	12·7	9·5	13·2	117	68	49	116
Newry and south-eastern streams																
Annalong Valley	8·6	22·2	7·5	15·9	9·8	10·1	6·1	6·1	6·2	13·6	8·4	15·8	130	80	54	134
Erne and western streams																
Belleek	4·5	8·8	6·6	6·1	7·7	10·9	5·9	4·8	8·6	6·0	9·0	16·6	95	51	44	95
Channel Islands																
Jersey Airport	16·1	10·4	2·6	14·1	6·9	7·7	4·7	8·9	4·7	18·7	12·0	16·1	123	79	47	125

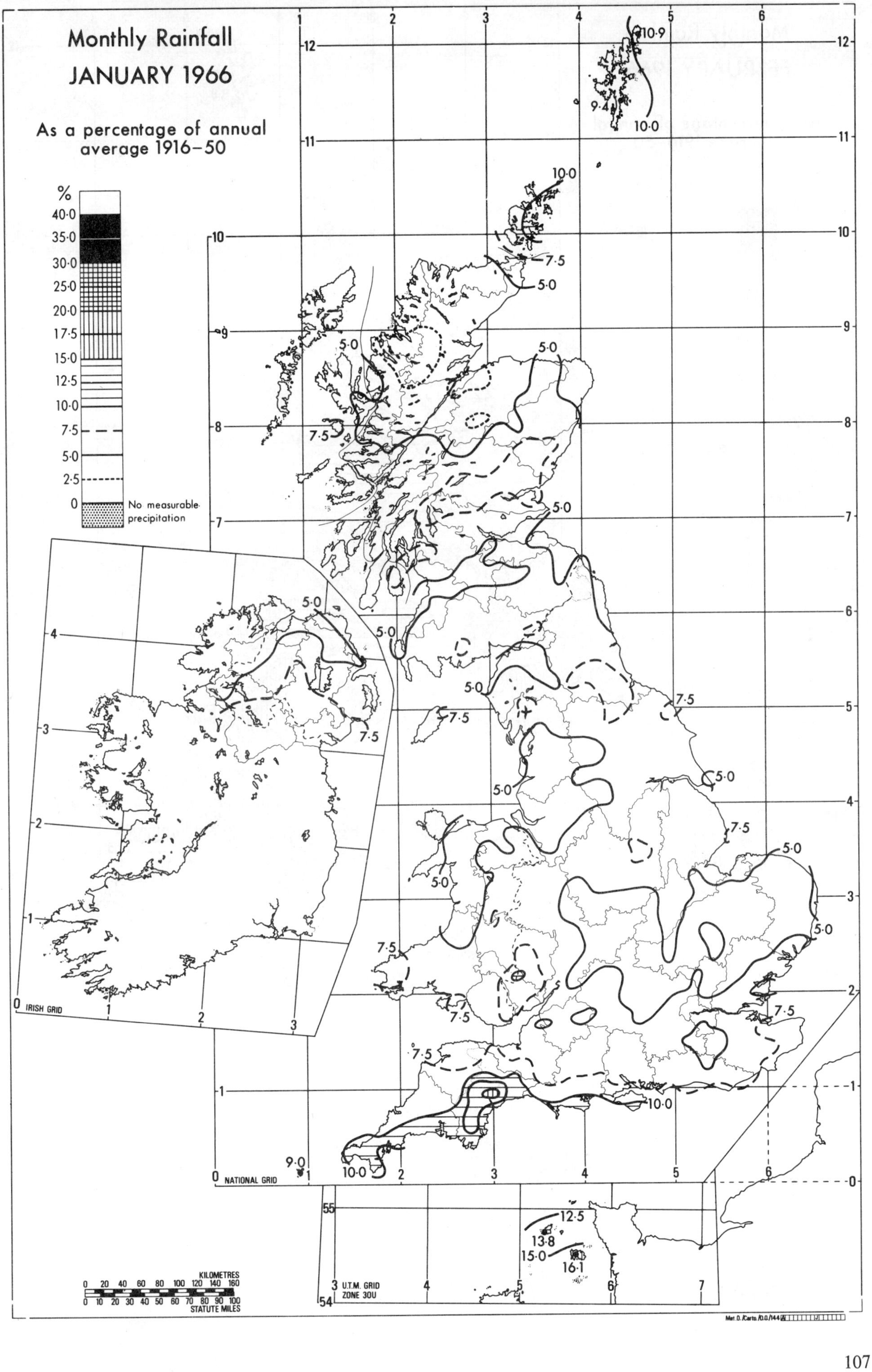
Monthly Rainfall
JANUARY 1966
As a percentage of annual average 1916–50
%
40·0
35·0
30·0
25·0
20·0
17·5
15·0
12·5
10·0
7·5
5·0
2·5
0
No measurable precipitation
IRISH GRID
NATIONAL GRID
U.T.M. GRID ZONE 30U
KILOMETRES
STATUTE MILES
10·9
9·4
10·0
7·5
5·0
9·0
12·5
13·8
15·0
16·1

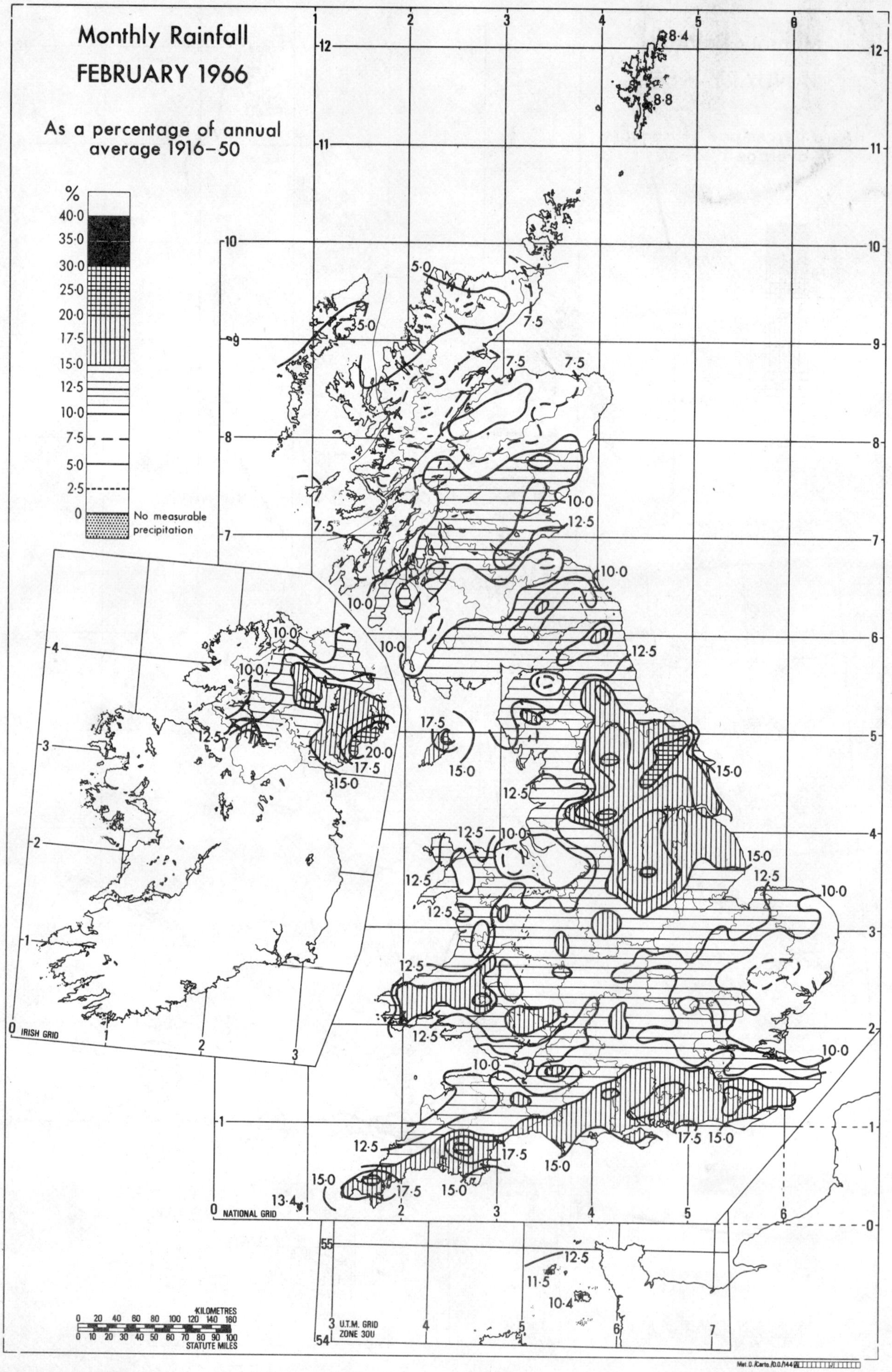
Monthly Rainfall
FEBRUARY 1966
As a percentage of annual average 1916–50
%
40·0
35·0
30·0
25·0
20·0
17·5
15·0
12·5
10·0
7·5
5·0
2·5
0
No measurable precipitation
IRISH GRID
NATIONAL GRID
U.T.M. GRID ZONE 30U
KILOMETRES
STATUTE MILES

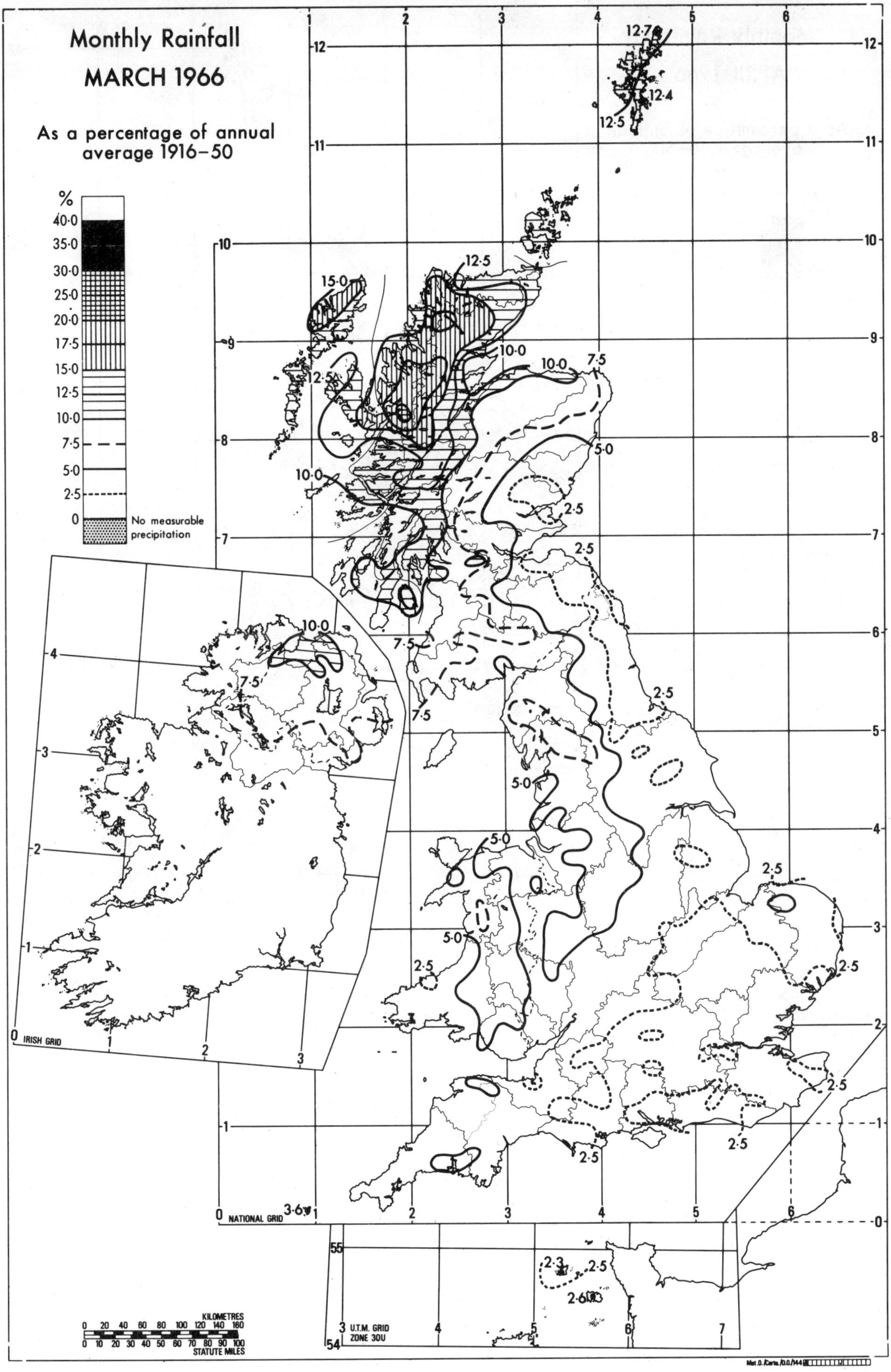
Monthly Rainfall
MARCH 1966
As a percentage of annual average 1916–50
%
40·0
35·0
30·0
25·0
20·0
17·5
15·0
12·5
10·0
7·5
5·0
2·5
0
No measurable precipitation
IRISH GRID
NATIONAL GRID
U.T.M. GRID ZONE 30U
KILOMETRES
STATUTE MILES

H

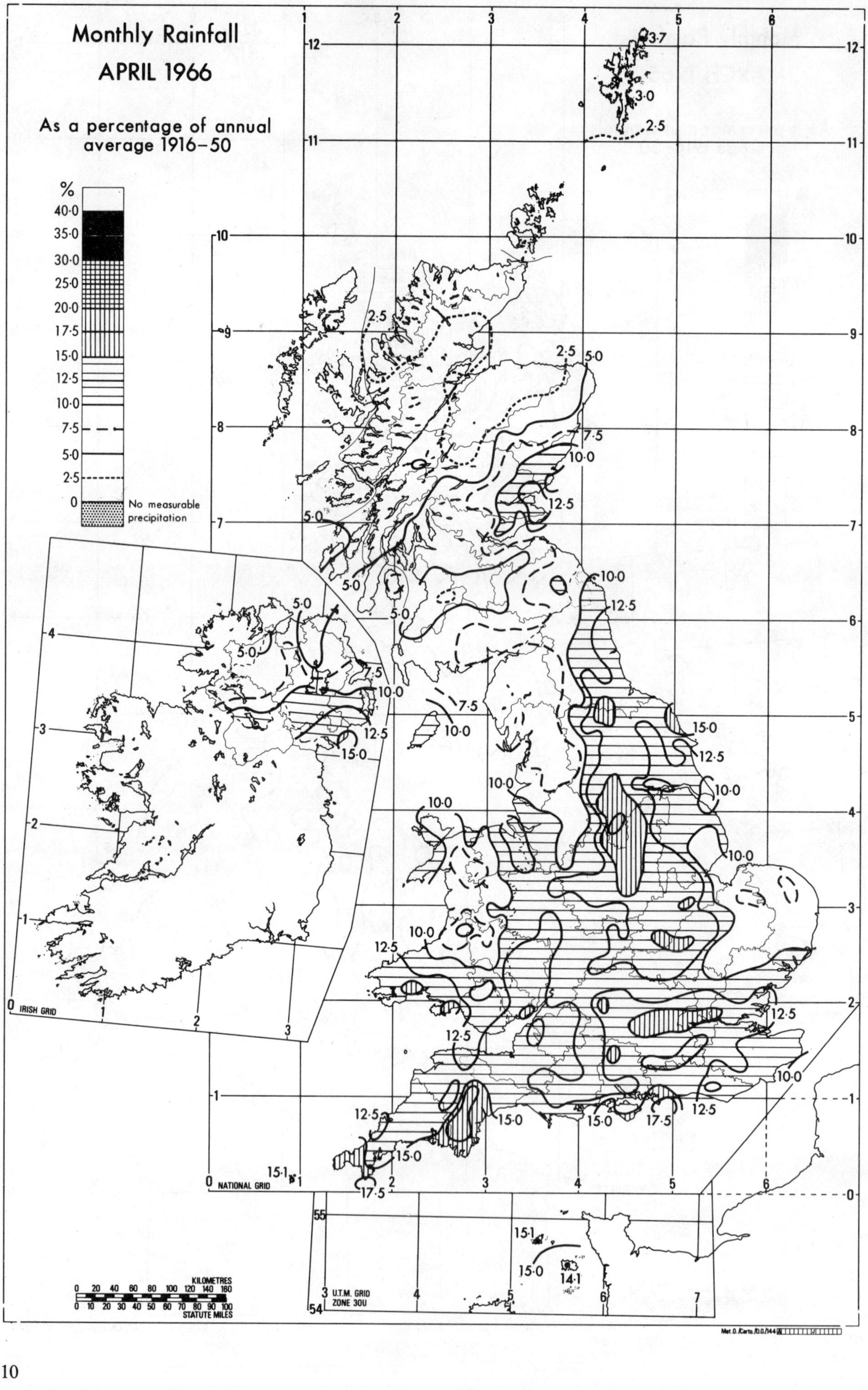
Monthly Rainfall
APRIL 1966
As a percentage of annual average 1916–50
%
40·0
35·0
30·0
25·0
20·0
17·5
15·0
12·5
10·0
7·5
5·0
2·5
0
No measurable precipitation
IRISH GRID
NATIONAL GRID
U.T.M. GRID ZONE 30U
KILOMETRES
STATUTE MILES

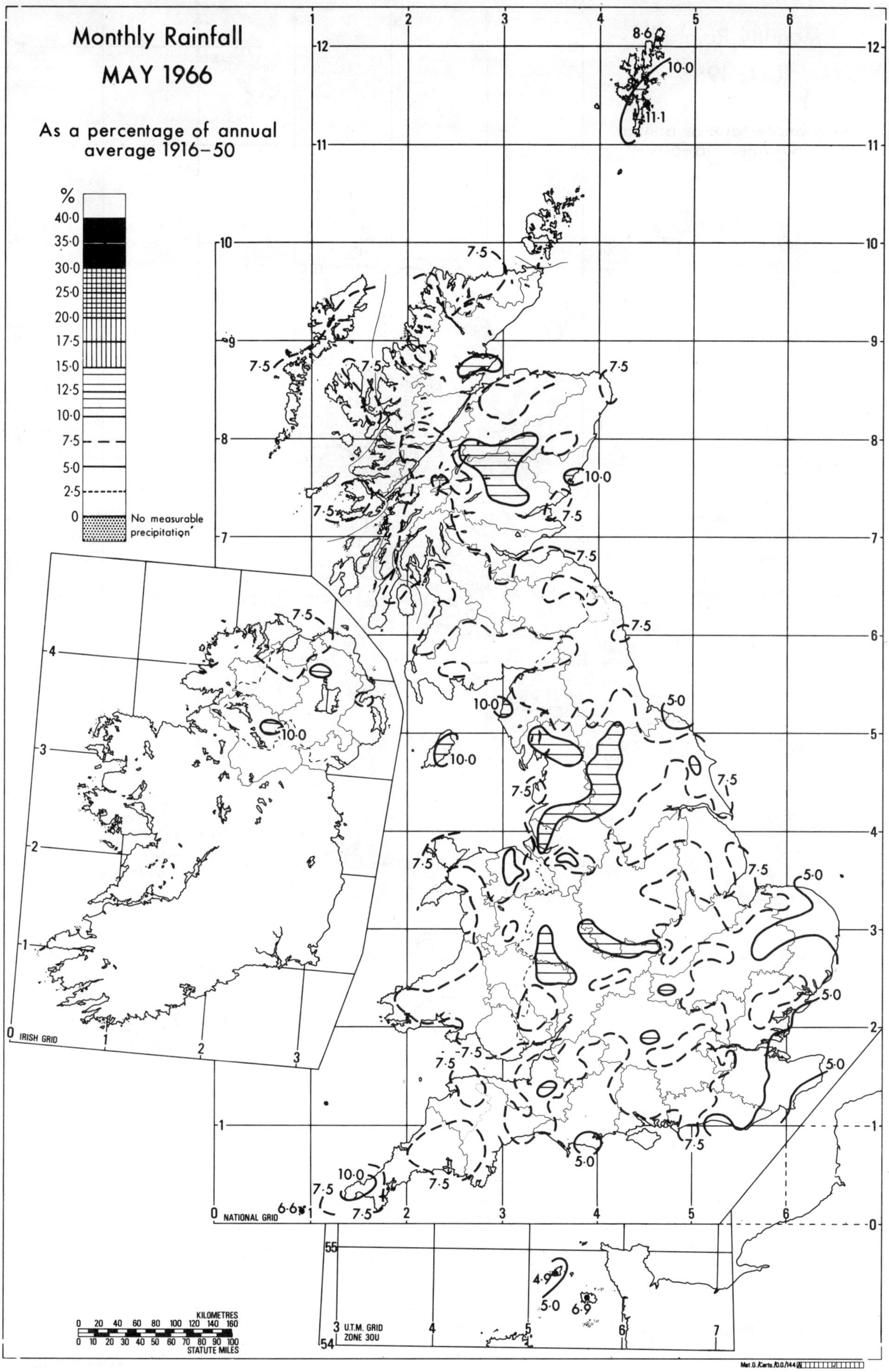
Monthly Rainfall
MAY 1966
As a percentage of annual average 1916–50
%
40·0
35·0
30·0
25·0
20·0
17·5
15·0
12·5
10·0
7·5
5·0
2·5
0
No measurable precipitation
IRISH GRID
NATIONAL GRID
U.T.M. GRID ZONE 30U
KILOMETRES
STATUTE MILES
10·0
11·1
8·6
7·5
5·0
4·9
6·9
6·6
Met.O./Carto./D.O./144

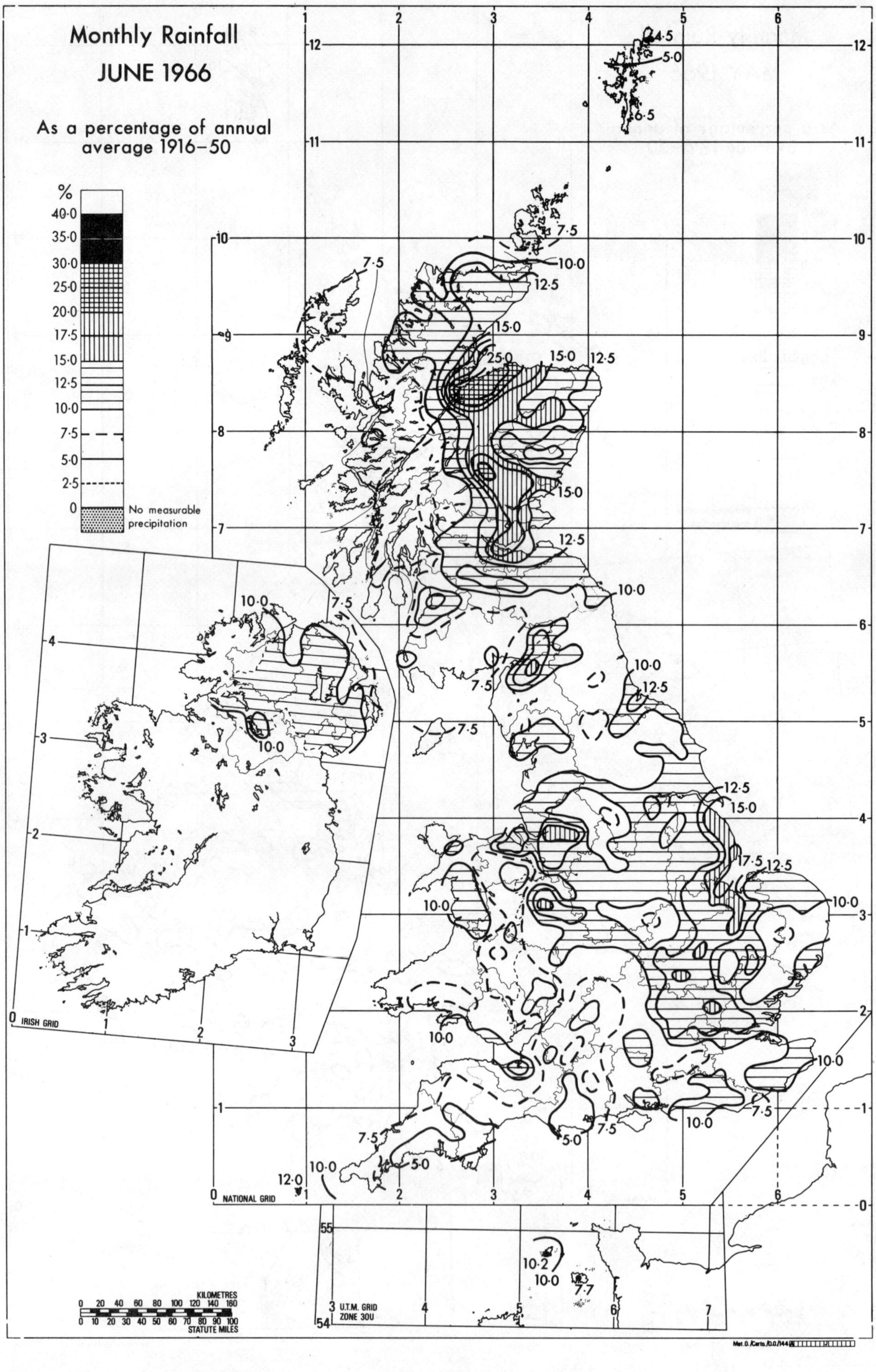
Monthly Rainfall
JUNE 1966
As a percentage of annual average 1916–50
%
40·0
35·0
30·0
25·0
20·0
17·5
15·0
12·5
10·0
7·5
5·0
2·5
0
No measurable precipitation
IRISH GRID
NATIONAL GRID
U.T.M. GRID ZONE 30U
KILOMETRES
STATUTE MILES
Met.O./Carto./D.O./144

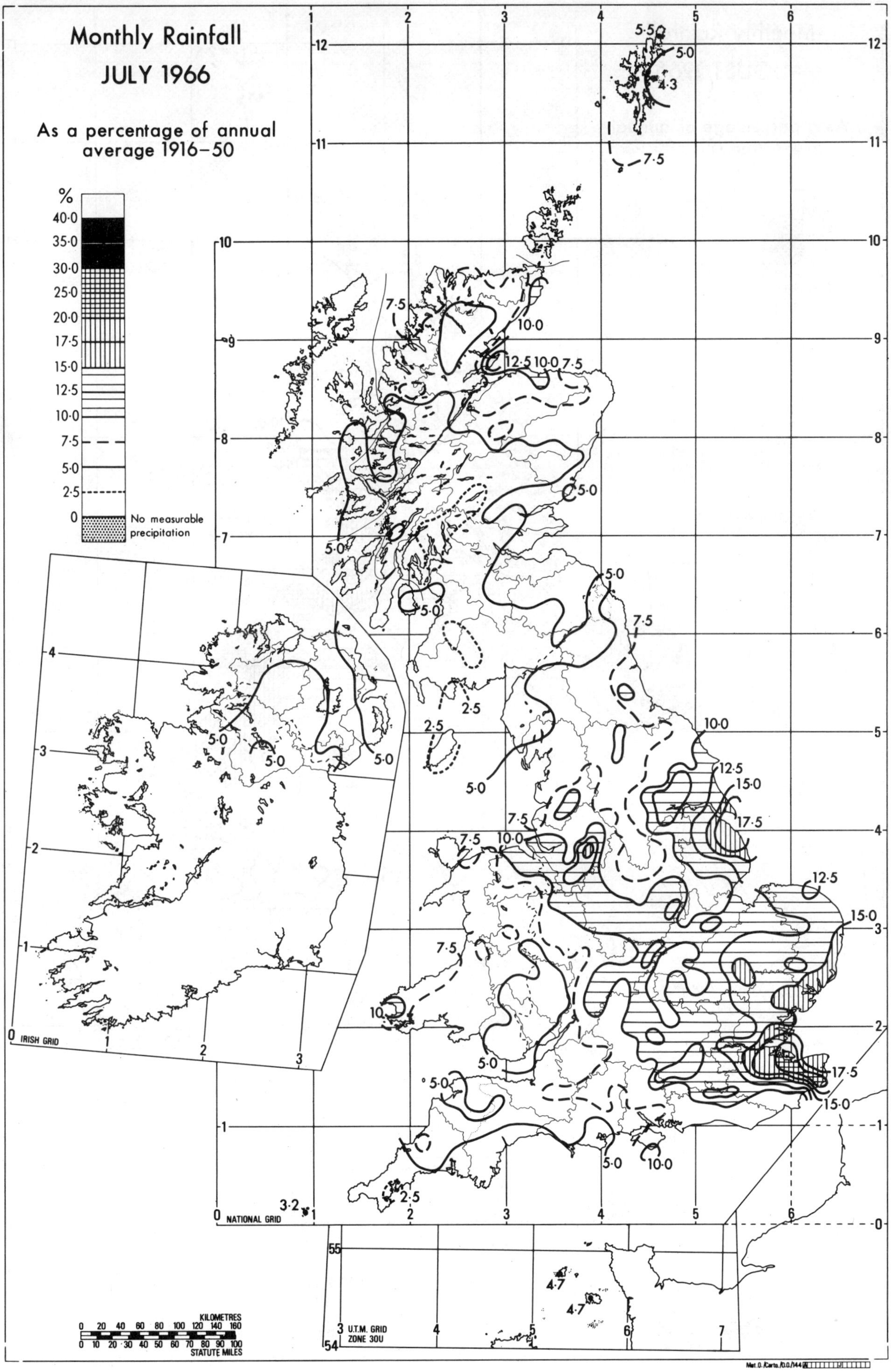
Monthly Rainfall
JULY 1966
As a percentage of annual average 1916–50
%
40·0
35·0
30·0
25·0
20·0
17·5
15·0
12·5
10·0
7·5
5·0
2·5
0
No measurable precipitation
IRISH GRID
NATIONAL GRID
U.T.M. GRID ZONE 30U
KILOMETRES
STATUTE MILES

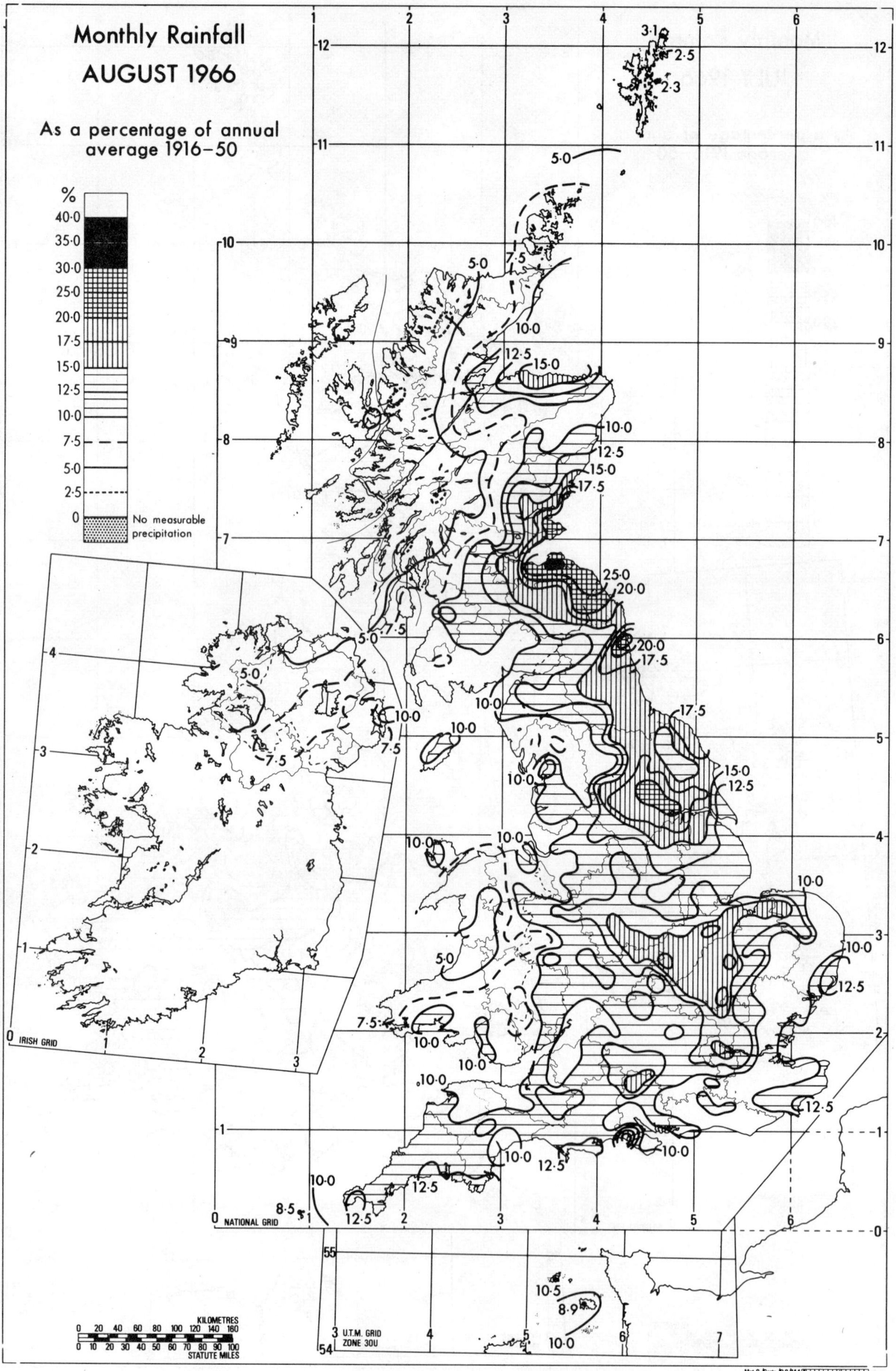
Monthly Rainfall
AUGUST 1966
As a percentage of annual average 1916–50
%
40·0
35·0
30·0
25·0
20·0
17·5
15·0
12·5
10·0
7·5
5·0
2·5
0
No measurable precipitation
IRISH GRID
NATIONAL GRID
U.T.M. GRID ZONE 30U
KILOMETRES
0 20 40 60 80 100 120 140 160
0 10 20 30 40 50 60 70 80 90 100
STATUTE MILES
Met.O./Carto./D.D./144

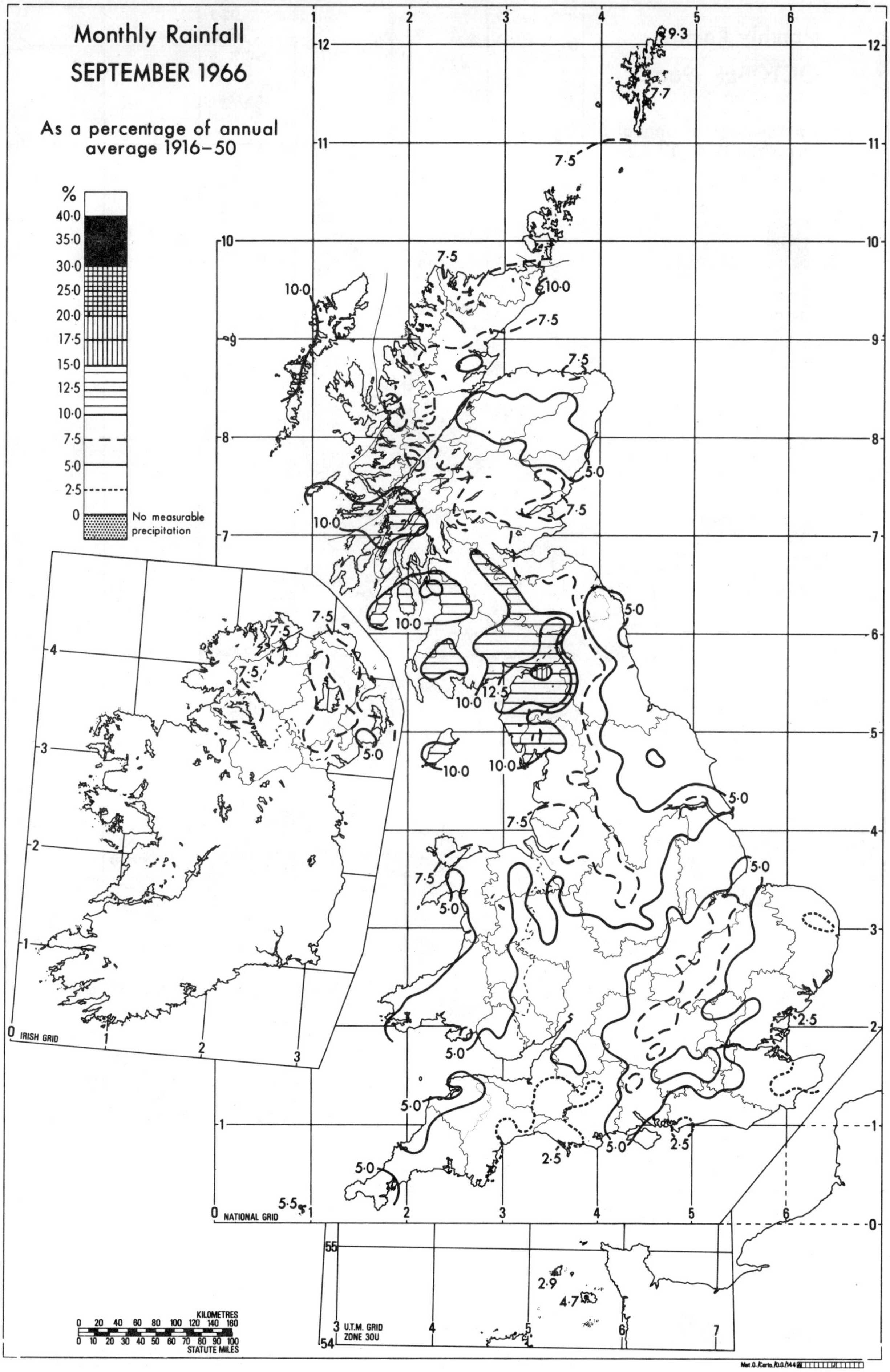

Monthly Rainfall
SEPTEMBER 1966
As a percentage of annual average 1916–50
%
40·0
35·0
30·0
25·0
20·0
17·5
15·0
12·5
10·0
7·5
5·0
2·5
0
No measurable precipitation
9·3
7·7
7·5
10·0
12·5
5·0
2·5
2·9
4·7
5·5
IRISH GRID
NATIONAL GRID
U.T.M. GRID ZONE 30U
KILOMETRES
0 20 40 60 80 100 120 140 160
0 10 20 30 40 50 60 70 80 90 100
STATUTE MILES

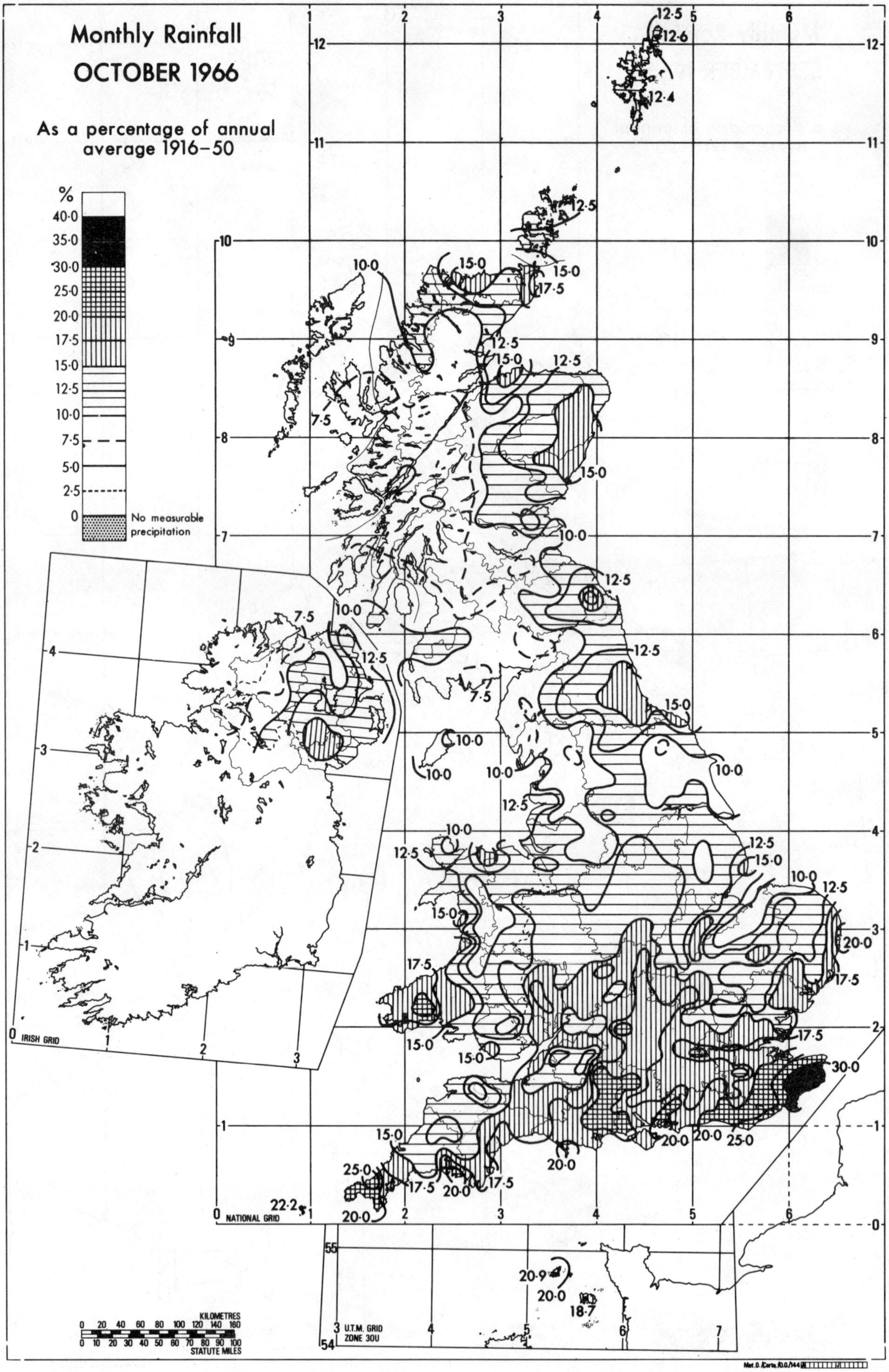
Monthly Rainfall
OCTOBER 1966
As a percentage of annual average 1916–50
%
40·0
35·0
30·0
25·0
20·0
17·5
15·0
12·5
10·0
7·5
5·0
2·5
0
No measurable precipitation
IRISH GRID
NATIONAL GRID
U.T.M. GRID ZONE 30U
KILOMETRES
0 20 40 60 80 100 120 140 160
0 10 20 30 40 50 60 70 80 90 100
STATUTE MILES

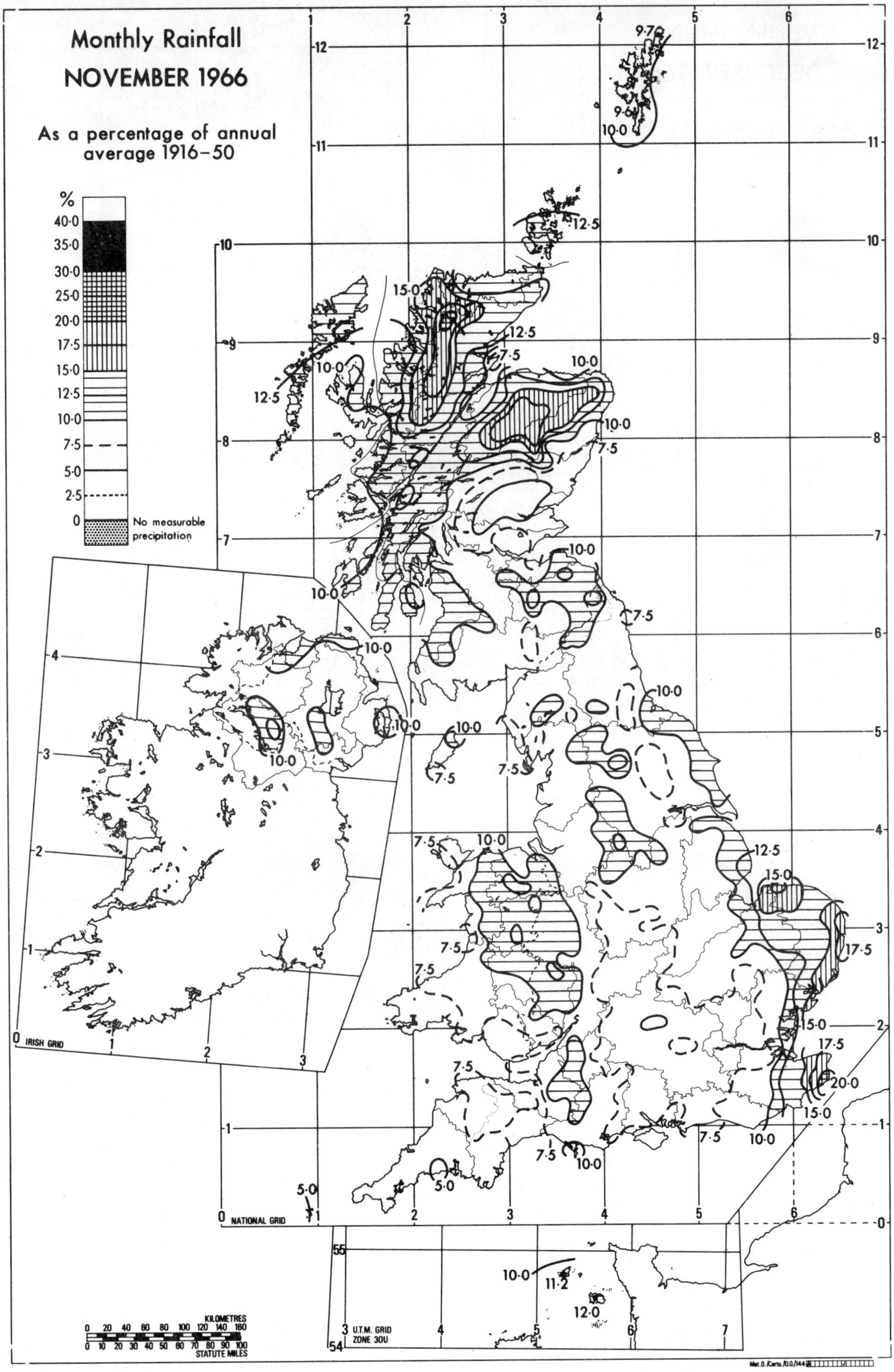
Monthly Rainfall
NOVEMBER 1966
As a percentage of annual average 1916–50
%
40·0
35·0
30·0
25·0
20·0
17·5
15·0
12·5
10·0
7·5
5·0
2·5
0
No measurable precipitation
IRISH GRID
NATIONAL GRID
U.T.M. GRID ZONE 30U
KILOMETRES
STATUTE MILES

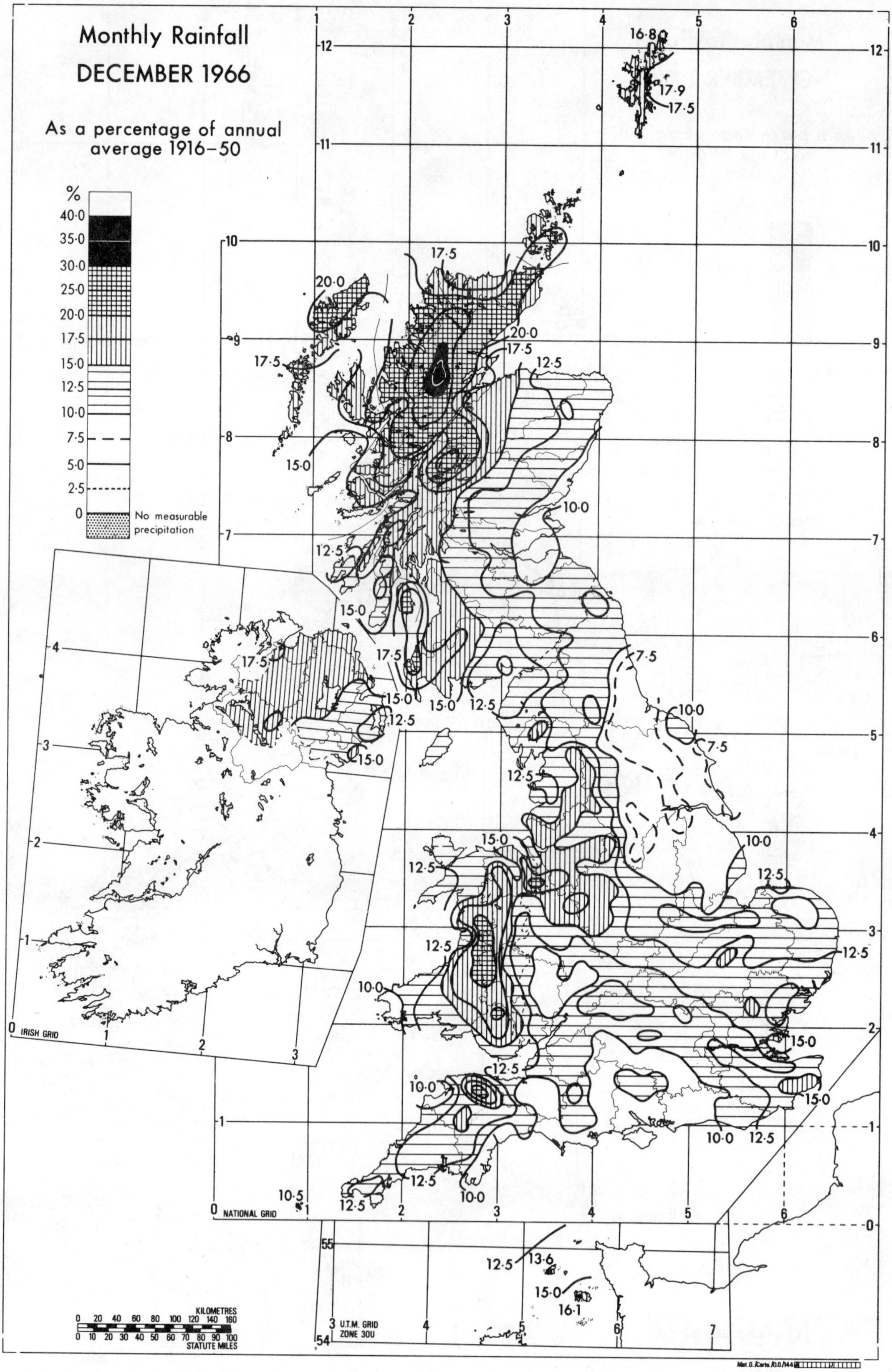
Monthly Rainfall
DECEMBER 1966
As a percentage of annual average 1916–50
%
40·0
35·0
30·0
25·0
20·0
17·5
15·0
12·5
10·0
7·5
5·0
2·5
0
No measurable precipitation
IRISH GRID
NATIONAL GRID
U.T.M. GRID ZONE 30U
KILOMETRES
STATUTE MILES
16·8
17·9
17·5
20·0
12·5
10·0
15·0
7·5
10·5
13·6
16·1

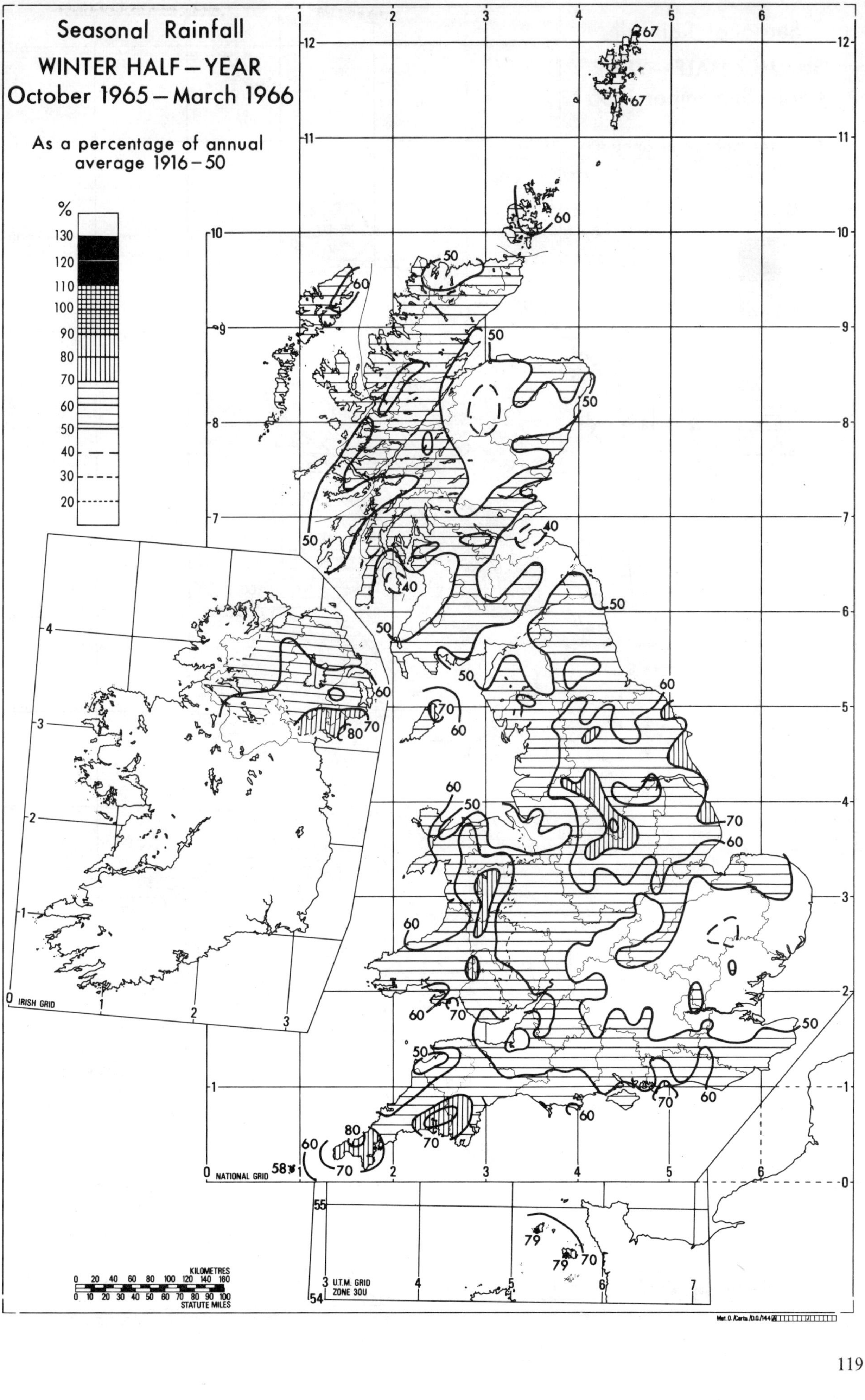
Seasonal Rainfall
WINTER HALF—YEAR
October 1965—March 1966
As a percentage of annual average 1916—50
%
130
120
110
100
90
80
70
60
50
40
30
20
IRISH GRID
NATIONAL GRID
U.T.M. GRID ZONE 30U
KILOMETRES
0 20 40 60 80 100 120 140 160
0 10 20 30 40 50 60 70 80 90 100
STATUTE MILES

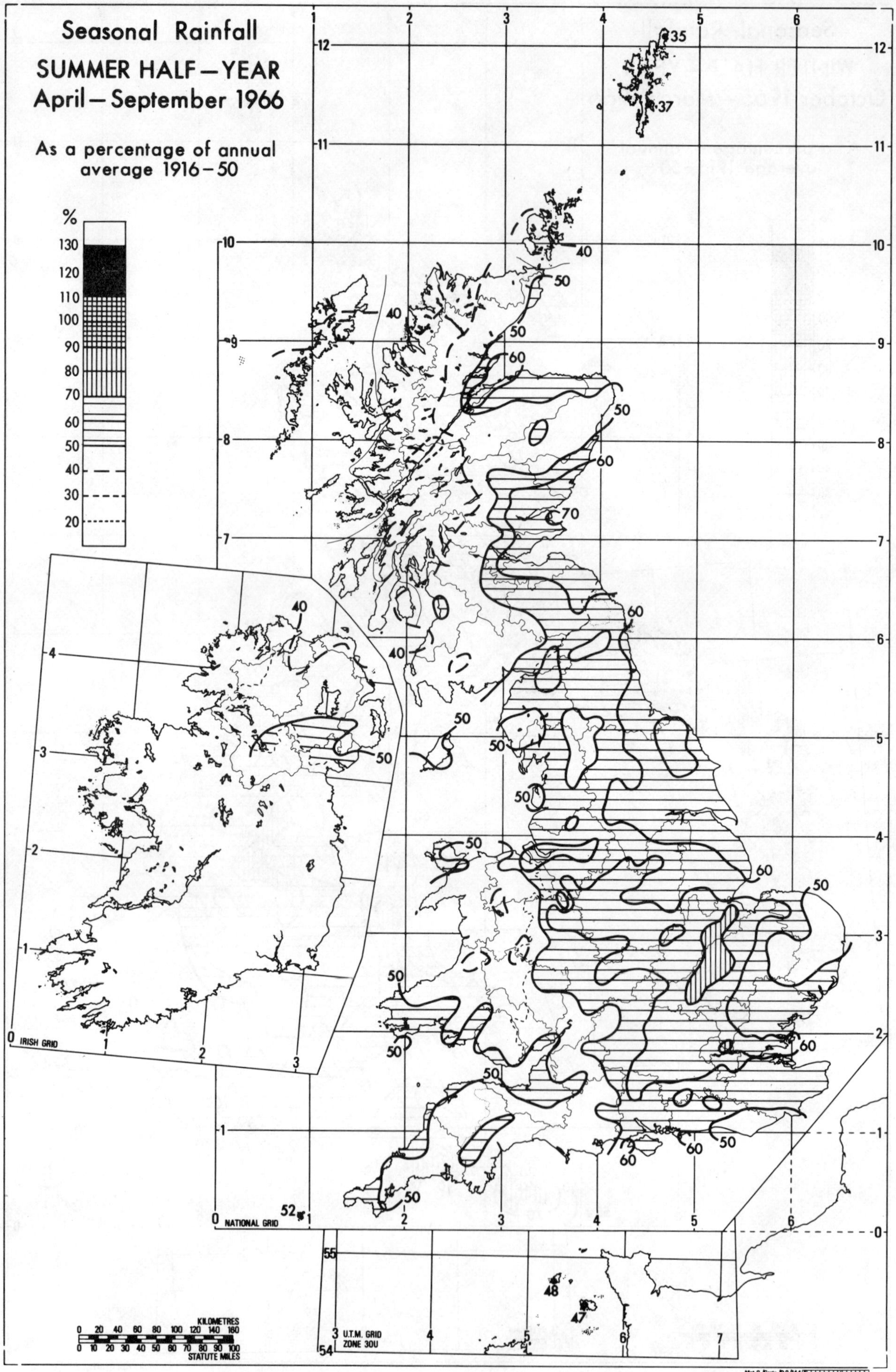

Seasonal Rainfall
SUMMER HALF—YEAR
April—September 1966
As a percentage of annual average 1916—50
%
130
120
110
100
90
80
70
60
50
40
30
20
IRISH GRID
NATIONAL GRID
U.T.M. GRID ZONE 30U
KILOMETRES
STATUTE MILES
Met.O./Carto./D.O./144

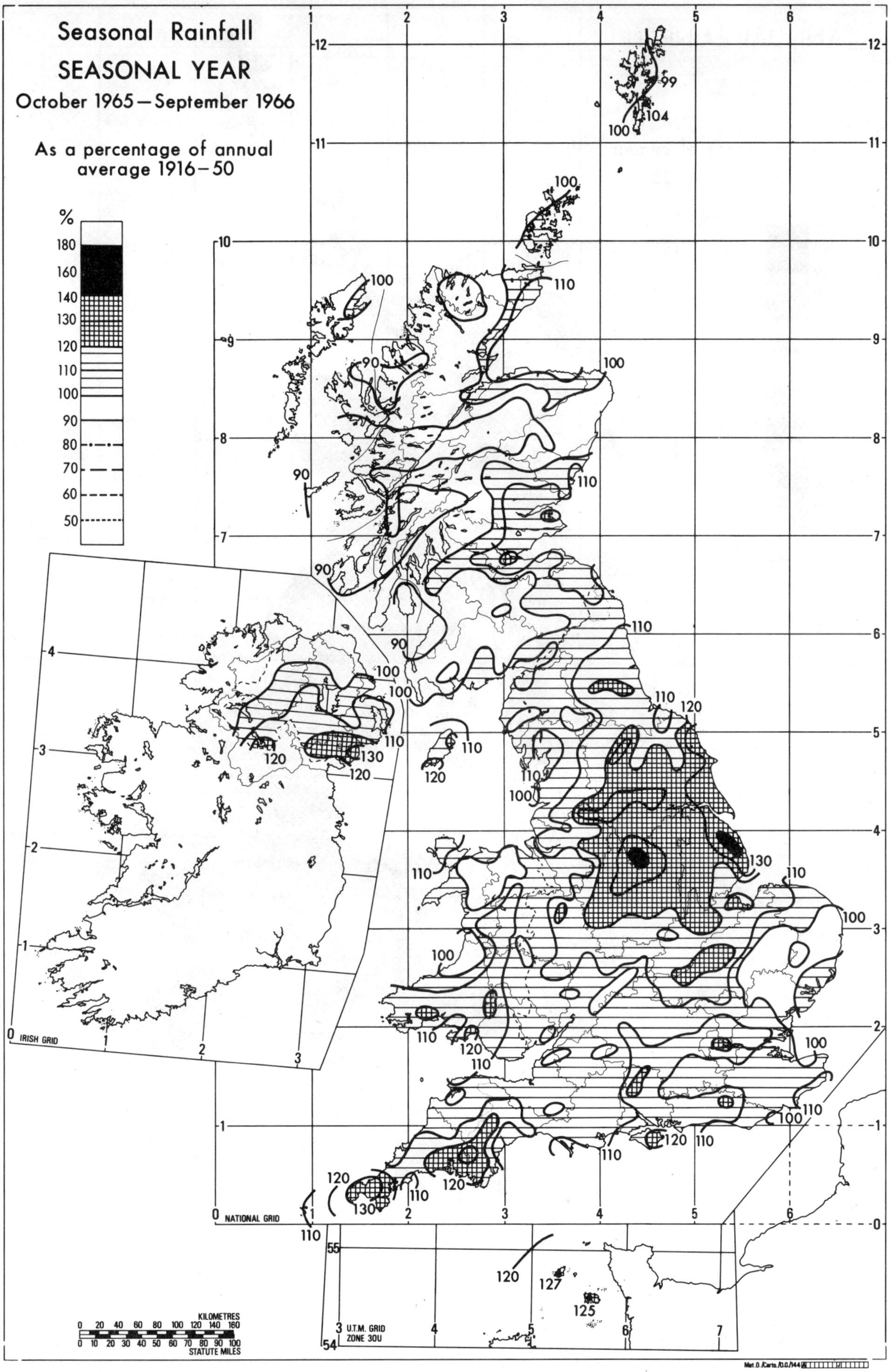
Seasonal Rainfall
SEASONAL YEAR
October 1965 — September 1966
As a percentage of annual average 1916–50
%
180
160
140
130
120
110
100
90
80
70
60
50
IRISH GRID
NATIONAL GRID
U.T.M. GRID ZONE 30U
KILOMETRES
0 20 40 60 80 100 120 140 160
0 10 20 30 40 50 60 70 80 90 100
STATUTE MILES
Met.O./Carto./D.O./144

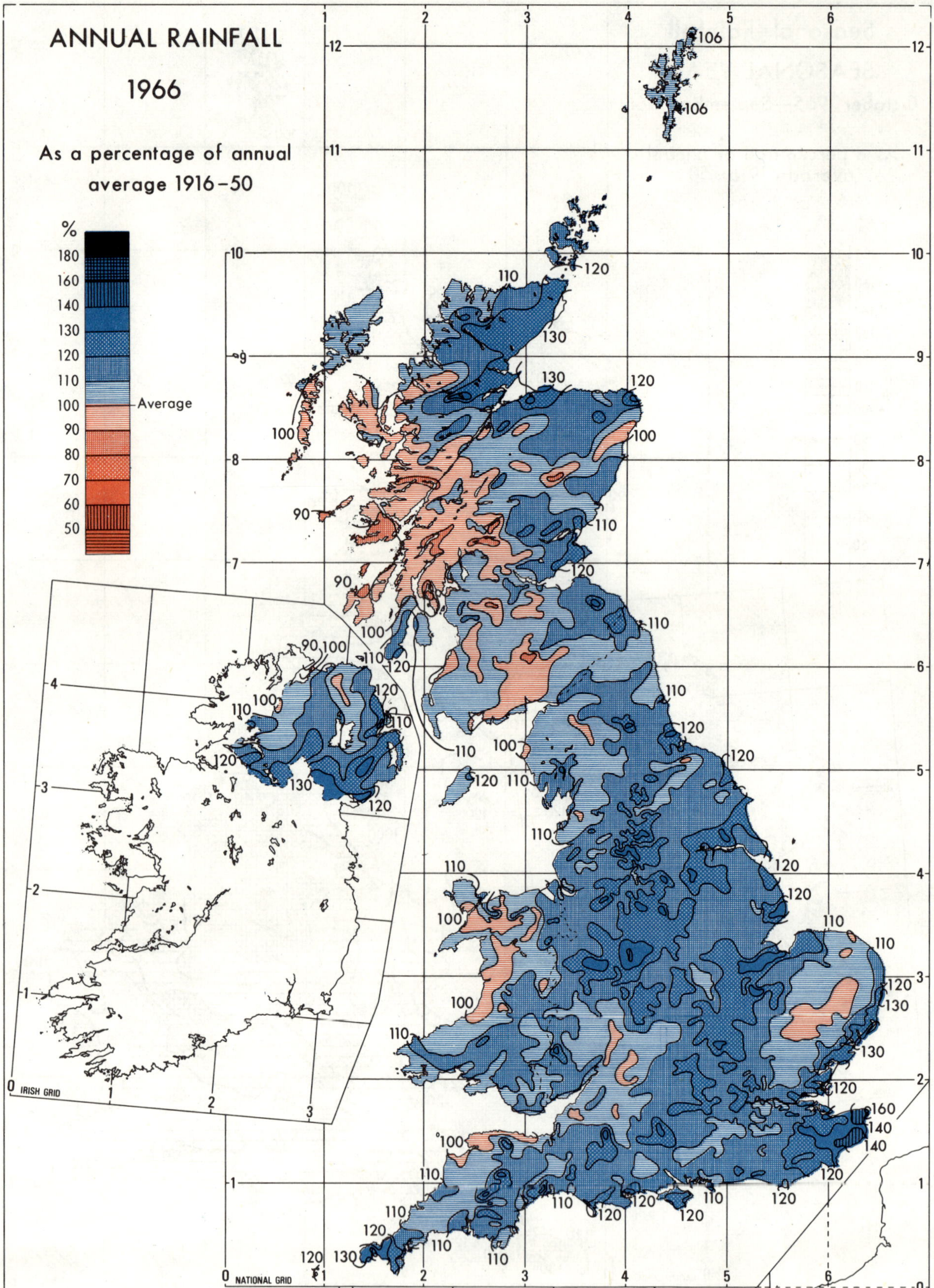
ANNUAL RAINFALL
1966
As a percentage of annual average 1916–50
%
180
160
140
130
120
110
100
Average
90
80
70
60
50
NATIONAL GRID
IRISH GRID
U.T.M. GRID ZONE 30U
KILOMETRES
STATUTE MILES
Met.O./Carto./D.O./144

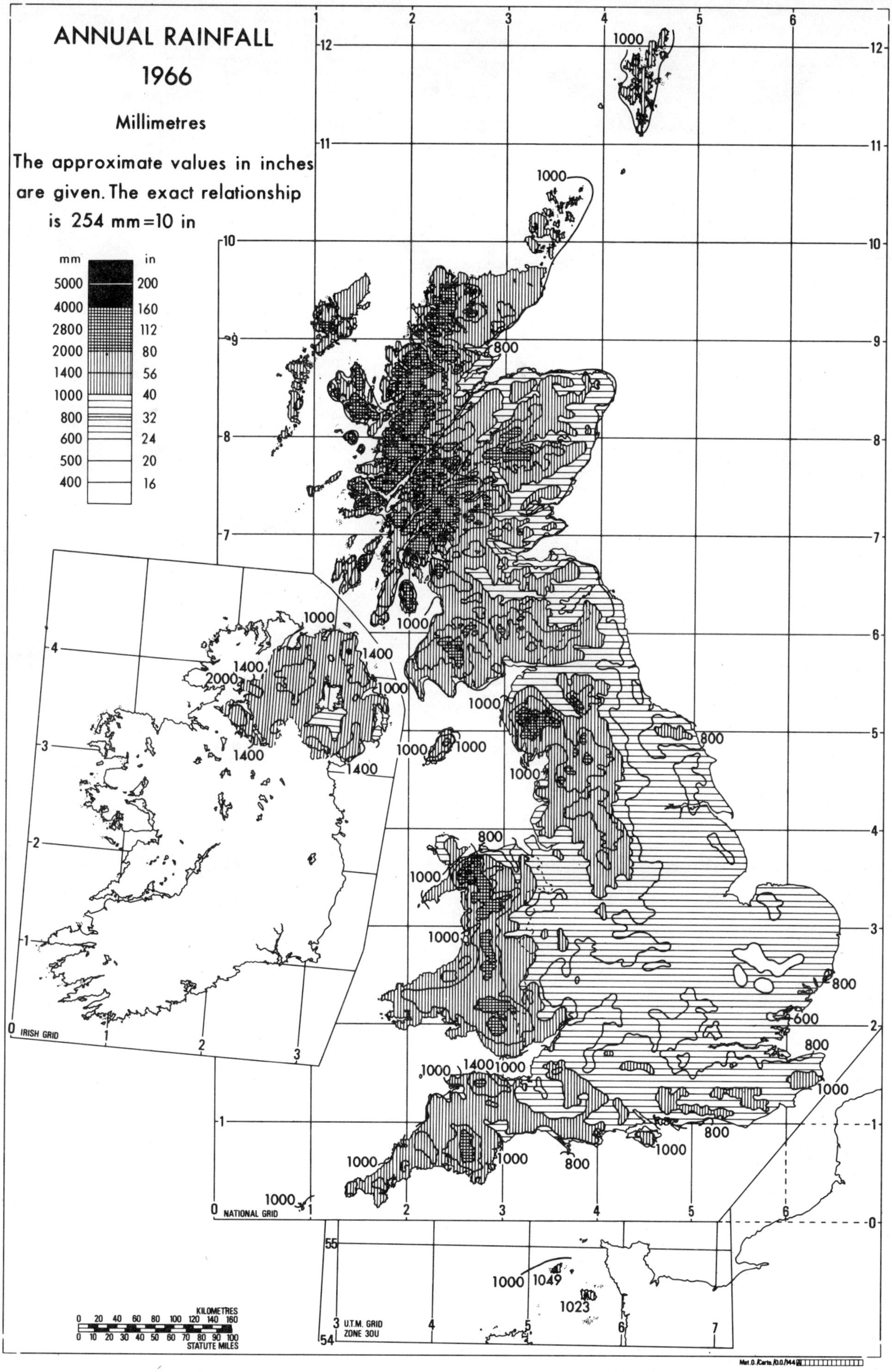
ANNUAL RAINFALL
1966
Millimetres
The approximate values in inches are given. The exact relationship is 254 mm=10 in
mm
in
5000
200
4000
160
2800
112
2000
80
1400
56
1000
40
800
32
600
24
500
20
400
16
0 IRISH GRID
0 NATIONAL GRID
U.T.M. GRID
ZONE 30U
KILOMETRES
0 20 40 60 80 100 120 140 160
0 10 20 30 40 50 60 70 80 90 100
STATUTE MILES
1049
1023

3 SPELLS OF RAINFALL DEFICIENCY AND EXCESS

The reasons why tabulations of occurrences of droughts and dry spells, or wet spells and rain spells based, prior to 1961, on a set of arbitrary criteria are considered unsatisfactory have been discussed in *British Rainfall* 1961. Accordingly, a method was introduced for examining spells of rainfall deficiency and excess without dependence on fixed or arbitrary definitions.

For any station, cumulative amounts of rainfall at intervals throughout the year are first expressed as percentages of average annual rainfall, and then as departures from the percentage amounts which would accumulate at a uniform rate corresponding to the average. The results may be conveniently plotted at five-day intervals (pentads) as in the graphs which follow. Beside each graph are included relevant details of rainfall during the year.

Approximately horizontal sections in these graphs correspond to rainfall at more or less the average rate for the station. Significantly descending sections represent rainfall deficiency spells, with a limiting slope corresponding to no rain at all. Ascending sections, for which there is no accurately assignable limiting slope, represent periods of excess; during thundery spells, particularly in areas with low average rainfall, the rise from one point to the next may be very steep. The zero line is arbitrary and a new start may be made at any point in the year by an appropriate zero shift without any change in the shape of the curve. Graphs for successive years may be joined together by simply joining the curves, with any appropriate adjustments of zero lines.

Used either singly or, if required, in combination, graphs of this type can help to overcome very largely some of the disadvantages of working constantly with the conventional divisions of the year into months and seasons. Though it is convenient and in some ways necessary to use these divisions, far more often than not it turns out that there is a meteorologically arbitrary element about the point of separation of one such period from another. In arranging the material given in the table and maps of Section 2, in particular, it is almost impossible to avoid imputing to individual months and seasons quite separate meteorological identities which in fact they rarely possess. The graphs provide a useful way of showing how the characteristics dominant at the end of one month or season often run on into the next. A sufficient body of data analysed on these lines would indicate when decisive changes most frequently occur, and would probably help, for instance, to determine whether there is any specially advantageous month to be chosen as the beginning of that 'seasonal year' which would be most meaningful in relation to rainfall and perhaps also to evaporation and other elements in the hydrological cycle.

In fact the most serious weakness of this form of presentation is that it refers to percentage rainfall in isolation. This is convenient for studies restricted to rainfall alone as it reduces to a common basis the stations with low average rainfall and those with high average rainfall. For some purposes concerned with actual quantities of water, rather than variations relative to average, it would be better to express rainfall in absolute measure and adopt some form of combined presentation with other elements, notably evaporation. This matter is receiving attention as a probable development for the future.

A minor objection to this form of presentation is that when there is any degree of regular seasonal variation of rainfall underlying more random irregularities, the graph for any individual year should be taken in conjunction with the corresponding curve for average rainfall through the year. The average curve will certainly be very much flatter than the separate graphs for almost all individual years, but will not usually offer a really close approximation to a horizontal straight line. The preparation of average curves has been undertaken and a small selection was published in the *British Rainfall Supplement* 1961–1965. A larger selection is given here with data for 1966.

As in previous years the data for 1966 are set out in Table 3A with the pentad numbers for ready reference in Table 3B. The corresponding graphs for these stations are given in the same order on pages 131–137 Heading the details given on the right-hand side of each graph is the station name followed on the next line by the appropriate river authority and the drainage area. Some of the details give information additional to that shown in Table 3A and in the graphs themselves, since the rainfall amounts for individual days within the pentads have been taken into consideration, where appropriate, to determine the full extent in days of any spell worthy of note.

Next, with each of the graphs for 1966, there is shown the corresponding graph for *average* rainfall (1916–50) calculated by pentads throughout the year. Most of these graphs fall into one of two classes which may be rather simply described, though there are one or two transitional cases which do not fall definitely into either class:

1. In the drier parts of the country, on the eastern side of Britain, seasonal variation of rainfall is not very marked, even on the average, and one of the summer months, typically July or August, is likely to be the wettest month of the year. (This rather flat seasonal variation is often completely swamped by erratic variations in individual years.) On the average there is a slight build-up of 1 or 2 per cent excess in the first 7 or 8 pentads of the year. From some point during the first half of February onwards there is a more or less steady development of rainfall deficiency (compared with a uniform average rate throughout the year), and this eventually reaches a total deficiency of up to 8 per cent of average annual rainfall by about pentad 37 or 38 (beginning of July). Thereafter, for the rest of the year, slight excess of rainfall makes good this deficiency. The total seasonal range from greatest excess of average rainfall to greatest deficiency does

not reach as much as 10 per cent of average annual rainfall and is usually about 6 to 9 per cent.

2. In the wetter parts of the country, especially the mountainous areas in the west and north, there is a much more pronounced seasonal variation of average rainfall, which may not be completely swamped by erratic variations in individual years. The build-up of excess rainfall (compared with a uniform average rate) in the early part of the year is larger, yielding an excess which by the first half of February may reach 5 per cent of average annual rainfall. The development of rainfall deficiency from then onwards goes on much longer, since in these areas July and August though on the average wet, are by no means the wettest months of the year; there is a late summer and autumn increase to a winter maximum. The total deficiency, which of course is *relative* to a much larger average annual rainfall, may exceed 8 per cent or even 10 per cent by early autumn, after which it is made good by a pronounced excess for the rest of the year. The total seasonal *range* of average excess and deficiency, measured from a uniform average rate, is usually about 10 to 14 per cent of average annual rainfall.

For some purposes it may be useful to compare the excess and deficiency curves for individual years with these corresponding curves for *average* rainfall, instead of with the horizontal base line, which represents a strictly uniform rate of average rainfall throughout the year. The degree of usefulness must of course depend on the stability of average seasonal distribution of rainfall, which in the drier parts of the country is poor. In the wetter parts of the country departure from 'expectation' rather than from the uniform average rate will certainly have a more definite meaning.

Twenty graphs have been selected to illustrate the incidence of rainfall over the British Isles during 1966. The graphs for 1966 are plotted with dots and the corresponding graphs of average rainfall are full lines joining plotted pentad values.

The graphs are largely self-explanatory and detailed written comment would merely put into words what is obvious from the graphs. However, one or two features are worthy of mention, the first being the marked difference in the incidence of rainfall that may occur at stations which are relatively short distances apart. This is illustrated by comparing the Armagh and Coleraine graphs, also the Eskdalemuir and Edinburgh graphs. Another unusual feature is the long period of deficiency at Tiree. Here the deficiency increased, albeit somewhat erratically, until as late as pentad 65, after which an appreciable excess of rainfall failed to restore the year's rainfall to average. The Llwydiarth Esgob rainfall, proceeding as it does, throughout the year at a rate so near the uniform rate corresponding to the average annual rainfall, is probably unusual enough to merit mention.

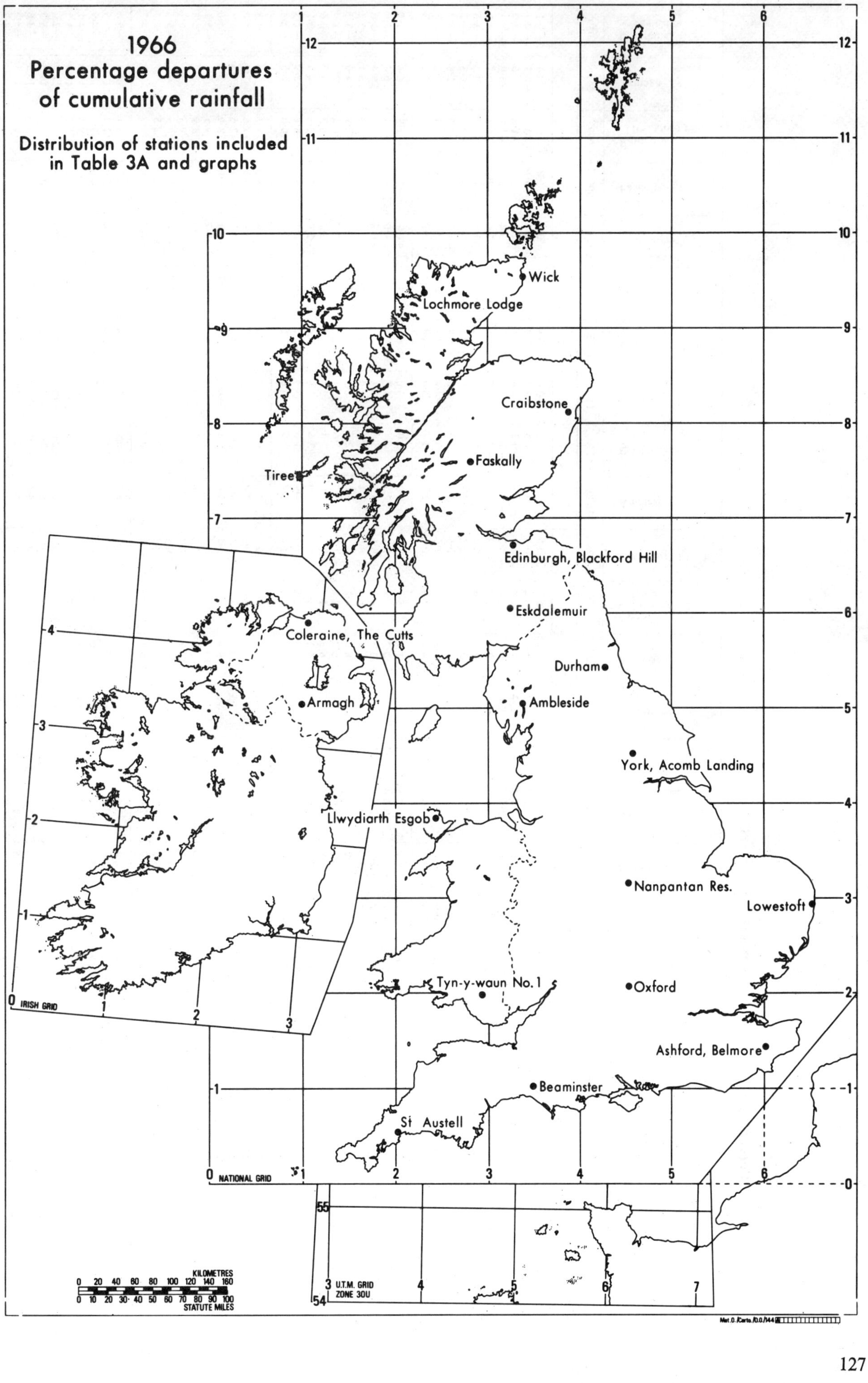
1966
Percentage departures
of cumulative rainfall
Distribution of stations included
in Table 3A and graphs
Wick
Lochmore Lodge
Craibstone
Faskally
Tiree
Edinburgh, Blackford Hill
Eskdalemuir
Coleraine, The Cutts
Durham
Armagh
Ambleside
York, Acomb Landing
Llwydiarth Esgob
Nanpantan Res.
Lowestoft
Tyn-y-waun No. 1
Oxford
Ashford, Belmore
Beaminster
St Austell
IRISH GRID
NATIONAL GRID
U.T.M. GRID
ZONE 30U
KILOMETRES
STATUTE MILES

TABLE 3A

Percentage departures of cumulative rainfall

Pentad	ending	Durham	York, Acomb Landing	Nanpantan Res.	Lowestoft	Oxford	Ashford, Belmore	Beaminster	St Austell	Treherbert, Tyn-y-Waun No. 1	Llwydiarth Esgob	Ambleside	Eskdalemuir	Tiree	Lochmore Lodge	Wick	Craibstone	Faskally	Edinburgh, Blackford Hill	Armagh	Coleraine, The Cutts
	A.a.r. (mm)	668	629	715	600	652	730	1018	1232	2377	1024	1890	1582	1117	1933	761	861	900	699	845	1039
		%	%	%	%	%	%	%	%	%	%	%	%	%	%	%	%	%	%	%	%
1	5·1	−0·3	0·0	−0·2	−0·6	0·0	0·3	0·4	2·1	0·5	0·6	1·3	2·3	−0·4	−0·9	−0·4	0·1	1·0	0·5	0·4	−0·1
2	10·1	−1·6	−1·4	−1·6	−1·9	−1·4	−1·0	0·7	2·6	−0·4	−0·6	0·0	1·0	−1·5	−2·3	−1·8	−1·1	−0·1	−0·6	−0·3	−0·7
3	15·1	0·1	−2·4	−2·7	−2·0	−2·7	−1·2	−0·7	1·3	−1·8	−1·9	−1·3	−0·3	−2·8	−3·6	−2·9	−2·3	−1·4	−1·8	−1·6	−2·0
4	20·1	−1·0	−3·3	−3·5	−2·0	−3·6	−1·8	−2·0	0·4	−3·1	−3·3	−2·7	−1·7	−4·2	−3·9	−3·1	−3·4	−2·7	−3·1	−3·0	−3·4
5	25·1	−1·2	−3·7	−3·1	−3·0	−3·5	−2·4	−1·1	2·4	−2·5	−3·3	−2·5	−2·5	−5·0	−4·9	−3·3	−3·0	−2·9	−4·1	−2·1	−4·3
6	30·1	−0·7	−2·1	−2·7	−2·5	−3·7	−2·1	−0·3	2·7	−1·4	−2·4	−0·3	−1·4	−2·9	−4·9	−2·3	−1·5	−1·7	−2·5	0·0	−3·6
7	4·2	−1·3	−2·3	−3·6	−3·5	−4·3	−2·4	−0·1	2·3	−0·4	−2·3	2·7	1·1	−2·1	−5·1	−1·3	−0·8	−0·1	−2·1	0·3	−3·2
8	9·2	3·3	2·7	0·0	−1·6	−2·1	0·3	3·0	3·9	2·2	−0·5	3·4	1·9	−3·0	−5·7	−0·8	−1·2	0·5	−1·7	2·8	−3·7
9	14·2	3·7	1·9	−0·6	−2·6	−3·3	0·0	1·7	4·5	0·9	−1·9	2·1	0·7	−4·4	−7·1	−2·2	−2·5	−0·9	−2·4	2·2	−4·8
10	19·2	4·7	5·6	2·6	−1·9	−3·0	0·2	3·1	9·6	3·0	0·1	2·3	0·7	−5·0	−8·3	−2·5	−2·9	−1·7	−1·9	4·6	−4·3
11	24·2	5·9	10·8	3·7	−1·1	−1·0	2·0	6·6	11·1	5·1	1·7	4·9	2·0	−3·0	−8·6	−1·7	−0·7	−1·1	−1·3	6·4	−2·7
12	1·3	6·0	10·0	4·1	−1·5	−0·2	2·7	7·0	10·7	5·7	2·2	7·3	4·7	−1·7	−8·9	−1·2	1·0	1·1	−1·3	6·4	−2·6
13	6·3	4·6	9·7	4·2	−2·7	−1·5	1·7	6·6	10·7	6·3	0·8	6·3	3·8	−2·1	−7·1	−2·2	−0·3	0·3	−2·7	5·4	−3·0
14	11·3	3·3	8·5	3·3	−3·7	−2·7	0·9	5·7	9·4	5·5	0·6	7·1	3·9	−1·9	−5·6	−1·5	−1·4	0·0	−3·8	5·1	−2·2
15	16·3	1·9	7·2	1·9	−5·0	−4·1	−0·5	4·8	8·0	4·1	−0·5	5·8	2·5	−2·5	−6·2	−2·8	−2·7	−1·3	−5·1	4·0	−3·2
16	21·3	0·5	5·8	0·5	−6·2	−5·4	−1·9	3·4	6·7	2·8	−1·7	4·7	1·6	−3·1	−3·5	−2·9	−3·7	−2·3	−6·5	3·1	−3·9
17	26·3	−0·2	5·2	0·9	−6·8	−6·2	−2·3	3·5	6·5	2·5	−0·2	5·7	3·4	−2·4	−2·1	−0·2	−3·7	−1·5	−6·5	4·2	−0·6
18	31·3	−0·9	3·9	−0·1	−7·5	−6·9	−3·0	2·2	5·3	1·3	−0·7	6·0	2·6	−2·8	0·0	0·7	−2·9	−2·2	−7·6	4·5	−0·1
19	5·4	1·8	8·0	2·8	−7·1	−6·8	−3·8	2·4	6·5	2·2	1·7	5·9	1·6	−3·9	−1·2	−0·4	−4·1	−3·5	−8·1	5·9	−1·3
20	10·4	4·1	9·6	4·6	−7·0	−6·8	−2·5	2·5	8·0	2·4	1·4	5·7	1·6	−5·0	−2·5	−1·6	−3·0	−3·7	−6·0	6·5	−2·3
21	15·4	4·7	9·0	4·1	−6·6	−5·2	−1·6	4·5	9·4	2·3	0·4	4·3	0·3	−6·4	−3·9	−2·6	−4·1	−5·0	−6·7	7·6	−3·5
22	20·4	4·6	9·8	7·4	−5·6	−2·7	0·5	6·3	9·9	2·0	0·2	3·2	−0·6	−7·1	−5·3	−4·0	−5·4	−6·1	−7·5	7·9	−4·0
23	25·4	4·1	10·0	8·6	−6·2	−1·5	0·5	7·5	10·3	3·7	1·1	4·7	0·3	−6·6	−4·9	−4·5	−4·8	−6·1	−7·6	9·3	−3·7
24	30·4	2·8	8·7	7·3	−7·6	−2·6	−0·9	6·8	9·9	3·2	0·1	3·9	−0·3	−6·7	−5·9	−5·7	−6·1	−6·6	−8·9	8·3	−4·6
25	5·5	2·0	8·5	8·5	−8·2	−2·1	−0·7	7·7	10·2	3·6	−0·3	3·5	−1·3	−7·1	−6·0	−5·8	−7·2	−7·1	−9·5	8·3	−5·1
26	10·5	2·3	9·3	9·4	−7·9	−1·9	−0·6	10·0	11·8	4·9	0·9	3·8	−2·0	−6·8	−7·0	−6·5	−7·9	−7·9	−9·9	8·0	−5·5
27	15·5	4·3	9·5	9·3	−9·0	0·0	−1·3	9·3	10·7	4·3	−0·4	4·3	−2·2	−8·0	−7·1	−6·1	−5·8	−4·3	−9·8	7·5	−6·1
28	20·5	5·0	10·5	8·8	−9·3	−0·6	−2·0	8·5	10·0	4·2	0·1	4·5	−2·2	−6·9	−6·9	−6·2	−7·0	−4·4	−10·5	9·2	−5·5
29	25·5	4·5	10·6	8·2	−9·7	−1·2	−3·0	8·0	9·7	4·6	0·3	7·0	0·5	−5·9	−4·9	−3·7	−4·8	−2·7	−10·5	9·7	−4·3
30	30·5	3·1	9·2	6·8	−11·0	−2·6	−4·3	6·6	8·3	3·3	−1·0	5·7	−0·9	−7·3	−6·3	−5·0	−6·2	−4·0	−11·9	8·3	−5·7
31	4·6	1·8	7·9	5·9	−12·4	−3·9	−5·7	5·3	7·1	2·7	−0·6	6·3	−0·8	−7·0	−4·4	−3·8	−6·3	−4·9	−13·1	7·5	−6·8
32	9·6	0·7	6·5	4·7	−13·8	−5·2	−7·1	5·2	6·9	3·0	0·2	5·4	−1·7	−7·7	−5·6	−4·8	−7·4	−5·0	−12·9	12·0	−5·5
33	14·6	0·0	5·5	3·7	−14·3	−4·0	−5·2	5·8	5·9	2·1	0·4	4·7	−2·5	−9·1	−6·7	−3·5	−8·5	−4·9	−13·2	11·9	−4·9
34	19·6	2·3	6·6	6·6	−11·9	−3·2	−1·7	5·0	5·8	1·6	−0·4	6·2	−1·8	−10·4	−6·7	0·0	−7·4	1·3	−9·4	11·6	−4·1
35	24·6	4·3	8·5	9·6	−12·0	−1·3	−1·0	4·6	5·7	2·6	−0·9	6·7	−1·4	−10·3	−4·6	0·4	−2·7	11·6	−5·1	12·3	−1·5
36	29·6	4·3	8·9	9·3	−11·3	−2·1	−1·3	3·7	4·9	2·7	0·2	7·0	−2·0	−10·0	−4·1	0·8	−0·4	11·4	−4·7	12·3	−2·0
37	4·7	4·3	7·6	7·9	−12·7	−3·4	−2·7	2·6	3·5	1·5	−1·0	5·7	−3·3	−10·5	−3·5	1·8	−0·9	10·0	−6·1	11·3	−3·3
38	9·7	3·2	6·5	9·0	−13·8	−2·1	−2·7	2·2	2·7	0·3	−0·7	4·6	−4·5	−10·7	−3·3	2·0	−1·4	9·5	−6·6	10·9	−3·6
39	14·7	4·1	6·7	8·5	−12·5	−2·6	−3·7	1·8	2·2	−0·1	−0·7	4·4	−5·1	−11·4	−2·3	1·9	−0·9	8·7	−6·5	11·1	−3·5
40	19·7	3·2	6·3	8·4	−9·6	−1·8	−2·1	2·3	1·1	−1·1	−1·7	3·2	−6·4	−12·5	−3·2	1·3	−2·1	7·4	−7·1	9·8	−3·3

		Durham	York, Acomb Landing	Nanpantan Res.	Lowestoft	Oxford	Ashford, Belmore	Beaminster	St Austell	Treherbert, Tyn-y-Waun No. 1	Llwydiarth Esgob	Ambleside	Eskdalemuir	Tiree	Lochmore Lodge	Wick	Craibstone	Faskally	Edinburgh, Blackford Hill	Armagh	Coleraine, The Cutts
41	24·7	2·8	5·5	8·3	−7·5	−2·4	−0·2	1·2	0·0	−1·8	−2·6	2·2	−7·7	−13·4	−3·9	0·1	−3·5	6·0	−8·5	8·8	−4·1
42	29·7	3·7	7·8	9·7	−7·4	0·3	3·4	0·2	−0·7	−1·5	−2·5	2·3	−6·8	−12·3	−3·0	3·6	−2·7	6·0	−8·2	8·0	−4·1
43	3·8	7·8	12·6	11·5	−4·8	1·4	7·1	1·4	1·6	−0·7	0·1	2·0	−7·0	−13·7	−3·1	2·8	0·2	7·7	−1·5	7·1	−5·4
44	8·8	8·0	15·7	12·2	−5·1	3·7	8·6	3·0	3·6	−0·9	−1·2	1·0	−7·8	−14·7	−3·7	6·4	0·3	7·4	0·5	6·8	−5·6
45	13·8	12·1	17·7	11·9	−5·9	3·9	7·5	4·1	4·1	2·1	2·4	4·8	−4·7	−14·8	−4·3	7·1	3·1	9·5	9·7	8·5	−5·6
46	18·8	11·2	16·3	10·5	−7·3	2·5	6·2	3·3	2·8	0·7	1·0	3·6	−5·6	−15·0	−3·3	8·2	2·4	8·3	8·4	7·5	−6·8
47	23·8	13·3	20·5	13·2	−5·7	1·1	6·4	1·9	1·5	−0·6	−0·3	3·8	−6·1	−16·3	−4·3	7·6	1·3	7·5	12·3	8·8	−7·7
48	28·8	12·0	19·1	11·8	−6·5	−0·2	5·3	3·1	0·7	−1·7	−1·5	2·4	−7·5	−17·7	−5·6	6·2	0·0	6·1	11·1	7·4	−9·1
49	2·9	13·8	20·7	15·2	−3·8	4·8	6·6	3·0	1·9	−0·8	−2·1	3·0	−6·6	−17·9	−6·2	8·8	0·3	5·8	11·0	8·9	−9·2
50	7·9	13·7	19·8	14·3	−4·7	4·1	5·9	3·0	2·1	−0·3	−1·0	6·1	−4·5	−17·6	−5·8	8·3	−0·9	6·4	10·7	9·7	−9·3
51	12·9	14·1	19·9	14·6	−5·9	3·0	4·6	1·8	1·1	−0·8	0·1	8·4	−2·3	−16·1	−5·5	9·9	−0·1	6·9	11·1	10·8	−8·3
52	17·9	13·1	18·9	15·1	−5·7	2·9	4·1	0·6	0·3	−0·7	−0·3	8·8	−2·3	−15·0	−4·1	10·7	−0·5	7·2	10·7	10·1	−8·6
53	22·9	11·7	17·5	13·7	−7·1	1·6	2·7	−0·7	−1·1	−2·1	−1·7	7·5	−3·7	−16·4	−3·9	9·5	−1·9	5·8	9·3	8·8	−10·0
54	27·9	10·3	16·2	12·3	−8·5	0·2	1·3	−2·1	−2·4	−3·4	−3·1	6·1	−5·1	−17·7	−5·2	8·1	−3·1	4·4	8·0	7·5	−10·9
55	2·10	10·1	16·4	12·8	−6·7	8·9	2·9	−0·1	−1·8	−3·2	0·4	6·5	−4·9	−16·7	−3·3	11·9	−1·5	6·5	8·5	7·4	−11·8
56	7·10	13·7	18·0	14·9	−2·9	9·4	7·5	0·5	−0·7	−3·4	0·8	7·6	−2·9	−16·7	−3·7	16·2	1·6	8·4	10·2	12·3	−10·5
57	12·10	12·6	16·7	14·5	−1·6	9·7	7·8	1·2	−0·2	−4·0	−0·4	6·6	−4·1	−17·5	−4·7	15·4	2·4	7·5	8·8	16·2	−10·9
58	17·10	16·8	19·0	18·6	−1·1	14·9	11·5	4·9	4·2	−3·5	−1·1	6·7	−3·8	−18·0	−5·5	14·5	2·9	9·4	8·6	16·9	−11·2
59	22·10	16·0	18·5	19·1	2·2	17·0	17·9	8·3	6·9	−0·2	−1·6	6·7	−3·8	−18·3	−4·7	16·5	3·1	9·1	7·7	17·3	−11·3
60	27·10	17·5	18·0	19·2	1·5	16·3	22·9	7·1	6·4	−1·5	−1·9	6·0	−4·9	−18·9	−4·1	16·7	3·0	7·8	6·4	16·4	−11·2
61	1·11	16·8	17·7	18·3	5·2	15·0	25·3	5·9	5·9	−2·8	−2·1	4·8	−5·4	−18·9	−3·1	17·1	4·0	6·6	5·9	15·4	−11·1
62	6·11	17·1	17·8	18·5	7·0	15·6	28·7	11·7	6·2	−2·8	−1·6	3·5	−6·1	−19·7	−3·7	16·6	4·5	5·7	7·5	17·7	−11·7
63	11·11	16·3	16·9	17·3	6·4	14·5	28·4	10·4	4·9	−3·5	−1·9	3·5	−6·9	−19·2	−3·1	16·0	3·4	4·6	6·5	17·1	−12·0
64	16·11	15·3	16·2	17·7	9·3	13·9	28·0	10·0	4·7	−3·3	−1·6	5·5	−6·1	−18·8	0·5	19·3	5·9	5·6	6·3	18·2	−10·1
65	21·11	16·9	15·8	16·7	14·4	12·6	29·3	8·7	3·4	−4·6	−3·0	4·1	−7·4	−19·7	−0·6	18·1	5·8	4·6	5·5	16·9	−11·4
66	26·11	15·7	15·2	15·5	14·8	11·4	29·1	7·5	2·7	−5·6	−4·0	3·7	−8·3	−19·5	1·7	17·9	4·7	3·8	4·4	15·9	−12·1
67	1·12	16·6	16·1	17·2	15·3	12·8	32·7	8·1	4·7	−2·3	−2·0	8·0	−6·2	−16·9	7·6	26·8	7·1	5·5	4·7	18·8	−9·0
68	6·12	15·8	15·0	16·4	16·1	12·4	34·4	7·7	4·8	−3·0	−1·5	7·5	−6·8	−16·6	8·8	27·2	7·9	4·8	3·3	18·0	−9·0
69	11·12	15·7	15·3	19·1	17·7	15·5	37·3	7·7	7·4	−0·7	2·1	9·1	−6·1	−15·1	10·3	29·7	7·9	5·5	3·5	19·5	−6·8
70	16·12	15·9	14·4	18·3	17·5	15·7	37·2	8·9	8·1	0·8	3·9	9·1	−6·1	−13·7	11·7	30·7	8·3	6·0	2·7	21·8	−5·0
71	21·12	15·5	13·5	17·4	16·5	14·3	36·4	7·6	7·5	0·9	3·0	12·2	−4·0	−11·5	15·0	32·4	8·3	8·7	5·3	21·6	−3·2
72	26·12	14·3	12·8	18·0	17·0	13·8	36·7	7·3	7·6	1·4	3·4	12·1	−4·2	−11·3	15·9	32·7	7·3	8·7	3·9	21·8	−3·4
73	31·12	14·5	14·7	18·9	18·6	15·2	39·1	8·6	9·6	5·4	3·3	13·1	−2·3	−10·9	17·1	32·1	6·5	8·9	4·5	22·2	−1·9

TABLE 3B

The calendar year divided into 73 five-day groups, or pentads

Date	Jan	Feb	Mar	Apr	May	June	July	Aug	Sept	Oct	Nov	Dec	Date
1			12					43			61	67	1
2		7				31	37		49	55			2
3	1			19	25								3
4			13								62	68	4
5									50	56			5
6								44					6
7		8				32	38						7
8	2			20	26								8
9			14								63	69	9
10									51	57			10
11								45					11
12		9				33	39						12
13	3			21	27								13
14			15								64	70	14
15									52	58			15
16								46					16
17		10				34	40						17
18	4			22	28								18
19			16								65	71	19
20									53	59			20
21								47					21
22		11				35	41						22
23	5			23	29								23
24			17								66	72	24
25									54	60			25
26								48					26
27		12				36	42						27
28	6			24	30								28
29		ϕ	18								67	73	29
30		×				37	43		55	61			30
31	7	×		×	31	×		49	×		×		31

ϕ In a leap year 29 February is counted as 28 February, i.e. the total rainfall is credited to the 28th.

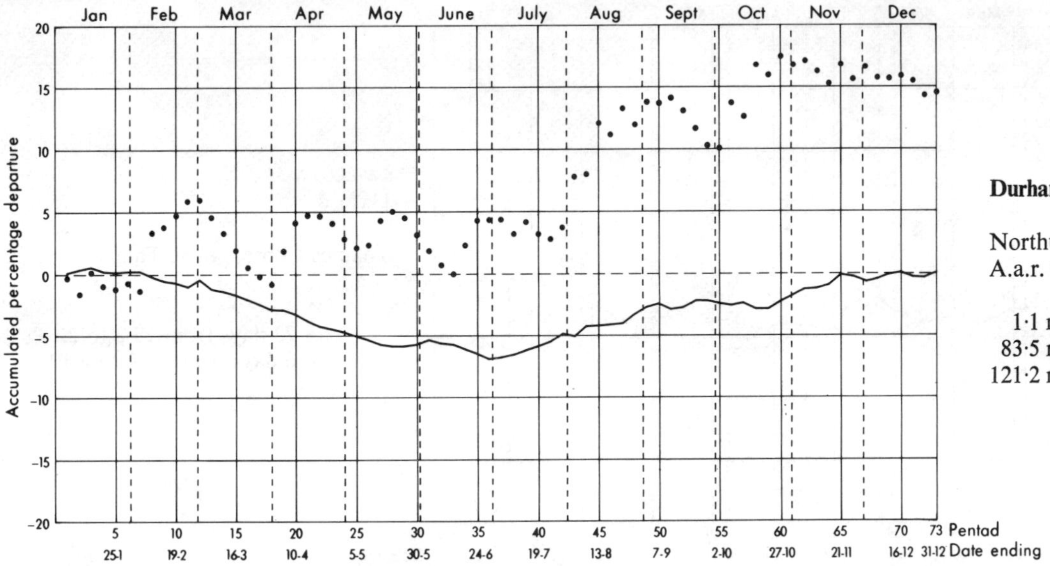

Durham

Northumbrian, Wear
A.a.r. 668 mm

1·1 mm 27 days from 27·2 to 25·3
83·5 mm 19 days from 1·10 to 19·10
121·2 mm 31 days from 22·7 to 21·8

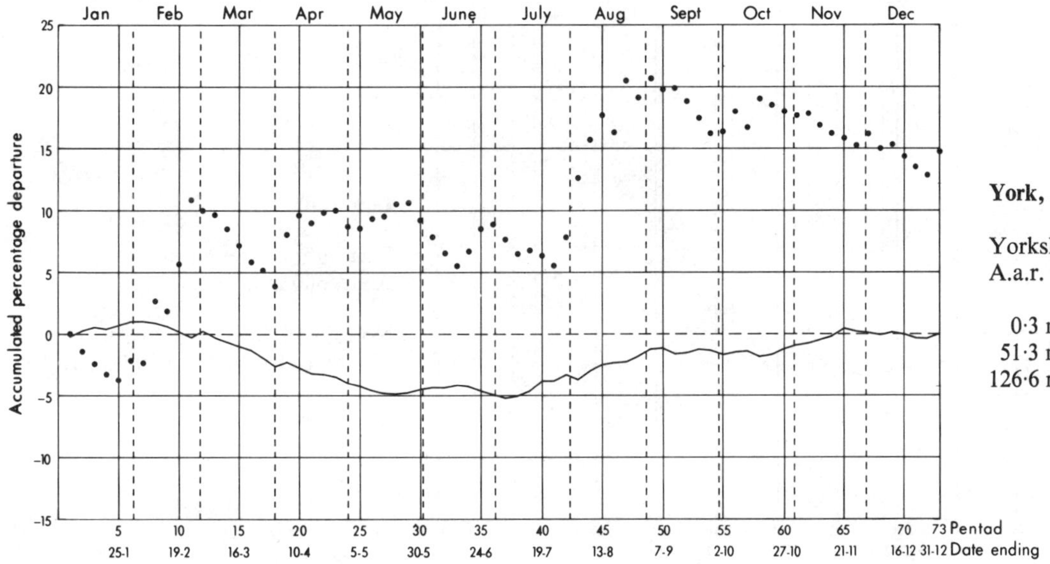

York, Acomb Landing

Yorkshire Ouse and Hull, Ouse
A.a.r. 629 mm

0·3 mm 16 days from 26·5 to 10·6
51·3 mm 10 days from 20·8 to 29·8
126·6 mm 25 days from 2·2 to 26·2

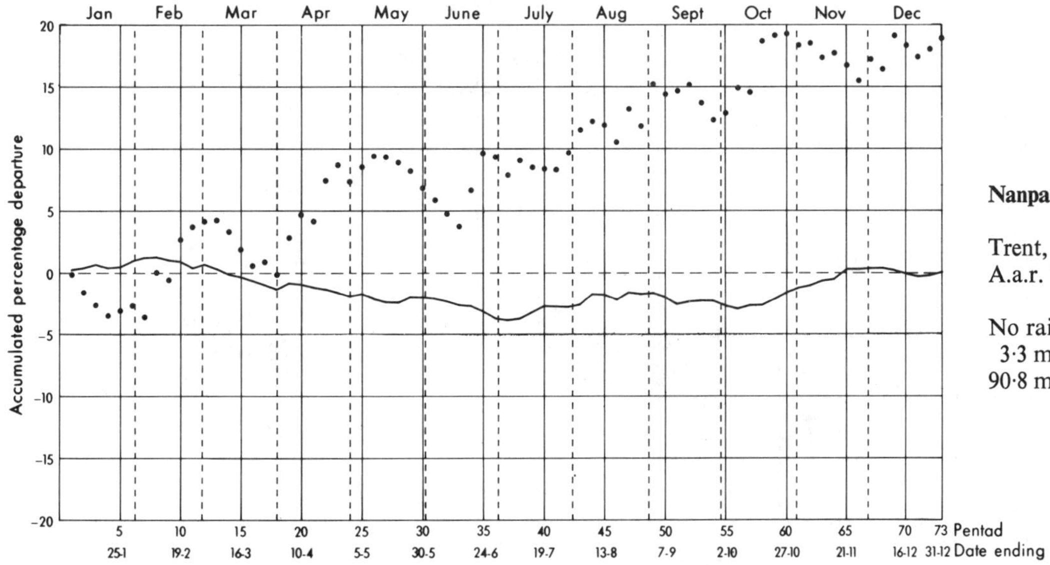

Nanpantan Res.

Trent, Soar (Trent)
A.a.r. 715 mm

No rain 15 days from 15·9 to 29·9
3·3 mm 19 days from 5·3 to 23·3
90·8 mm 18 days from 2·10 to 19·10

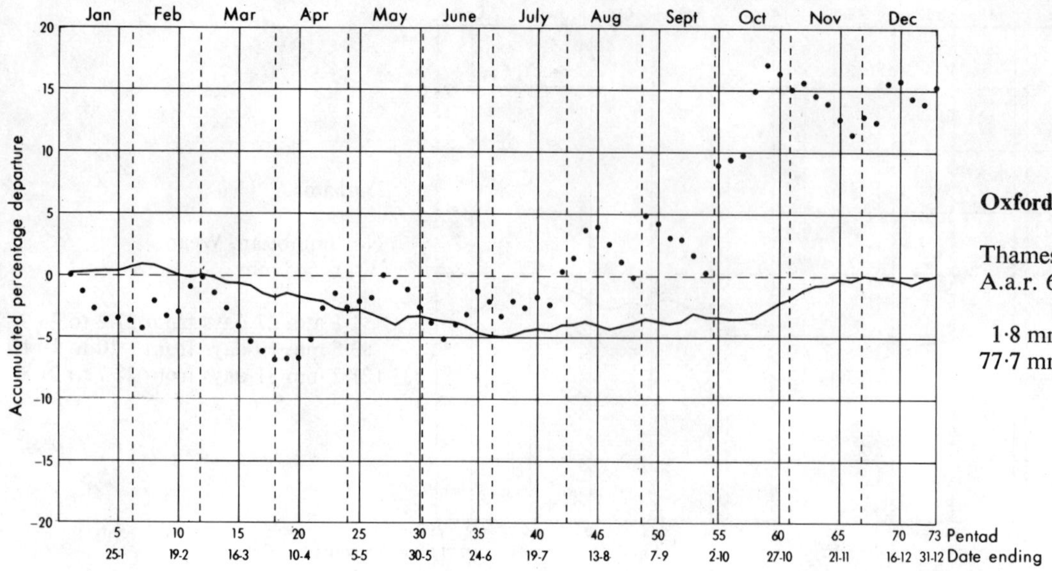

Oxford

Thames Conservancy, Thames
A.a.r. 652 mm

1·8 mm 23 days from 28·2 to 22·3
77·7 mm 6 days from 29·9 to 4·10

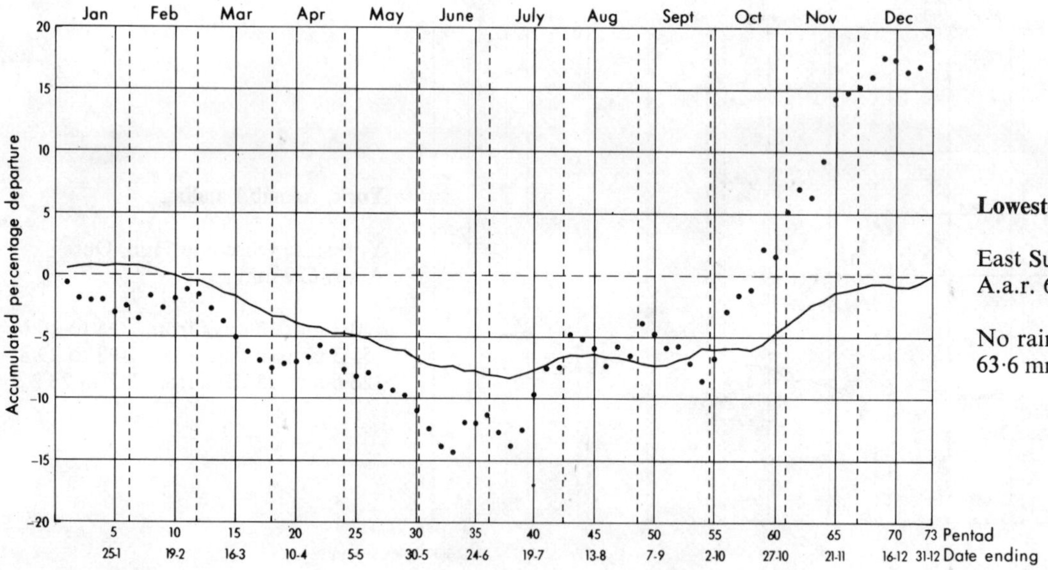

Lowestoft

East Suffolk and Norfolk, Coastal
A.a.r. 600 mm

No rain 16 days from 26·5 to 10·6
63·6 mm 11 days from 30·9 to 10·10

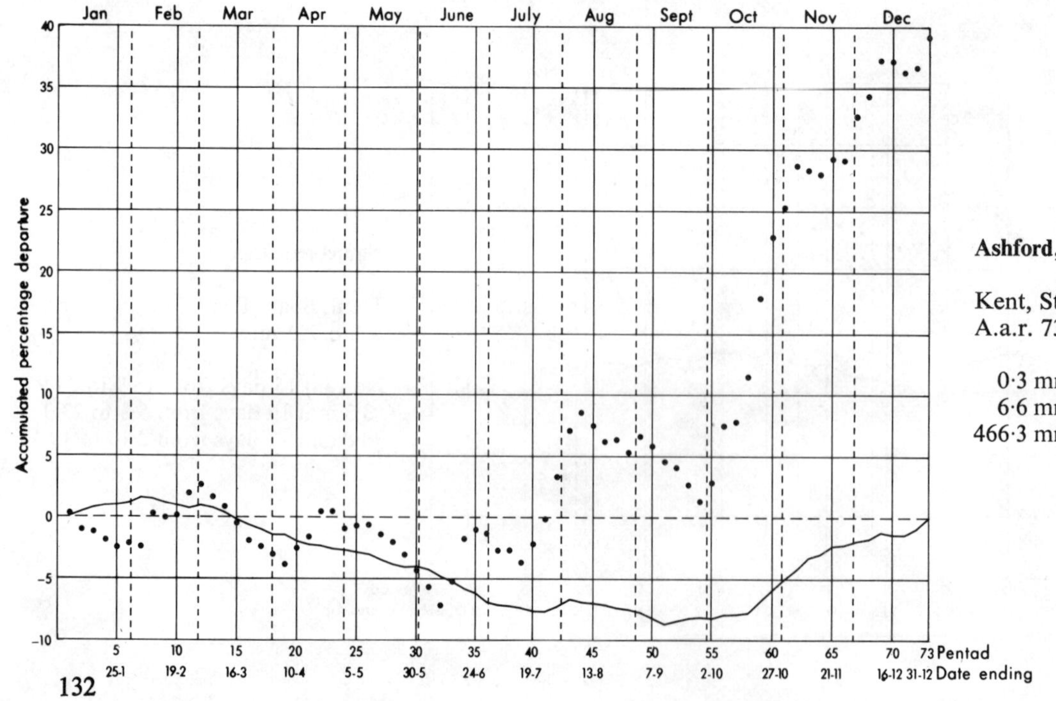

Ashford, Belmore

Kent, Stour
A.a.r. 730 mm

0·3 mm 15 days from 26·5 to 9·6
6·6 mm 24 days from 5·9 to 28·9
466·3 mm 93 days from 29·9 to 31·12

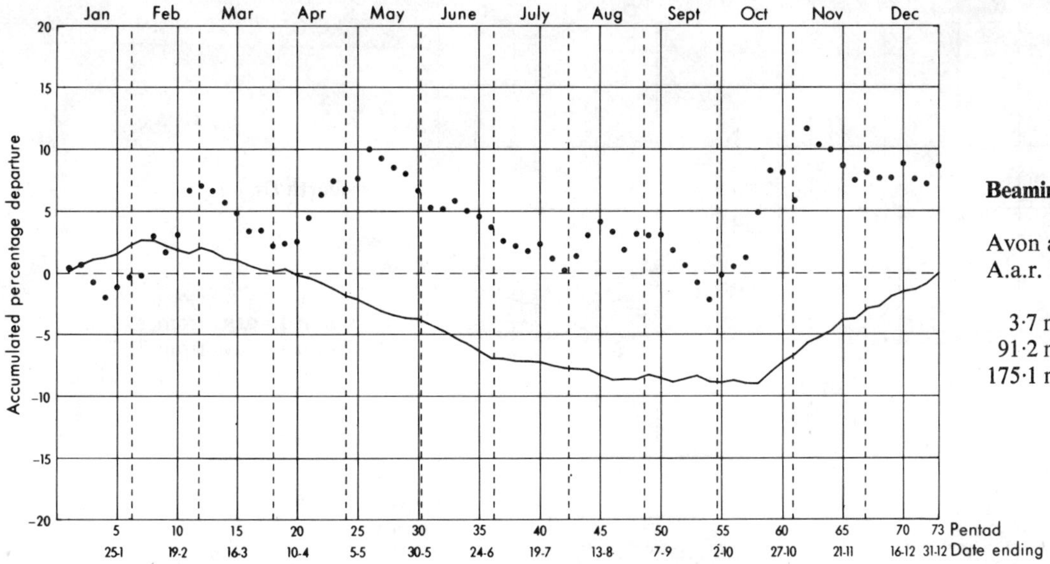

Beaminster

Avon and Dorset, Coastal
A.a.r. 1018 mm

3·7 mm 25 days from 5·9 to 29·9
91·2 mm 9 days from 18·2 to 26·2
175·1 mm 23 days from 30·9 to 22·10

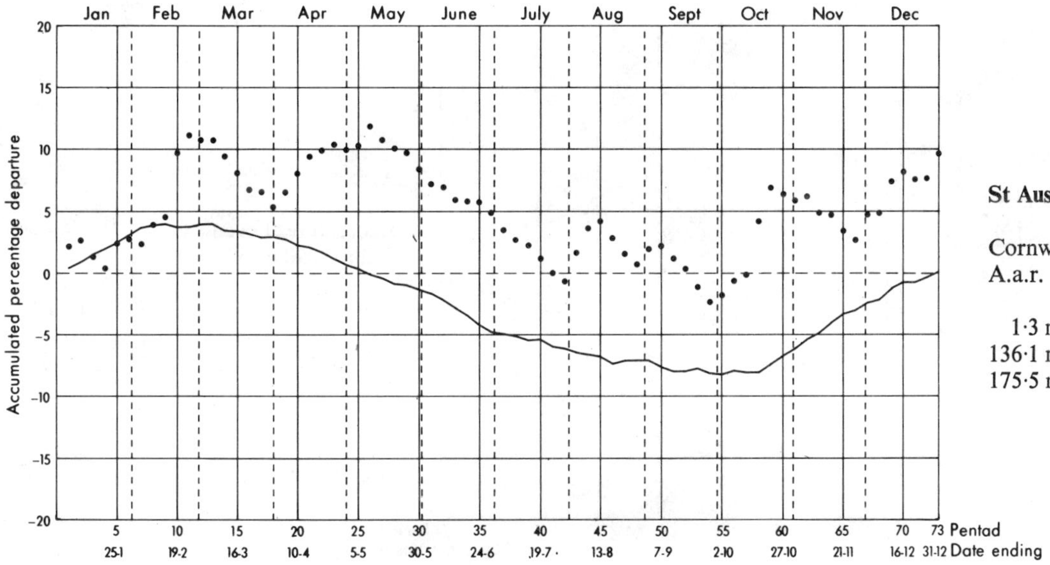

St Austell

Cornwall, Coastal
A.a.r. 1232 mm

1·3 mm 20 days from 4·3 to 23·3
136·1 mm 12 days from 14·2 to 25·2
175·5 mm 19 days from 4·10 to 22·10

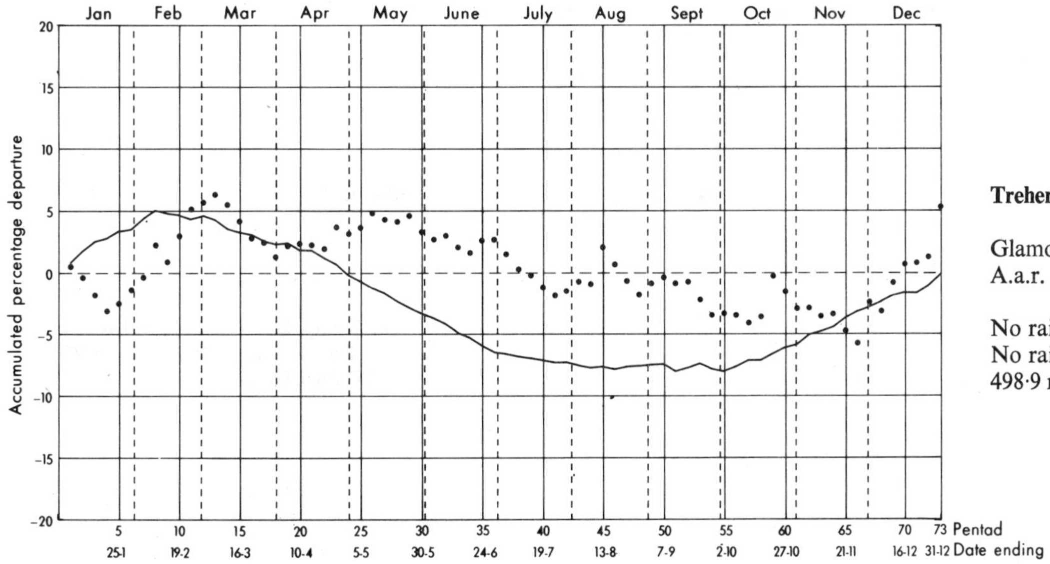

Treherbert, Tyn-y-Waun No. 1

Glamorgan, Rhondda (Taff)
A.a.r. 2377 mm

No rain 14 days from 14·8 to 27·8
No rain 14 days from 16·9 to 29·9
498·9 mm 38 days from 24·11 to 31·12

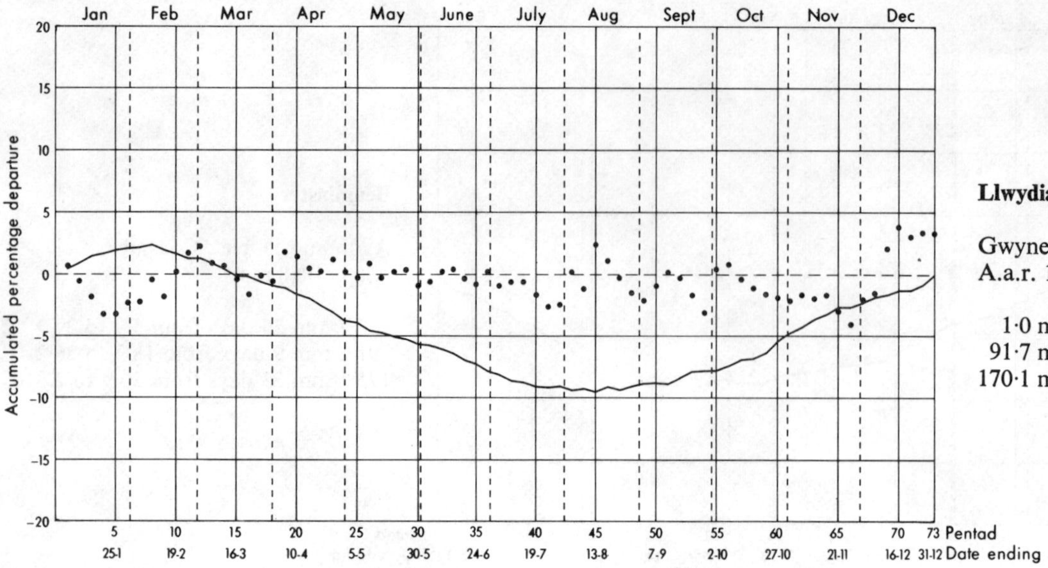

Llwydiarth Esgob

Gwynedd, Coastal
A.a.r. 1024 mm

1·0 mm 15 days from 15·9 to 29·9
91·7 mm 14 days from 31·7 to 13·8
170·1 mm 32 days from 29·11 to 30·12

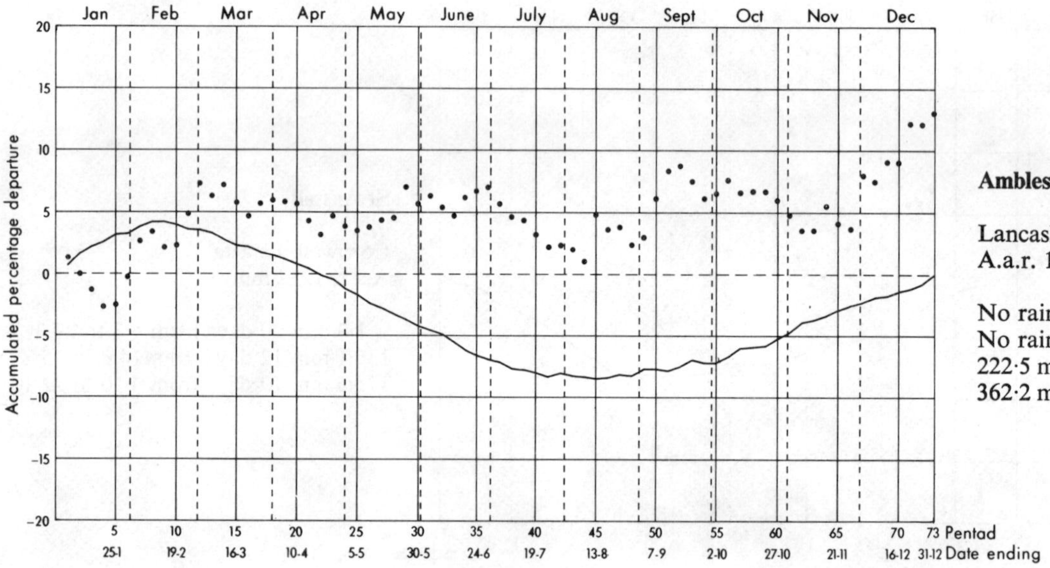

Ambleside

Lancashire, Leven
A.a.r. 1890 mm

No rain 15 days from 15·9 to 29·9
No rain 18 days from 7·1 to 24·1
222·5 mm 14 days from 1·9 to 14·9
362·2 mm 34 days from 25·1 to 27·2

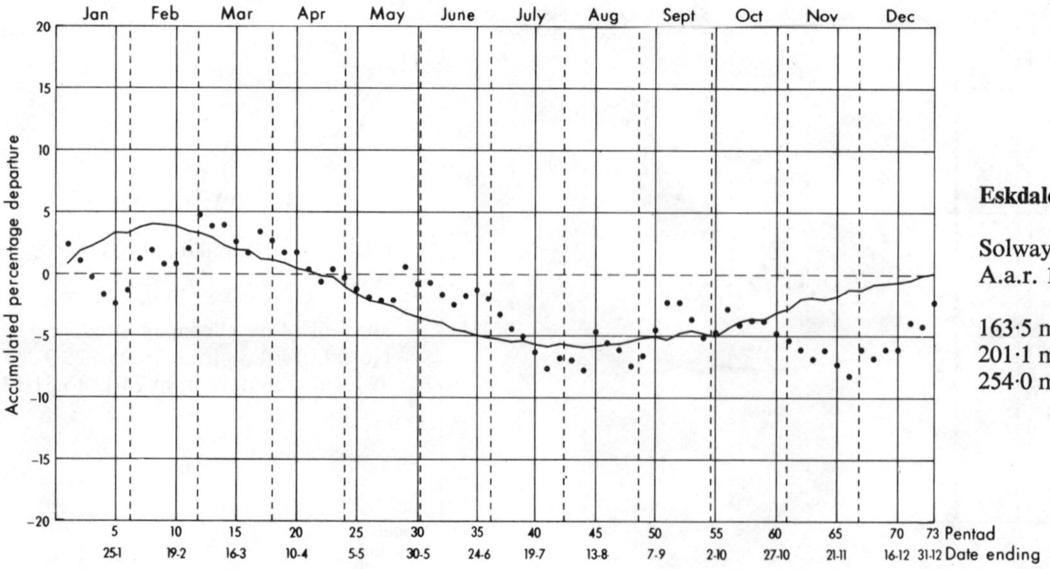

Eskdalemuir

Solway, Esk
A.a.r. 1582 mm

163·5 mm 14 days from 1·9 to 14·9
201·1 mm 27 days from 1·2 to 27·2
254·0 mm 38 days from 24·11 to 31·12

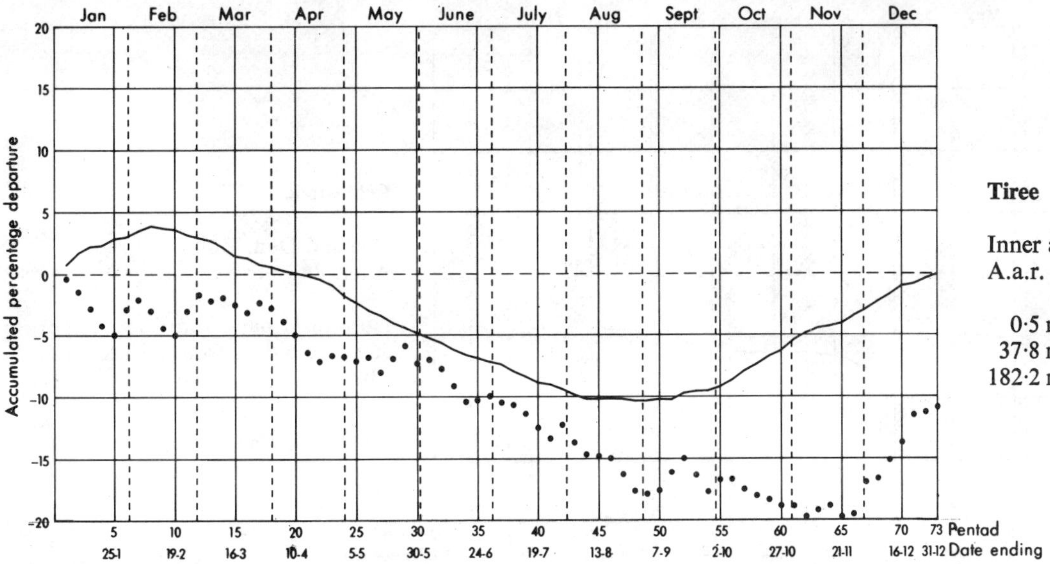

Tiree

Inner and Outer Hebrides, Coastal
A.a.r. 1117 mm

0·5 mm 14 days from 9·1 to 22·1
37·8 mm 3 days from 27·1 to 29·1
182·2 mm 28 days from 23·11 to 20·12

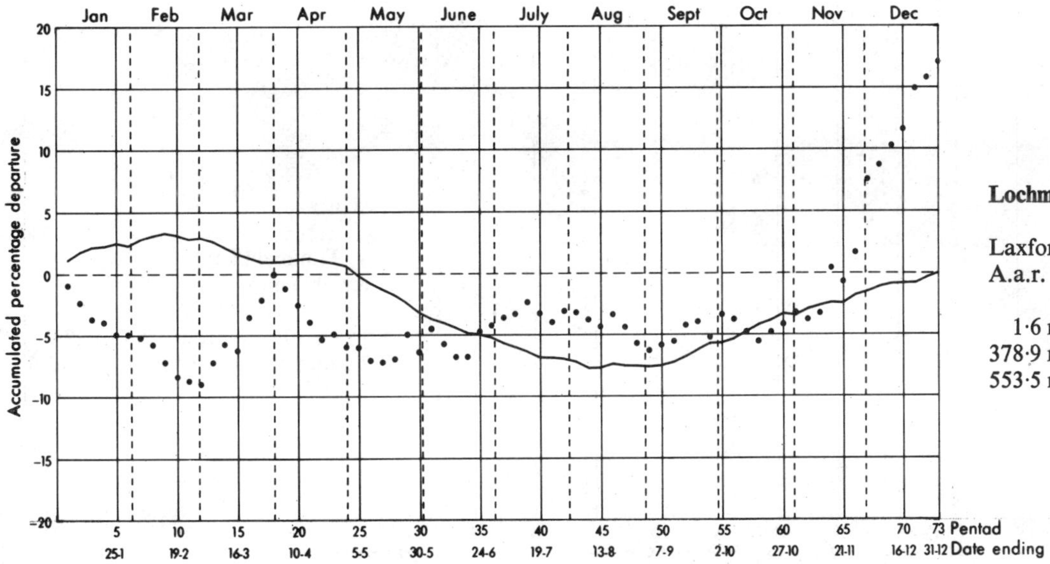

Lochmore Lodge

Laxford, Laxford
A.a.r. 1933 mm

1·6 mm 20 days from 2·4 to 21·4
378·9 mm 42 days from 19·2 to 1·4
553·5 mm 39 days from 23·11 to 31·12

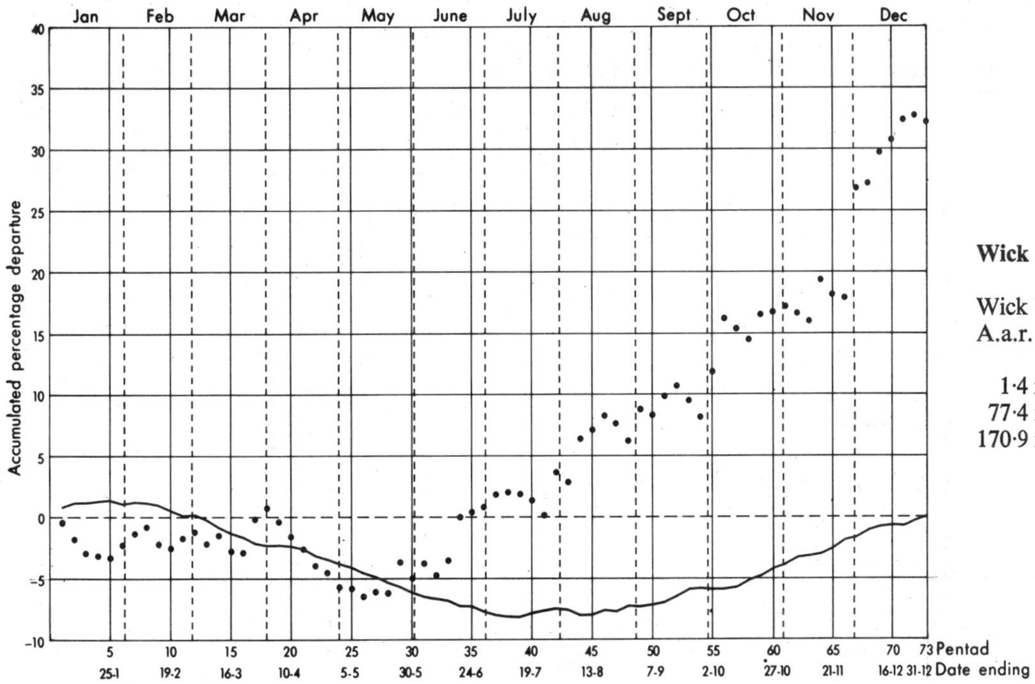

Wick

Wick Water to Conon, Wick
A.a.r. 761 mm

1·4 mm 12 days from 7·2 to 18·2
77·4 mm 7 days from 1·10 to 7·10
170·9 mm 27 days from 27·11 to 23·12

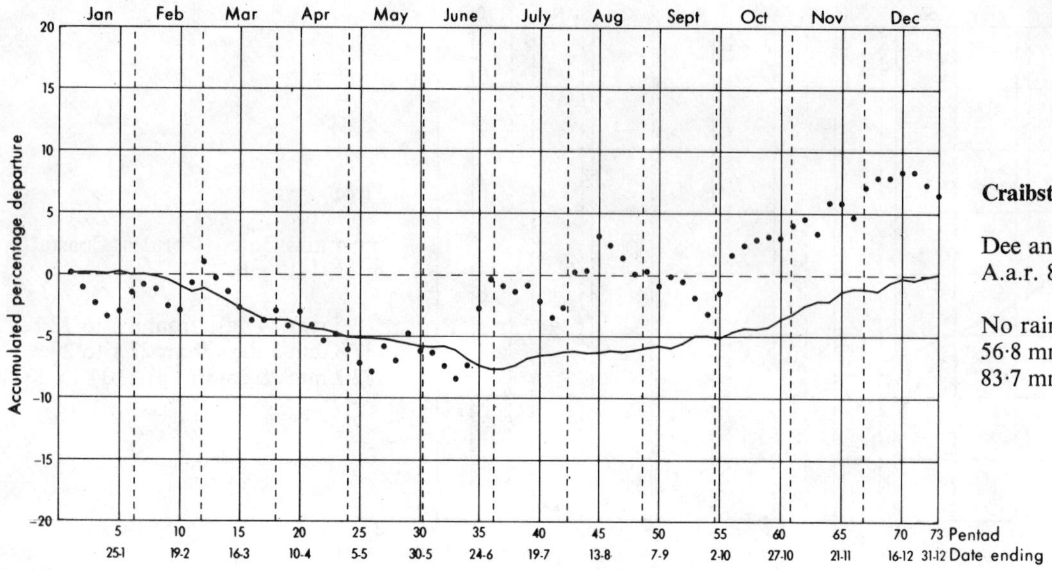

Craibstone

Dee and Don, Don
A.a.r. 861 mm

No rain 11 days from 16·9 to 26·9
56·8 mm 6 days from 5·10 to 10·10
83·7 mm 6 days from 22·6 to 27·6

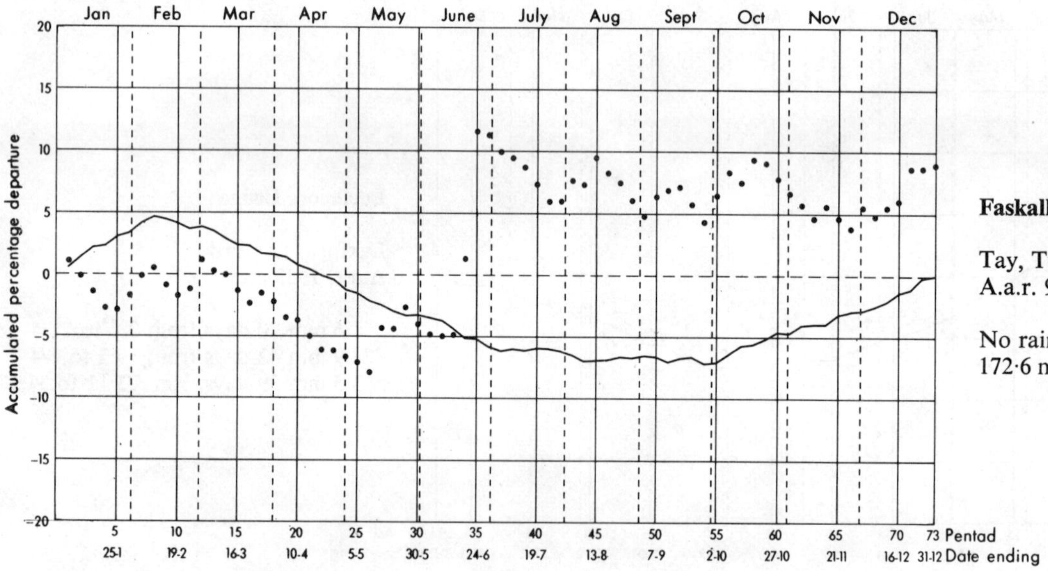

Faskally

Tay, Tummel (Tay)
A.a.r. 900 mm

No rain 10 days from 18·9 to 27·9
172·6 mm 10 days from 14·6 to 23·6

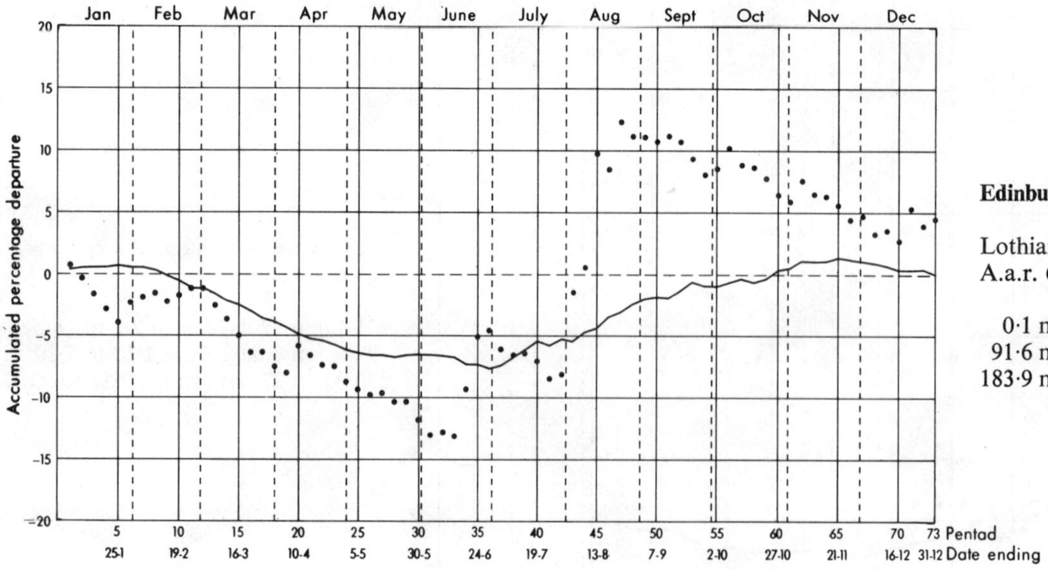

Edinburgh, Blackford Hill

Lothians, Figgate
A.a.r. 699 mm

0·1 mm 13 days from 17·9 to 29·9
91·6 mm 14 days from 14·6 to 27·6
183·9 mm 19 days from 3·8 to 21·8

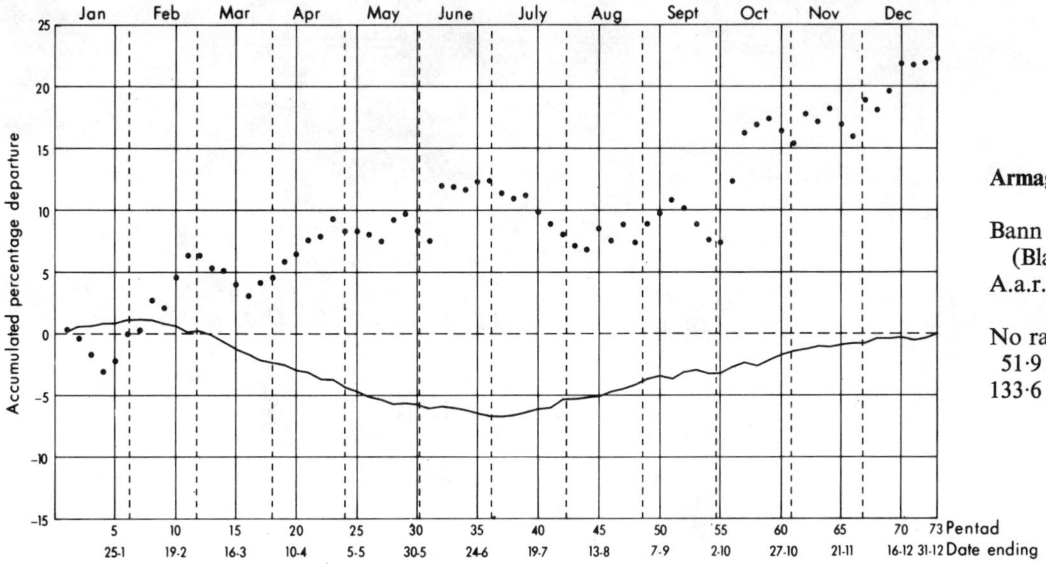

Armagh

Bann and Lough Neagh, Callan (Blackwater)
A.a.r. 845 mm

No rain 13 days from 11·1 to 23·1
51·9 mm 2 days from 9·6 to 10·6
133·6 mm 17 days from 2·10 to 18·10

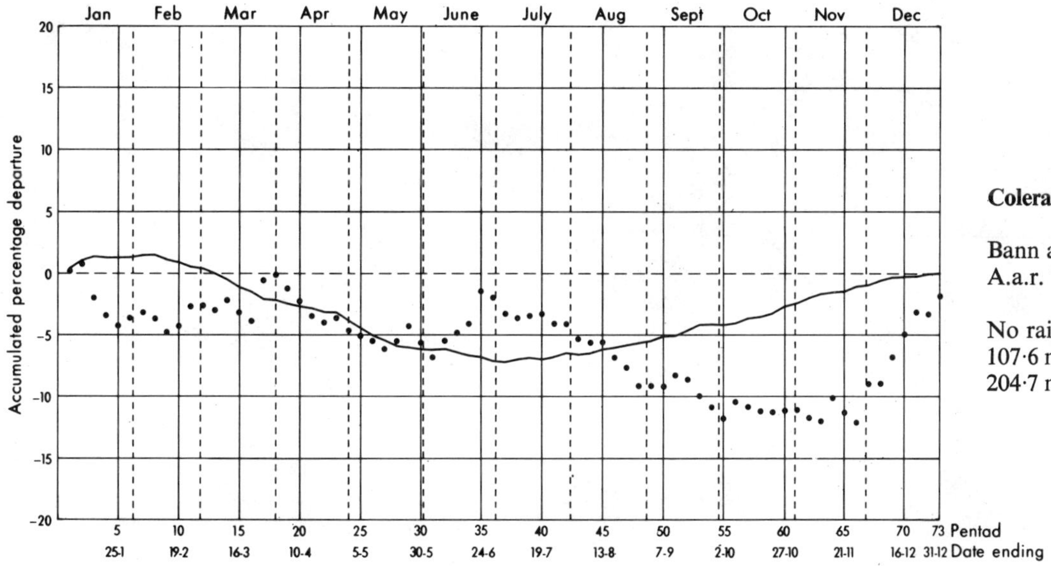

Coleraine, The Cutts

Bann and Lough Neagh, Bann
A.a.r. 1039 mm

No rain 15 days from 9·1 to 23·1
107·6 mm 16 days from 9·6 to 24·6
204·7 mm 35 days from 27·11 to 31·12

4 FREQUENCY DISTRIBUTIONS OF DAILY AMOUNTS OF RAIN

Table 4 gives seven-point frequency distributions of daily amounts for the two seasonal six-month periods and for the calendar year. Nine maps are also provided, giving isopleths of the number of rainfall days with 1·0 millimetre or more for the periods mentioned and similarly for the values 2·5 millimetres and 10·0 millimetres. The stations listed in Table 4 are a selection of the many stations used in compiling the data for the maps.

As 0·25 millimetre is relatively small, the values for this threshold are more haphazard than for those of the larger amounts. This is mainly due to the difficulties of observing small amounts of rain within narrow percentage limits and partly due to purely meteorological reasons. For these reasons it has never been considered really satisfactory to try to map frequencies for those days with 0·25 millimetre or more. However, with the reservations mentioned the 0·25-millimetre threshold has been included in Table 4.

The 0·25-millimetre threshold has been used in Table 4 since the publication of the first metric version of *British Rainfall*, namely that for 1964, even though daily values of rainfall in millimetres are measured to one place of decimals only. This is because 0·25 millimetre is within less than 2 per cent of 0·01 inch and this is the value which has to be used in extracting much of the data for Table 4, most stations in which were still reporting in inches in 1966. A threshold value of 0·3 millimetre is used for the small number of stations reporting in millimetres.

TABLE 4

Frequency distributions of daily amounts

Numbers of days with rainfall amounts reaching or exceeding specified values (in millimetres)

Area and station	October 1965–March 1966									April–September 1966									Calendar year 1966						
								Max. fall									Max. fall								
	0·25	1·0	2·5	5	10	25	50	Amount	Date	0·25	1·0	2·5	5	10	25	50	Amount	Date	0·25	1·0	2·5	5	10	25	50
								mm									*mm*								
Northumbrian																									
Acklington	91	66	43	23	9	0	0	18·7	17 Nov.	87	69	39	21	11	2	1	92·3	13 Aug.	179	134	79	40	17	2	1
Whickhope	119	88	67	41	21	3	1	50·8	31 Oct.	110	84	65	44	18	2	0	33·8	3 Aug.	236	182	131	87	39	4	0
Hexham Res.	103	78	59	22	7	1	0	34·0	19 Nov.	105	71	45	28	12	2	0	29·0	13 Aug.	214	151	99	50	19	3	0
Durham	101	73	46	27	9	0	0	24·6	26 Nov.	105	72	48	27	12	0	0	22·4	11 May	212	144	97	50	17	1	0
Hury Res.	122	92	54	40	14	0	0	24·4	31 Oct.	108	81	53	30	15	2	0	26·2	13 Aug.	241	183	115	66	29	4	0
Yorkshire Ouse and Hull																									
Mulgrave Castle	109	85	55	27	7	0	0	19·0	{8 Feb. / 17 Nov.}	96	73	54	30	12	1	0	25·7	21 Aug.	211	154	106	56	23	3	0
Scarborough	106	81	56	31	9	0	0	20·1	7 Feb.	93	66	50	31	14	0	0	22·9	21 Aug.	195	143	101	52	17	0	0
Hull	102	79	47	32	11	1	0	31·2	7 Feb.	88	63	42	29	11	1	0	27·7	21 Aug.	192	140	88	55	17	2	0
Richmond, Green Howard Rd	104	87	53	32	15	1	0	31·2	7 Feb.	98	73	54	30	15	3	0	35·6	11 May	212	160	107	62	29	5	0
Ramsgill	101	94	80	54	32	4	0	45·0	31 Oct.	89	82	61	42	20	2	0	36·3	11 May	197	183	139	95	45	8	0
York, Acomb Landing	98	66	44	33	10	1	0	28·4	7 Feb.	82	65	45	29	10	1	0	28·2	21 Aug.	184	131	82	50	16	2	0
Ampleforth	90	79	58	31	11	1	0	30·7	7 Feb.	93	75	55	39	15	0	0	20·6	21 Aug.	195	157	108	61	23	1	0
Malham Tarn	115	97	81	57	33	8	0	44·4	23 Feb.	104	84	72	54	34	5	0	32·8	11 May	236	194	155	111	62	12	0
Bradford	115	87	58	37	21	1	0	25·4	7 Feb.	102	77	56	37	14	3	1	53·6	20 Aug.	224	164	107	71	30	4	1
Goole S. Wks	104	68	39	27	9	1	0	36·6	7 Feb.	95	70	49	36	11	1	0	26·2	29 Aug.	198	144	89	56	15	2	0
Thorne	91	61	41	24	10	1	0	31·7	7 Feb.	81	68	50	28	11	1	0	26·4	15 June	180	133	91	45	17	2	0
Trent																									
Keele	106	72	52	35	10	0	0	21·1	9 Dec.	100	77	50	35	14	2	0	26·9	20 Aug.	220	163	112	67	24	2	0
Edgbaston	93	74	49	29	12	1	0	28·6	18 Dec.	87	70	48	29	12	2	0	30·2	10 June	188	152	101	62	24	2	0
Nanpantan Res.	89	71	49	30	13	0	0	22·6	9 Dec.	86	70	48	32	13	1	0	26·7	21 Aug.	187	149	99	60	23	1	0
Mansfield	86	63	50	37	18	1	0	26·4	9 Dec.	80	67	45	31	14	0	0	22·6	21 Aug.	180	145	99	65	26	0	0
Lincolnshire																									
Manby	88	63	47	31	15	0	0	22·3	7 Feb.	75	62	46	27	8	1	1	50·3	22 June	175	132	92	50	14	1	1
Revesby Res.	92	70	47	31	10	0	0	24·1	7 Feb.	83	69	51	23	6	1	0	41·7	22 June	193	155	99	46	13	1	0
Welland and Nene																									
Eye Brook Res.	84	61	40	24	7	0	0	15·2	{28 Nov. / 22 Dec.}	78	57	41	24	9	2	0	35·3	21 Aug.	178	121	82	45	16	2	0
Ravensthorpe	93	64	43	21	6	0	0	14·5	18 Dec.	90	67	50	29	8	2	0	49·3	29 Aug.	196	141	94	53	15	2	0
Great Ouse																									
Foxcote P. Sta.	89	65	46	27	7	0	0	16·0	22 Dec.	78	65	46	31	9	1	0	25·9	22 June	176	133	96	56	17	1	0
Cardington	75	58	36	22	3	0	0	11·1	22 Dec.	70	55	40	27	8	2	0	42·4	22 June	161	118	74	48	15	2	0
Odsey	93	59	42	22	1	0	0	13·2	28 Nov.	86	61	45	19	9	3	0	37·8	21 Aug.	192	125	84	37	15	3	0
Cambridge, Botanic Garden	81	56	31	14	1	0	0	10·9	10 Feb.	66	55	42	25	7	2	0	29·7	21 Aug.	172	119	77	36	13	2	0
Mildenhall	71	51	33	15	0	0	0	9·7	18 Feb.	78	56	38	19	5	2	0	31·9	29 Aug.	171	118	73	35	10	2	0
Sandringham House	101	73	51	31	8	0	0	19·8	9 Dec.	83	64	43	29	12	2	0	48·8	30 Aug.	203	152	102	61	19	2	0
East Suffolk and Norfolk																									
Bressingham Hall	93	66	39	21	3	0	0	19·0	9 Dec.	80	58	39	19	6	0	0	24·6	29 Aug.	198	138	84	41	9	0	0
Sprowston	101	68	38	28	5	0	0	24·1	9 Dec.	86	64	39	21	7	0	0	18·0	16 June	206	147	88	45	10	0	0
Lowestoft	90	67	36	16	1	0	0	17·8	9 Dec.	79	52	41	22	7	0	0	16·3	21 Aug.	198	136	96	51	15	0	0
Belstead Hall	104	60	37	19	2	0	0	14·0	9 Dec.	90	59	45	28	9	1	0	25·1	19 July	215	136	89	49	14	1	0
Essex																									
Felsted School	82	63	43	26	4	0	0	14·2	18 Nov.	82	62	44	27	8	0	0	22·1	19 July	185	138	88	48	11	0	0
Shoeburyness	73	58	35	13	2	0	0	11·1	9 Dec.	68	54	37	22	9	1	0	39·0	19 July	160	122	79	45	14	1	0

Lee																									
Meesden	85	63	49	24	2	0	0	15·7	26 Feb.	90	66	48	25	6	2	0	33·3	22 June	196	136	96	45	12	2	0
Darnicle Hill P. Sta. . .	89	61	49	27	7	0	0	15·7	28 Nov.	84	68	47	31	13	3	0	33·8	22 June	179	135	92	56	20	3	0
London, left bank																									
St James's Park . . .	79	66	45	23	7	0	0	12·7	9 Feb.	75	61	41	27	11	2	0	34·3	29 Aug.	162	133	88	52	19	2	0
Thames																									
Cirencester, Chesterton Grove .	84	60	37	30	14	0	0	22·9	18 Dec.	93	67	43	28	12	1	0	44·2	29 Aug.	180	126	83	56	24	1	0
Oxford	78	61	37	24	2	0	0	14·5	22 Dec.	91	61	41	28	12	1	0	34·5	29 Aug.	181	128	84	54	22	1	0
Maidenhead S. Wks . .	89	62	40	22	5	0	0	17·0	28 Nov.	84	61	48	27	12	1	0	33·3	22 June	184	130	93	51	20	1	0
Hindhead W. Wks . . .	95	72	55	37	18	2	0	31·2	7 Nov.	84	70	47	30	13	2	0	34·0	22 June	185	152	104	64	32	4	0
Mickleham	87	68	46	29	9	0	0	23·6	28 Nov.	81	65	42	30	9	1	0	25·1	22 June	182	138	91	59	21	2	0
London, right bank																									
Kew	81	61	40	23	5	0	0	15·3	9 Feb.	69	58	43	28	10	2	0	30·1	29 Aug.	147	120	87	49	17	2	0
Bromley	80	70	50	25	4	0	0	19·0	28 Nov.	78	62	43	27	12	2	0	34·5	29 Aug.	175	134	94	56	22	2	0
Kent																									
Falconhurst	90	66	50	31	11	0	0	23·1	9 Feb.	84	61	44	33	12	1	0	40·4	22 June	196	146	100	67	26	2	0
Goudhurst	88	66	46	33	16	2	0	27·7	9 Dec.	90	69	52	36	13	0	0	21·6	6 Aug.	204	154	119	76	32	1	0
East Malling	88	66	42	27	5	0	0	22·9	9 Feb.	78	59	47	30	10	0	0	23·1	18 July	188	135	101	62	20	1	0
Herne Bay	86	63	42	18	3	0	0	10·7	28 Nov.	65	46	33	23	9	0	0	20·3	21 Aug.	175	128	85	53	20	2	0
Ashford, Belmore . . .	88	67	46	32	8	0	0	18·0	28 Nov.	76	67	44	32	12	0	0	23·1	10 June	192	160	107	73	32	1	0
Dover	94	74	48	34	8	0	0	21·3	9 Feb.	80	62	41	31	15	0	0	24·6	16 June	194	157	109	79	36	4	0
Sussex																									
Seaford, Newlands . . .	77	62	46	35	14	0	0	19·8	19 Nov.	73	54	42	25	6	2	0	32·5	22 June	173	132	97	64	23	3	0
Knepp Castle	90	67	47	35	17	2	0	26.9	18 Dec.	84	65	48	33	14	1	0	26·2	6 Aug.	185	139	102	64	30	2	0
Worthing	82	62	48	33	11	0	0	22·4	20 Nov.	79	64	44	26	5	2	0	35·1	6 Aug.	178	135	95	61	19	3	0
Hampshire																									
Farlington Res. . . .	85	65	50	31	14	1	0	32·5	7 Nov.	80	63	50	30	10	1	0	41·9	6 Aug.	177	139	104	59	27	2	0
Southampton	96	69	52	34	16	0	0	22·9	28 Nov.	88	66	48	28	12	1	0	29·2	6 Aug.	200	139	97	59	26	3	0
Wootton House . . .	106	74	50	35	10	0	0	24·1	28 Nov.	98	74	54	34	14	2	0	27·7	22 June	215	157	108	63	26	2	0
Andover S. Wks . . .	113	66	48	33	12	1	0	26·9	7 Nov.	83	61	42	30	10	1	1	59·2	29 Aug.	184	126	89	60	23	2	1
Isle of Wight																									
Ryde	93	70	52	37	18	3	0	32·0	7 Nov.	80	65	49	26	12	1	0	34·8	6 Aug.	188	143	105	59	28	2	0
Avon and Dorset																									
Stourhead	94	77	57	40	21	1	0	26·7	28 Nov.	90	67	53	36	17	1	0	29·2	6 Aug.	196	149	112	77	38	3	1
Blandford, Kings Road . .	100	74	57	36	15	3	0	33·3	7 Nov.	88	61	41	25	8	0	0	24·9	28 Aug.	190	141	101	62	23	6	0
Weymouth	95	80	55	39	18	2	0	28·4	10 Jan.	76	65	41	21	12	1	0	34·3	6 Aug.	181	145	100	57	31	4	1
Beaminster	88	75	59	44	25	5	0	30·2	28 Nov.	88	69	54	31	13	2	0	28·4	6 Aug.	192	154	117	71	35	5	1
Devon, south																									
Exeter	93	71	49	33	15	5	1	57·0	10 Jan.	73	61	44	26	9	0	0	20·7	8 Apr.	167	133	94	64	26	3	1
Totnes	117	93	74	56	32	7	0	35·6	18 Dec.	95	73	50	37	19	2	0	29·0	6 Aug.	211	161	118	87	45	8	0
Cornwall																									
Mount Batten	103	87	68	50	27	3	0	29·3	16 Nov.	89	73	49	30	14	1	0	33·9	6 Aug.	195	154	109	72	39	4	0
St Austell	111	101	78	52	29	5	0	35·3	24 Jan.	104	79	61	40	14	1	0	28·4	2 Aug.	228	182	144	92	40	6	0
Tresco Abbey	104	88	66	38	11	1	0	26·7	28 Nov.	81	64	48	29	13	1	0	39·4	8 June	195	153	105	68	30	3	0
Rosewarne	119	98	73	51	26	6	0	40·9	16 Nov.	107	74	54	35	20	3	0	30·0	8 Apr.	235	178	123	82	43	10	1
Newquay	104	79	60	37	14	1	0	41·4	16 Nov.	91	72	48	26	11	1	0	26·9	10 Aug.	202	157	107	59	27	4	0
Bude	104	80	58	32	11	0	0	21·8	16 Nov.	96	84	62	31	9	2	0	31·0	17 June	210	164	118	61	19	4	0
Devon, north																									
Okehampton, Pleasure Gardens	113	97	79	58	37	7	1	55·9	16 Nov.	110	87	68	46	24	1	0	29·2	3 Aug.	235	190	158	107	53	5	0
Chivenor	91	74	50	34	13	0	0	18·7	16 Nov.	88	74	60	32	12	0	0	24·9	10 Aug.	187	152	111	64	27	0	0
Somerset																									
Taunton, Vivary Park . .	82	71	49	33	12	0	0	20·8	28 Nov.	77	64	48	25	8	1	0	28·4	10 June	172	138	97	54	22	2	0
Street S. Wks	91	71	53	30	7	0	0	17·0	6 Nov.	82	70	52	31	11	0	0	20·3	{8 May, 28 Aug.	180	143	108	62	16	1	0
Bristol Avon																									
West Lavington . . .	99	81	56	35	10	0	0	16·5	7 Nov.	101	70	48	28	11	0	0	19·3	29 Aug.	210	149	99	57	22	3	0
Long Ashton	94	70	50	36	15	2	0	31·5	17 Dec.	93	73	52	29	14	0	0	19·8	6 Aug.	194	146	99	60	28	2	0

J

TABLE 4 (*continued*)

Frequency distributions of daily amounts

Numbers of days with rainfall amounts reaching or exceeding specified values (in millimetres)

Area and station	October 1965–March 1966									April–September 1966									Calendar year 1966						
								Max. fall									Max. fall								
	0·25	1·0	2·5	5	10	25	50	Amount	Date	0·25	1·0	2·5	5	10	25	50	Amount	Date	0·25	1·0	2·5	5	10	25	50
								mm									*mm*								
Severn																									
Lower Hazel	99	78	50	33	13	2	0	48·8	18 Dec.	92	69	49	29	11	0	0	16·8	29 Aug.	197	146	96	58	24	3	0
Lake Vyrnwy	129	103	85	64	48	7	2	70·4	17 Dec.	107	85	72	52	21	2	0	34·3	1 Apr.	246	198	162	115	66	6	1
Shawbury	91	69	51	32	9	0	0	20·3	18 Dec.	92	69	45	29	10	1	0	48·0	17 June	193	147	99	57	21	1	0
Anchor	120	97	67	49	21	1	0	34·8	9 Dec.	104	73	53	34	10	0	0	24·4	8 May	241	183	125	82	30	4	0
Oakly Park	102	77	50	27	12	0	0	21·6	28 Nov.	87	72	46	26	11	2	0	31·7	20 Aug.	198	159	102	51	21	2	0
Dingle Leys	104	86	61	42	14	0	0	22·1	18 Feb.	93	76	60	36	11	0	0	22·1	22 June	211	172	129	79	22	1	0
Finham S. Wks	91	70	46	28	7	0	0	20·8	9 Dec.	84	68	51	29	11	2	0	35·1	29 Aug.	188	147	98	57	19	3	0
Wellesbourne	99	64	40	19	4	0	0	14·7	22 Dec.	90	61	44	26	8	1	0	30·5	29 Aug.	194	129	87	48	15	1	0
Cheltenham	98	63	41	21	7	0	0	23·6	28 Nov.	91	61	48	26	5	0	0	16·5	8 May	199	129	92	50	17	0	0
Wye																									
Llandrindod Wells	108	91	69	45	19	2	0	29·5	9 Dec.	101	76	52	33	14	1	0	25·9	8 May	221	172	121	76	34	2	0
Glasbury	105	86	61	36	16	0	0	23·1	9 Dec.	88	75	54	24	9	1	0	35·8	8 May	205	167	116	59	24	2	0
Ross-on-Wye	93	71	45	29	11	0	0	19·2	28 Nov.	78	60	44	25	6	0	0	21·0	20 Aug.	170	134	88	51	16	1	0
Usk																									
Blaenau-isaf	118	102	84	69	41	13	2	58·7	17 Dec.	99	93	77	57	35	2	0	32·0	23 Apr.	227	204	170	129	77	13	1
Gilwern	109	97	72	52	33	5	0	34·8	18 Dec.	95	80	60	41	20	0	0	23·4	8 Apr.	217	183	129	90	50	5	0
Glamorgan																									
Treherbert, Tyn-y-Waun No. 1	108	97	85	66	50	19	6	111·0	17 Dec.	106	98	85	68	39	12	0	37·6	23 June	230	208	179	136	86	30	3
South West Wales																									
Golden Grove Farm	111	93	79	61	32	7	0	43·2	17 Dec.	102	94	74	43	27	1	0	27·7	5 Apr.	226	200	153	103	62	5	0
Cwmoernant Res.	108	96	83	55	30	7	0	36·6	22 Dec.	99	90	66	48	24	1	0	27·9	21 Apr.	202	178	137	97	51	4	0
Brawdy	108	96	69	47	25	4	0	29·0	16 Nov.	86	73	55	39	18	4	1	72·1	31 July	201	170	121	84	39	4	1
Gwynedd																									
Aberdovey	109	89	64	38	16	1	0	34·0	1 Oct.	94	73	49	37	15	1	0	48·5	26 June	224	172	115	83	30	3	0
Llwydiarth Esgob	111	87	64	47	21	3	1	59·2	1 Oct.	86	64	49	32	18	1	0	25·7	2 Aug.	207	155	107	69	33	2	0
Bryn Adda	105	88	64	42	17	2	0	27·4	1 Oct.	85	73	55	37	18	4	0	48·8	1 Apr.	209	173	124	77	35	6	0
Bodnant	123	89	62	42	16	3	0	35·3	24 Feb.	100	72	48	31	9	1	0	39·4	1 Apr.	241	179	115	67	24	4	1
Dee and Clwyd																									
Bala Sluices	115	100	77	55	31	3	1	58·4	17 Dec.	103	79	53	36	12	2	0	33·3	26 June	238	188	134	95	42	5	0
Alwen	120	105	80	54	22	3	0	37·6	9 Dec.	106	90	61	39	13	1	0	31·0	1 Apr.	246	208	149	93	35	4	0
Pack Saddle Res.	106	87	62	45	12	1	0	27·9	9 Dec.	94	86	65	36	11	0	0	22·9	1 Apr.	206	177	131	76	25	3	0
Mersey and Weaver																									
Heaton Res.	110	83	68	49	25	4	0	45·5	7 Feb.	102	83	62	41	18	2	0	39·9	26 June	235	185	144	86	38	6	0
Knott Hill Res.	107	88	61	42	22	4	0	41·7	7 Feb.	98	78	61	43	19	3	0	36·6	20 Aug.	224	184	127	83	36	4	0
Ringway	86	64	54	30	13	0	0	24·6	7 Feb.	88	76	56	36	16	3	0	31·6	5 July	191	151	109	67	27	3	0
Lancashire																									
Stonyhurst	93	84	64	47	20	3	0	44·7	17 Dec.	89	79	66	43	19	4	1	50·8	26 June	203	176	137	97	39	6	1
Squires Gate	89	71	50	32	12	1	0	26·2	7 Feb.	84	65	51	32	10	0	0	24·8	26 June	190	150	110	68	23	1	0
Ambleside	105	90	79	63	42	13	0	48·8	4 Feb.	92	83	68	54	30	9	2	65·5	3 Sept.	216	185	154	119	71	24	2
Poaka Beck	99	81	67	52	26	2	0	34·0	4 Feb.	89	78	59	48	26	5	0	46·0	26 June	204	174	133	98	52	8	0
Cumberland																									
Dale Head Hall	124	98	81	71	44	16	1	54·6	31 Oct.	130	93	69	56	34	10	1	64·0	3 Sept.	279	208	158	131	82	28	2
Newton Rigg	103	73	50	34	12	1	0	27·2	31 Oct.	97	69	50	30	16	2	0	32·5	13 Aug.	214	157	105	64	29	3	0
Carlisle S. Wks	92	64	46	26	7	1	0	27·9	31 Oct.	94	70	56	33	16	1	0	45·0	11 Sept.	194	146	103	59	25	1	0
Isle of Man																									
Douglas	109	90	69	47	30	2	0	33·5	7 Feb.	88	73	49	33	17	5	0	37·8	3 Sept.	201	167	125	79	42	6	0

Solway																									
Eskdalemuir	106	90	76	55	36	4	0	38·1	4 Feb.	102	80	56	37	20	6	0	43·8	3 Sept.	224	176	135	99	51	10	0
Dumfries	96	83	62	36	12	1	0	25·9	31 Oct.	88	67	50	33	17	4	0	39·9	13 Aug.	192	156	114	74	29	4	0
Glenlee	105	90	79	60	33	5	0	39·9	31 Oct.	99	79	54	36	17	4	1	51·3	13 Aug.	222	186	141	98	48	11	1
Stranraer Filters	116	88	68	45	19	2	0	31·0	4 Dec.	89	68	50	33	18	3	0	30·5	21 Apr.	213	169	131	87	47	5	0
Ayrshire																									
Prestwick	97	80	58	31	8	0	0	19·2	4 Feb.	86	64	41	31	15	2	1	66·7	13 Aug.	192	151	105	68	28	3	1
Clyde																									
Camps Res.	117	91	64	43	22	2	0	32·8	31 Oct.	109	79	56	36	12	3	1	66·0	13 Aug.	238	173	123	85	33	5	1
Sloy Power Sta.	120	102	92	76	53	21	3	62·7	4 Oct.	109	94	78	55	34	9	1	59·9	3 Sept.	236	208	178	139	92	31	2
Lephinmore	115	100	82	61	30	7	0	48·5	31 Oct.	113	90	63	42	24	2	0	41·1	23 June	245	208	156	113	62	9	0
Kintyre and south-western islands																									
Largie Castle	110	107	83	54	21	0	0	24·1	18 Feb.	96	83	53	32	15	2	0	28·2	9 Sept.	217	201	147	85	38	2	0
Lochy and Linnhe																									
Fersit	130	99	78	62	39	8	0	43·2	27 Feb.	106	87	66	39	17	2	0	29·7	3 Sept.	246	199	159	113	63	11	1
Ardgour House	116	105	87	67	40	16	1	100·8	26 Oct.	113	98	78	63	34	7	0	38·9	21 May	243	215	181	142	85	24	1
Sheil, Alsh and Maree																									
Polloch House	128	109	94	71	43	8	1	52·8	31 Oct.	111	102	83	57	33	2	0	45·7	22 May	245	223	186	142	87	9	0
Rhubana	119	105	80	51	30	5	1	50·8	26 Oct.	115	90	69	48	26	3	0	27·7	{16 Aug. 3 Sept.	247	205	167	111	58	9	0
Ratagan	124	111	91	70	43	10	4	81·8	26 Oct.	128	107	74	47	25	2	0	30·0	14 Sept.	263	232	184	138	79	11	5
Poolewe	109	97	84	54	23	4	0	38·6	5 Mar.	93	85	61	39	20	1	0	33·3	27 June	218	200	158	109	50	5	1
Inner and Outer Hebrides																									
Tiree	110	91	71	40	12	3	0	37·4	31 Oct.	91	63	45	32	14	0	0	17·3	25 July	219	170	122	78	24	0	0
Portree	117	104	80	55	34	10	0	47·8	26 Oct.	109	91	67	45	19	3	0	35·3	21 May	246	215	168	120	53	12	0
Benbecula	123	93	69	43	20	1	0	29·2	28 Jan.	101	75	49	28	10	1	0	28·0	1 Sept.	229	182	135	82	28	2	0
Laxford																									
Lochmore Lodge	126	115	94	65	33	6	0	42·7	21 Mar.	126	112	88	63	24	4	0	45·7	22 May	272	246	199	146	75	14	1
Naver and Thurso																									
Strathy	133	97	68	45	10	0	0	16·0	23 Nov	119	85	52	28	13	2	0	35·6	14 June	269	206	147	83	31	4	0
Orkney and Shetland Islands																									
Lerwick	132	115	80	47	24	3	0	36·3	28 Jan.	102	77	51	25	9	1	0	49·8	22 May	247	195	138	76	36	3	0
Kirkwall	123	102	81	45	14	1	0	25·6	28 Jan.	103	77	48	31	8	1	0	27·2	4 Aug.	245	195	144	87	26	5	0
Wick Water to Conon																									
Wick	117	91	63	30	8	0	0	15·0	22 Nov.	89	72	48	28	12	1	0	29·6	4 Aug.	217	177	115	59	25	3	0
Achany	141	95	67	46	14	0	0	17·5	20 Mar.	134	81	51	29	10	0	0	19·6	23 June	283	193	136	87	30	4	0
Geanies House	89	64	42	21	8	0	0	14·2	2 Dec.	96	75	47	28	11	1	0	41·9	23 June	208	159	90	44	15	2	0
Fairburn House	98	76	64	42	13	1	0	25·9	30 Oct.	90	73	50	27	12	1	0	36·3	17 June	196	163	123	76	37	4	0
Beauly and Ness																									
Kingie Camp	112	102	86	69	48	12	1	57·7	26 Oct.	113	91	68	44	20	4	0	31·2	14 Sept.	239	205	166	126	76	18	2
Inverness	78	60	39	23	6	0	0	15·7	31 Oct.	91	67	42	21	7	3	1	58·4	14 June	181	140	85	46	13	5	1
Banff, Moray and Nairn																									
Cluny Castle	122	80	57	35	19	0	0	21·1	28 Oct.	120	83	54	25	11	1	0	25·1	22 May	252	184	127	74	32	4	0
Achnagoichan	97	76	47	27	4	0	0	19·0	9 Dec.	101	80	54	22	8	1	0	29·0	23 June	216	175	119	59	19	3	0
Gordon Castle	97	69	44	25	7	0	0	22·1	27 Nov.	92	68	44	23	9	3	0	49·8	23 June	202	151	94	49	17	5	0
Dee and Don																									
New Deer School	117	92	59	32	10	0	0	20·3	2 Dec.	98	72	50	25	9	2	0	48·3	23 June	235	175	117	61	22	2	0
Craibstone	122	83	51	30	8	0	0	18·0	20 Feb.	111	73	46	24	8	4	0	45·7	23 June	232	161	101	57	19	5	0
Braemar	117	86	59	26	8	1	0	25·4	27 Nov.	100	73	49	26	7	3	0	30·7	23 June	238	169	111	63	25	4	0
North and South Esk																									
Inchgrundle	147	92	70	43	20	3	0	30·2	25 Feb.	132	81	52	29	14	5	0	46·0	4 Aug.	282	180	130	75	40	10	1
Tay																									
Dundee	103	67	50	26	9	0	0	22·6	31 Oct.	100	66	48	32	14	4	1	50·0	9 Apr.	196	133	97	56	23	5	1
Ardtalnaig	125	99	71	46	19	4	0	29·2	22 Nov.	115	82	56	39	15	0	0	23·4	26 Apr.	252	186	128	83	34	3	0
Faskally	113	81	52	29	8	0	0	23·9	4 Oct.	105	76	48	31	12	5	0	40·1	20 June	227	170	104	61	22	5	0
Perth	108	84	54	28	8	1	0	43·2	31 Oct.	106	75	50	32	17	1	0	31·5	3 Aug.	215	157	105	58	25	2	0

TABLE 4 (*continued*)

Frequency distributions of daily amounts

Numbers of days with rainfall amounts reaching or exceeding specified values (in millimetres)

Area and station	October 1965–March 1966									April–September 1966									Calendar year 1966						
								Max. fall									Max. fall								
	0·25	1·0	2·5	5	10	25	50	Amount	Date	0·25	1·0	2·5	5	10	25	50	Amount	Date	0·25	1·0	2·5	5	10	25	50
								mm									*mm*								
Forth																									
Stirling	103	82	56	38	15	1	0	39·6	31 Oct.	100	73	56	34	14	1	0	31·2	3 Sept.	209	156	109	69	28	2	0
Lothians																									
Edinburgh, Blackford Hill .	87	58	38	17	5	0	0	22·2	31 Oct.	79	67	45	26	11	3	1	55·3	13 Aug.	158	121	84	41	16	3	1
Tweed																									
Lauder	119	75	47	31	13	1	0	26·4	31 Oct.	104	78	50	35	12	2	0	37·8	3 Aug.	232	161	105	69	21	3	0
Wauchope	107	74	48	31	9	0	0	20·8	31 Oct.	103	74	53	35	15	3	0	35·1	3 Aug.	219	161	105	69	25	3	0
Pallinsburn	93	55	36	22	7	2	0	26·2	17 Nov.	82	61	39	17	6	2	1	56·4	13 Aug.	180	125	88	42	17	3	1
Foyle and north-western streams																									
Edenfel	130	106	75	49	19	0	0	24·4	31 Oct	116	88	63	38	10	0	0	18·8	9 June	259	200	137	85	31	0	0
Lough Foyle, east																									
Banagher Filters . . .	139	112	87	63	27	1	0	30·5	31 Oct.	118	86	65	40	15	1	0	28·7	9 June	266	206	157	109	48	2	0
Bann and Lough Neagh																									
Armagh	123	93	65	38	12	1	0	25·7	24 Nov.	113	83	52	29	10	1	0	48·3	9 June	245	178	120	67	28	4	0
Stoneyford Res. No. 1 . .	122	102	73	43	22	2	0	30·2	16 Nov.	103	79	52	32	12	1	0	36·8	9 June	229	179	126	75	35	3	0
Coleraine, The Cutts . .	124	98	75	48	11	1	0	26·9	31 Oct.	101	73	58	30	9	0	0	23·4	9 June	236	185	143	81	25	0	0
Lagan and eastern streams																									
Belfast P. Sta.	118	97	72	46	21	0	0	23·9	31 Oct.	101	72	42	29	10	2	1	55·9	9 June	223	160	109	73	30	4	1
Newry and south-eastern streams																									
Annalong Valley . . .	123	101	75	58	34	9	1	50·0	7 Feb.	113	93	68	41	20	5	0	44·4	21 Apr.	245	192	139	90	50	13	2
Erne and western streams																									
Belleek	141	115	84	43	15	1	0	44·7	6 Oct.	126	95	62	42	14	2	0	27·7	19 June	269	213	153	87	25	3	0
Channel Islands																									
Jersey Airport . . .	106	91	73	42	18	4	0	34·4	19 Jan.	82	64	44	26	11	0	0	20·3	22 June	201	164	123	62	24	4	0

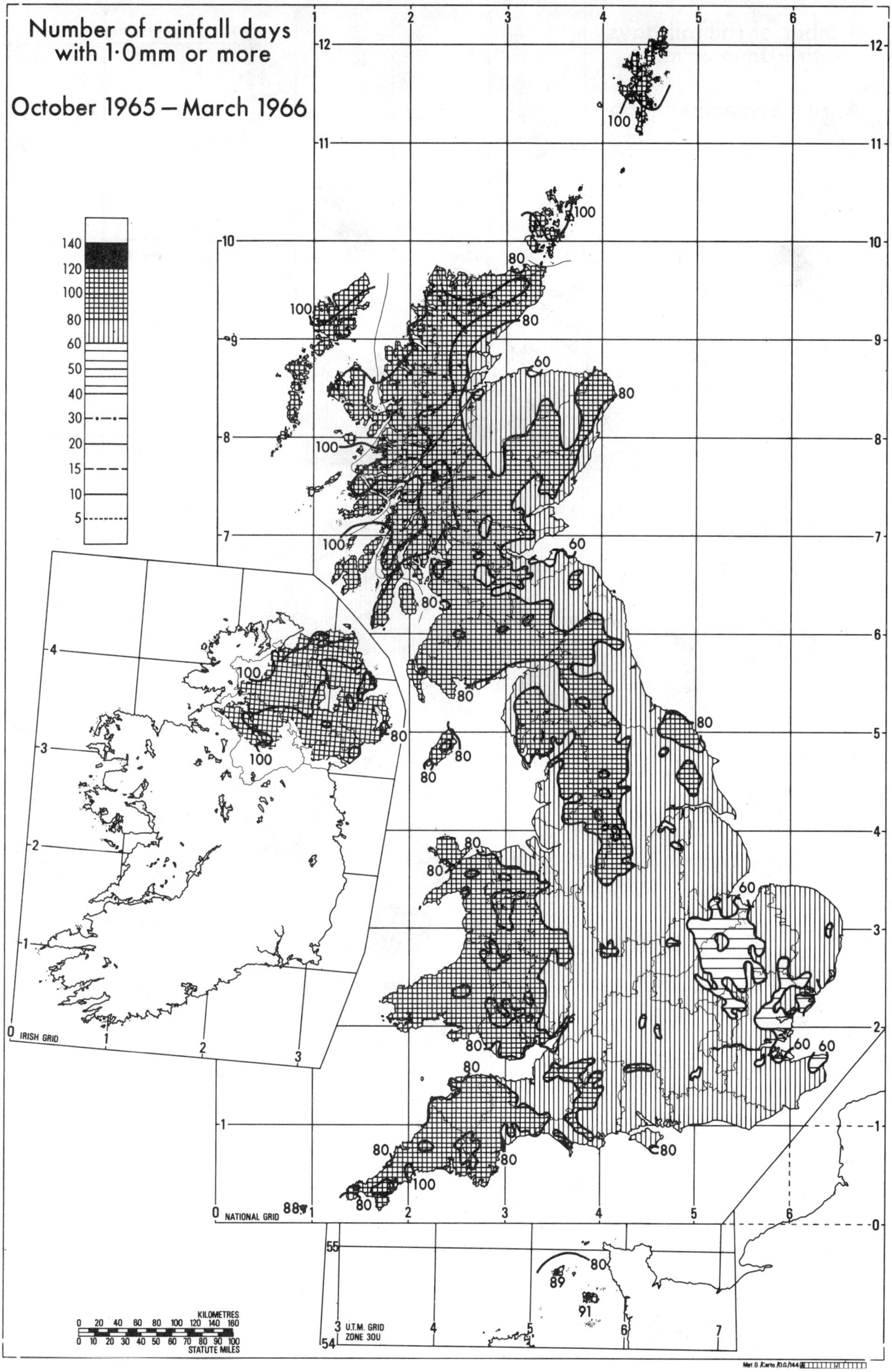
Number of rainfall days
with 1·0 mm or more
October 1965 – March 1966
140
120
100
80
60
50
40
30
20
15
10
5
IRISH GRID
NATIONAL GRID
U.T.M. GRID
ZONE 30U
KILOMETRES
STATUTE MILES

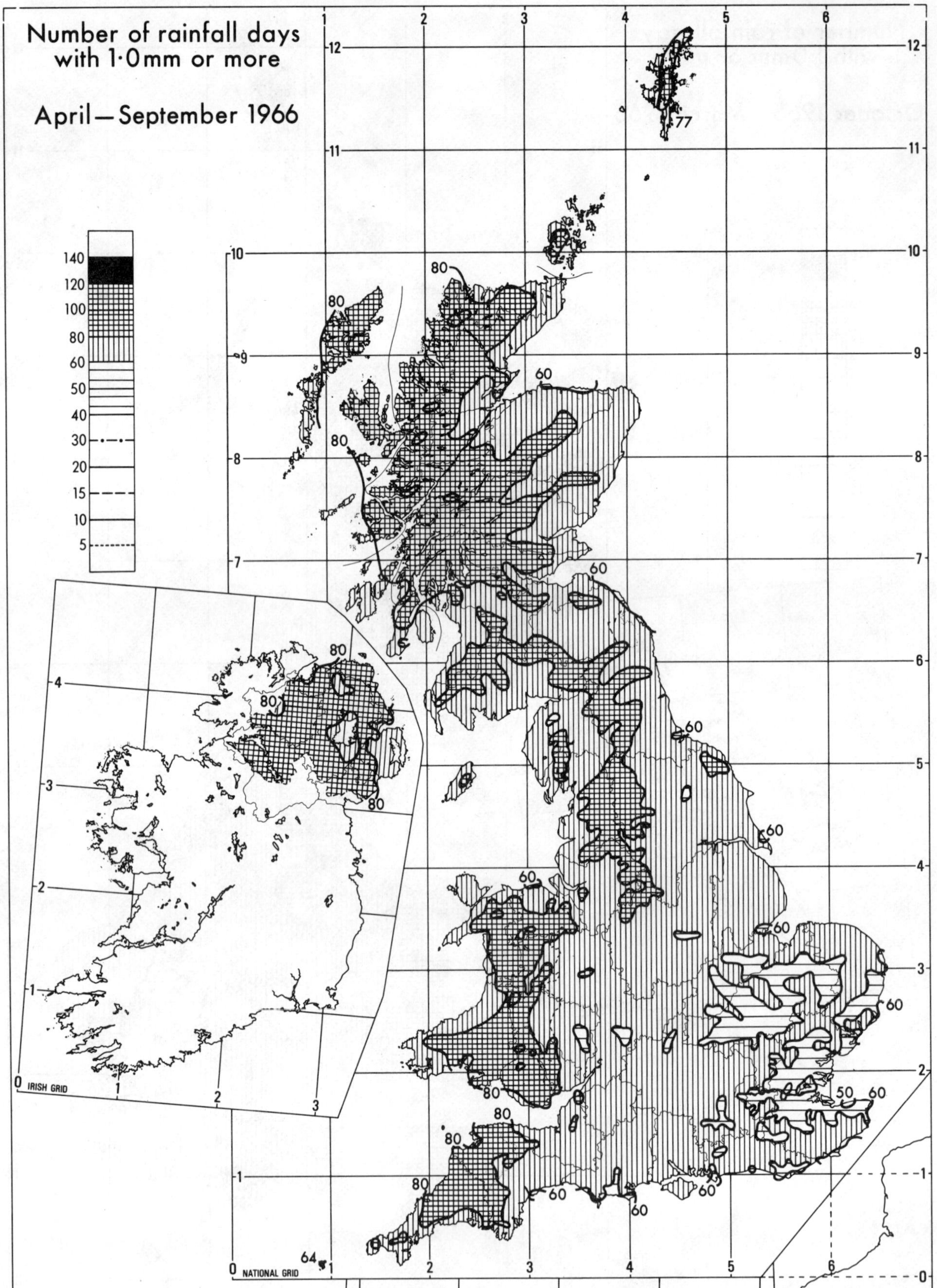
Number of rainfall days
with 1·0mm or more
April—September 1966
140
120
100
80
60
50
40
30
20
15
10
5
IRISH GRID
NATIONAL GRID
U.T.M. GRID
ZONE 30U
KILOMETRES
0 20 40 60 80 100 120 140 160
0 10 20 30 40 50 60 70 80 90 100
STATUTE MILES
Met.O./Carto./D.O./144

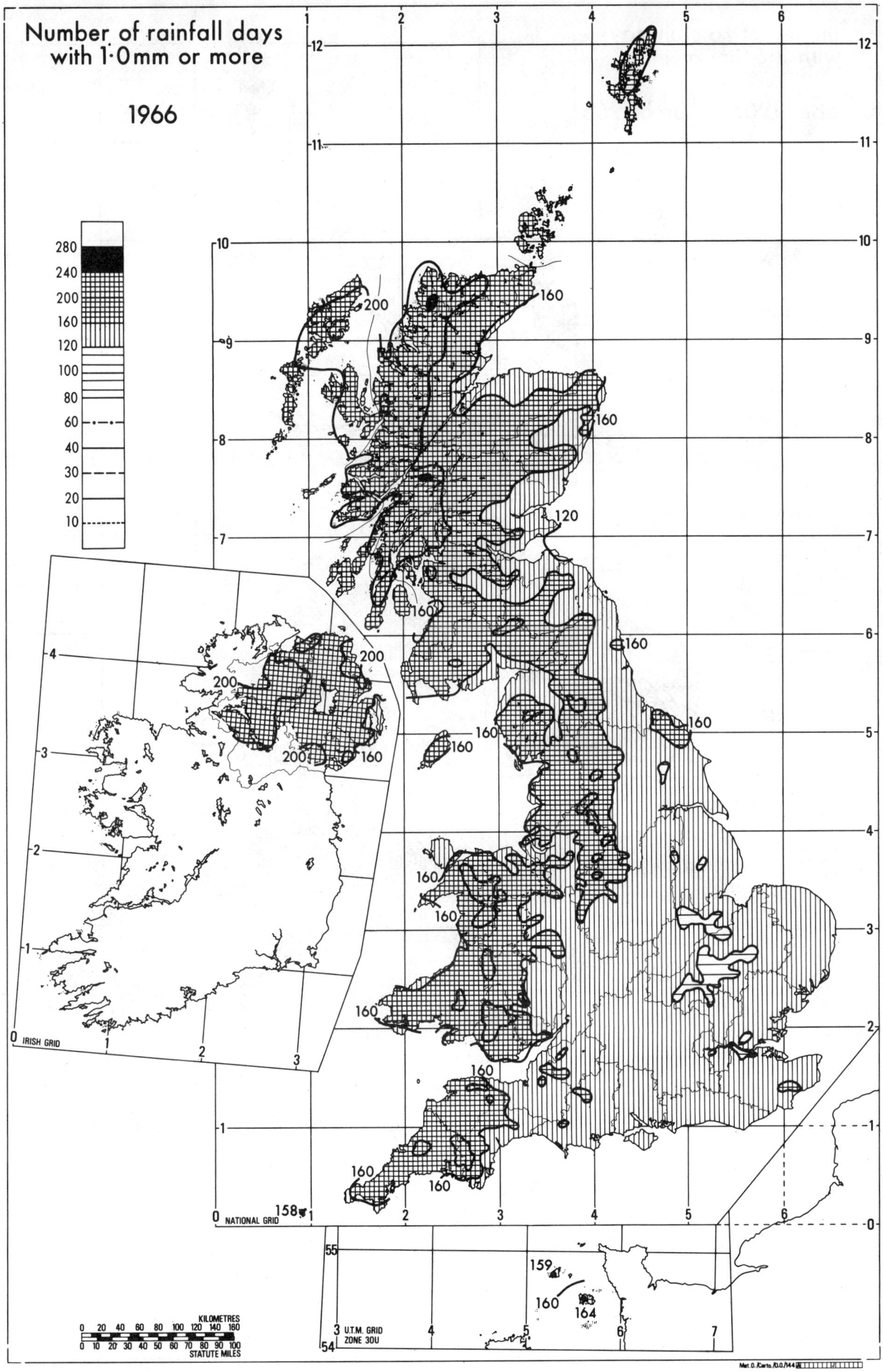
Number of rainfall days
with 1·0 mm or more
1966
280
240
200
160
120
100
80
60
40
30
20
10
200
160
120
160
200
200
200
160
160
160
160
160
160
160
160
160
160
160
158
159
160
164
IRISH GRID
NATIONAL GRID
U.T.M. GRID
ZONE 30U
KILOMETRES
0 20 40 60 80 100 120 140 160
0 10 20 30 40 50 60 70 80 90 100
STATUTE MILES

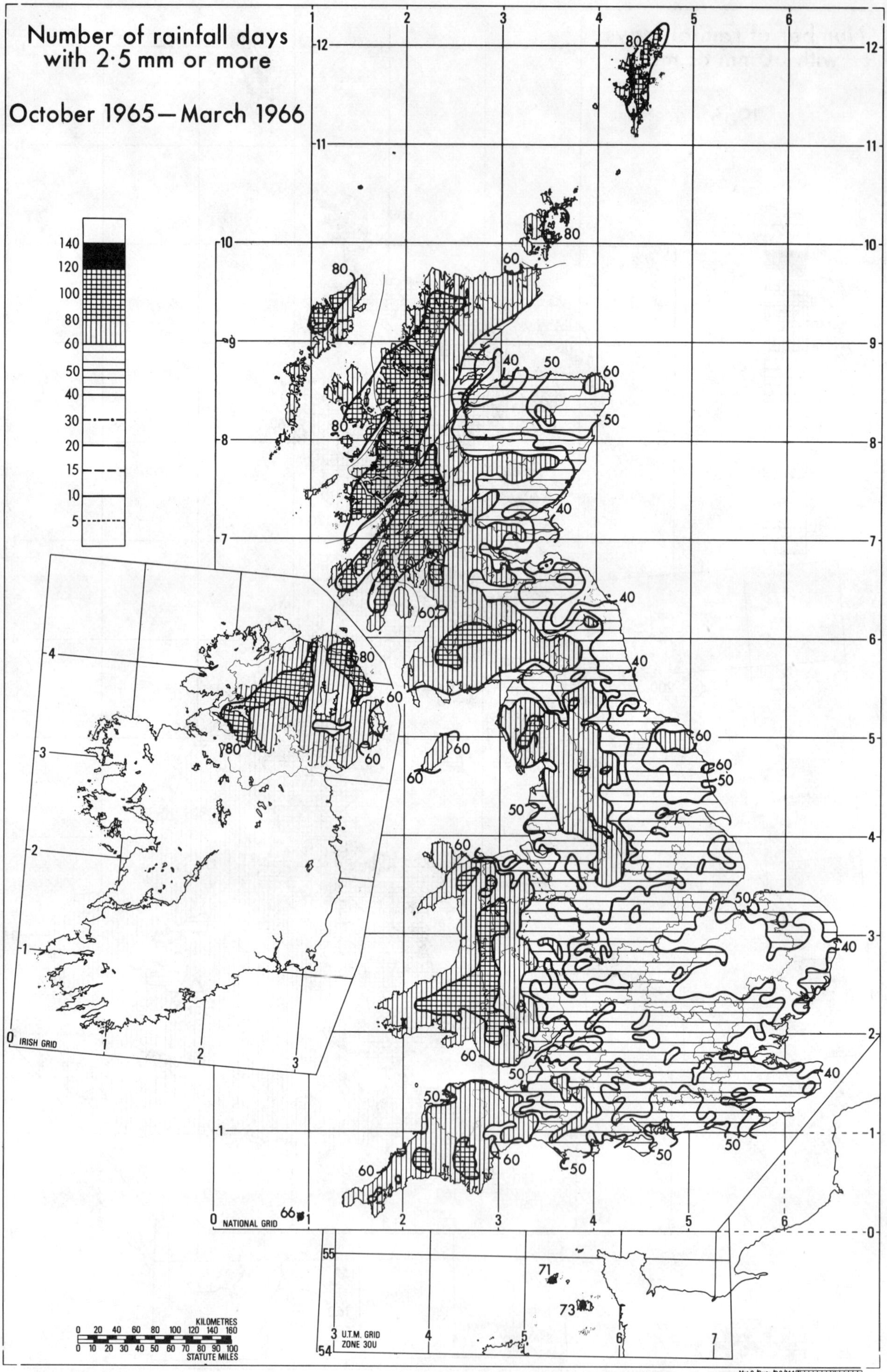
Number of rainfall days
with 2·5 mm or more
October 1965—March 1966
140
120
100
80
60
50
40
30
20
15
10
5
IRISH GRID
NATIONAL GRID
U.T.M. GRID
ZONE 30U
KILOMETRES
0 20 40 60 80 100 120 140 160
0 10 20 30 40 50 60 70 80 90 100
STATUTE MILES
Met.O./Carto./D.O./144

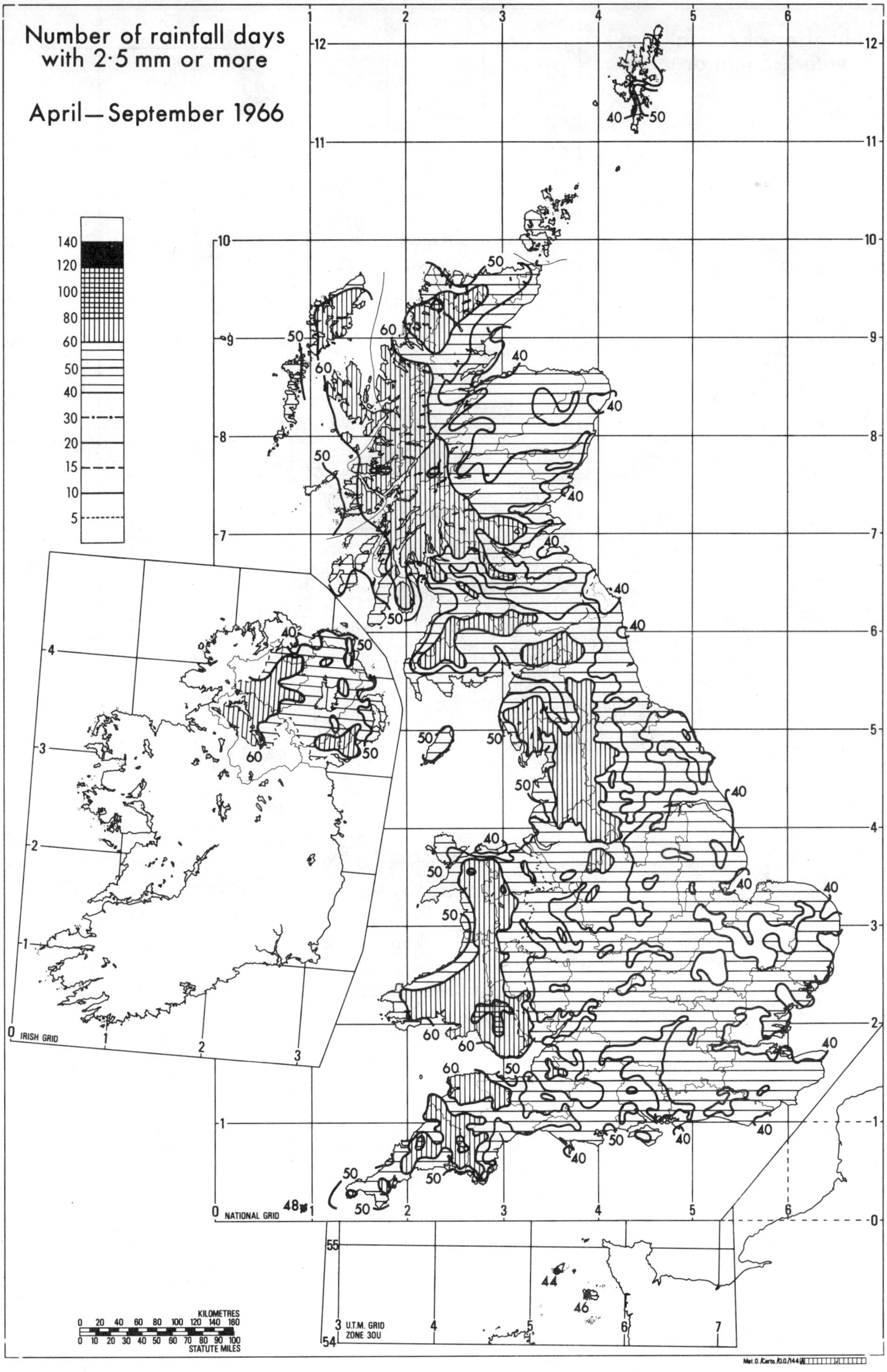
Number of rainfall days with 2·5 mm or more
April—September 1966
140
120
100
80
60
50
40
30
20
15
10
5
IRISH GRID
NATIONAL GRID
U.T.M. GRID ZONE 30U
KILOMETRES
STATUTE MILES

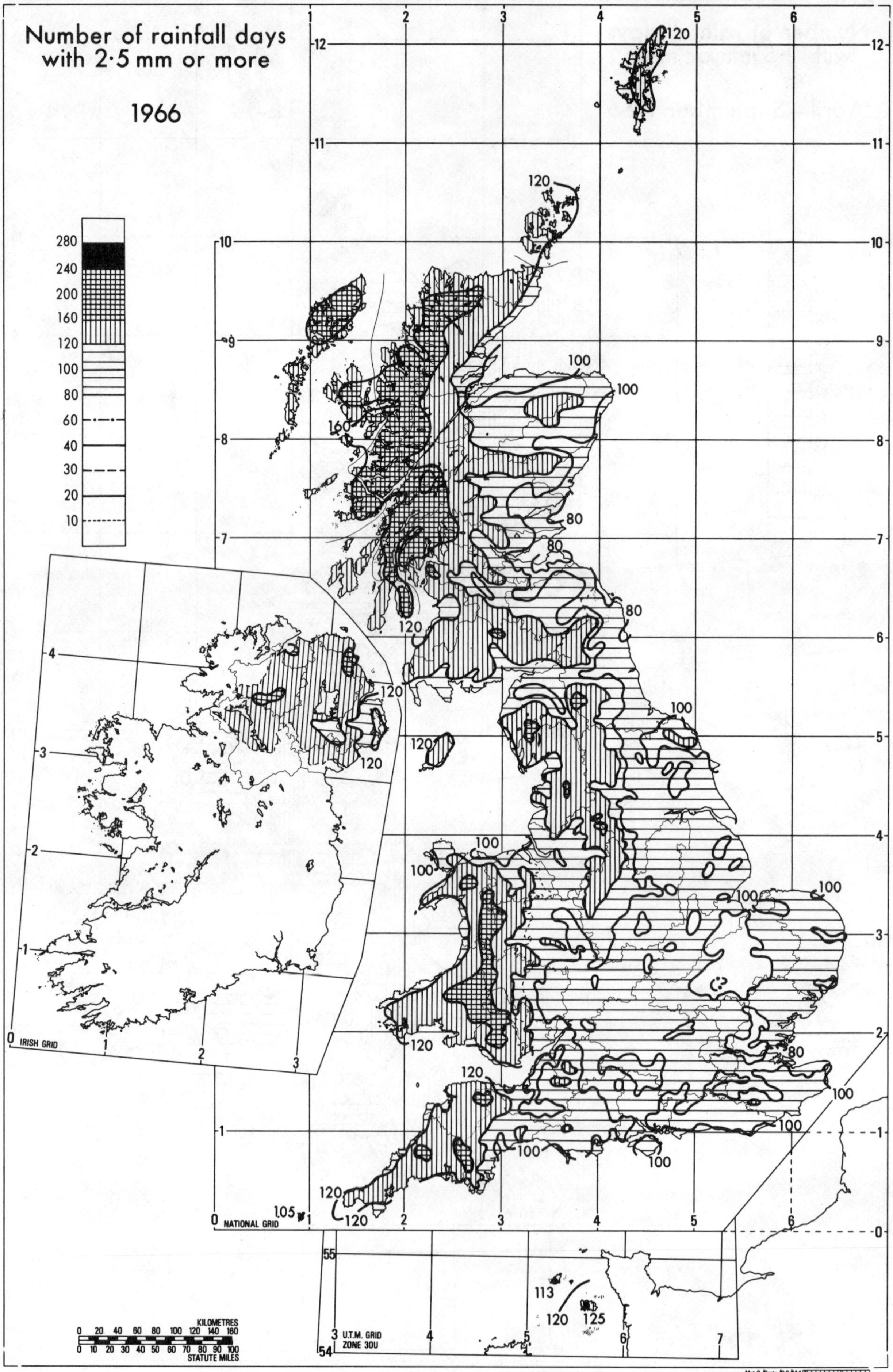
Number of rainfall days with 2·5 mm or more
1966
280
240
200
160
120
100
80
60
40
30
20
10
IRISH GRID
NATIONAL GRID
U.T.M. GRID ZONE 30U
KILOMETRES
STATUTE MILES
Met. O./Carto./D.O./144

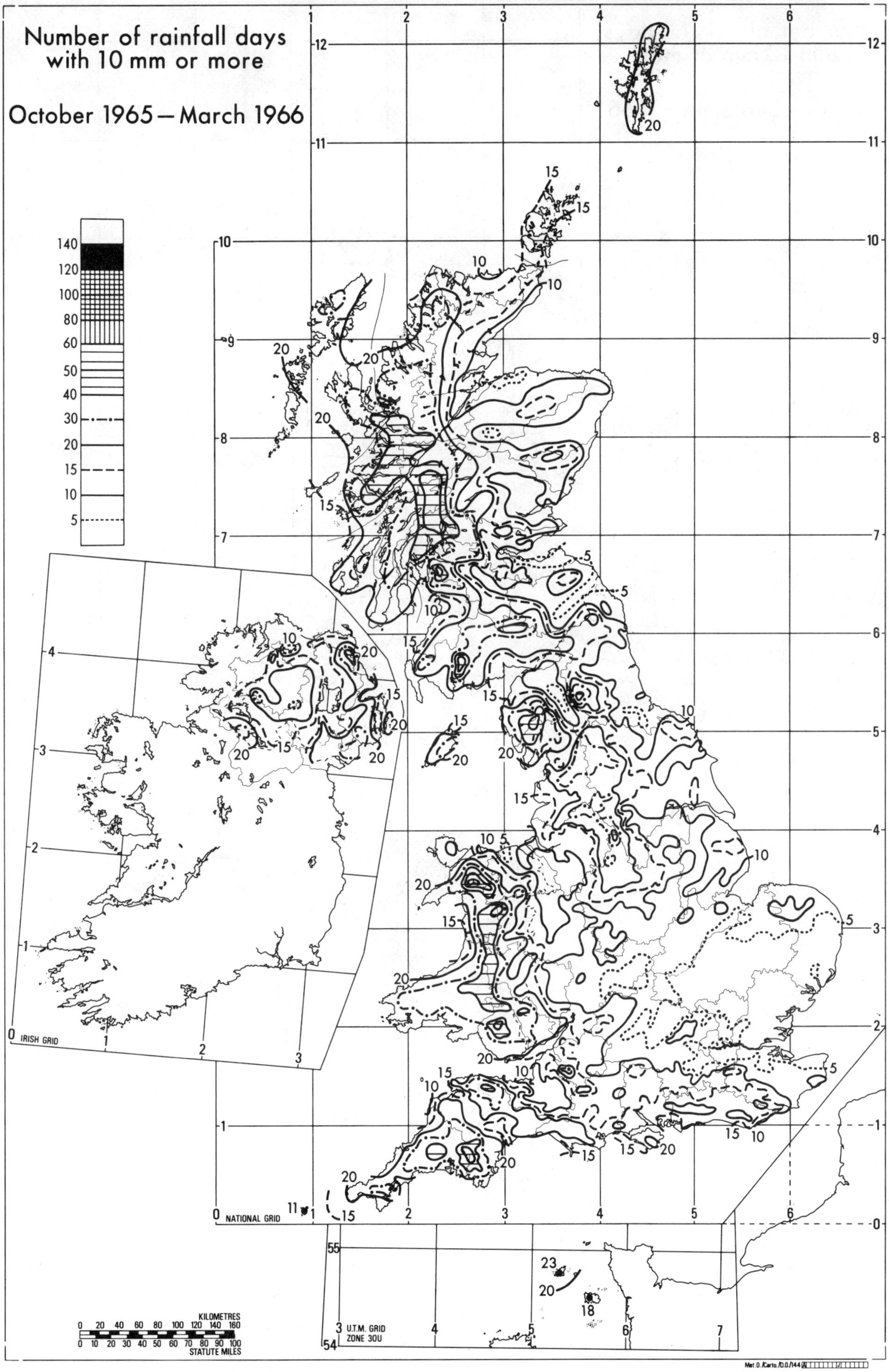
Number of rainfall days
with 10 mm or more
October 1965—March 1966
140
120
100
80
60
50
40
30
20
15
10
5
IRISH GRID
NATIONAL GRID
U.T.M. GRID
ZONE 30U
KILOMETRES
0 20 40 60 80 100 120 140 160
0 10 20 30 40 50 60 70 80 90 100
STATUTE MILES

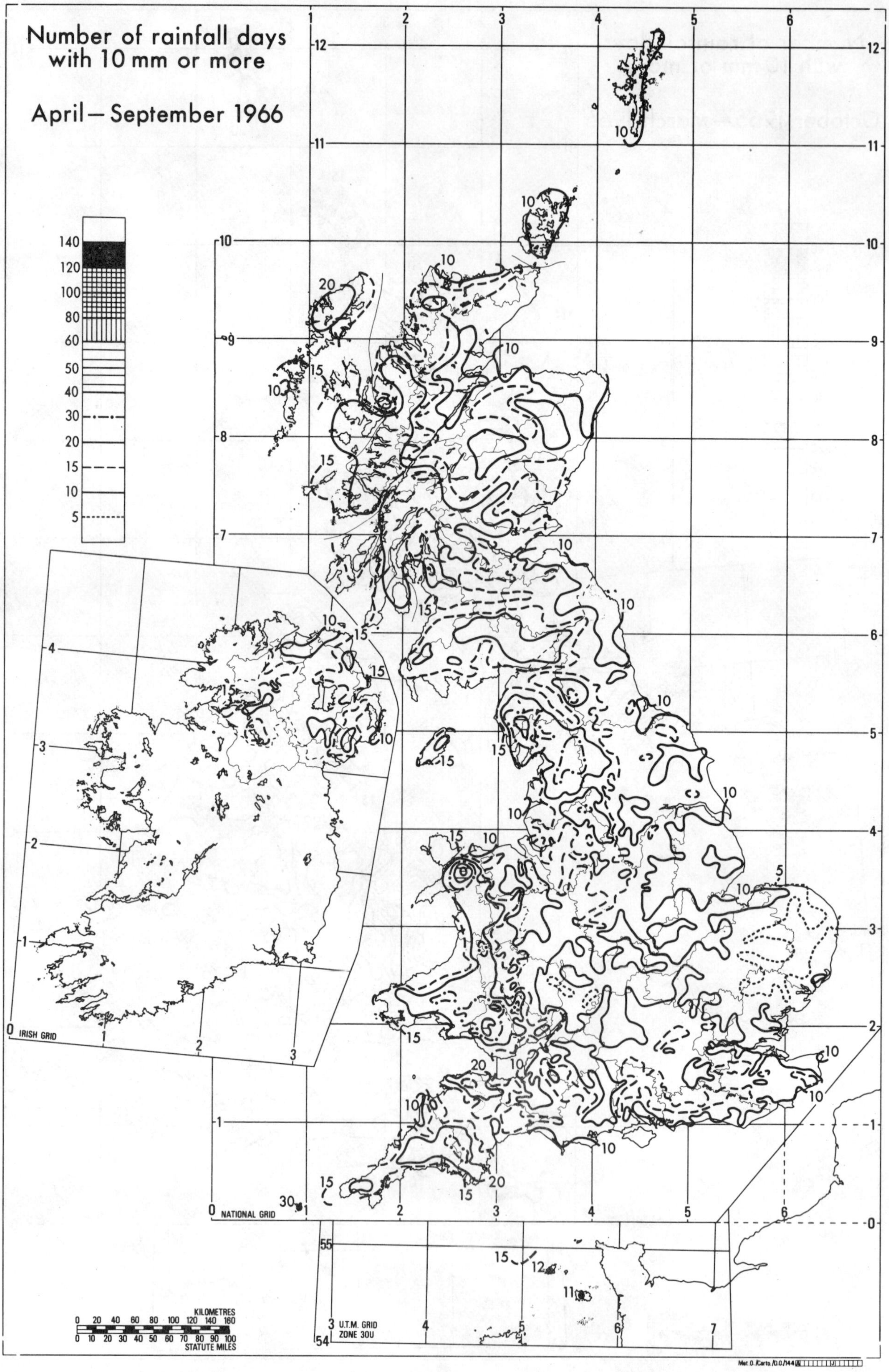
Number of rainfall days
with 10 mm or more
April—September 1966
140
120
100
80
60
50
40
30
20
15
10
5
IRISH GRID
NATIONAL GRID
U.T.M. GRID
ZONE 30U
KILOMETRES
0 20 40 60 80 100 120 140 160
0 10 20 30 40 50 60 70 80 90 100
STATUTE MILES

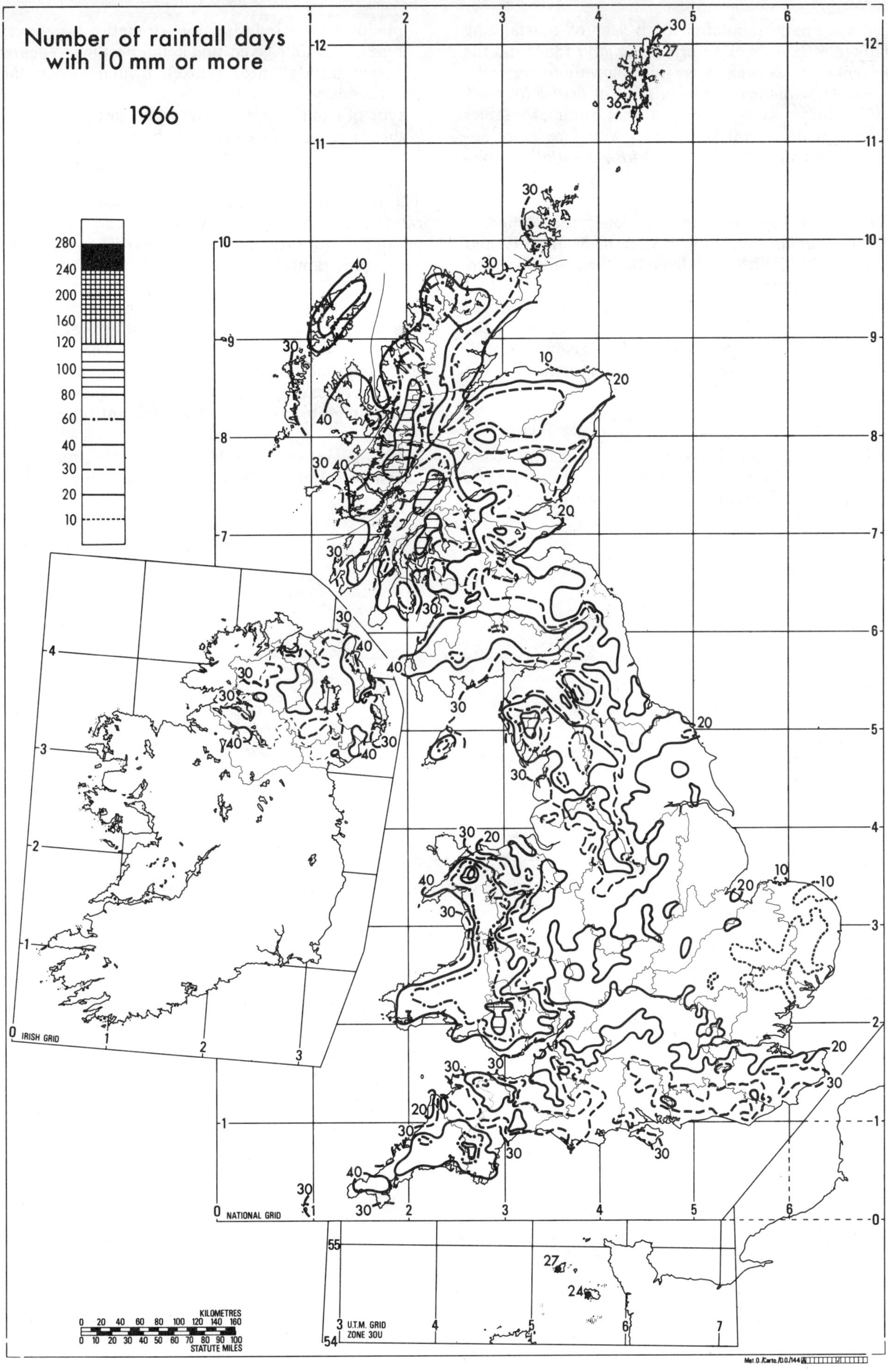
Number of rainfall days with 10 mm or more
1966
280
240
200
160
120
100
80
60
40
30
20
10
IRISH GRID
NATIONAL GRID
U.T.M. GRID ZONE 30U
KILOMETRES
STATUTE MILES
Met.O./Carto./D.O./144

5 HEAVY FALLS ON RAINFALL DAYS

The systematic tabulation each year of outstanding daily falls dates back to *British Rainfall* 1864, and the present title has been in use since the volume for 1895. However, with the new presentation of *British Rainfall*, 1961 edition, it was decided that summary tables dealing with several years' data would be more appropriately included in the *British Rainfall* supplements.

Adjustments to criteria for inclusion in this section of *British Rainfall* publications were made in 1919 and again in 1961. With effect from the 1962 edition it was considered desirable to modify the criteria so that all falls of 50 millimetres or more and at least 4 per cent of the annual average rainfall are published in the chronicle. In addition to this, all falls of 100 millimetres or more are included, irrespective of the percentage value.

An attempt has been made to introduce a means of characterizing rainfall types on the occasions listed in the chronicle by means of brief notes covering in regular order some or all of the following:

predominant wind direction, strength;
type of rainfall given under the headings—depression (rainfall near centre), frontal, orographic, thundery;
type of front—warm, cold, occlusion;
direction of movement of front, depression or frontal wave.

The broad distinctions between the various types of rainfall can be gauged by considering the falls in relation to locality and season, bearing in mind that orographic rainfall is generally characterized by large amounts but percentage values are not very high and often range from 2 to 5. On the other hand, rainfall of a thundery type has typical percentage values much closer to and sometimes exceeding 10. There is always the likelihood, however, of a combination of the different types of rainfall with the consequent difficulty of deciding objectively the predominant type.

Maps showing the areal distributions of rainfall for 11 occasions selected from the chronicle follow Table 5.

TABLE 5

Chronicle of heavy falls on rainfall days

10 January
Winds SE moderate becoming light; frontal; occlusion moving slowly east.

Devon, south	*mm*	*%*
Clyst St Lawrence	57·1	6·4
Exeter	57·0	6·7
Ide	61·2	6·9
Shillingford St George	52·1	5·7
Yellam Cottage	53·6	4·4
Clifford Bridge Farm	53·6	4·7
Fursdon, Moretonhampstead	67·8	4·9

24 January
Winds S to SE light; frontal; warm fronts moving north-east.

Devon, south		
Clifford Bridge Farm	52·3	4·6
Cornwall		
Stenalees	60·5	4·7

25 January
Wind SE moderate; frontal; occlusion moving slowly north-east over Scotland.

Ayrshire		
Middleton Filters	55·1	4·2

28 January
Winds SE strong to gale veering to S; frontal; occlusion followed by a warm front moving north-east.

Orkney and Shetland Islands		
Hamnavoe	51·6	5·3

30 January
Wind SW moderate becoming light; thundery; unstable westerly air-stream.

Northern Ireland		
Killylane Res.	58·9	4·7

7 February
Winds SW light, backing to SE becoming moderate; frontal; occlusion moving north-east.

Yorkshire Ouse and Hull		
Calf Lee	53·3	4·2
Whiteholme	51·8	4·1
Rishworth Drain	53·3	4·1
Green Withens Res.	55·4	4·4
Great Wolden Edge No. 1	57·1	4·3
Great Wolden Edge No. 2	60·2	4·5
Rishworth Moor Lodge No. 2	59·4	4·7
Rishworth Moor Lodge No. 1	61·0	4·9
Ringstone Res.	57·1	4·6
Marsden Res.	55·9	4·1
Mersey		
Ox Hey, New Site	51·3	4·0
Gin Hall	53·1	4·4
Blackstone Edge	50·8	4·0
Broadhead	55·1	4·4
Brushes Clough	53·8	4·3
Whitefield Park	50·3	4·2
Smithills Hall	51·6	4·1
Northern Ireland		
Tollymore Park	53·6	4·2
Annalong Valley	50·0	4·1

19 February
Winds SE moderate; frontal; occlusion moving north.

Northern Ireland		
Castlewellan Forest	50·8	4·5
Tollymore Park	56·4	4·4

5 March
Winds SW moderate to strong; frontal; warm front moving slowly east followed by a slow-moving cold front.

Inner and Outer Hebrides		
Broadford	97·0	4·9

1 April
Winds SW moderate becoming E veering to N later; depression, frontal; frontal depression to the west of southern Ireland moving east to the Midlands.

Dee and Clwyd	*mm*	*%*
Gwytherin	52·8	4·0

5 April
Winds SE moderate to strong; frontal; occlusion moving slowly north-east.

Ayrshire		
Hunterston	63·2	5·4
Largs	50·8	4·1

9 April
Winds E moderate; frontal; occlusion moving slowly north.

Tay		
Dundee	50·0	6·3
Stobsmuir Res.	52·6	6·9
Newport, Tayfield	54·6	7·1

21 April
Winds light W becoming SE moderate; frontal; occlusion approaching from the west.

Isle of Man		
Ballure Res.	52·1	4·0

22 May
Winds E moderate backing to NE; depression, frontal; depression west of south-west Scotland moving slowly east with associated occlusion slow moving.

Orkney and Shetland Islands		
Uyeasound, New Site	50·0	4·7

9 June
Winds light variable becoming N moderate; frontal; almost stationary occlusion over Northern Ireland.

Northern Ireland		
Spelga Dam	78·0	4·9
Hilltown Police Sta.	75·9	6·6
Rathfriland Police Sta.	55·9	4·9
Whitecross Police Sta.	55·4	5·3
Glenanne	56·5	5·6
Ballylane W. Wks	54·6	5·4
Tandragee S. Wks	51·6	6·2
Gosford Castle Forest	53·3	5·7
Newtownabbey	57·9	5·1
Belfast P. Sta.	55·9	5·8
Dromara Police Sta.	58·4	5·7
Belmont	52·6	5·5
Drumkeeragh Forest	63·5	7·6
Downpatrick, la Salle	62·2	7·2
Downpatrick, Technical School	51·1	5·9
Seaforde	52·3	4·9
Castlewellan Forest	68·8	6·1
Newcastle, Murlough Upper	82·5	7·3
Foffany Res. No. 1	94·0	6·2
Tollymore Park	95·2	7·5
Newcastle, Slievenamaddy Avenue	55·6	4·4
Poyntzpass Police Sta.	50·8	5·7
Bessbrook	56·6	5·5
Bessbrook No. 2	57·1	5·5
Newry Abbey	60·5	6·3
Newry, Bridge Street House	59·7	6·1
Newry Urban S. Wks	60·2	6·2
Narrow Water Forest	69·1	7·0

10 June
Winds light becoming NE moderate; frontal, thundery; depression over Biscay with associated occlusion moving slowly north over area.

Somerset		
Syndercombe No. 1	53·1	4·7
Bridgwater, Blake Gardens	50·8	6·5
Brymore	51·8	6·5
Cannington	58·7	7·7
Wells S. Wks	50·3	5·9

14 June
Winds S light; orographic, thundery; unstable air.

Beauly and Ness		
Inverness	58·4	8·9

TABLE 5 (*continued*)

Chronicle of heavy falls on rainfall days

	mm	%
Forth		
Forrestburn Res.	54·6	4·7

16 June
Winds light variable; thundery; unstable air.

Kent		
Margate	51·6	9·5

18 June
Winds SW light; thundery; unstable air.

East Sussex		
Race Hill Res.	55·9	6·5
Brighton, Lewes Road	54·6	7·8

20 June
Winds light variable; depression; shallow depression over area.

Northern Ireland		
Runkerry	50·8	5·0

22 June
Winds mainly light variable becoming W moderate; depression, frontal, thundery; small depression and associated cold front moving quickly north over east and south-east England.

Lincolnshire		
Cleethorpes P. Sta.	50·8	9·0
Louth, Technical School	51·1	7·9
Manby	50·3	8·7
Fordington W. Wks	50·0	7·6
Boston, Cooks Lock	52·6	8·6
Twentylands P. Sta.	51·3	8·5
Boston, Church Road P. Sta.	50·5	8·3
Tytton Hall	50·3	8·3
Wyberton	51·3	8·4
Eastville	52·1	9·3
Leverton	50·5	8·9
Welland		
Irby Hall Farm	54·9	10·1
Nene		
Sutton Bridge	55·6	9·5
Sutton Bridge, Experimental Sta.	52·8	9·0
Wingland Grange	51·3	9·8
Great Ouse		
Terrington St Clement	53·6	10·3
Potton	53·3	9·1
Potton S. Wks	51·3	9·0
St Neots P. Sta.	59·7	11·0
St Neots	58·9	10·8

23 June
Wind E light, becoming NE moderate; depression, frontal, thundery; depression and associated cold front slow moving over east coast of Scotland.

Wick Water to Conon		
Ardross, Glensax	54·6	6·9
Beauly and Ness		
Balblair	53·3	7·3
Banff, Moray and Nairn		
Cawdor Castle	61·7	7·7
Firhall	54·9	7·8
Kintessack	55·4	8·3
Drynachan Lodge	74·9	7·1
Glenferness House	74·4	8·0
Lochindorb Lodge	50·3	4·8
Forres	58·4	9·3
Kinloss	59·8	9·7
Dallas Lodge	77·7	8·7
Glenlatterach Filters	65·3	8·3
Elgin, The Bield	60·5	9·1
Elgin S. Wks	58·4	8·5
Blackhills House	56·6	7·0
Glenmore Lodge	66·8	6·6
Forglen House	61·5	7·4
Dee and Don		
Auchterless	65·3	7·7
Fyvie Castle	53·3	6·1
Udny	55·9	5·8
Glenbuchat Lodge	52·3	5·1
Littlewood Park	83·3	8·3
Insch	63·5	7·1
Aberdeen	50·5	6·0
Craigiebuckler	53·6	6·0
Northern Ireland		
Ballyhanna Forest	86·9	8·1
Cam Forest	52·1	4·5

26 June
Winds light becoming SW strong; frontal, orographic; warm front moving west.

South West Wales		
Lletty-Evan-hên	61·2	4·3
Gwynedd		
Ceinws	82·8	4·2
Abergynolwyn	80·8	4·5
Llyn Stwlan	125·0	4·8
Blaenau Ffestiniog	101·6	4·0
Pont Newydd	91·1	4·7
Trawsfynydd	88·9	4·7
Cwm Dyli	103·1	2·9
Mersey		
Springs	67·1	4·5
Smithills Hall	57·1	4·5
Lancashire		
Stonyhurst	50.8	4·2
Earnsdale Res.	57·7	4.4
Sunnyhurst Hey Res.	52·1	4·0

27 June
Wind NW strong, moderating later; frontal; occlusion moving south.

Dee and Don		
Strachan	52·1	5·5

5 July
Winds light variable becoming N light; frontal, thundery; shallow wave moving south-east over area.

Thames		
Hilfield Park Res.	63·2	9·1

31 July
Winds light variable becoming moderate NW; depression, frontal, thundery; shallow frontal depression moving east across Wales.

South West Wales		
Brawdy	72·1	6·2

3 August
Winds light variable becoming NW; depression, frontal, thundery; depression and associated occlusion moving north-east over northern England to east of Scotland with trough affecting south-west England.

Devon, south		
Throwleigh	57·9	4·5
North and South Esk		
Arbroath	56·1	8·4
Tay		
Cupar	57·1	7·7
Leuchars	62·5	9·2
St Andrews	64·8	9·1
St Andrews W. Wks, New Site	60·2	8·2
Cameron W. Wks	60·2	7·3
Forth		
Isle of May Lighthouse	51·3	9·9
Belliston	63·2	8·1
Ovenstone W. Wks	55·4	7·8
Lothians		
Greenhall School	56·1	7·6
Edinburgh, The Drum	53·1	7·7
Inveresk, Wedderburn House	60·5	9·7
St Germains	65·0	9·7
Longniddry	63·2	9·6
Elvingston	70·9	10·0
Gosford House	51·3	7·8

TABLE 5 (*continued*)

Chronicle of heavy falls on rainfall days

	mm	%
East Fortune Hospital	59·7	9·5
Hagbrae	78·7	10·7
Ford	78·5	11·3
Humbie House	53·3	6·7
Stobshiel Filters	59·7	7·3
Samuelston	77·7	11·3
West Hopes	76·2	8·2
Skedsbush	63·5	8·1
Haddington	69·8	10·7
Hailes Castle	66·5	9·8
East Linton	61·7	9·3
Whittingehame, West Lodge	81·3	11·0
Belton House	56·4	8·2
Dunbar	71·4	12·0
Broxmouth	57·4	9·0
Banns Ness Lighthouse	52·1	8·2
Tweed		
Dunglass	65·3	9·2
Lochton	51·3	7·8
Hungry Snout	77·2	9·2
Whitchester	61·5	7·2
Manderston House	51·3	6·5
Duns Castle	59·2	7·5

4 August
Winds NW moderate; depression, frontal, thundery; almost stationary depression with occlusion east of Scotland.

Banff, Moray and Nairn		
Dallas Lodge	64·0	7·2
Aultmore Lodge	76·7	9·4
Banff	62·7	8·4
Forglen House	79·0	9·5
Dee and Don		
Strichen	61·5	6·5
Fedderate Res.	77·5	8·3
Tay		
Crombie Res.	61·0	6·7
Monikie Res.	52·8	6·1
Clatto Res.	67·3	8·2
Forth		
Ovenstone W. Wks	65·3	9·2

6 August
Winds light variable; depression, frontal; shallow warm-sector depression, occluding and moving east over south coast of England.

West Sussex		
North End	52·1	5·7
West Dean Park	57·9	6·1
Chilgrove House	53·3	5·7
Isle of Wight		
Shanklin, Victoria Avenue	52·8	5·7
Shanklin	52·8	5·9
Devon, south		
Hazelwood House	52·6	4·2

9 August
Winds S light, backing to W, becoming strong; depression, frontal; frontal depression occluding and moving east-north-east to west of Scotland.

Northern Ireland		
Tollymore Park	57·7	4·5

13 August
Winds E light, backing to NW, becoming moderate to strong; depression, frontal, orographic, thundery; wave depression off south-west England, deepening and moving north-north-east.

	mm	%
Northumbrian		
Newlands	92·5	14·1
Seahouses	56·4	8·5
Newton House	65·5	9·5
Stamford	78·2	11·1
Howick Hall	51·3	6·8
Callaly Castle	54·6	6·7
Acklington	92·3	13·8
Allenheads	67·1	5·4
Whitburn	58·4	8·9
Cumberland		
Summergrove	63·5	5·0
St Bees Head	64·5	6·6
Workington, Coke Ovens	57·4	5·9
Braithwaite	74·7	4·4
High Lorton	62·2	4·4
Cockermouth Moor Res.	52·8	4·7
Stainburn Res.	57·1	4·3
Quarry Hill Filter Wks	53·6	5·5
Bromfield	58·9	6·6
Brookfield	51·1	5·9
Hutton John	50·5	4·1
Geltsdale	62·0	5·8
Isle of Man		
West Baldwin Res.	68·6	4·5
Point of Ayre	53·7	6·8
Solway		
Ae	64·8	5·1
Bank Colliery	64·0	4·8
Kirkconnel	54·6	4·3
Euchan Filters	78·2	4·1
Eliock	101·9	8·2
Drumlanrig Castle	61·0	4·4
Kettleton Filters	78·0	5·4
Dalmacallan	81·0	5·7
Maxwelton House	76·2	6·0
Glenkiln Res.	62·7	4·3
Lochrutton W. Wks	59·7	4·7
Pullaugh Burn	83·3	4·4
Drumjohn	96·3	6·8
Drumjohn, New Site	94·5	6·7
Glenlochar Barrage	50·8	4·2
Threave	55·9	4·4
Kirkcudbright	56·4	4·5
Penninghame Prison	64·0	4·8
Palnure	59·4	4·4
Ayrshire		
Barr	62·2	4·2
Kirkbride House	54·6	4·3
Garpel Burn	83·8	5·1
Mossdale Filters	88·4	6·7
Kerse Filters	64·8	4·6
Otterden House, New Site	53·1	4·6
Doonholm	55·6	4·9
Alloway, Doonbrae	49·5	4·4
Cumnock	63·5	5·4
Hillhead Water Tank	53·6	4·7
Auchincruive	64·3	6·4
Ayr Cemetery	55·9	6·1
Ladykirk Filters	66·8	6·9
Prestwick	66·7	7·1
Collenan Res.	51·3	5·3
Darvel, Burgh Yard	58·4	5·0
Arnlaird Filters	60·2	4·7
Stewarton Filters, New Site	54·4	4·2
Clyde		
New Kirkhope	95·2	4·7
New Crookburn	91·4	4·3
Sweetshaw Burn	106·7	5·9
Garls Craig	82·5	4·6
Camps Res.	66·0	5·7
Leadhills	85·3	5·1
Abington	68·1	5·4
Loch Lyoch Res.	57·1	4·5
Monksfoot	90·4	7·1
Bonnington Weir	55·6	6·3
Dunside Res.	64·3	4·7
Glengavel Res.	81·5	5·8
West Browncastle	83·1	5·9
Kype Res.	73·7	5·4
Glassford Filters	51·8	4·2
Airdrie, Broomknoll Street	56·6	6·1

TABLE 5 *(continued)*

Chronicle of heavy falls on rainfall days

	mm	%
Coatbridge, Whifflet Park	50·8	5·5
Kilsyth, Colzium House	63·5	4·8
Condorrat	55·1	5·1
South Moorhouse, New Site	58·4	4·1
Thorntonhall	51·3	4·4
Forth		
Craigannet	64·8	4·1
Cumbernauld	57·7	4·5
Forrestburn Res.	50·3	4·3
Stonerigg Filters	53·8	5·8
Lothians		
Harperrig	50·3	4·9
Harperrig, New Site	50·8	4·8
Harlaw Res.	54·9	6·2
Colinton, Firhill Tank	55·9	7·5
Riccarton	51·8	6·7
Edinburgh, Hillview Road	53·6	7·6
Clubbiedean Res.	65·8	8·0
Edinburgh, Blackford Hill	55·3	7·9
Liberton, The Inçh	51·8	6·9
Liberton, Alnwick Hill	59·4	8·1
North Esk Res.	56·9	5·3
Newhall House	69·6	6·2
Penicuik	81·8	8·6
Loganlea	69·8	7·1
Glen Cottage	69·3	7·2
Glencorse Filters	67·3	7·4
Bush House	57·1	6·3
Gladhouse Res.	57·1	6·0
Rosebery	66·8	7·8
Portmore Res.	58·9	6·2
Greenhall School	51·8	7·0
Gourlaw Filters	63·5	7·6
Edinburgh, The Drum	56·1	8·2
Inveresk, Wedderburn House	50·8	8·2
St Germains	53·6	8·0
Longniddry	59·7	9·1
Elvingston	53·3	7·5
Gosford House	56·6	8·6
East Fortune Hospital	55·6	8·8
Hagbrae	57·1	7·7
Ford	57·9	8·3
Haddington	53·8	8·3
Hailes Castle	52·1	7·7
Belton House	54·6	8·0
Tweed		
Dunglass	57·1	8·0
Fruid Dam	52·8	4·1
Victoria Lodge No. 2	53·8	4·2
Victoria Lodge	54·1	4·3
Baddinsgill Res.	56·1	4·7
Blyth Bridge	50·8	5·5
Roberton Filters	51·8	5·1
Linhope	56·6	5·5
Chillingham Barns	53·3	6·8
Middleton Hall	56·6	7·0
Mindrum	51·1	7·1
Hethpool	82·3	9·7
Pallinsburn	56·4	9·0
Tweedhill	63·0	9·2
Hungry Snout	59·4	7·1
Whitchester	55·9	6·5
Manderston House	60·7	7·7
Duns Castle	51·8	6·5

20 August

Wind light variable; depression, frontal, thundery; shallow area of low pressure with stationary front weakening over northern England.

	mm	%
Yorkshire Ouse and Hull		
Horton Park	53·3	6·0
Bradford, Peel Park	54·9	6·5
Bradford	53·6	6·3
Ringstone Res.	51·1	4·1
Halifax, Stannary	51·1	5·1
Trent		
Great Wyrley	63·0	8·6
Longford Road S. Wks	66·8	9·4
Milford W. Wks	67·8	9·2
Rugeley, Brinley Bank	87·1	12·6
Slitting Mill	80·8	10·9
Blithfield Res.	78·7	10·8
Gentleshaw	88·6	11·1
Willenhall S. Wks No. 2	67·3	10·2
Darlaston S. Wks	57·1	8·5
Bilston, The Lunt Outfall	62·5	8·9
Brownhills, Combe House	57·4	7·9
Walsall, Aldridge	50·0	7·0
Cannock Res.	55·4	7·7
Severn		
Harescombe Grange	83·8	10·2
Pitchcombe	56·4	6·7
Dudbridge	65·3	8·9

21 August

Wind light mainly NE; depression, frontal, thundery; shallow frontal depression over south-west England, occluding and moving east.

	mm	%
Great Ouse		
Potton	52·8	9·0
Boxworth, Manor House	50·0	8·3
Bassingbourn	72·9	12·4
Lee		
Welwyn, Mill Lane	84·3	13·0
Welwyn S. Wks	74·7	11·5
Digswell P. Sta.	63·0	9·8
Avon and Dorset		
Maiden Bradley House	77·7	7·8
Bristol Avon		
Bradford-on-Avon, Frome Road	54·6	6·7

29 August

Winds SE light becoming variable; depression, frontal, thundery; shallow depression over southern Ireland moving slowly east. Associated occlusion extending south-eastwards to France moving slowly north-east.

	mm	%
Trent		
Solihull, Station Road	50·0	6·6
Highters Heath Res.	59·4	7·8
Stechford	57·1	8·1
Elmdon	60·0	8·8
Chilcote	54·6	8·6
Frolesworth	59·4	9·1
Welland		
Market Harborough, Northampton Road	57·9	9·1
Market Harborough, Burnmill Road	60·7	9·6
Market Harborough, Grammar School	57·7	9·3
Great Bowden	54·6	8·7
Nene		
Hollowell	57·9	8·9
Kelmarsh Hall	64·3	8·9
Wingland Grange	55·1	10·5
Lee		
Hadham Hall	50·0	7·8
London, north bank		
Mill Hill Cemetery	53·8	8·0
Thames		
Baunton W. Wks	54·6	6·6
Purton	71·6	10·4
Shipton Lee, Middle Farm	50·3	7·9
Shipton Lee, Woodlands Farm	57·7	8·5
Grendon Underwood, Lower Greatmoor Farm	63·5	9·7
Finmerhill House	65·0	9·7
Edgcott, Prune Farm	55·9	8·4
Wantage, Kitford Gardens	52·3	7·1
Compton	66·8	9·1
Peasemore House	64·8	9·1
Bushey Heath Res.	53·6	7·9
Hilfield Park Res.	50·8	7·3
Watford, Aldenham Road	55·4	8·3

TABLE 5 (*continued*)

Chronicle of heavy falls on rainfall days

	mm	%
Watford W. Wks	54·9	8·0
Little Gaddesden P. Sta.	61·0	7·7
Ashridge Park	53·1	7·0
Hawridge P. Sta.	51·3	6·5
Batchworth P. Sta.	52·1	7·5
London, south bank		
Raynes Park P. Sta.	50·0	8·1
Rock Hill Res.	50·3	7·9
Hampshire		
Fosbury Manor	55·4	6·2
Linkenholt Manor	53·1	5·9
Conholt Park	56·6	6·3
Andover W. Wks	55·1	7·5
Andover S. Wks	59·2	7·9
Andover, Clarendon Avenue	63·2	8·3
Leckford	71·9	8·9
Sparsholt	53·8	6·3
Severn		
Stanford Res.	56·9	8·7
Kenilworth	70·6	10·0
Staverton	57·4	7·9
Dunchurch, Winton Hall	52·1	7·6
Southam, Banbury Road	52·6	8·1
Southam, Council Office	58·2	8·8
Warwick, St Nicholas Park	80·5	12·4
Heathcote S. Wks	78·5	11·7
Hatton	63·8	9·0
Kineton High School	58·4	8·7
Henley-in-Arden P. Sta.	60·5	8·7
Wroxall	105·4	14·0
Blakeney Hill Lodge	63·5	7·6

30 August
Winds light variable mainly N to NW; depression, thundery; shallow depression over eastern England moving slowly north-east.

Welland		
Tongue End	50·0	8·6
Nene		
Thorney Toll	50·0	9·0
Clough's Cross	53·1	9·8
London, south bank		
Battersea Park	81·8	13·5

1 September
Winds SE becoming light variable; depression, frontal; depression west of Ireland moving north-east and becoming complex with associated occlusion moving slowly north-east.

Solway		
Drumlanrig Castle	62·5	4·5

3 September
Wind W moderate backing to SSW becoming strong; frontal, orographic; warm front moving east followed by cold front with minor waves.

Lancashire		
Barrow-in-Furness, Hill Road	52·1	4·4
Cumberland		
Wastwater Hotel	114·3	5·0
Anthorn	51·3	5·9
Solway		
Newcastleton Forest	59·7	4·5

9 September
Winds SE light becoming SW moderate; frontal; warm front moving east followed by cold front, with wave, moving south-east.

Ayrshire		
Montgreenan	51·6	4·3

11 September
Winds light, mainly SW; frontal, orographic; cold front, with wave, moving east.

	mm	%
Cumberland		
Ennerdale	74·9	4·2

30 September
Winds light variable; depression, thundery; shallow depression with stationary occlusion along west coast of England and Wales.

Great Ouse		
Abbots Ripton Hall	50·8	9·2

1 October
Winds light becoming SE over south-west England; frontal; occlusion moving north-east over south-west England. Winds light becoming NW over Scotland; depression, frontal, thundery; frontal depression east of Scotland, moving north.

Devon, south		
Slade	65·3	4·6
Banff, Moray and Nairn		
Cawdor Castle	50·8	6·3
Kintessack	53·3	8·0
Forres	59·2	9·5
Kinloss	52·6	8·5
Dallas Lodge	53·1	5·9

3 October
Winds NE moderate in north England but S light over rest of area; frontal; slow-moving occlusion over England and Wales.

Northumbrian		
Boltshope	52·6	5·2
Presser Villa	55·9	5·8
Birkside	52·1	5·9
Buckshott	50·0	5·6
Hisehope Res.	53·3	5·9
Waskerley	50·0	5·5
Yorkshire Ouse and Hull		
Harper Hill	50·0	5·1
Kent		
Sandling Park	53·1	6·8
Fairlight Res.	57·4	7·5
Brede Sluice	50·5	6·9
Fairlight	53·1	7·2
East Sussex		
Halton Res.	59·7	7·9
Hastings, Wykeham Road	52·8	7·3
Hastings	51·6	7·1
Hastings, Newgate	52·6	7·1
Filsham W. Wks	52·1	7·1
South West Wales		
Letterston	54·1	4·2
Gwynedd		
Bodnant	51·6	5·3
Lancashire		
Heysham Harbour	57·1	5·9
Urswick	52·1	4·4
Little Urswick	57·7	4·7
Northern Ireland		
Lowtown	52·6	4·1
Annalong Valley	50·3	4·1

4 October
Winds NE light; frontal; occlusion moving slowly south-east.

Devon, south		
Stoodleigh Court	51·1	4·5
Templeton	59·7	5·7
Devon, north		
Lapford Cross	53·3	6·1

5 October
Winds SE strong; frontal, orographic; occlusion moving slowly from the west with warm front developing over England and moving north-east.

Dee and Don		
Bridge of Dye	52·6	4·6
Strachan	81·3	8·7
North and South Esk		
Inchgrundle	66·0	5·3
Invermark Lodge	61·7	5·3

TABLE 5 (*continued*)

Chronicle of heavy falls on rainfall days

	mm	%
Whitehillocks	60·7	5·0
Fettercairn	50·0	5·1
Tay		
Pitcarmick	50·3	4·8
Cardney House	51·3	5·2
Strathearn Hydro	50·5	4·9
Forth		
Glensherup Res. No. 1, New Site	50·0	4·8

12 October

Winds SE moderate to strong; frontal; occlusion moving slowly north-east.

Northern Ireland		
Bonds Mountain Forest	51·6	4·8

14 October

Winds mainly light variable; depression, frontal; northward moving wave depression developing on a cold front, and becoming slow moving over south-west England.

Cornwall		
Hendra P. Sta.	50·0	4·1
St Michael's Mount	60·5	7·0
Gulval	55·6	5·5
Penzance	55·1	5·2
Trengwainton House	61·7	5·3
Drift W. Wks	57·4	5·3
Trewidden	60·7	5·3
Bussow Moor	61·0	5·6
St Ives, Halsetown Vicarage	65·0	6·0
Rosewarne	60·2	6·2
Boswyn Res.	54·4	4·5
St Agnes	55·4	5·5

19 October

Winds S moderate becoming SW light; depression; depression over Irish Sea moving slowly northwards.

Glamorgan		
Llanishen Res.	52·1	4·9
Rhiwbina Res.	53·1	4·2
St Fagans Castle	55·4	4·5
Cogan P. Sta.	51·6	4·9
Cowbridge, Pwllwy	57·1	5·0
Tair Onen	53·3	4·7
St Mary Church	52·1	5·1
Merthyr Mawr Schwyll	53·6	5·1

22 October

Winds light variable; depression, frontal; shallow wave depression developing and moving east on almost stationary front along the English Channel.

	mm	%
Hampshire		
Holmsley	55·1	6·1
Avon and Dorset		
Corfe Mullen Plantation	53·1	6·1
Hurnbridge	57·9	7·6
Alderney Res.	51·3	6·4
Branksome, Bourne Valley	57·1	7·2
Bournemouth, Rothesay Road	54·4	6·9
Bournemouth	52·1	6·6
Bournemouth, East Avenue	58·2	7·4
Parkstone, Balmoral Road	56·4	7·2
Poole, Gasworks	54·1	7·0
Poole, Henry Harbin School	51·6	6·6
Poole S. Wks	53·3	6·5
Bloxworth Heath	57·4	6·7
Bere Regis	54·6	6·3
Trigon	54·4	6·3
Dorchester W. Wks	54·6	5·6
Kingston Maurward	53·3	5·6
Martinstown	50·0	5·1
Dorchester, Queens Avenue	56·6	6·0
Dorchester, Alfred Road	56·6	5·9
Empool P. Sta.	55·9	6·1
Friar Waddon	62·7	7·0
Weymouth	56·6	7·4
Devon, south		
Halwell	53·1	4·2
Yealmpton, Torr	50·0	4·5
Little Woodford	51·6	4·6

29 October

Winds N light to moderate; thundery; unstable air.

Kent		
Stone House	50·8	8·9
Broadstairs	52·1	8·9
Ramsgate	53·3	9·1
Sandwich, Jacobs Churchyard	54·9	8·8

4 November

Winds SE moderate becoming light; depression, frontal, thundery; cold front moving slowly north-east with depression moving over area later.

Avon and Dorset		
Fonthill House	51·8	5·4
Swallowcliffe	54·6	6·1
Maiden Bradley House	54·6	5·5
Stourhead	55·6	5·5
Penselwood	58·2	6·3
Gillingham	53·8	6·8
Stour Provost	59·7	7·8
Wincanton	66·0	7·6
Horsington	65·8	8·2
Fifehead Magdalen	55·6	7·3
Glanvilles Wootton Manor	54·9	6·1
Little Westrow	58·9	6·8
Stock House	56·6	6·7
Sturminster Newton	57·1	6·8
Boyne Hollow W. Wks	57·1	6·6
Shaftesbury	55·9	6·7
Okeford Manor	62·0	7·0
Okeford Fitzpaine P. Sta.	59·4	6·8
Shillingstone	51·3	5·9
Shillingstone, Green Hills	55·4	6·2
Bryanston	56·9	6·5
Bryanston School	52·1	5·5
Winterborne Whitchurch	57·4	5·9
Piddlehinton S. Wks	58·4	5·7
Puddletown	5·9	5·6
Milton Abbas	55·9	5·5
Bere Regis	52·6	6·1
Cattistock P. Sta.	67·1	6·5
Minterne House	52·1	5·1
Birkin House	50·3	5·3
Empool P. Sta.	52·3	5·7
Winfrith	55·9	6·3
Portland Bill	50·9	7·5
Portland	53·6	7·3
Abbotsbury, East Cottage	58·4	7·2
Abbotsbury, West End	64·0	7·9
Litton Cheney	63·0	7·3
Beaminster	55·9	5·5
Netherbury	56·6	5·7
Devon, south		
Drimpton	59·4	5·8
Somerset		
Chedington	57·7	6·1
Henley	55·4	6·3
Crewkerne Res.	58·7	6·7
Crewkerne, Lyme Road	57·7	6·5
Crewkerne School	50·8	6·2
Little Norton	53·1	6·7
Sherborne, Leafield Cottage	62·2	7·7
Sherborne, Manor House	60·5	7·8
Sherborne S. Wks	59·2	7·9
Yetminster	63·5	7·6
Sutton Bingham Res.	58·9	7·2
Sutton Bingham Res., New site	59·4	7·3
Yeovil, Preston Grove	54·9	6·9
Yeovil, Summerleaze Park School	54·1	6·9
Pen Mill S. Wks	51·6	6·2

TABLE 5 (*continued*)

Chronicle of heavy falls on rainfall days

	mm	%
Woolston	55·9	6·5
West Camel	53·3	7·0
Castle Cary	55·4	6·6
Folly Farm	73·4	7·6
Hardway	68·6	7·5
Shepton Mallet, Whitstone School	65·5	6·4
Bristol Avon		
Witham Friary Vicarage	61·0	6·9
Maiden Bradley	55·4	5·7
Gurney Slade	67·3	5·8
Pondsmead	52·3	4·7
Egford P. Sta.	54·1	5·9
Chilcompton	62·5	5·7
Downside Abbey	62·5	5·8
Ammerdown House	62·2	5·7
Hinton Blewett Manor	62·0	5·9
Southstoke	55·4	6·5
Colerne	55·8	6·7
Monkswood Res., New Site	52·1	5·8
Batheaston Res.	56·1	6·8
Charlcombe	57·4	6·9
Bath City Boys School	54·1	6·8
Bath, Priory Close	53·3	6·3
Weston Res.	57·1	6·8
Bitton	58·4	7·3
Chewton House	66·0	5·8
Sherborne Res.	55·9	5·1
Sutton Court	61·0	6·3
Conham S. Wks	55·4	6·6
Tytherington	51·3	5·8
Filton	55·8	7·1
Bristol, Hopewell Hill	57·7	7·1
Horfield	52·1	6·3
Severn		
Sharpness	52·1	6·6
Knighton, Town Hall	51·8	5·3
Cradley	56·9	7·7
Bosbury	52·1	7·3
Putley, The Sladd	52·1	7·3
Upton Bishop	51·6	7·0
Wye		
Coleford, Marions Lodge	54·9	6·3
Monmouth, Cornwall House	52·8	6·7
Llantilio Crossenny	51·8	5·7
Redbrook P. Sta.	52·1	5·6
Usk		
Crumbland Plantation	56·4	5·5
Northern Ireland		
Spelga Dam	64·8	4·1
Hilltown Police Sta.	55·9	4·9
Foffany Res. No. 1	66·0	4·3
Slieve Gullion Forest	56·1	5·8

5 November
Winds E moderate, becoming light variable; depression, frontal; slow-moving occlusion weakening with minor depression developing over south-west England later.

Bristol Avon		
Kingsfield School	53·3	6·6

11 November
Winds S moderate veering SW; frontal; warm front moving east with following cold front moving slowly south-east over area.

Clyde		
South Moorhouse	56·9	4·0

24 November
Winds SW light to moderate veering NW; frontal, thundery; secondary cold front moving south-east over area.

Clyde		
Coatbridge, Dunbeth Park	66·0	6·7

29 November
Winds SW strong veering NW later; frontal, thundery; occlusion moving quickly east over Scotland.

Clyde	mm	%
Coatbridge, Dunbeth Park	91·4	9·3

1 December
Winds SE light backing to N becoming strong; frontal, orographic; depression moving east across northern England with associated occlusion moving north.

Dee and Clwyd		
Ruabon	56·4	6·3
Penycae	54·4	5·3
Llanarmon, Dyffryn Ceiriog	52·3	4·3
Cae Llwyd Res.	56·4	5·5
Ty-Mawr Res.	55·6	5·7
Shiel, Alsh and Maree		
Applecross House	62·7	4·1
Kinlochewe	96·3	4·9
Laxford		
Glackour	69·8	4·8
Ullapool	54·9	4·4
Elphin	66·8	4·0
Kylestrome	63·5	4·3
Wick Water to Conon		
Suisgill Lodge	61·5	6·5
Uppat	50·8	6·0
Ospisdale House	54·4	6·8
Rosehall	56·4	5·3
Rosskeen	52·6	6·7
Ardross, Glensax	71·6	9·1
Assynt Filters, New Site, No. 2	63·0	6·7
Fannich Lodge	73·7	4·3
Aultdearg	68·6	5·0
Loch Luichart Manse	56·6	5·0
Bridgend, New Site	59·7	4·7
Inchbae Lodge	86·9	7·4
Beauly and Ness		
Muchrachd	62·7	4·7
Banff, Moray and Nairn		
Drynachan Lodge	53·3	4·8

2 December
Winds N strong; thundery; unstable airstream.

Laxford		
Ullapool	51·1	4·1

9 December
Winds W strong becoming light in north of area; frontal, orographic; small depression and associated warm front moving quickly south-east across area.

Severn		
Hafren Lodge	91·4	5·9
Llanidloes, Abernant	55·9	5·0
Carno	61·7	4·6
Cefn Coch	63·0	5·5
Cwm Derwen	81·3	4·7
Abercannon Farm	94·0	5·4
Wye		
Manod	120·6	5·5
Pant Mawr	121·9	5·5
Bod-talog	102·6	5·6
Abergwngu	110·0	6·0
Tynant	99·1	5·6
Nantgwillt	78·7	4·9
Tyrmynydd	76·2	4·8
Elan Village	79·2	5·3
South West Wales		
Cwmystwyth	78·7	4·5
Gwynedd		
Llanymawddwy	88·1	4·1
Dylife	83·1	4·4
Rhydymain	83·8	4·3

12 December
Winds NW moderate to strong; depression, frontal; depression extending north-west to south-east England, moving east with associated occlusion moving south.

TABLE 5 (*continued*)

Chronicle of heavy falls on rainfall days

	mm	%
Somerset		
Lower Thorne	84·8	5·3
South West Wales		
Abergorlech	63·8	4·1
Swyddffynnon	55·4	4·2
Lampeter, Grammar School	62·7	5·1
Bwlch-y-Graig	82·0	6·6

16 December

Winds SW light becoming strong; frontal, orographic; warm fronts moving east across Scotland.

	mm	%
Sheil, Alsh and Maree		
Kinlochewe	119·4	6·1
Glackour	61·7	4·3
Naver and Thurso		
Gobernuisgach Lodge	86·1	4·3
Wick Water to Conon		
Cassley Power Sta.	105·4	5·9
Luib	74·2	4·3
Aultdearg	73·7	5·4
Loch Luichart Manse	81·3	7·2
Corriefeol	59·2	4·2
Inchbae Lodge	53·8	4·6

17 December

Winds SW strong veering to NW; frontal, orographic; cold front moving south-east over the British Isles.

	mm	%
Yorkshire Ouse and Hull		
Busker Beck	61·5	4·5
Lee Gate House	59·2	4·5
Ayrshire		
Stewarton Filters, New Site	58·4	4·5
Dunlop	61·2	4·9
Beith Filters	66·0	4·5
Dalry Filters	57·1	4·1
Clyde		
Lochwinnoch S. Wks	67·3	4·1
Glentyan House	80·0	5·1
South Moorhouse	69·6	4·9
South Moorhouse, New Site	63·0	4·4
Neilston Filters	60·5	4·1
Barrhead	52·1	4·5
Paisley	50·0	4·5
Parkhill Filters	54·6	4·1
Add, Awe and Etive		
Dalness	199·1	6·5
Lochy and Linnhe		
Blackwater Dam South	171·7	7·1
Achnacarry House	88·9	4·3
Mucomir Generating Sta.	77·0	4·3
Fersit	118·1	6·9
Sheil, Alsh and Maree		
Achnashellach	92·2	4·5
Inner and Outer Hebrides		
Broadford	150·6	7·6
Wick Water to Conon		
Cassley Power Sta.	89·9	5·0
Luib	76·2	4·4
Loch Luichart Manse	82·3	7·3
Corriefeol	63·2	4·5
Inchbae Lodge	63·2	5·4
Beauly and Ness		
Affric Lodge	92·2	5·7
Cannich	65·8	5·5
Mullardoch Dam	121·4	7·0
Muchrachd	84·3	6·4
Kingie Camp	97·3	4·5
Forth		
Cumbernauld	51·6	4·1

19 December

Winds SW strong veering later to N; frontal, orographic; cold front associated with wave depression moving south-east over Britain.

	mm	%
Northumbrian		
Fishlake	72·6	4·8
Lancashire		
Stocks No. 10	71·1	4·5
Buttock	70·1	4·7

28 December

Wind S light becoming SW strong; frontal; warm front moving east with associated cold front moving slowly south-east.

	mm	%
Cornwall		
Caradon Hill	67·1	4·3

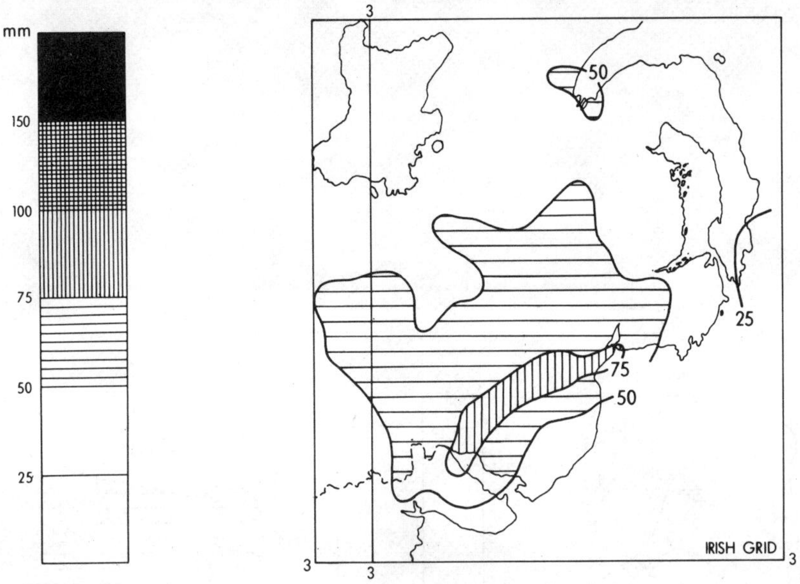

Rainfall (mm) for the 24 hours beginning 09 GMT, 9 June 1966: County Down, Northern Ireland

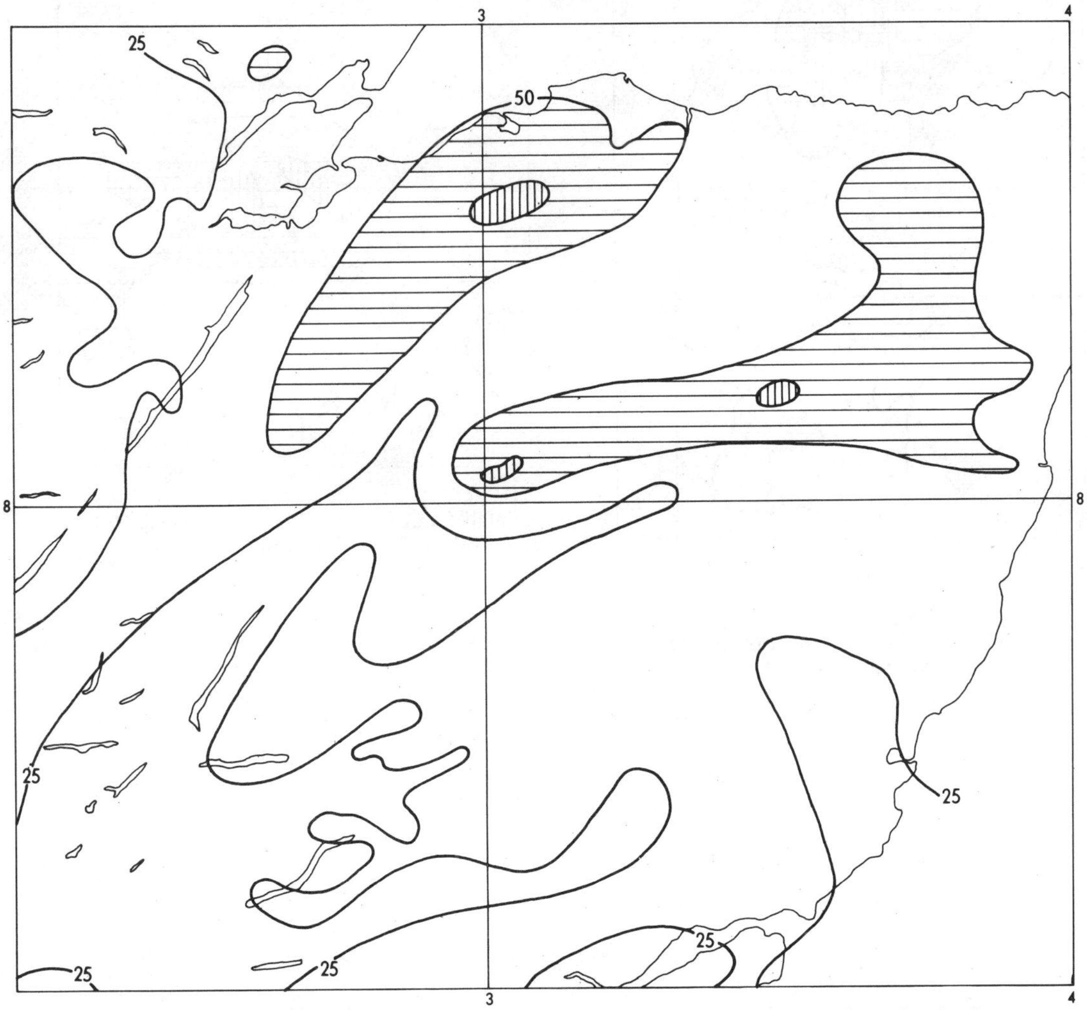

Rainfall (mm) for the 24 hours beginning 09 GMT, 23 June 1966: central and east Scotland

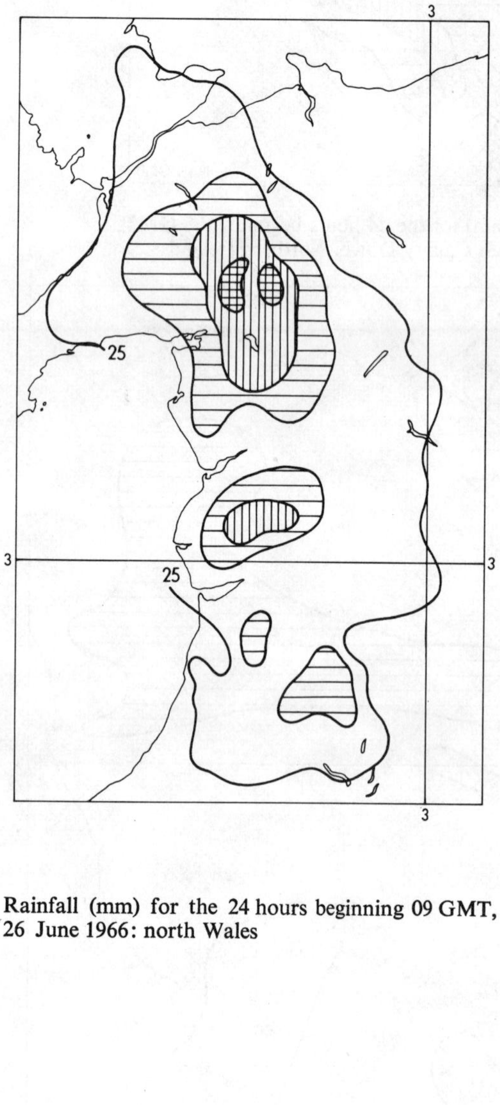

Rainfall (mm) for the 24 hours beginning 09 GMT, 26 June 1966: north Wales

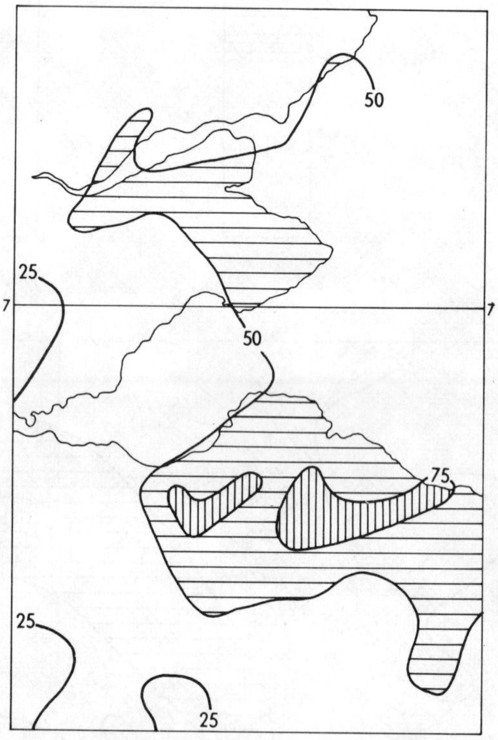

Rainfall (mm) for the 24 hours beginning 09 GMT, 3 August 1966: east Scotland

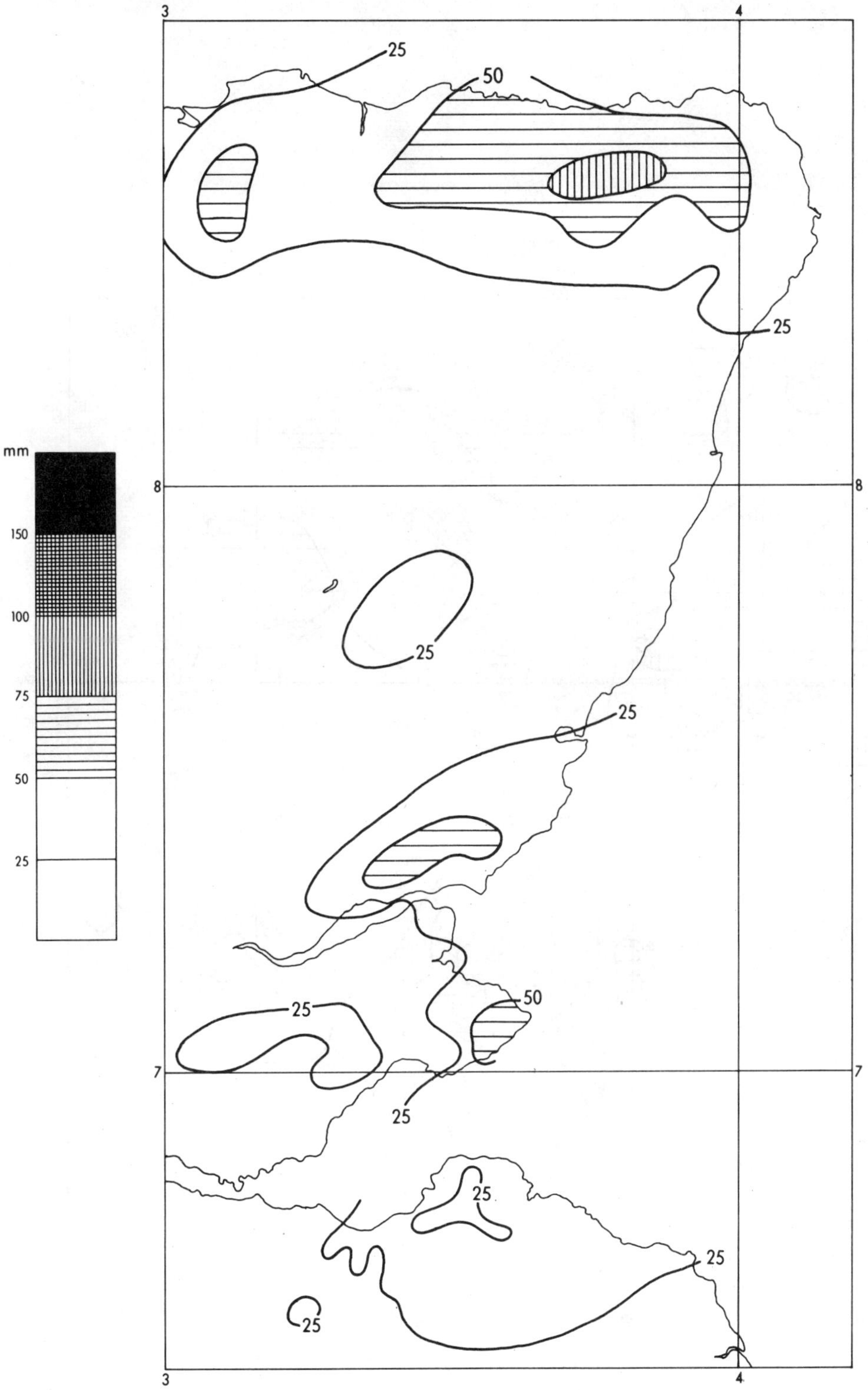

Rainfall (mm) for the 24 hours beginning 09 GMT, 4 August 1966: east Scotland

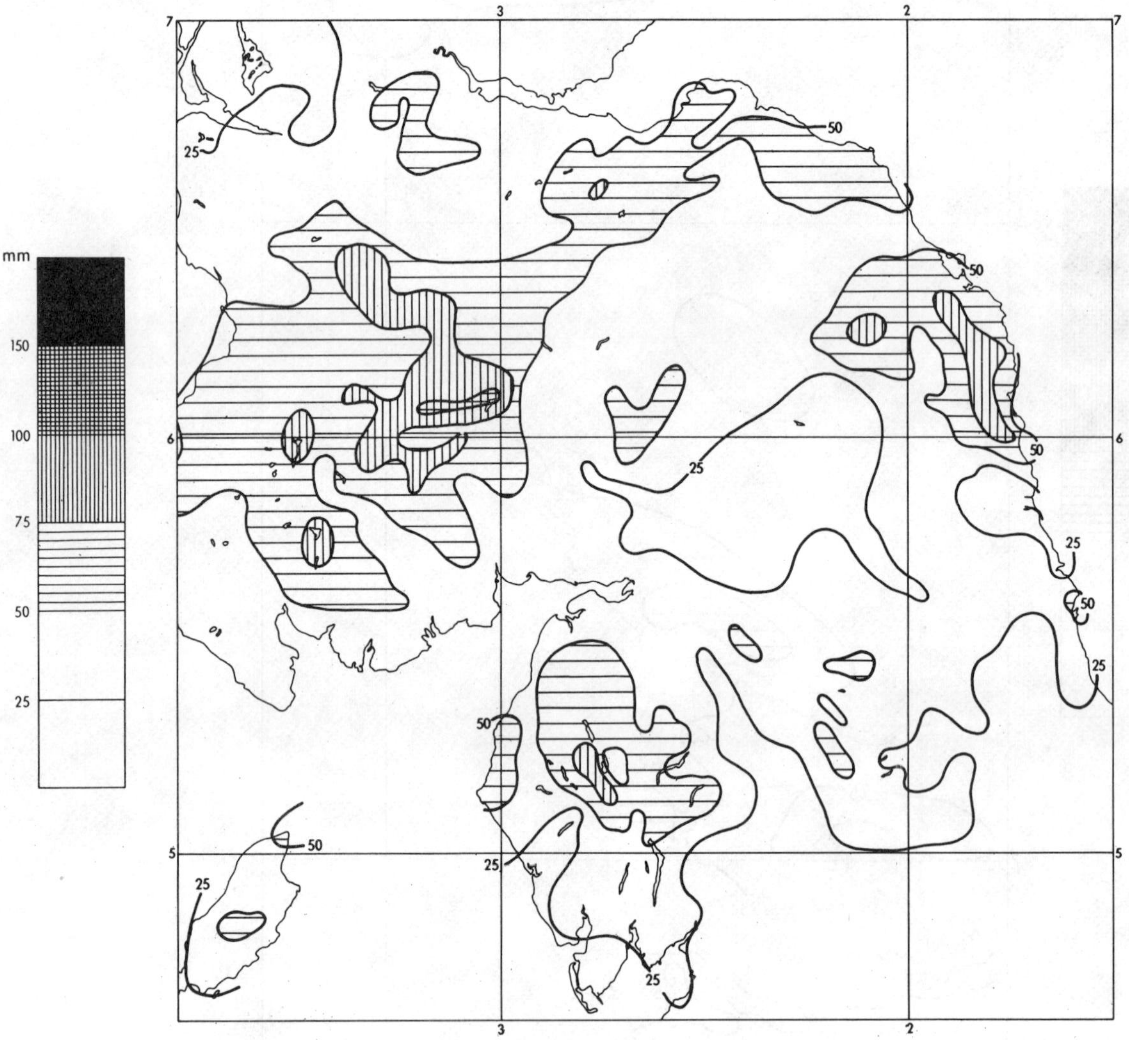

Rainfall (mm) for the 24 hours beginning 09 GMT, 13 August: north England and south Scotland

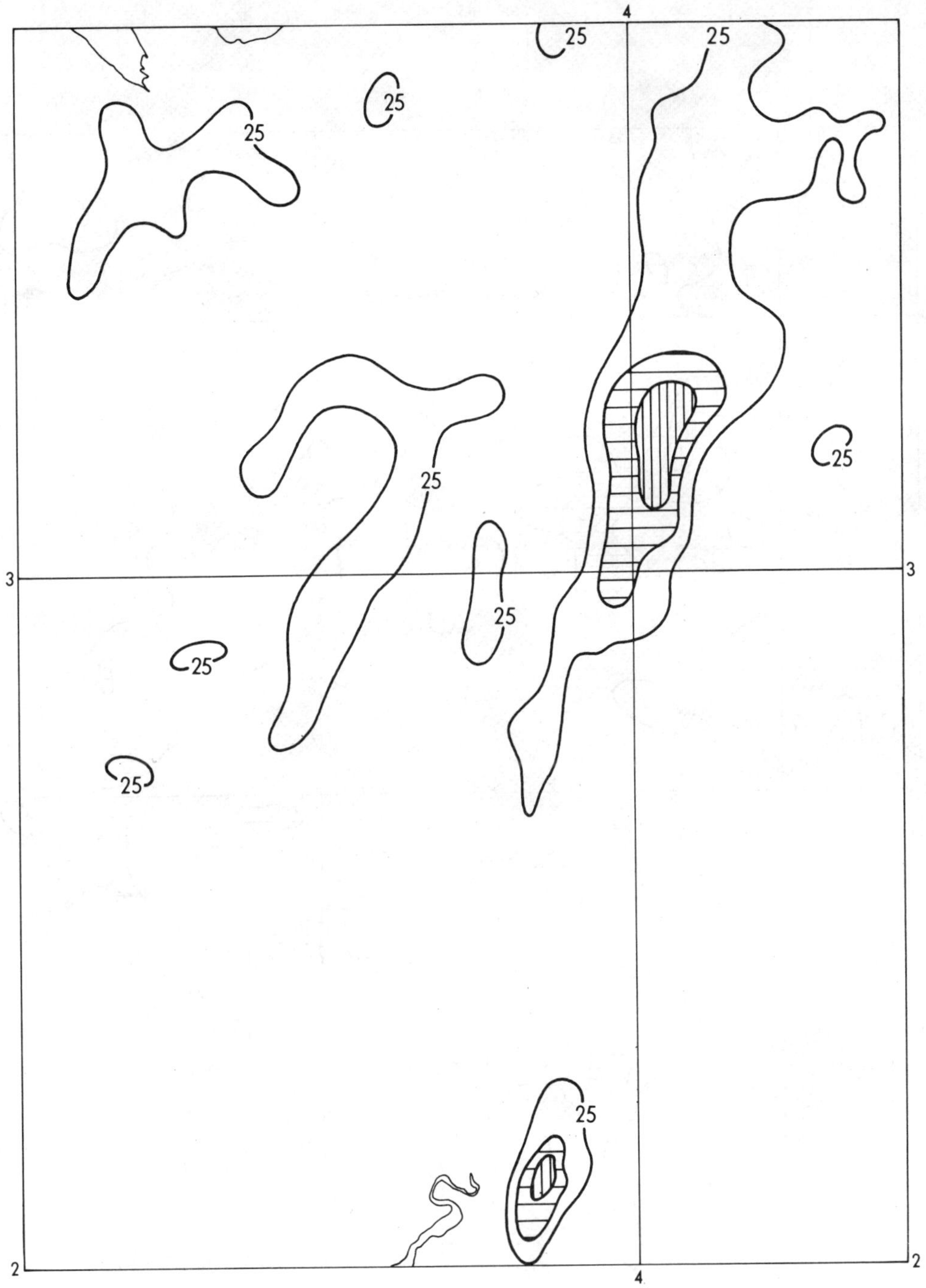

Rainfall (mm) for the 24 hours beginning 09 GMT, 20 August 1966: Midlands

Rainfall (mm) for the 24 hours beginning 09 GMT, 29 August 1966: Midlands to south coast

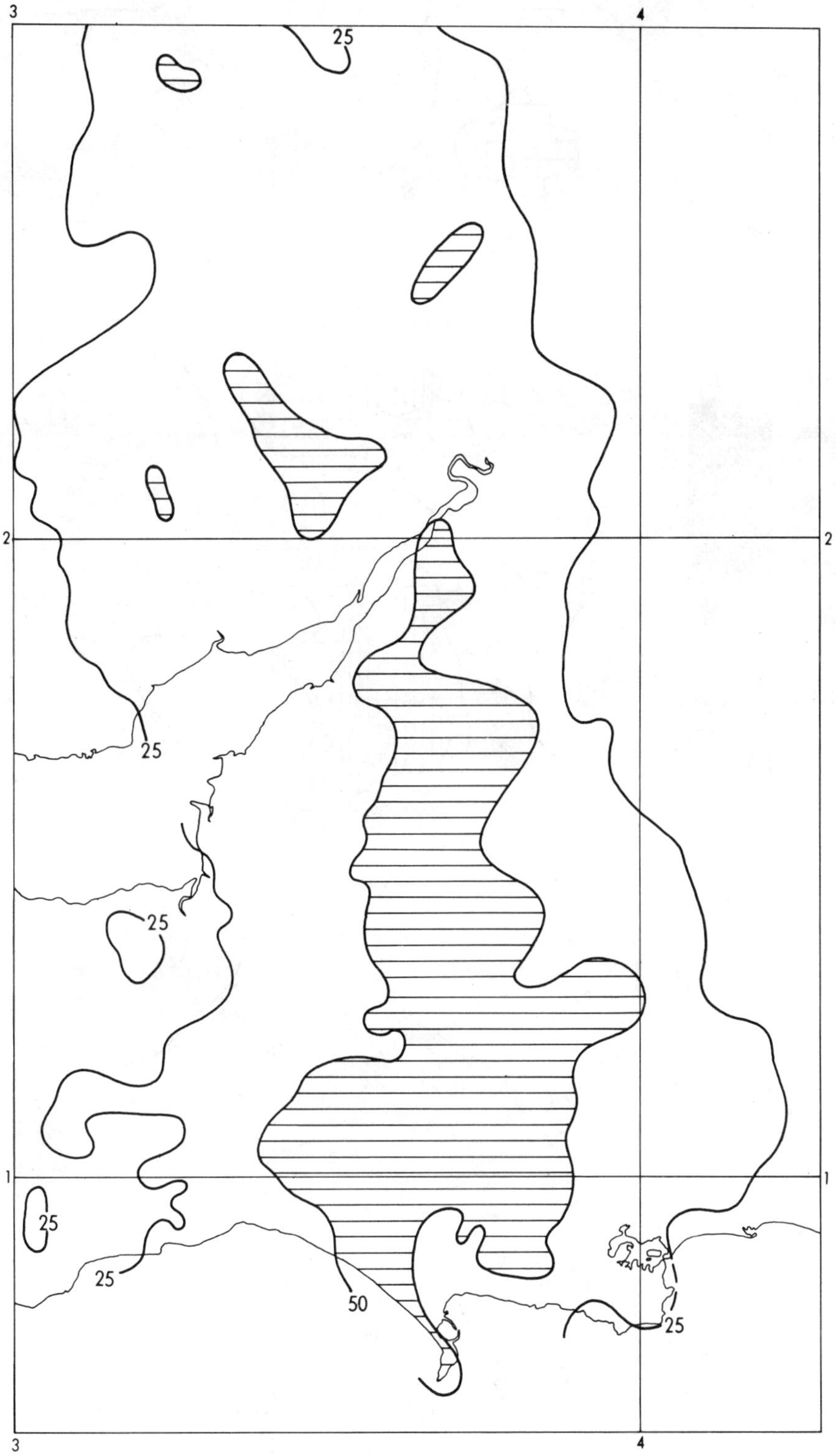

Rainfall (mm) for the 24 hours beginning 09 GMT, 4 November 1966: south-west Wales to English Channel

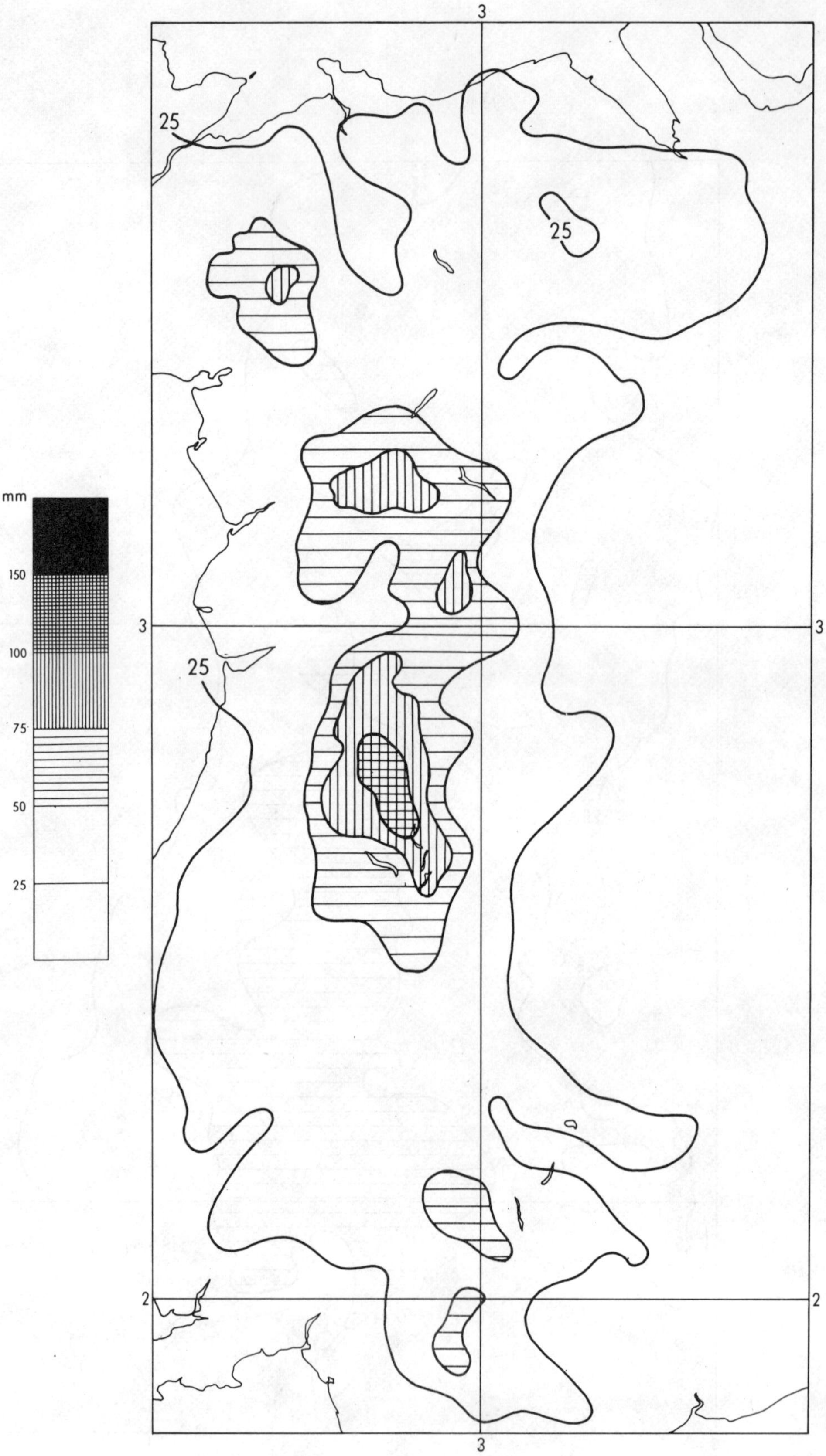

Rainfall (mm) for the 24 hours beginning 09 GMT, 9 December 1966: Wales

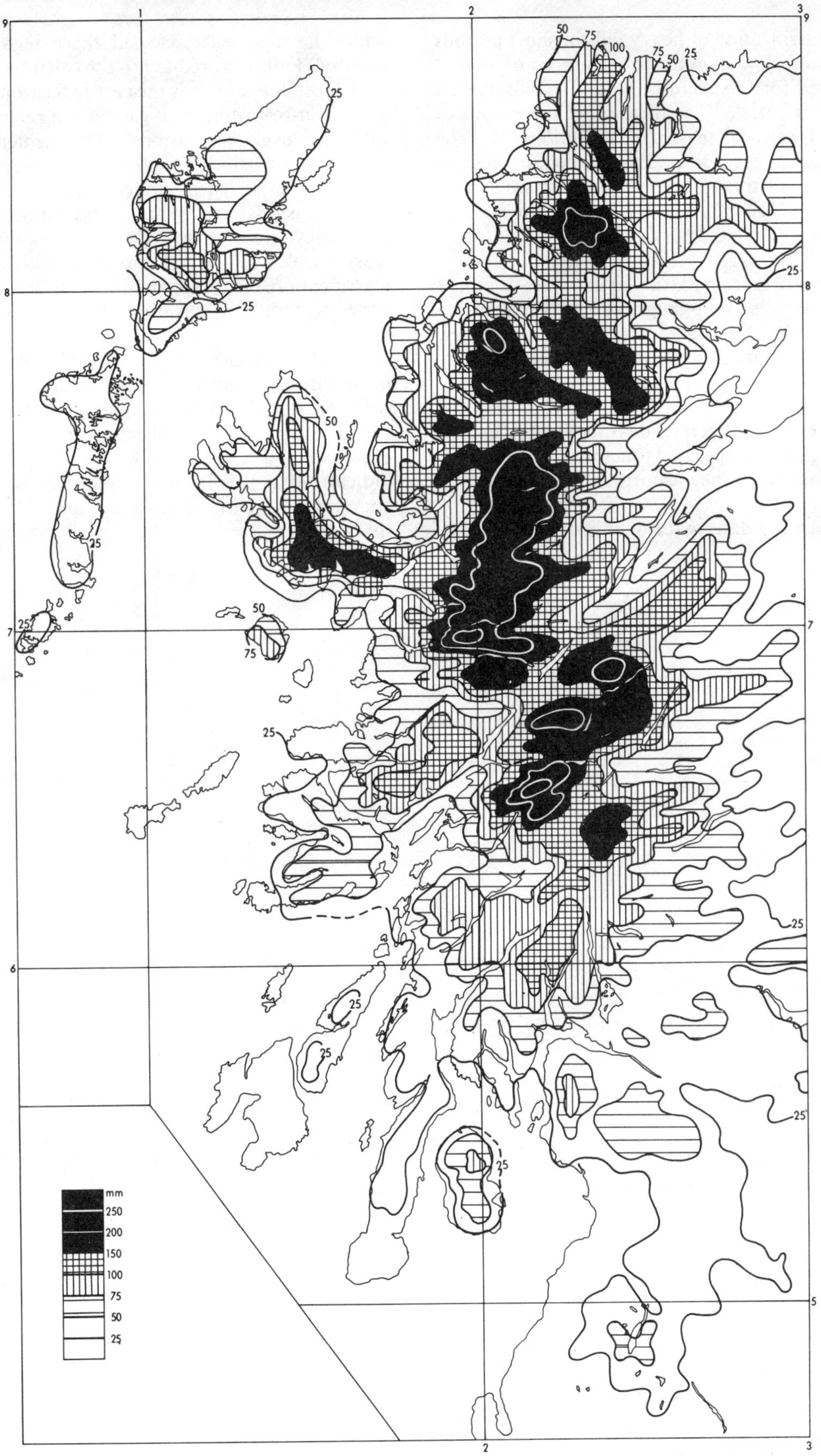

Rainfall (mm) for the 48 hours beginning 09 GMT, 16 December 1966: north and west Scotland

6 HEAVY FALLS IN SHORT PERIODS

A basis of classification of heavy falls in short periods (ranging from a few minutes to a number of hours) was developed for this country by E. G. Bilham and published in an article 'Classification of heavy falls of rain in short periods' in *British Rainfall* 1935. The current position regarding the frequency analysis of intense falls of this type has been discussed in an article by D. J. Holland in Part III of *British Rainfall* 1961.

Details of intense falls on particular dates are obtained either from recording rain-gauges or from appropriate multiple readings of standard non-recording gauges, particularly those intended primarily for daily observations.

Table 6A gives the numbers of days in which specified amounts of rain fell in specified times without reference to the dates on which they occurred.

Table 6B gives the dates with details of known falls which have at least reached the category of 'noteworthy'. Falls are credited to the calendar day and not to the 'rainfall day' as is the case in Section 5. Readings from non-recording rain-gauges are referred to in the table as 'eye observations'. To augment these eye observations, similar information from recording rain-gauges is listed under 'autographic records'. The occasions are summarized under the broad categories 'noteworthy' (N), 'remarkable' (R) and 'very rare' (VR) in accordance with the Bilham classification. The information is listed in ascending order according to duration of fall.

Table 6C is supplementary to Table 6B and contains information concerning which there is some doubt with regard either to the magnitude or to the duration of fall, or again because the duration is known to have exceeded two hours. The letter '*c*' against any item indicates that it is in doubt. Falls for which times of commencement are not known are also included in this table.

TABLE 6A

Frequency distributions of intense falls

Number of days with the threshold values reached within the specified durations

Area and station	5 mm within			10 mm within				15 mm within					20 mm within					25 mm within				
	5 min	10 min	15 min	5 min	15 min	30 min	1 hour	15 min	30 min	1 hour	2 hours	4 hours	30 min	1 hour	2 hours	4 hours	8 hours	1 hour	2 hours	4 hours	8 hours	16 hours
Northumbrian																						
Acklington	0	0	1	0	0	0	1	0	0	0	0	0	0	0	0	0	2	0	0	0	0	1
Yorkshire Ouse and Hull																						
Leconfield	0	1	2	0	0	0	0	0	0	0	0	2	0	0	0	0	2	0	0	0	1	2
Leeming	0	0	0	0	0	0	0	0	0	0	0	1	0	0	0	0	1	0	0	0	1	2
Trent																						
Keele	0	1	1	0	0	1	1	0	0	1	1	2	0	0	0	1	1	0	0	0	1	1
Elmdon	0	0	1	0	0	0	1	0	0	0	1	2	0	0	0	1	1	0	0	1	1	1
Watnall	1	3	3	0	1	1	1	0	1	1	1	1	1	1	1	1	1	1	1	1	1	3
Finningley	3	4	7	0	1	3	4	1	1	1	1	1	0	1	1	1	1	0	0	0	0	1
Lincolnshire																						
Manby	0	3	4	0	1	1	3	0	0	1	1	4	0	0	0	0	0	0	0	0	0	0
Waddington	1	2	4	0	0	0	0	0	0	0	0	1	0	0	0	1	2	0	0	0	1	1
Cranwell	1	1	3	1	1	1	1	0	0	0	0	2	0	0	0	1	2	0	0	0	1	1
Great Ouse																						
Cardington	0	0	1	0	0	0	2	0	0	1	2	3	0	0	1	1	3	0	0	1	1	2
Mildenhall	0	1	2	0	1	1	3	0	1	1	3	3	0	0	0	2	3	0	0	1	2	3
East Suffolk and Norfolk																						
West Raynham	0	0	1	0	0	0	1	0	0	1	2	2	0	0	0	2	2	0	0	0	2	3
Essex																						
Shoeburyness	0	0	1	0	0	0	1	0	0	0	1	1	0	0	1	1	1	0	0	1	1	1
Lee																						
Stansted	0	1	3	0	0	2	3	0	0	1	1	2	0	0	0	1	3	0	0	0	2	3
London, left bank																						
Northolt	0	0	1	0	0	0	0	0	0	0	1	4	0	0	0	2	2	0	0	1	2	2
Heathrow	0	1	2	0	0	1	4	0	0	0	2	4	0	0	1	3	3	0	0	0	2	2
Thames																						
Oxford	0	2	4	0	0	0	2	0	0	0	0	3	0	0	0	0	1	0	0	0	1	1
Benson	0	0	1	0	0	1	3	0	0	2	2	6	0	0	0	0	2	0	0	0	0	2
Little Rissington	0	1	1	0	0	0	0	0	0	0	0	0	0	0	0	0	0	0	0	0	0	1
South Farnborough	0	1	4	0	0	0	4	0	0	0	3	4	0	0	1	2	3	0	0	1	1	1
Bracknell	0	2	2	0	0	0	1	0	0	0	2	4	0	0	1	1	1	0	0	1	1	2
Bovingdon	0	0	2	0	0	0	2	0	0	0	1	2	0	0	0	1	3	0	0	0	1	2
Gatwick	0	2	5	0	1	1	4	0	0	1	2	3	0	0	1	2	3	0	0	2	2	3
London, right bank																						
Greenwich	0	3	4	0	0	2	2	0	0	1	1	3	0	0	0	1	2	0	0	0	1	2
Kew	1	4	6	0	0	1	5	0	0	1	2	3	0	0	0	1	1	0	0	1	1	2
Kent																						
Canterbury	0	1	1	0	0	0	1	0	0	0	0	3	0	0	0	0	0	0	0	0	0	0
Scots Float Sluice	0	0	2	0	0	0	1	0	0	0	1	1	0	0	0	1	1	0	0	0	1	1
Manston	0	2	2	0	1	2	3	0	0	2	3	3	0	0	0	2	5	0	0	1	1	3
West Malling	0	1	1	0	0	0	3	0	0	0	0	1	0	0	0	1	1	0	0	0	1	1
Sussex																						
Thorney Island	0	1	3	0	0	0	1	0	0	1	1	2	0	0	1	1	2	0	0	1	2	2
Hampshire																						
Havant	0	0	1	0	0	0	2	0	0	1	2	3	0	0	1	3	3	0	1	2	3	3
Avon and Dorset																						
Boscombe Down	0	0	2	0	0	0	2	0	0	0	1	3	0	0	0	2	3	0	0	0	0	0
Porton	1	2	2	0	0	0	2	0	0	0	1	6	0	0	0	1	3	0	0	0	0	0
Hurn	2	4	5	0	0	3	5	0	0	1	4	7	0	0	1	3	6	0	1	2	3	4
Larkhill	0	0	0	0	0	0	1	0	0	0	1	2	0	0	0	1	2	0	0	0	1	2
Devon, south																						
Exeter	0	2	4	0	0	0	1	0	0	0	0	6	0	0	0	1	4	0	0	0	3	4
Cornwall																						
Mount Batten	0	2	3	0	0	0	2	0	0	0	3	5	0	0	1	3	8	0	0	2	4	6
St Mawgan	0	0	2	0	0	1	2	0	0	0	2	5	0	0	1	2	5	0	0	1	3	4
Devon, north																						
Chivenor	0	1	1	0	0	0	1	0	0	0	0	2	0	0	0	1	1	0	0	0	0	0
Bristol Avon																						
Lyneham	0	1	2	0	1	1	3	0	1	1	2	4	0	1	1	1	3	1	1	1	1	2
Filton	0	0	1	0	0	0	1	0	0	0	1	4	0	0	1	2	4	0	0	1	2	2
Severn																						
Caersws	0	0	0	0	0	0	1	0	0	0	0	0	0	0	0	0	0	0	0	0	0	0
Shawbury	0	1	1	0	1	1	2	0	1	1	2	2	1	1	1	1	2	1	1	1	1	2
Pershore	0	2	3	0	0	1	2	0	0	0	2	2	0	0	0	2	2	0	0	0	0	0

TABLE 6A (*continued*)

Frequency distributions of intense falls

Number of days with the threshold values reached within the specified durations

	5 mm within			10 mm within				15 mm within					20 mm within					25 mm within				
Area and station	5 min	10 min	15 min	5 min	15 min	30 min	1 hour	15 min	30 min	1 hour	2 hours	4 hours	30 min	1 hour	2 hours	4 hours	8 hours	1 hour	2 hours	4 hours	8 hours	16 hours
Glamorgan																						
Rhoose	0	2	2	0	0	1	2	0	0	1	2	5	0	0	1	2	3	0	1	1	2	3
South West Wales																						
Aberporth	1	1	1	0	0	1	1	0	0	1	1	3	0	0	0	1	4	0	0	0	1	3
Aberystwyth	2	2	3	0	1	2	3	0	1	1	1	2	0	1	1	1	3	0	0	0	2	3
Gwynedd																						
Valley	0	2	3	0	0	0	0	0	0	0	1	4	0	0	0	0	3	0	0	0	0	2
Mersey and Weaver																						
Ringway	1	4	4	0	1	2	2	0	0	1	1	3	0	1	1	1	3	0	1	1	2	4
Speke	1	1	4	0	1	2	2	1	1	1	1	2	1	1	1	1	2	0	0	0	1	2
Lancashire																						
Squires Gate	1	2	3	0	0	1	1	0	0	1	1	2	0	0	1	1	2	0	0	1	1	3
Cumberland																						
Spadeadam	0	0	0	0	0	0	0	0	0	0	0	6	0	0	0	2	5	0	0	0	1	6
Carlisle	0	1	3	0	0	1	3	0	0	0	2	5	0	0	0	2	4	0	0	1	3	3
Isle of Man																						
Ronaldsway	0	0	1	0	0	0	0	0	0	0	0	3	0	0	0	1	3	0	0	0	1	4
Solway																						
West Freugh	0	0	0	0	0	0	0	0	0	0	0	2	0	0	0	1	1	0	0	0	1	3
Eskdalemuir	0	0	3	0	0	0	1	0	0	0	2	7	0	0	0	2	7	0	0	1	2	10
Ayrshire																						
Prestwick	0	1	2	0	1	1	2	0	1	1	2	4	0	0	1	1	2	0	0	1	1	2
Kilmarnock	0	0	0	0	0	0	0	0	0	0	0	1	0	0	0	0	2	0	0	0	1	1
Clyde																						
Abbotsinch	0	1	1	0	1	1	2	0	0	0	1	6	0	0	0	2	5	0	0	0	4	5
Kintyre and south-western islands																						
Machrihanish	0	0	1	0	0	0	0	0	0	0	0	1	0	0	0	0	2	0	0	0	0	0
Shiel, Alsh and Maree																						
Kinlochewe	0	0	0	0	0	0	1	0	0	0	3	7	0	0	0	4	8	0	0	3	5	8
Inner and Outer Hebrides																						
Isle of Rhum	0	1	1	0	0	1	6	0	0	0	3	21	0	0	1	8	18	0	0	2	12	16
Benbecula	0	0	0	0	0	0	3	0	0	0	2	3	0	0	0	0	4	0	0	0	0	2
Stornoway	0	0	1	0	0	1	1	0	0	0	0	0	0	0	0	0	0	0	0	0	0	0
Orkney and Shetland Islands																						
Kirkwall	0	0	0	0	0	0	0	0	0	0	0	1	0	0	0	0	2	0	0	0	0	4
Lerwick	0	0	0	0	0	0	1	0	0	0	0	0	0	0	0	0	2	0	0	0	0	2
Wick Water to Conon																						
Wick	1	2	4	0	1	2	3	0	0	0	2	2	0	0	0	2	4	0	0	0	0	4
Beauly and Ness																						
Dalcross	0	0	0	0	0	0	0	0	0	0	0	2	0	0	0	0	3	0	0	0	1	3
Banff, Moray and Nairn																						
Kinloss	0	1	3	0	1	1	1	0	0	1	1	4	0	0	0	1	3	0	0	0	2	4
Dee and Don																						
Dyce	0	1	1	0	0	0	1	0	0	0	1	3	0	0	0	2	3	0	0	1	2	4
Tay																						
Leuchars	0	0	0	0	0	0	0	0	0	0	1	2	0	0	0	1	2	0	0	1	2	6
Forth																						
Cumbernauld	0	0	0	0	0	0	1	0	0	0	0	6	0	0	0	3	3	0	0	1	3	4
Stronachlachar	1	1	1	0	0	0	1	0	0	0	3	12	0	0	0	4	13	0	0	2	5	15
Lothians																						
Haddington	0	0	0	0	0	0	0	0	0	0	0	3	0	0	0	1	3	0	0	0	2	4
Turnhouse	2	3	4	0	1	2	3	0	1	1	1	3	0	0	0	0	3	0	0	0	2	2
Whitburn	0	0	0	0	0	0	0	0	0	0	0	4	0	0	0	0	2	0	0	0	1	1
Foyle and north-western streams																						
Creggan Res. No. 2	0	2	2	0	0	0	1	0	0	0	0	4	0	0	0	0	2	0	0	0	0	1
Lough Foyle, east																						
Ballykelly	0	1	2	0	0	1	1	0	0	0	1	1	0	0	0	1	1	0	0	0	0	1

TABLE 6A (*continued*)

Frequency distributions of intense falls

Number of days with the threshold values reached within the specified durations

Area and station	5 mm (0·2 in) within			10 mm (0·4 in) within				15 mm (0·6 in) within					20 mm (0·8 in) within					25 mm (1·0 in) within				
	5 min	10 min	15 min	5 min	15 min	30 min	1 hour	15 min	30 min	1 hour	2 hours	4 hours	30 min	1 hour	2 hours	4 hours	8 hours	1 hour	2 hours	4 hours	8 hours	16 hours
Bann and Lough Neagh																						
Loughgall	0	0	0	0	0	0	0	0	0	0	0	2	0	0	0	0	2	0	0	0	0	2
Aldergrove	0	0	1	0	0	0	0	0	0	0	0	1	0	0	0	0	2	0	0	0	0	2
Portadown S. Wks	0	0	1	0	0	0	1	0	0	0	1	1	0	0	0	0	1	0	0	0	0	2
Armagh	0	0	0	0	0	0	0	0	0	0	0	2	0	0	0	0	2	0	0	0	0	3
Lurgan Cemetery	0	1	2	0	1	1	1	0	0	1	1	2	0	0	0	1	2	0	0	1	1	2
Toomebridge	0	0	0	0	0	0	0	0	0	0	0	0	0	0	0	0	0	0	0	0	0	2
Coleraine, The Cutts	0	1	1	0	0	0	0	0	0	0	0	0	0	0	0	0	0	0	0	0	0	0
Lagan and eastern streams																						
Hare Island	0	0	1	0	0	0	1	0	0	0	0	3	0	0	0	1	2	0	0	0	1	3
Newry and south-eastern streams																						
Newry Urban S. Wks	0	1	1	0	0	0	0	0	0	0	0	2	0	0	0	0	2	0	0	0	2	2
Erne and western streams																						
Castle Archdale Forest	0	1	1	0	1	1	1	0	0	0	0	1	0	0	0	0	1	0	0	0	0	1

TABLE 6B

Heavy falls in short periods

Area	Station	Date	Amount *mm*	Began GMT	Duration *min*	Rate *mm/h*	Category
	(a) EYE OBSERVATIONS						
Banff, Moray and Nairn	Dalrachney Beg	21 June	13·7	1700	10	82·2	N
Somerset	Folly Farm	10 June	18·3	1743	12	91·5	N
Lee	Hornsey, Creighton Avenue	5 July	23·0	1245	30	46·0	N
Cornwall	Trengwainton House	10 June	22·6	1800	30	45·2	N
Hampshire	Lyndhurst	29 Aug.	22·4	0630	35	38·4	N
Somerset	Nether Stowey	10 June	31·7	1730	40	47·6	N
Lincolnshire	Lade Bank Bridge	17 June	34·5	1600	60	34·5	N
Banff, Moray and Nairn	Dallas Lodge	14 June	43·7	1100	60	43·7	R
Hampshire	Grateley	9 June	28·4	2100	75	22·7	N
Hampshire	Lyndhurst	14 Aug.	34·3	1425	75	27·4	N
Somerset	Brymore	10 June	38·6	1830	75	30·9	N
Severn	Dudbridge	20 Aug.	65·3	1200	75	52·2	VR
Wear and Tees	Whitburn	13 Aug.	31·5	0020	90	21·0	N
Somerset	Bridgwater	10 June	43·9	1845	90	29·3	R
Somerset	Wembdon Hill	10 June	36·6	1830	90	24·4	N
Severn	Harescombe Grange	20 Aug.	83·8	2320	90	55·9	VR
Cheshire	Red Bull S. Wks	5 July	29·0	1700	90	19·3	N
Banff, Moray and Nairn	Lochindorb Lodge	14 June	37·3	1145	120	18·7	N
	(b) AUTOGRAPHIC RECORDS						
Trent	Finningley	15 June	19·6	1755	18	65·3	N
Mersey	Speke	17 June	20·4	1450	20	61·2	N
Trent	Willenhall S. Wks No. 2	21 Aug.	41·4	0400	40	62·1	R
Yorkshire Ouse	Halifax, Stannary	21 Aug.	25·4	0500	52	29·4	N
Trent	Watnall	20 June	29·9	1410	60	29·9	N
Severn	Shawbury	17 June	25·0	1750	60	25·0	N
Bristol Avon	Lyneham	21 Aug.	40·9	1918	120	20·5	N

TABLE 6C

Heavy falls in short periods. Supplementary list

Area	Station	Date	Amount *mm*	Duration *min*	Category
Nene	Sudborough	11 June	19·3	15	N
Yorkshire Ouse	Ings Lane P. Sta.	15 June	32·0	20	R
Yorkshire Ouse	Coxwold	17 June	28·4	*c* 30	N
Yorkshire Ouse	Thorpe Hesley	15 June	25·4	30	N
Devon, south	Hittisleigh	16 Apr.	37·9	60	N
Kent	Fairseat Manor	21 Aug.	29·0	*c* 75	N
Somerset	Ashford W. Wks	10 June	34·5	*c* 120	N
West Sussex	Littlehampton	30 Sept.	31·0	120	N
Bristol Avon	Bradford-on-Avon, Frome Road	21 Aug.	54·6	*c* 120	R

'*c*' against a value indicates that it is doubtful.

N Noteworthy R Remarkable VR Very rare

7 EVAPORATION AND PERCOLATION

This section brings together groups of data having a bearing on the subsequent history of water that reaches the earth's surface as precipitation. These data, derived from measurements of evaporation, percolation and drainage, were discussed in *British Rainfall* 1951 and 1952, particular attention being drawn to the nature of the instruments used and to the limitations of the measurements made. Further, in Part III of *British Rainfall* 1961, various aspects concerning evaporation were treated comprehensively in a group of papers by D. J. Holland.

The presentation of data is generally similar to that in the 1959–60 and earlier editions but with a few alterations. The 1916–50 averages have been replaced by 1956–62 averages, because averages for the latter period are available for most of the stations. Further, in the volume for 1962 the names of the evaporation stations were brought into line with their associated rainfall stations. For this reason, the names of some evaporation stations may now differ slightly from those used in volumes prior to 1962 but as National Grid references have been published for a number of years no ambiguity should arise.

Table 7A gives evaporation data from standard tanks. Any value prefixed by a negative sign indicates probable condensation.

In Table 7B the data from 12 percolation gauges stem from four stations whilst in Table 7C the data from 5 drainage gauges stem from two stations. Both the tables include appropriate rainfall values and also details concerning the gauges in use.

Severe weather in the winter months may cause some of the evaporation tanks and the percolation and drainage gauges to be frozen. The number of days for which data were available is indicated in brackets beside the monthly value concerned. Totals for the year are totals of the available data and are included because they have some comparative value.

TABLE 7A

Evaporation from standard tanks (values in millimetres)

Station	Harlow Hill Res.	Lower Laithe Res.	Ardsley Res., Upper	Upper Linacre Filters	Frankley	Bromwich Wood
Altitude	167 m	236 m	120 m	159 m	187 m	186 m
Grid ref.	SE44 289542	SE44 015367	SE44 287248	SK43 338726	SP42 005803	SO32 999810
January	1 (8 days)	0 (8 days)	9	1	14	15 (16 days)
February	0 (15 days)	2 (15 days)	12	4 (27 days)	15	24 (23 days)
March	37	17	32	24	30	35
April	24	20	29	20	30	35
May	63	61	68	66	83	50
June	72	56	70	66	81	57
July	88	70	76	72	77	42
August	51	46	53	56	63	53
September	40	37	38	42	53	42 (26 days)
October	14	15	19	19	13	37
November	11	−3 (29 days)	12	10	10	24 (27 days)
December	7 (20 days)	−2 (23 days)	12	2	6	25 (27 days)
Year	408	319	431	383	475	439
Average 1956–62	455	364	430	474		422

Station	Revesby Res.	Lound P. Sta.	Ormesby St Michael	Farmoor	Wallingford	Rothamsted
Altitude	38 m	10 m	7 m	63 m	48 m	128 m
Grid ref.	TF53 303634	TG63 501008	TG63 468152	SP42 452064	SU41 618898	TL52 132134
January	0	4	5 (20 days)	3	3	1 (6 days)
February	2	3	9	19	14	6 (22 days)
March	28	27	28	37	37	42
April	32	28	58	35	35	33
May	70	73	73	78	75	78
June	74	76	69	95	92	88
July	72	71	70	81	75	81
August	69	67	63	66	74	76
September	44	50	41	49	53	62
October	17	12	24	24	26	26
November	6	3		12	13	15 (27 days)
December	0	1	16	1	2	4 (23 days)
Year	415	415		499	498	513
Average 1956–62	416	491	402			540

TABLE 7A (*continued*)

Evaporation from standard tanks (values in millimetres)

Station	Kempton W. Wks	Kew	Weir Wood	Otterbourne W. Wks	Rosewarne	Barrow Gurney Res.
Altitude	12 m	5 m	69 m	34 m	76 m	84 m
Grid ref.	TQ51 111710	TQ51 171757	TQ51 406354	SU41 469233	SW10 643412	ST31 537679
January	4	11 (23 days)	−1 (13 days)	1	19	11
February	17	17	−1	6	20	23
March	44	41	35	34	44	33
April	37	36	29	31	61	35
May	87	78	76	67	102	90
June	97	101	96	88	83	94
July	92	88	77	74	109	88
August	82	75	76	62	86	75
September	59	53	45	42	64	53
October	21	23	15 (30 days)	17	53	27
November	18	17	9	11	17	11
December	−1	12	0 (29 days)	8	19	−2
Year	559	553	457	443	676	538
Average 1956–62	602	605	452	503		578

Station	Lake Vyrnwy	Wellesbourne	Stocks No. 10	Stocks No. 21D
Altitude	303 m	47 m	201 m	192 m
Grid ref.	SJ33 016192	SP42 271565	SD34 719550	SD34 717548
January	10	0	5 (11 days)	5 (11 days)
February	20	5	15 (15 days)	14 (15 days)
March	28	33	35 (29 days)	37 (29 days)
April	33	25	33 (23 days)	30 (26 days)
May	71	74	67	60
June	65	83	82	75
July	70	76	94	87
August	58	67	61	61
September	42	53	49	45
October	26	22	24	22
November	10	6	22	19
December	1	−1	29	25
Year	434	449	516	481
Average 1956–62	426	536*	558	455

* Average 1957–62

TABLE 7B

Rainfall, percolation and inferred evaporation (values in millimetres)

STATION	Fulbourn (Cambs.)			Lound P. Sta. (Suffolk)				Rothamsted (Herts.)			Farlington Res. (Hants)					
Grid ref.	TL52 539549			TG63 501008				TL52 132134			SU41 679060					
Gauge size		4·05 m²			38-cm diameter				4·05 m²			91-cm diameter				
Core		Bare soil on chalk	Turfed soil on chalk		Turfed soil	Turfed soil	Gravel		Bare soil	Bare soil		Chalk marl under 31 cm top soil and turf			Sand under 15 cm top soil and turf	Chalk under 15 cm top soil and turf
Depth (cm)		91	91		28	91	91		51	152		70	91	122	91	91
	Rainfall			Rainfall				Rainfall			Rainfall					
January	26	41	41	36	14	33	39	33	33	34	47	47	49	48	49	45
February	40	37	38	59	58	58	56	100	89	90	109	93	97	98	101	98
March	9	4	4	16	1	9	5	18	1	1	21	3	4	8	10	9
April	53	21	16	48	46	37	35	83	44	49	102	65	69	59	58	57
May	30	1	1	29	0	1	3	67	21	23	53	4	8	5	6	5
June	75	25	0	56	2	0	3	79	25	22	86	0	0	1	0	0
July	99	45	20	76	1	0	14	79	19	15	72	0	0	0	0	0
August	71	37	20	46	1	0	9	81	43	41	83	19	33	1	6	18
September	20	5	0	17	0	0	4	40	4	4	21	0	0	0	0	0
October	52	21	10	104	60	39	80	90	65	63	154	76	97	71	71	78
November	42	28	33	81	103	106	93	50	30	27	60	44	37	37	37	37
December	73	75	75	61	69	68	64	88	74	75	73	68	67	65	65	65
Year	591	342	258	629*	356	351	406	808	447	443	881	420	459	394	403	413
Inferred evaporation		249	333		273	278	223		361	365		461	422	487	478	468

* Gauge faulty—probably reading low

TABLE 7C

Rainfall, drainage observations and inferred evaporation (value in millimetres)

Station	Lake Vyrnwy (Montgomeryshire)					Prescot (Lancs.)	
Grid ref.	SJ33 016192					SJ33 473936	
Gauge size		Seepage gauge 40-cm diameter		Percolation gauge 46-cm diameter			Seepage gauge 40-cm diameter
Core		No. 1 Bare soil	No. 2 Turfed soil	No. 3 Grass cover over chalk	No. 4 Grass cover over chalk		Turfed soil
Depth (cm)		23	23	23	23		28
	Rainfall					Rainfall	
January	114	109	114	117	117	39	40
February	282	272	256	257	271	90	83
March	123	104	87	93	88	42	20
April	158	155	135	135	136	104	92
May	139	95	92	94	92	83	35
June	110	60	51	56	55	113	61
July	100	47	37	35	33	109	30
August	107	69	67	63	65	95	48
September	100	75	71	70	70	57	11
October	150	129	132	119	124	95	55
November	189	181	181	175	175	77	62
December	289	305	308	304	305	140	149
Year	1862	1602	1533	1520	1531	1044	686
Inferred evaporation		260	329	342	331		358

8 POTENTIAL EVAPOTRANSPIRATION

The estimates of potential evapotranspiration supplied in Table 8 are calculated values using the Penman formula. This formula, since its first publication,[1] has received wide acclaim as one of the most soundly based methods of calculating evaporation by using readily available meteorological data.

The merit of Penman's method lies in the combination of two of the classical approaches to calculating evaporation, the energy-budget and the aerodynamic, thereby eliminating the requirement in either approach for the temperature of the evaporating surface, a quantity which is rarely measured on a routine basis and is usually difficult to obtain.

Although the formula has a sound physical basis a good deal of empiricism is also inherent, particularly in deriving values for incoming and outgoing radiation in the energy-budget component. Much of this empiricism can be eliminated if measured amounts of global radiation or net radiation are available but these are quantities which are obtainable for very few stations in the United Kingdom. Over the years since the first publication of the formula, the empirical constants in the basic equations have been changed from time to time and in quoting values derived from the formula it is important to state precisely what constants have been used.

The version of the formula used in the derivation of the values in the table is, with one important amendment, that published by Penman.[2] The calculations are carried out by computer and relate to evapotranspiration from a vegetated surface, the albedo of which is taken to be 0·25. This albedo is generally considered representative of grass, most agricultural crops in most phases of their development and deciduous woodlands in leaf (but not conifers); it is not appropriate to an open water surface and the values in the table cannot be considered representative of such a surface.

The formula is

$$E = \frac{\Delta H + \gamma E_a}{\Delta + \gamma},$$

where Δ is the slope of the saturation vapour pressure curve at air temperature and γ is the hygrometric constant, taken as 0·49, temperature being in degrees Celsius and vapour pressure in millimetres of mercury.*

* Conversions compatible with SI units are as follows: 1 mmHg = 1·333 mb; 1 cal = 4·1868 J; 1 mile/day = 1·609 km/day; 1 mm of water evaporated/day = 2·858 mW/cm².

H is given by the equation

$$H = 0{\cdot}75R_a\left(0{\cdot}18 + 0{\cdot}55\frac{n}{N}\right) - 0{\cdot}95\sigma T_a^4\left(0{\cdot}10 + 0{\cdot}90\frac{n}{N}\right)(0{\cdot}56 - 0{\cdot}092\sqrt{e_d}).$$

Here, R_a is the amount of short-wave radiation reaching the outside of the earth's atmosphere expressed in millimetres of water evaporated per day (the simple latent-heat conversion is 59 cal/cm² = 1 mm water evaporated), n/N is the ratio of observed hours of sunshine to possible hours, σT_a^4 is the theoretical black-body radiation in equivalent millimetres per day at mean air temperature T_a (expressed in kelvins) and e_d is actual vapour pressure in mmHg at T_a. The coefficient 0·95 is the important amendment referred to earlier and is intended to allow for vegetation not radiating as a perfect black body, following Budyko.[3] E_a is given by the equation

$$E_a = 0{\cdot}35(e_a - e_d)\left(1 + \frac{U_2}{100}\right),$$

where e_a is the saturation vapour pressure at the mean air temperature T_a and e_d the actual vapour pressure (in mmHg), and U_2 is the run of wind in miles a day. The data required for the calculation of evapotranspiration in the form adopted in the programme are: sunshine (actual hours per month), air temperature (in degrees Celsius) and vapour pressure (in mmHg), run of wind (derived from a frequency table of ranges of speeds), and latitude and time of year (for the calculation of R_a and N).

All the values in Table 8 are based on meteorological elements measured four times a day (at 03, 09, 15 and 21 GMT), averaged to give a mean value for the month. Values are also calculated, but not published, for stations which measure the required meteorological elements less frequently than four times a day and even for stations which measure only once a day at 09 GMT.

References

1. PENMAN, H. L.; Natural evaporation from open water, bare soil and grass. *Proc R Soc, London*, A, **193**, 1948, pp. 120–145.
2. PENMAN, H. L.; Woburn irrigation, 1951–59. I. Purpose, design and weather. *J Agric Sci, Cambridge*, **58**, 1962, pp. 343–348.
3. BUDYKO, M. I.; The heat balance of the earth's surface. Leningrad, Gidrometeoizdat, 1956. (Translation available—Washington, U.S. Weather Bureau, 1958.)

TABLE 8

Calculated potential evapotranspiration (values in millimetres)

Area and station	Grid reference	Altitude *metres*	Jan.	Feb.	Mar.	Apr.	May	June	July	Aug.	Sept.	Oct.	Nov.	Dec.	Year
Northumbrian															
Acklington	NU 46 225007	42	7	12	45	49	81	72	85	61	41	16	8	2	479
Tynemouth	NZ 45 374694	27	16	22	48	54	79	76	92	62	46	24	17	4	541
Yorkshire Ouse and Hull															
Leconfield	TA 54 026438	7	7	14	41	48	84	81	88	69	44	17	10	5	507
Huddersfield Oakes	SE 44 113177	232	5	9	33	35	88	80	81	65	41	13	8	0	458
Trent															
Elmdon	SP 42 170835	97	9	17	39	44	98	98	93	78	45	14	9	8	553
Watnall	SK 43 503456	112	2	11	34	38	88	83	80	67	39	12	5	4	462
Finningley	SK 43 658995	10	5	14	43	43	86	93	90	74	47	16	10	9	530
Lincolnshire															
Binbrook	TF 53 196958	114	6	15	39	45	88	88	90	78	52	18	9	8	537
Manby	TF 53 391869	17	6	13	36	44	84	83	88	74	45	15	8	4	498
Waddington	SK 43 984650	68	2	13	38	44	90	96	88	75	49	15	7	3	520
Welland and Nene															
Wittering	TF 53 048032	80	4	14	41	46	92	96	(91)	(76)	53	17	10	5	545
Great Ouse															
Cardington	TL 52 083464	23	4	15	39	47	83	96	87	77	51	19	11	9	538
Mildenhall	TL 52 683778	5	5	14	38	46	88	99	84	83	51	17	6	4	534
Marham	TF 53 739087	23	7	14	38	48	86	96	86	81	51	18	7	(1)	534
East Suffolk and Norfolk															
West Raynham	TF 53 847245	73	5	11	32	42	83	94	85	80	52	16	8	1	509
Coltishall	TG 63 262229	17	9	18	39	47	88	91	88	82	55	20	14	8	558
Gorleston	TG 63 534037	1	14	13	32	42	86	90	88	84	57	24	16	9	556
Wattisham	TM 62 026514	89	4	10	37	42	84	99	91	83	53	16	7	3	530
Lee															
Stansted	TL 52 551226	101	3	13	40	45	84	97	88	83	55	17	9	7	540
Essex															
Walton on Naze	TM 62 259225	5	11	14	36	44	92	105	96	85	58	28	3	4	576
London, left bank															
Heathrow	TQ 51 077768	25	8	19	42	50	100	115	98	90	56	22	13	9	622
Thames															
Little Rissington	SP 42 209191	225	1	10	36	40	90	93	83	76	47	15	7	5	503
Abingdon	SU 41 482990	64	5	14	34	46	88	95	84	77	49	18	8	5	524
South Farnborough	SU 41 867544	69	6	14	39	46	91	107	91	84	51	19	10	8	566
Bovingdon	TL 52 007038	157	4	13	34	43	86	92	84	75	49	17	10	8	514
Gatwick	TQ 51 265407	58	4	14	35	44	85	101	87	78	45	17	8	5	524
Kent															
Manston	TR 61 335666	44	9	14	37	45	89	106	94	88	58	23	14	7	584
Deal	TR 61 377536	5	10	14	36	42	82	96	86	95	62	26	12	5	566
Sussex															
Thorney Island	SU 41 756027	3	4	13	33	42	84	106	93	81	56	27	12	10	563
Avon and Dorset															
Boscombe Down	SU 41 172393	127	5	15	36	47	89	88	90	78	51	19	9	8	535
Hurn	SZ 40 116978	10	4	15	37	48	91	103	100	76	48	20	5	6	552
Devon, south															
Exeter	SX 20 995937	32	7	18	38	45	90	91	95	76	46	18	8	10	543
Cornwall															
Mount Batten	SX 20 492529	27	9	18	37	46	90	92	103	77	51	24	12	11	570
St Mary's, Scilly	SV 00 913121	48	15	18	45	54	95	92	101	83	49	35	27	22	636
St Mawgan	SW 10 871642	102	7	18	41	52	95	84	99	82	56	27	19	17	597
Devon, north															
Hartland Point	SS 21 231277	91	9	20	36	47	93	86	94	82	59	31	30	13	601
Chivenor	SS 21 494347	6	9	22	38	50	94	86	93	75	53	24	17	13	573
Bristol Avon															
Lyneham	SU 41 012786	139	3	13	34	42	89	93	87	77	45	17	6	5	512
Filton	ST 31 598802	58	4	16	35	44	91	91	92	76	47	18	9	8	531
Severn															
Shawbury	SJ 33 553220	72	6	16	37	46	87	82	82	66	40	15	10	7	495
Pershore	SO 32 973495	40	7	17	40	46	94	99	93	79	49	18	10	13	565
Gloucester	SO 32 851177	27	9	19	39	47	96	97	89	80	49	19	10	13	568
Glamorgan															
Mumbles Head	SS 21 626871	36	20	20	41	53	99	92	103	91	63	40	24	13	659
Rhoose	ST 31 060670	67	6	11	36	44	95	89	96	76	46	23	10	7	540
South West Wales															
Aberporth	SN 22 247516	115	12	20	35	42	84	77	84	72	50	28	24	17	545
Gwynedd															
Valley	SH 23 310758	10	11	16	34	54	97	84	95	75	47	29	28	14	584
Mersey and Weaver															
Ringway	SJ 33 818850	76	13	30	44	65	102	99	91	74	50	19	15	5	609
Speke	SJ 33 415839	23	13	20	44	55	96	83	98	73	47	20	16	9	573

TABLE 8 (*continued*)

Calculated potential evapotranspiration (values in millimetres)

Area and station	Grid reference	Altitude *metres*	Jan.	Feb.	Mar.	Apr.	May	June	July	Aug.	Sept.	Oct.	Nov.	Dec.	Year
Lancashire															
Squires Gate	SD 34 316317	10	5	12	31	43	84	86	104	73	40	16	10	5	510
Cumberland															
Carlisle	NY 35 384603	26	9	17	36	50	83	80	88	59	42	14	7	4	490
Isle of Man															
Ronaldsway	SC 24 279688	17	22	22	40	53	81	78	98	69	44	27	20	17	570
Point of Ayre	NX 25 467043	9	13	17	34	47	85	71	93	71	43	29	19	6	529
Solway															
Eskdalemuir	NT 36 235026	242	4	8	30	35	70	68	79	50	33	12	4	1	393
Mull of Galloway . . .	NX 25 157304	78	7	11	28	43	75	65	90	67	41	26	18	7	479
Ayrshire															
Prestwick	NS 26 396261	16	7	16	33	54	82	71	90	64	39	16	9	7	488
Clyde															
Renfrew	NS 26 480667	5	8	18	34	51	81	69	91	60	38	13	7	3	473
Kintyre and south-western islands															
Rhuvaal	NR 16 426792	20	15	16	27	51	75	65	77	75	37	23	11	10	482
Inner and Outer Hebrides															
Tiree	NL 07 999446	9	16	23	31	53	76	66	77	74	37	25	17	14	508
Stornoway	NB 19 465332	3	2	15	22	53	66	66	70	60	32	12	7	5	411
Benbecula	NF 08 784556	5	7	15	27	51	76	71	81	70	37	21	17	14	487
Laxford															
Cape Wrath	NC 27 259747	112	12	14	33	45	56	58	66	66	35	26	8	(4)	422
Orkney and Shetland Islands															
Lerwick	HU 411 453397	82	6	4	22	45	57	58	66	57	27	11	14	14	380
Kirkwall	HY 310 483076	26	10	10	30	49	69	(57)	74	57	37	17	19	12	442
Wick Water to Conon															
Wick	ND 39 366523	36	14	18	28	52	70	57	70	55	36	18	13	9	439
Banff, Moray and Nairn															
Kinloss	NJ 38 069625	5	9	9	38	56	85	69	85	66	45	20	16	10	509
Dee and Don															
Rattray Head	NK 48 093582	26	14	17	42	57	76	61	88	64	39	21	11	9	499
Dyce	NJ 38 873125	58	7	10	37	46	77	63	83	54	45	15	11	4	451
Tay															
Leuchars	NO 37 468208	11	5	12	46	52	82	69	96	58	44	14	9	4	493
Lothians															
Turnhouse	NT 36 159739	35	5	15	43	51	83	68	90	59	42	17	10	7	491
Lough Foyle, east															
Ballykelly	IC 24 624238	1	7	18	32	51	76	66	74	64	39	18	10	10	465
Bann and Lough Neagh															
Aldergrove	IJ 33 147798	69	8	15	26	47	82	72	81	61	38	16	8	4	458

Part III Special articles

1 POTENTIAL EVAPOTRANSPIRATION DATA, 1966

by F. H. W. Green, M.Sc., Nature Conservancy, Monks' Wood, Huntingdon

Irrigated-lysimeter observations have been made at a number of locations in the British Isles for several years. The observational techniques and the use of the observations to determine evapotranspiration were discussed in *British Rainfall* 1958, the first volume of this publication to include evapotranspiration data. Further, D. J. Holland discussed potential evapotranspiration observations at some length in Part III of *British Rainfall* 1961.

When observational difficulties have led to short gaps in the monthly returns, values for the gaps have been interpolated and the total potential evapotranspiration for the month is given in brackets.

Commencing with *British Rainfall* 1959–60, potential water deficit (PWD) and potential water surplus (PWS) have been assessed by wet and dry periods (i.e. periods when rainfall (R) exceeds potential evapotranspiration (PE) and when PE exceeds R) of five days or more. These figures are ecologically and hydrologically more realistic than figures for the calendar months.

The data for 1966 are presented in the same form as for previous years. Attention is drawn to a few anomalous cases:

(i) it is not clear why there should, in 1966, in contrast to the preceding years, be a significant difference in PE between the values for the adjoining stations Monks' Wood and Wood Walton Fen.

(ii) It is not clear why the PE values for Cawood should be significantly less than those for High Mowthorpe and Westlands Farm.

(iii) The PE values for Overton Down are, as usual, high and this can be attributed to the oasis effect;[1] it has been suggested that the 'dryness' (lack of transpiration) of the summit areas of the Marlborough Downs may be due not so much to paucity of water in the soil as to the inability of nitrogen-starved vegetation to extract it (Bunting and Elston[2]).

A table of some similar observations made in the Republic of Ireland was included in the 1965 volume, with acknowledgements to the Irish Meteorological Service and the Irish Agricultural Institute. The longest-maintained records are in fact from the former's Valentia Observatory, where observations commenced in 1952 (Guerrini[3] and Morgan[4]).

References

1. GREEN, F. H. W.; Potential evapotranspiration data, 1962. *Brit Rainf* 1962, *London*, 1971, Part III, pp. 3–7.
2. BUNTING, A. D. and ELSTON, J.; Water relations of crop and grass on chalk soil. *Scient Hort, Canterbury*, **18**, 1966, pp. 116–120.
3. GUERRINI, V. H.; Evaporation and transpiration in the Irish climate. *Tech Notes, Met Serv Eire, Dublin*, No. 14, 1953.
4. MORGAN, W. A.; Potential evapotranspiration as measured at Valentia Observatory over the period August 1952 to February 1962, and a comparison with values as computed by the Penman formula. *Tech Notes, Met Serv Eire, Dublin*, No. 29, 1962.

TABLE 1A

Potential evapotranspiration and rainfall 1966 (values in inches; 1 inch = 25·4 millimetres)

PE = Potential evapotranspiration, R = Rainfall, PWD = Potential water deficit, PWS = Potential water surplus

Note: The rainfall amount in those months when PE exceeds R is shown in *italics*

		Achnag-oichan	Cwm Dyli	Glenlatt-erach	Harris	Inverpolly	Isle of Rhum	Killie-chonate	Kin-lochewe
January	PE	0·3	(0·1)	0·2	0·2	(0·2)	(0·4)	1·1	0·5
	R	0·8	7·5	1·3	4·9	1·5	7·5	4·7	2·2
	PWD	0·1	0·0	0·1	—	(0·1)	0·2	0·5	0·4
	PWS	0·6	7·4	1·2	—	1·4	7·3	4·1	2·1
February	PE	0·5	(0·1)	0·5	0·5	0·2	(0·3)	1·4	0·7
	R	2·0	14·1	1·8	4·1	2·0	6·7	5·4	5·1
	PWD	0·2	0·0	0·2	—	0·1	0·1	0·6	0·2
	PWS	1·7	14·0	1·5	—	1·9	6·5	4·6	4·6
March	PE	1·2	0·4	0·7	1·5	0·3	(0·4)	1·9	2·0
	R	2·9	9·4	3·4	6·2	7·1	12·2	10·6	13·6
	PWD	0·0	0·0	0·1	—	0·0	0·0	0·0	0·0
	PWS	1·7	9·0	2·8	—	6·8	11·8	8·7	11·6
April	PE	1·0	1·3	1·0	1·4	0·5	1·1	1·9	1·4
	R	*0·5*	10·9	*0·2*	2·1	1·2	3·7	2·3	2·0
	PWD	0·7	0·1	0·8	—	0·2	0·5	1·2	0·6
	PWS	0·2	9·7	0·0	—	0·9	3·1	1·6	1·2
May	PE	2·0	1·9	3·2	1·6	1·6	1·5	1·3	(1·6)
	R	2·8	10·5	*2·0*	4·3	4·0	8·2	4·4	5·9
	PWD	0·4	0·4	1·7	—	0·3	0·3	0·2	0·2
	PWS	1·2	9·0	0·5	—	2·7	7·0	3·3	4·5
June	PE	2·0	1·9	2·3	2·0	2·3	2·1	0·7	1·7
	R	6·4	13·3	8·5	4·3	5·3	4·9	4·7	5·9
	PWD	0·0	0·2	0·0	—	0·4	0·4	0·1	0·2
	PWS	4·4	11·6	6·2	—	3·4	3·2	4·1	4·4
July	PE	2·2	2·5	3·3	1·7	2·1	2·3	1·9	1·4
	R	3·1	8·4	*3·2*	3·1	3·9	4·6	2·3	4·6
	PWD	0·6	0·4	0·8	—	0·4	0·5	0·6	0·2
	PWS	1·5	6·3	0·7	—	2·2	2·8	1·0	3·4
August	PE	1·6	2·3	3·2	0·6	1·8	1·9	1·7	1·6
	R	2·5	8·1	4·4	1·9	2·1	2·9	2·3	2·2
	PWD	0·4	1·0	0·8	—	0·9	0·5	0·7	0·3
	PWS	1·3	6·8	2·0	—	1·2	1·5	1·3	0·9
September	PE	1·3	1·9	2·3	0·5	1·8	0·8	1·5	0·6
	R	1·7	9·4	*1·7*	5·0	4·5	8·8	5·1	5·9
	PWD	0·3	0·7	0·8	—	0·2	0·2	0·4	0·0
	PWS	0·7	8·2	0·2	—	2·9	8·2	4·0	5·3
October	PE	0·6	0·7	1·1	0·5	0·8	1·1	0·9	(0·0)
	R	3·7	9·2	3·8	3·4	4·9	5·8	4·6	5·3
	PWD	0·0	0·0	0·1	—	0·0	0·0	0·0	0·0
	PWS	3·1	8·5	2·8	—	4·1	4·7	3·7	5·3
November	PE	0·5	0·0	0·4	0·4	0·3	(1·3)	1·0	(0·0)
	R	6·5	13·5	4·4	5·2	6·5	9·0	7·5	11·5
	PWD	0·0	0·0	0·0	—	0·0	0·0	0·1	0·0
	PWS	6·0	13·5	4·0	—	6·2	7·7	6·6	11·5
December	PE	0·5	0·0	0·1	0·6	0·1	(0·7)	0·6	(0·0)
	R	5·4	20·5	4·3	8·5	10·5	13·3	15·1	22·3
	PWD	0·0	0·0	0·0	—	0·0	0·0	0·0	0·0
	PWS	4·9	20·5	4·2	—	10·4	12·6	14·5	22·3
April to September inclusive	PE	10·1	11·8	15·3	7·8	10·1	9·7	9·0	8·3
	R	17·0	60·6	20·0	20·7	21·0	33·1	21·1	26·5
	PWD	2·4	2·8	4·9	—	2·4	2·4	3·2	1·5
	PWS	9·3	51·6	9·6	—	13·3	25·8	15·3	19·7
Year	PE	13·7	13·1	18·3	11·5	12·0	13·9	15·9	11·5
	R	38·3	134·8	39·0	53·0	32·5	87·6	69·0	86·5
	PWD	2·7	2·8	5·4	—	2·6	2·7	4·4	2·1
	PWS	27·3	124·5	26·1	—	23·1	76·4	57·5	77·1

TABLE 1A (*continued*)

		Lossie-mouth	Monks' Wood	Moor House	Morden Bog	Overton Down	Prabost	Rannoch Rly. Sta.	Strachan	Strachan, Kerloch
January	PE	0·4	0·0	1·1	0·6	0·0	(0·2)	—	0·3	0·0
	R	1·1	1·1	4·1	3·6	2·1	4·2	—	2·2	2·4
	PWD	0·2	0·0	0·1	(0·2)	0·0	0·0	—	0·0	0·0
	PWS	0·9	1·1	3·2	(3·2)	2·1	4·0	—	1·9	2·4
February	PE	0·4	0.3	2.3	0·7	0·9	0·4	—	0·2	0·8
	R	2·9	3·1	9·5	5·8	4·3	3·8	—	5·2	5·5
	PWD	0·1	0·0	0·0	(0·1)	0·1	0·2	—	0·0	0·0
	PWS	2·6	2·8	7·2	(5·2)	3·5	3·6	—	5·0	4·7
March	PE	1·1	1·4	2·8	0·7	(2·9)	1·0	0·6	1·0	1·0
	R	2·4	0·3	7·3	*0·7*	*0·5*	10·3	7·5	1·3	1·4
	PWD	0·5	1·2	0·1	(0·3)	2·5	0·0	0·0	0·3	0·3
	PWS	1·8	0·1	4·6	(0·3)	0·1	9·3	6·9	0·6	0·7
April	PE	2·2	1·8	3·5	1·3	(3·2)	1·5	0·5	1·3	1·0
	R	*0·4*	3·4	5·0	4·8	7·0	2·9	2·3	2·9	3·1
	PWD	1·9	0·4	0·3	(0·3)	0·3	0·8	0·2	0·1	0·0
	PWS	0·1	2·0	1·8	(3·8)	4·1	2·2	2·0	1·7	2·1
May	PE	2·9	3·6	2·6	2·4	(3·6)	2·2	1·3	2·5	1·9
	R	*2·2*	*1·7*	6·7	*2·0*	*2·9*	5·3	4·5	3·0	3·2
	PWD	0·9	2·1	0·8	(1·4)	1·5	0·3	0·1	0·6	0·0
	PWS	0·2	0·2	4·9	(1·0)	0·8	3·4	3·3	1·1	1·3
June	PE	3·1	3·8	2·3	3·3	3·0	2·1	2·2	2·5	2·4
	R	6·9	*2·7*	6·4	*1·9*	*2·9*	5·3	6·2	5·7	5·3
	PWD	0·5	2·1	0·6	(1·7)	0·6	0·1	0·0	0·3	1·0
	PWS	4·3	1·0	4·7	(0·3)	0·5	3·3	4·0	3·5	3·9
July	PE	4·2	3·8	3·2	3·4	3·3	2·0	2·3	3·0	2·5
	R	*2·3*	*2·6*	4·8	*0·4*	*2·9*	4·2	*1·3*	*2·1*	*2·1*
	PWD	2·5	0·5	0·9	3·0	1·6	0·5	1·2	1·0	0·4
	PWS	0·6	1·7	2·5	0·0	1·2	2·7	0·2	0·1	0·0
August	PE	3·5	3·3	1·8	2·4	3·7	1·7	1·7	2·2	2·2
	R	*2·0*	4·4	6·7	4·1	6·1	2·6	2·5	4·2	3·9
	PWD	3·1	0·8	0·4	1·1	1·4	0·6	0·5	0·8	0·7
	PWS	1·6	1·9	5·3	2·8	3·8	1·5	1·3	2·8	2·4
September	PE	1·8	2·5	1·8	1·9	2·6	1·1	1·1	2·0	1·9
	R	2·2	*0·7*	7·1	2·4	*1·1*	5·8	5·8	*1·3*	2·6
	PWD	0·7	2·1	0·7	1·3	1·8	0·0	0·3	1·3	0·6
	PWS	1·1	0·3	6·0	1·8	0·3	4·7	5·0	0·6	1·3
October	PE	0·6	0·8	0·4	0·7	2·6	0·7	0·1	0·9	1·4
	R	3·0	3·4	7·1	8·3	7·1	4·6	2·5	7·6	5·9
	PWD	0·1	0·0	0·0	0·1	0·5	0·0	0·0	0·0	0·0
	PWS	2·5	2·6	6·7	7·7	5·0	3·9	2·4	6·7	4·5
November	PE	0·7	0·5	1·4	0·4	1·8	0·3	—	0·2	1·2
	R	3·3	1·7	8·1	2·8	2·4	8·3	—	2·7	2·4
	PWD	0·1	0·1	0·0	0·1	0·3	0·0	—	0·0	0·0
	PWS	2·7	1·3	6·7	2·5	0·9	8·0	—	2·5	1·2
December	PE	(0·0)	0·1	4·4	0·2	1·8	0·3	—	0·0	0·7
	R	3·5	2·2	11·7	3·2	3·8	12·6	—	4·4	6·0
	PWD	0·0	0·0	0·0	0·0	0·0	0·0	—	0·0	0·0
	PWS	3·5	2·1	7·3	3·0	2·0	12·3	—	4·4	5·3
April to September inclusive	PE	17·7	18·8	15·2	14·7	19·4	10·6	9·1	13·5	11·9
	R	16·0	15·5	36·7	15·6	22·9	26·1	22·6	19·2	20·2
	PWD	9·6	8·0	3·7	8·8	7·2	2·3	2·3	2·1	2·7
	PWS	7·9	4·7	25·2	9·7	10·7	17·8	15·8	7·8	11·0
Year	PE	20·9	21·9	27·6	18·0	29·4	13·5	—	16·1	17·0
	R	32·2	27·3	84·5	40·0	43·1	69·9	—	42·6	43·8
	PWD	10·6	9·3	3·9	9·6	10·6	2·5	—	4·4	3·0
	PWS	21·9	14·7	60·8	31·6	24·3	58·9	—	30·9	29·8

TABLE 1A (*continued*)

		White-hillocks	Wood Walton Fen	Yarner Wood	Alice Holt Lodge	Blacklaw Moss	Cawood	Edgbaston	Everton (Efford)
January	PE	1·0	0·1	0·4	0·1	0·1	—	0·0	—
	R	3·8	1·2	7·5	2·0	1·2	—	1·4	—
	PWD	0·2	0·0	0·1	—	—	—	0·0	—
	PWS	3·0	1·1	7·2	—	—	—	1·4	—
February	PE	1·2	0·1	0·4	1·0	0·3	—	0·2	—
	R	7·1	1·1	12·2	5·4	3·5	—	4·3	—
	PWD	0·0	0·0	0·0	—	—	—	0·0	—
	PWS	5·9	1·0	11·8	—	—	—	4·1	—
March	PE	1·4	1·3	1·6	1·2	0·5	0·9	1·1	1·3
	R	1·6	(*0·3*)	2·2	*0·5*	1·9	*0·7*	1·4	*0·5*
	PWD	0·5	1·0	0·9	—	—	—	0·4	—
	PWS	0·7	0·0	1·5	—	—	—	0·7	—
April	PE	1·4	1·6	1·5	1·6	0·8	1·0	0·9	1·1
	R	4·4	2·7	10·6	3·9	1·5	3·4	3·2	4·4
	PWD	0·2	0·5	0·2	—	—	—	0·2	—
	PWS	3·2	1·6	9·3	—	—	—	2·5	—
May	PE	2·0	2·4	2·8	2·9	1·9	2·1	2·4	2·7
	R	4·3	*1·6*	3·7	*2·8*	2·3	*2·0*	3·2	*1·5*
	PWD	0·4	1·0	0·8	—	—	—	0·9	—
	PWS	2·7	0·2	1·7	—	—	—	1·7	—
June	PE	2·1	2·8	3·5	3·4	3·2	2·6	2·6	4·4
	R	4·4	3·2	*2·6*	*2·4*	5·5	*1·8*	3·4	*2·2*
	PWD	0·5	1·3	1·2	—	—	—	0·5	—
	PWS	2·8	1·7	0·3	—	—	—	1·3	—
July	PE	2·5	2·8	4·0	3·6	3·0	3·0	2·3	4·1
	R	*0·9*	(*0·9*)	*1·4*	*2·6*	*2·3*	*2·6*	2·9	*1·9*
	PWD	1·8	(2·1)	2·8	—	—	—	0·3	—
	PWS	0·2	(0·2)	0·2	—	—	—	0·9	—
August	PE	2·2	1·8	2·5	2·7	2·3	1·6	2·4	3·2
	R	4·7	4·7	6·0	*2·6*	4·6	4·0	3·7	5·3
	PWD	0·8	0·5	0·7	—	—	—	0·9	—
	PWS	3·3	3·4	4·2	—	—	—	2·2	—
September	PE	1·3	1·8	1·7	2·2	1·9	1·5	1·6	2·0
	R	2·1	*0·8*	*1·4*	*1·3*	4·5	*1·0*	*1·1*	2·1
	PWD	0·3	1·2	0·8	—	—	—	1·0	—
	PWS	1·1	0·2	0·5	—	—	—	0·5	—
October	PE	0·9	0·4	1·1	0·8	0·8	—	0·7	0·6
	R	7·7	3·1	8·3	4·4	2·7	—	4·3	4·9
	PWD	0·0	0·1	0·3	—	—	—	0.0	—
	PWS	6·8	2·8	7·5	—	—	—	3·6	—
November	PE	0·7	0·3	0·6	0·9	0·4	—	0·4	—
	R	5·8	1·9	3·4	2·1	2·4	—	2·1	—
	PWD	0·2	0·0	0·1	—	—	—	0·0	—
	PWS	5·3	1·6	2·9	—	—	—	1·7	—
December	PE	0·7	0·2	0·2	0·7	0·1	—	0·4	—
	R	5·5	2·4	6·1	3·6	4·9	—	3·9	—
	PWD	0·0	0·0	0·0	—	—	—	0·0	—
	PWS	4·8	2·2	5·9	—	—	—	3·5	—
April to September inclusive	PE	11·5	13·2	16·0	16·4	13·1	11·8	12·2	17·5
	R	20·8	13·9	25·7	15·6	20·7	15·5	17·5	17·4
	PWD	4·0	6·6	6·5	—	—	—	3·8	—
	PWS	13·3	7·3	16·2	—	—	—	9·1	—
Year	PE	17·4	15·6	20·3	21·1	15·3	—	15·0	—
	R	41·0	23·9	65·4	33·6	37·3	—	34·9	—
	PWD	4·9	7·7	7·9	—	—	—	4·2	—
	PWS	28·5	16·0	53·0	—	—	—	24·1	—

TABLE 1A (*continued*)

		High Mowthorpe	Hollins-clough	Mount St Bernard Abbey	Stocks No. 21D	Sutton Bonington	Wallingford	Westlands Farm
January	PE	—	0·0	0·0	0·2	0·1	0·5	—
	R	—	3·5	2·3	2·7	1·6	1·1	—
	PWD	—	0·0	0·0	0·1	0·0	—	—
	PWS	—	3·5	2·3	2·6	1·5	—	—
February	PE	—	0·1	0·5	0·7	0·2	0·5	—
	R	—	6·4	4·5	9·3	2·8	2·9	—
	PWD	—	0·0	0·0	0·1	0·0	—	—
	PWS	—	6·3	4·0	8·7	2·6	—	—
March	PE	—	0·8	1·0	0·9	1·0	1·3	—
	R	—	3·8	1·1	4·5	1·0	*0·4*	—
	PWD	—	0·0	0·4	0·0	0·4	—	—
	PWS	—	3·0	0·5	3·6	0·4	—	—
April	PE	1·8	1·0	1·1	1·4	1·1	2·9	1·3
	R	4·1	4·3	4·9	4·3	3·9	*2·8*	2·9
	PWD	—	0·0	0·2	0·3	0·3	—	—
	PWS	—	3·3	4·0	3·2	3·1	—	—
May	PE	2·7	2·1	2·5	2·4	2·0	2·8	2·4
	R	*2·3*	4·2	*2·3*	5·4	*1·5*	*1·8*	*2·0*
	PWD	—	0·4	1·0	0·8	0·6	—	—
	PWS	—	2·5	0·8	3·8	0·1	—	—
June	PE	3·1	2·5	2·9	2·8	3·0	3·5	2·9
	R	3·5	5·5	*2·2*	5·7	*2·3*	*2·7*	*2·1*
	PWD	—	0·3	0·8	0·7	1·2	—	—
	PWS	—	3·3	0·1	3·6	0·5	—	—
July	PE	3·4	2·8	2·3	2·9	2·7	2·8	3·7
	R	*3·1*	3·8	3·3	3·8	*2·0*	3·1	*3·4*
	PWD	—	1·0	0·8	0·4	1·6	—	—
	PWS	—	2·0	1·8	1·3	0·9	—	—
August	PE	2·3	2·2	2·2	2·1	2·2	2·4	2·4
	R	4·9	5·2	4·6	5·7	3·7	*2·3*	4·1
	PWD	—	0·3	1·0	0·6	0·5	—	—
	PWS	—	3·3	3·4	4·2	2·0	—	—
September	PE	1·5	2·1	1·3	1·5	1·7	2·5	1·6
	R	*1·1*	4·1	*1·1*	4·8	*1·1*	*1·8*	1·8
	PWD	—	0·7	0·7	0·7	0·8	—	—
	PWS	—	2·7	0·5	4·0	0·2	—	—
October	PE	—	1·1	0·6	0·8	0·5	1·9	1·1
	R	—	4·2	4·3	5·1	3·6	6·1	3·2
	PWD	—	0·0	0·1	0·2	0·0	—	—
	PWS	—	3·1	3·8	4·5	3·1	—	—
November	PE	—	0·8	0·2	0·8	0·0	0·5	0·3
	R	—	5·7	1·8	6·6	1·9	1·6	2·2
	PWD	—	0·0	0·1	0·1	0·0	—	—
	PWS	—	4·9	1·7	5·9	1·9	—	—
December	PE	—	0·7	0·2	0·2	0·0	0·3	0·2
	R	—	9·7	3·4	9·3	2·7	2·5	1·7
	PWD	—	0·0	0·0	0·0	0·0	—	—
	PWS	—	9·0	3·2	9·1	2·7	—	—
April to September inclusive	PE	14·8	12·7	12·3	13·1	12·7	16·9	14·3
	R	19·0	27·1	18·4	29·7	14·5	14·5	16·3
	PWD	—	2·7	4·5	3·5	5·0	—	—
	PWS	—	17·1	10·6	20·1	6·8	—	—
Year	PE	—	15·2	14·8	16·7	14·5	21·9	—
	R	—	60·4	35·8	67·2	28·1	29·1	—
	PWD	—	2·7	5·1	4·0	5·4	—	—
	PWS	—	47·9	26·1	54·5	19·0	—	—

TABLE 1B
Appendix for Irish stations

Potential evapotranspiration and rainfall 1966 (values in inches, 1 inch = 25·4 millimetres)
PE = Potential evapotranspiration, R = Rainfall
Note: the rainfall amount in those months when PE exceeds R is shown in *italics*

		Ballinamore Co. Leitrim	Glenamoy Co. Mayo	Johnstown Castle Co. Wexford	Valentia Co. Kerry
January	PE	0·1	1·3	0·2	0·5
	R	4·0	4·4	4·7	6·9
February	PE	1·2	0·9	0·7	1·0
	R	6·8	5·7	6·2	10·1
March	PE	1·6	1·3	0·1	1·5
	R	4·0	3·9	1·3	1·8
April	PE	1·7	1·7	0·9	1·5
	R	5·2	3·9	8·3	5·9
May	PE	3·3	3·3	3·5	2·5
	R	4·6	4·7	*2·7*	5·0
June	PE	3·1	3·4	3·3	2·4
	R	5·5	4·9	4·3	3·6
July	PE	4·2	2·7	4·4	2·6
	R	*1·9*	3·1	*1·7*	*1·9*
August	PE	2·5	2·6	2·7	2·1
	R	3·1	*2·5*	3·1	3·3
September	PE	1·8	1·7	2·9	1·5
	R	3·4	4·6	3·4	2·8
October	PE	1·2	1·6	0·4	0·8
	R	4·0	2·6	8·0	7·3
November	PE	0·6	0·8	0·9	0·5
	R	4·5	6·0	4·3	2·8
December	PE	*	0·4	0·1	0·4
	R	7·9	7·8	2·3	3·8
April to September	PE	16·6	15·4	17·7	12·6
	R	23·7	23·7	23·5	22·5
Year	PE	—	21·7	20·1	17·3
	R	54·9	54·1	50·3	55·2

* Not available.

Note. The Valentia observations are provided by courtesy of the Irish Meteorological Service and the remainder are by courtesy of the Irish Agricultural Institute.

TABLE 2

Station information 1966

Station	Observer	Grid reference	Height a.o.d. *feet**
	(1) Nature Conservancy and allied stations		
Achnagoichan	D. F. Holland	NH 913082	1 000
Cwm Dyli	Superintendent, Power Station, CEGB	SH 654542	310
Glenlatterach	Filter Station Superintendent, Laich of Moray Water Board	NJ 200545	600
Harris	J. Lamb	NM 337957	50
Inverpolly	R. Tweddle	NC 074134	45
Isle of Rhum	J. Lamb	NM 402996	15
Killiechonate	T. Hill	NN 243813	280
Kinlochewe	R. Balharry and A. Christie	NH 024630	76
Lossiemouth	Senior Meteorological Officer, RN Air Station	NJ 214698	15
Monks' Wood	K. Mellanby, C.B.E.	TL 202797	130
Moor House	M. Rawes	NY 757328	1 825
Morden Bog	M. T. Horwood	SY 914916	50
Overton Down	N. King	SU 129707	760
Prabost	Mr and Mrs R. M. Murray	NG 418501	220
Rannoch Railway Sta.	J. Michie	NN 424578	975
Strachan	A. Walker	NO 696917	400
Strachan, Kerloch Moor	G. R. Miller	NO 700903	475
Whitehillocks	W. E. C. Macdonald and I. Moir	NO 449797	845
Wood Walton Fen	G. Mason	TL 239848	0
Yarner Wood	R. V. Russell	SX 785782	650
	(2) Other stations		
Alice Holt Lodge	Forestry Commission	SU 804429	378
Blacklaw Moss	Hill Farming Research Organization	NS 925488	730
Cawood	Experimental Husbandry Farm	SE 561366	20
Edgbaston	A. L. Kelly (for the Trent River Authority)	SP 046864	535
Everton (Efford)	Experimental Horticultural Station	SZ 303937	51
High Mowthorpe	Experimental Husbandry Farm	SE 888688	574
Hollinsclough	Miss W. M. Leonard (for the Trent River Authority)	SK 067666	954
Mount St Bernard Abbey	Father Augustine (for the Trent River Authority)	SK 459158	620
Stocks No. 21D	Fylde Water Board	SD 717548	629
Sutton Bonington	Agricultural College, Nottingham University (for the Trent River Authority)	SK 503262	157
Wallingford	Institute of Hydrology	SU 618898	157
Westlands Farm	Department of Geography, Hull University	TA 181411	40

* 100 ft = 30·48 metres

2 SNOW SURVEY OF GREAT BRITAIN

Season 1965–66

by R. E. Booth

The basic material for this report has been obtained, as in previous years, from observers who have provided, month by month, daily records of snowfall and of snow-cover within sight of their stations. These records from a network of stations distributed over the country have been augmented by data extracted from the monthly returns of official stations manned by Meteorological Office staff, and also of climatological stations and rainfall stations all of which voluntarily report to the Meteorological Office. Without the ready co-operation of all those responsible for voluntary observations, this report could not have been prepared in such detail. The measurement of snow depth in the following pages refers in general to observations taken at 09 GMT or thereabouts.

Summary of general weather during the season

Table 1 gives for each month of the season and for the season as a whole the monthly mean temperature and the difference from the 1931–60 average, also the monthly and seasonal amounts of precipitation with percentages of the 1916–50 averages for England and Wales and for Scotland. It shows that the season September 1965 to May 1966 was colder than usual, especially in Scotland where temperature was below the average in every month except October and March. Over the season as a whole mean temperature showed an accumulated deficit of 2·5 degC in England and Wales and 6·6 degC in Scotland, giving mean monthly anomalies of −0·3 degC and −0·7 degC respectively. Rainfall was above average in England and Wales and almost average in Scotland. October and January were dry months generally and November and April were dry in Scotland but wet in England and Wales.

TABLE 1

	1965				1966					Season 1965–66
	Sept.	Oct.	Nov.	Dec.	Jan.	Feb.	Mar.	Apr.	May	
Temperature (degrees Celsius)										
England and Wales										
Mean	12·8	11·6	5·4	5·4	3·6	6·2	7·1	7·7	11·6	7·9
Diff. from average	−1·5	+0·8	−2·1	−0·2	−0·7	+1·7	+0·7	−1·1	−0·1	−0·3
Scotland										
Mean	11·7	9·9	3·8	3·1	3·1	3·9	6·2	5·5	9·9	6·3
Diff. from average	−0·7	+0·5	−2·5	−1·6	−0·4	−0·4	+0·8	−2·0	−0·3	−0·7
Precipitation (inches)*										
England and Wales										
Total	5·3	1·3	4·3	6·5	2·2	4·8	1·5	4·1	2·8	32·8
Per cent of average	178	37	115	189	60	186	65	173	113	121
Scotland										
Total	6·5	5·5	4·2	6·1	3·1	4·9	5·5	2·7	4·4	42·9
Per cent of average	129	88	75	109	52	117	156	77	127	99

* 1 inch = 25·4 millimetres.

Comparison of snowfall during 1965/66 with that of previous years

Data for 10 representative stations* in Great Britain at altitudes between 400 and 1200 ft (which have been used, with slight changes in stations, for seasonal comparisons since the survey of 1946/47) give for the season 1965/66 a mean of 31 days with snow lying at the morning observation. This is six days more than for the previous season and is equal to the average for the 20 successive years since 1946/47.

Number of days with snow falling during the season

At altitudes of about 1000 ft or above, snow fell on 80–90 days in the Cairngorms and over the Cross Fell Edge, 60–70 days in the extreme north of Scotland and 40–50 days in the Grampians, the Peak District, the North-west Highlands and the Southern Uplands. At lower levels it fell on 30–40 days in Snowdonia, on the Midland Plateau and on the Lincolnshire Edge. Some of the lower-lying ground in Norfolk also had 30–40 days. From sea level to 300 ft snow fell on 10–20 days over much of England and Wales south of a line drawn east–west through Birmingham.

* These stations are: Fort Augustus, Braemar, West Linton, Eskdalemuir, Huddersfield Oakes, Buxton, Luton, Little Rissington, Hawkridge and Llandrindod Wells.

Notes on the weather of individual months

September. September was dull, wet and rather cold, but there were no reports of snow.

October. The first snows of the winter season were reported from Scotland on the 28th–31st as a wave on a cold front deepened as it approached Scotland, bringing stormy weather and periods of heavy rain to most of the United Kingdom. The rain gave place to thundery showers late on the 28th which were of snow in parts of the Highlands.

November. Snow during November was mainly confined to the second half of the month although it was reported from a few places in Scotland during the first week, and in northern England during the latter part of the second week.

Strong northerly winds in the rear of a depression which moved across Scotland on the 1st brought frequent snow showers to Scotland and northern England on the 2nd and 3rd; isolated showers were reported as far south as Tavistock. During the second week, easterly winds associated with a developing anticyclone in the region of northern Scandinavia spread over the British Isles on the 11th and this proved to be the beginning of an exceptionally cold spell which lasted in many areas until the 16th. Though there was a good deal of sunshine, eastern districts had frequent snow showers and during the period snow accumulated to a level depth of 5 inches on the moors and on the higher ground of north-east England.

One of the most noteworthy features of November was the heavy snowfall in north-east England during the last week of the month, brought by a series of vigorous and often complex depressions which became established over the British Isles and the North Sea from the 27th. Severe blizzards swept across northern England during the nights of the 26th–27th and 28th–29th. At Ushaw, level snow lay 22 inches deep, and more generally in Co. Durham snowdrifts up to 15 feet deep blocked many roads.

Over the month as a whole snow fell on about 20 days in the Cairngorms and ground was snow-covered for about the same length of time. Snow fell on 10–20 days in the Grampians, the North-west Highlands, the Southern Uplands, and over most of the Pennines, including the Peak District, and ground was snow-covered for a rather shorter time, about 10 days in the Grampians and Southern Uplands and about 8 days in the Peak District. Elsewhere for the most part snowfall occurred on fewer than 5 days, though slightly more frequently in the Lake District and north-east England and on the higher ground in Lincolnshire, in north Wales and in parts of the Midlands.

December. December snowfall was mainly confined to the first week and last few days of the month. Many parts of northern England lay under a foot of snow at the beginning of the month, and at Ushaw level snow was reported to be lying to a depth of 19 inches on the 2nd after a night of snow. The snow gradually gave place to thundery showers as a depression moved slowly south-east across Scotland to the North Sea. A northerly airstream in the rear of a following depression brought thundery showers to most districts on the 6th and 7th, the showers being of snow in parts of Scotland. The depth of snow lying was not increased much after the 2nd and most of the snow had gone by the 11th when heavy rain led to a rapid thaw and widespread floods.

A few snow flurries were reported from northern England on the 20th–22nd, but appreciable snow fell in the northern isles of Scotland early on Christmas morning and spread southwards. On Boxing Day (26th) snow and sleet* showers affected Scotland, Northern Ireland and north Wales; 9 inches of level snow were recorded in the Shetland Islands and much of northern England was under 1–2 inches of snow on the 29th, but most of it had gone by the 30th.

Over the month as a whole snow fell on 20–30 days in the Cairngorms and much of the ground was snow-covered for at least 25 days. The Grampians, the North-west Highlands, the extreme north of Scotland and the Southern Uplands had snow on 10–20 days. The Pennines reported snow on 5–10 days but snow was recorded at most other places on fewer than 5 days and most of south-east England and central southern England was free of snow.

January. Snow and sleet fell mainly during the 10th–21st but the northern isles of Scotland reported falls during the first and last weeks.

On the 9th an anticyclone became established over Scandinavia and associated easterly winds, bringing very cold air from eastern Europe, dominated the weather over the British Isles for the next week. On the 10th a small depression developed in the western English Channel, later moving into northern France. Heavy rain associated with the depression turned to snow in south Wales and central southern England. Snow was 15 inches deep at Tredegar on the 11th (and did not clear until the 26th), 4–6 inches deep in the Salisbury Plain area and 4 inches deep as far east as Long Sutton, Lincolnshire. Snow showers continued daily in the easterly airstream, especially in eastern England, and by the 15th level snow had accumulated to a depth of 12 inches in parts of Kent. A small depression which moved westwards from northern Germany brought periods of more continuous snow to east and south-east England, but this slowly died out on the 17th. Alderney reported an unusual fall of 8 inches on the 15th. On the 18th a trough off Cornwall brought slight snow to south-west England, but the next day pressure fell rapidly in the western English Channel and by the 20th a vigorous depression moved eastwards along the Channel, bringing severe gales in the south while sleet and snow spread across southern England during the morning and turned to freezing rain later in the day. Level snow was 5–10 inches deep over much of southern England on the 20th but most of it had dispersed by the 22nd.

During the month snow fell on about 15 days in the Cairngorms and on 10–15 days in the North-west Highlands, the extreme north of Scotland including the northern isles and the Southern Uplands. Over most of the remainder of the country snow fell on fewer than 10 days, being rather more frequent in the east than in the west.

February. As in January, snowfall in February was mainly during the middle of the month. During most of the first week weather was generally mild and wet,

* The term sleet is commonly used in this country to describe precipitation of snow and rain (or drizzle) together, or of snow melting as it falls, but it has no agreed international meaning.

but on the 6th a belt of snow moved south over the north of Scotland followed by strong easterly winds with temperatures near freezing-point. By the evening of the 7th the cold easterlies had spread southwards to northern England and there were periods of snow in the east and rain in the west. On the 10th a small depression moving eastwards along the Channel coast gave considerable rain in southern England and snow in East Anglia, the Midlands and south-east England. By the 11th the wintry weather had extended to the whole country except the extreme south-west, and with pressure remaining high to the north of Scotland, the cold easterly winds became established over the country for about a week. Snow fell in most districts and was moderate to heavy at times in the eastern half of the country. By the 14th it lay 4–6 inches deep in many parts of north-east England, and 12 inches deep around Ushaw. Troughs associated with a deep depression approaching the south-west of the British Isles from the Atlantic brought rain to southern England on the 18th and substantial falls of snow farther north, but the snow soon turned to rain. As the cold easterly winds were gradually replaced by a southerly airstream, temperatures rose over most of England and Wales on the 19th and this milder weather extended to much of Scotland on the 20th. Periods of heavy rain accompanied the onset of the milder air and this, coupled with rapidly thawing snow, led to some flooding in the Midlands and Peak District on the 20th and 21st and in parts of northern England and Scotland on the 22nd and 23rd.

Over the month as a whole snow fell on 10–15 days in the Cairngorms, the Lammermuir Hills and the northern Pennines, and on 5–10 days in the Grampians, the North-west Highlands, the extreme north of Scotland, the Southern Uplands, north-east England and the Peak District and on the higher ground in Lincolnshire. Elsewhere snowfall was mainly on fewer than 5 days and much of the south coast of England, south-west England and the south coast of Wales was free of snow.

March. Snow fell widely in Scotland on the 10th–13th in a north-westerly airstream in the rear of a depression moving eastwards towards southern Scandinavia. As the wind veered more towards the north on the 11th, scattered snow showers reached the Midlands and East Anglia, but the snow did not lie in England. In England and Wales snow was fairly widespread on the 24th–26th in a north-westerly airstream associated with a deepening depression over southern Scandinavia. Snow lay an inch deep at a number of places in the Midlands on the 25th. On the 26th a trough of low pressure moved eastwards across the country accompanied by sleet and snow, and behind this trough winds again veered to the north-west, bringing snow showers to northern districts on the 27th and as far south as Bournemouth on the 28th. Snow lay 3 inches deep at Buxton on the 26th and 2 inches deep at Halesowen on the 29th. There were light falls of snow in Scotland on the 30th and 31st as a small depression moved eastward skirting the north coast.

Snow was most frequent in the Cairngorms, falling on 15–20 days during the month. In most other places snow fell on fewer than 10 days except in the extreme north of Scotland and the northern isles where there were 10–15 days of snowfall. The extreme south-west of Cornwall had no snow during the month.

April. A shallow depression moving from south-west Ireland to the southern North Sea on the 1st brought more than 12 inches of level snow to parts of north-west England and north Wales with drifts up to 5 feet deep. Occasional snow and sleet continued in eastern England on the 2nd, but only a few isolated places reported snow on the 3rd as the depression moved away eastwards into Germany. Easterly winds brought a few isolated snow showers to the east coast on the 4th ahead of more general rain and snow on the 5th which continued well into the following night, accompanying a trough as it spread from the Atlantic north-eastwards over the country. After a spell of rather milder weather, temperature fell sharply on the 11th as easterly winds became established over the country, heralding about 10 days of wintry showers. These were mainly confined to the eastern half of the country on the 12th and 13th, but on the 14th troughs associated with a deep depression approaching from the Atlantic brought rain to the south-west and snow to most of the southern half of the country. The snow continued in most places for about 48 hours and covered the ground to a depth of 4–6 inches in many places, but it moved slowly northwards and gave way to rain as temperatures, which had been near freezing-point in many southern districts, slowly increased. Snow showers continued in Scotland for another 3 or 4 days, but the cold spell finally came to an end on the 21st as a ridge of high pressure moved across the country ahead of freshening winds from the south.

Snow fell on about 10–15 days over the Cairngorms, the Lammermuir Hills, the Cheviots and the northern Pennines. It fell on 5–10 days in the extreme north of Scotland and the northern isles, the Southern Uplands, over much of the Lake District, north-east England, the Peak District and north Wales. Most resorts on the south coast had only one day of snow during the month and the extreme south-west of Cornwall had none.

May. Snow or sleet fell on the higher peaks of the Cairngorms on 6 days during the month. Elsewhere snow was reported on 2–3 days in the North-west Highlands and on 1–2 days in the northern isles, the Lammermuir Hills and the Cheviots; it lay for 3–4 days in parts of the Cairngorms.

The summer months June–August

June. At *Achnagoichan* snow lay down to 3000 ft until the 4th and down to 3500 ft until the 8th; thereafter no snow was reported below 4000 ft.

At *Derry Lodge* snow was reported down to 3000 ft until the 11th, down to 3500 ft until the 15th and down to 4000 ft until the 26th. There were large fields (but not 50 per cent cover) down to 2500 ft until the 15th and on higher slopes after the 26th.

Factory (near Fort William). Patches on the highest peaks throughout the month.

Glenshee Lodge. Isolated patches on highest range on 4th, 5th and 8th.

July. *Achnagoichan.* Snow patches at 3000 ft throughout the month.
Derry Lodge. Snow patches (but less than 50 per cent cover) down to 3000 ft throughout the month.

August. *Achnagoichan.* Small patches still visible at 4000 ft.
Derry Lodge. Large snowfields still seen down to 3000 ft. Thirty feet of snow still in Garth Corrie.

Number of days with snow lying during the season
The seasonal duration of snow-cover in the mountainous regions of Great Britain is summarized in Tables 2 and 3 and is illustrated for a selected number of stations in Figure 1. At altitudes of 1000 ft or above, snow lay on about 70–90 days in the Cairngorms (on 115 days at Derry Lodge), in the Shetlands and on Cross Fell Edge; on 50–60 days in the Grampians, the North-west Highlands and the central Pennines, and on 40–50 days in the Peak District. At lower levels it lay on 30–40 days in north Wales, away from the coast, and over much of north-east England. From sea level to 300 ft the ground was snow-covered on 10–20 days south-east of the line from Bridlington to Aberporth, except in the Peak District, and generally on less than 10 days in central southern England, south-west England and most coastal areas from the Solway Firth in the west to Manby in the east.

Table 2 gives, where possible, four values for each month of the snow season, and for the season as a whole, for 100 stations. These values are the number of days with snow falling at the station, the number of days with snow lying at the station, the greatest depth (in inches) of undrifted snow lying at the station and the first date when this depth was attained. The values are arranged in a set pattern as in the following example:

10	6
5	8

This means that snow occurred at the station on 10 days during the month and that half or more of the ground in the immediate neighbourhood was snow-covered on 6 days; that 5 inches was the greatest depth of accumulated and undrifted snow observed and that this depth was first measured on the 8th.

Where no values are given for a particular month this means that no observations were made, in which case no data are given for the season as a whole. It will be noted that snow depths are not always available.

Duration of snow-cover in British mountains
Diagrams showing the distribution of snow-cover relative to height for 11 areas in the British mountains are given in Figure 1. They are based very largely on reports received from stations in Table 3 in which the heights quoted for various mountain ranges are the heights of the highest peaks visible from the observing stations. It was found necessary in a few cases to supplement these reports with reports from other stations in the same area, but the names of these additional stations have not been included in the table.

TABLE 2

Number of days with snow falling and snow lying and maximum depth in inches, with date, during the month/season

Station	Height *feet*	Oct.		Nov.		Dec.		Jan.		Feb.		Mar.		Apr.		May		Season	
Ollaberry	750	0	0	17	13	20	25	14	19	10	16	15	19	5	12	0	0	81	104
		—		8	29	7	29	7	24	6	11	5	31	6	1	—		8	29/11
RAF Mossy Hill	750	0	0	10	16	4	22	6	13	8	12	10	7	8	1	1	0	47	71
		—		7	30	5	1	3	19	2	13	1	24	—		—		7	30/11
Stenness	75	0	0	8	8	7	4			5	12	5	4	3	1	0	0		
		—		4	22	3	28			2	20	2	25	1	1	—			
Inverpolly	50	0	0	12	10	11	5	3	0	6	0	6	2	3	1	—		41	18
		—		3	22	5	28	—		—		2	25	1	1	—		5	28/12
Glackour	80	0	0	10	10	3	17	1	4	2	1	5	6	0	0	0	0	21	38
		—		11	28	2	26	1	3	½	7	2	24	—		—		11	28/11
Shin	316	0	0	17	11	14	24	4	6	12	17	8	4	2	2	3	0	60	64
		—		4½	27	12½	8	1	22	4	20	3	26	1	1	—		12½	8/12
Kinlochewe	76	0	0	14	9	8	11	6	3	6	1	7	3	3	0	2	0	46	27
		—		4	29	4½	29	—		—		—		—		—		4½	29/12
Fairburn	500	0	0	11	11	11	21	3	4	5	20	7	5	4	0	0	0	41	61
		—		5	30	6	2	4	1	10	21	3	24	—		—		10	21/2
Strathconon	350	0	0	14	9	10	6	7	6	4	5	8	9	6	6	0	0	49	41
		—		5	22	6	27	—		10	19	5	24	—		—		10	19/2
Glenferness	700	0	0	19	20	17	31	10	20	14	16	13	8	10	1	0	0	83	96
		—		12	29	16	3	½	14	5	19	½	22	—		—		16	3/12
Drummuir	500	0	0	18	18	18	14			11	16	11	3	10	1	2	0		
		—		7	29	7½	9			5	20	4	25	—		—			
Achnagoichan	1000	0	0	19	15	22	20	10	13	11	14	11	7	13	3	2	0	88	72
		—		14	30	14	1	—		—		—		—		—		14	30/11
Isle of Rhum	15	0	0	7	3	5	0	0	0	2	0	6	2	1	0	0	0	21	5
		—		1	22	—		—		—		2	25	—		—		2	25/3
Glenshero Lodge	884	0	0	20	10	19	20	5	0	4	6	13	9	4	2	0	0	65	47
		—		12	28	11	1	—		1	6	8	24	¼	18	—		12	28/11
Derry Lodge	1400	4	2	19	19	20	23	17	28	14	20	20	9	15	16	6	0	115	117
		—		—		4	2	3	24	4	20	6	24	2½	13	—		6	24/3
Adit 3	750	0	0	5	8	6	18	0	0	2	2	9	7	2	0	0	0	24	35
		—		8	29	—		—		½	7	—		—		—		8	29/11
Fort William	90	0	0	5	6	6	4	1	1	2	0	9	6	2	0	0	0	25	17
		—		2½	29	—		½	23	—		1¼	28	—		—		2½	29/11
Fersit	800	0	0	9	10	11	18	3	5	4	5	9	9	3	1	0	0	39	48
		—		16	28	20	5	1	24	1	7	7	25	1	1	—		20	5/12
Glenshee Lodge	1100	0	0	0	0	16	29	13	22	11	16	13	6	13	5	0	0	66	78
		—		—		11	1	2	25	7	20	4	26	1	16	—		11	1/12
Blair Atholl	400	0	0	7	13	8	22	2	5	4	12	5	2	6	1	0	0	32	55
		—		6½	29	8	4	½	20	1	8	3	24	½	20	—		8	4/12
Ardtalnaig	427	0	0	10	4	6	8	5	1	11	12	6	1	7	1	0	0	45	27
		—		3	23	4	2	½	21	4	10	—		1	15	—		4	2/12
Glengyle	380	0	0	5	4	5	18	3	8	9	6	6	3	6	3	0	0	34	42
		—		3	29	10	4	1	4	5	19	3½	26	½	20	—		10	4/12
Stronachlachar	376	0	0	6	5	6	14	3	2	8	5	6	1	7	1	0	0	36	28
		—		2½	29	4	5	⅛	24	2½	19	2¼	26	½	20	—		4	5/12
Loch Arklet	480	0	0	4	7	5	15	4	3	6	15	11	4	2	4	0	0	32	48
		—		1½	29	7	3	1	4	5	19	1½	26	1	6	—		7	3/12
Brig o' Turk	275	0	0	6	6	3	16	0	0	4	11	1	1	0	0	0	0	14	34
		—		¾	23	8	5	—		4	19	1½	26	—		—		8	5/12

1 inch = 25·4 millimetres 100 feet = 30·48 metres

TABLE 2 (*continued*)

Number of days with snow falling and snow lying and maximum depth in inches, with date, during the month/season

Station	Height *feet*	Oct.	Nov.	Dec.	Jan.	Feb.	Mar.	Apr.	May	Season
Loch Venachar	275	0 0	5 2	4 17	1 0	3 6	3 2	0 0	0 0	16 27
		—	½ 30	11 5	—	5 19	2 26	—	—	11 5/12
Couligarton	450	2 0	11 3	4 18	1 3	8 12	5 1	11 2	0 0	42 39
		—	2 30	8 2	1 1	6 19	4 26	¼ 19	—	8 2/12
Rhuvaal	63	0 0	11 0	1 0	0 0	3 1	0 0	0 0	0 0	15 1
		—	—	—	—	—	—	—	—	—
Cramond	80	0 0	10 6	2 6	6 6	11 7	2 0		0 0	
		—	3 15	1 1	½ 16	3 19	—		—	
Hungry Snout	715	0 0	16 16	9 13	9 14	12 18	3 3	13 8	1 0	63 72
		—	7½ 16	5 2	4 16	10 19	½ 26	1 15	—	10 19/2
Hopes	736	0 0	4 6	5 4	9 12	14 14	6 1	12 12	0 0	50 49
		—	4 28	1 29	1 16	11½ 13	½ 24	3 14	—	11½ 13/2
Portmore Res.	1000	0 0	19 19	13 13	15 13	20 17	15 15	17 17	2 0	101 94
		—	4 26	2 1	½ 9	6 19	½ 27	1 9	—	6 19/2
Broughton	741	0 0	7 9	4 4	5 10	7 9	6 1	8 0	0 0	37 33
		—	3 29	3½ 2	—	2 19	—	—	—	3½ 2/12
Sourhope	900	0 0	13 16	6 11	14 9	10 13	2 0	5 2	0 0	50 51
		—	5 4	6 2	3 16	—	—	4 15	—	6 2/12
Leadhills	1270	0 0	11 12	9 10	7 0	10 7	6 2	9 3	0 0	52 34
		—	9 27	8 2	—	3 19	1 26	1 16	—	9 27/11
Lowther Hill	2377	1 0	13 20	15 21	12 19	10 14	11 10	11 10	0 0	73 94
		—	8 24	—	—	—	—	—	—	8 24/11
Eskdalemuir	772	1 0	18 9	12 10	15 3	17 12	10 1	15 1	0 0	88 36
		—	7 27	9 2	¼ 14	5 19	1 26	⅓ 16	—	9 2/12
Forrest Lodge	500	0 0	5 8	3 5	0 0	7 7	1 1	1 1	0 0	17 22
		—	9½ 29	10 1	—	4 18	1 23	½ 15	—	10 1/12
Catcleugh	800	0 0	7 15	7 14	7 14	9 17	3 0	10 2	0 0	43 62
		—	6½ 27	9 2	2½ 22	12½ 19	—	2 15	—	12½ 19/2
Burradon	220	0 0	9 11	4 7	9 8	9 13	1 0	8 4	0 0	40 43
		—	9 27	9 2	—	5 14	—	7 2	—	9 27/11
Gosforth	170	0 0	13 7	4 8	9 8	10 12	2 0	9 2	0 0	47 37
		—	10 30	8 1	2 15	5 14	—	6 2	—	10 30/11
Alston	1070	1 0	15 18	6 11	11 9	13 12	3 4	14 13	0 0	63 67
		—	10½ 27	2¾ 2	—	7½ 19	—	—	—	10½ 27/11
Patterdale	520	0 0	5 7	3 1	4 4	8 13	1 1	4 3	0 0	25 29
		—	3 26	1½ 29	½ 16	2½ 19	2 26	4 2	—	4 2/4
Copt Howe	392	0 0	6 6	6 6	7 0	11 4	5 1	8 2	0 0	43 19
		—	7 27	10 2	—	½ 16	3½ 26	8 2	—	10 2/12
Kildale Hall	575	0 0	14 12	1 12	10 14	1 12	7 0	12 6	0 0	45 56
		—	10½ 30	10½ 1	6 16	6½ 19	—	8 2	—	10½ 30/11
Moorland Cottage	1125	0 0	20 15	7 8	15 12	13 13	12 1	14 10	0 0	81 59
		—	—	—	—	—	—	—	—	—
High Nibthwaite	170	0 0	0 0	3 3	5 1	6 5	3 1	5 0	0 0	22 10
		—	—	3 29	—	—	—	—	—	3 29/12
Ribblehead	1025	0 0	10 10	1 9	2 6	1 8	1 1	3 1	0 0	18 35
		—	9 30	7 1	1 15	3 9	2 26	8 2	—	9 30/11
Litton	820	0 0	11 13	5 9	7 12	11 13	6 2	12 7	0 0	52 56
		—	15 29	16 2	4 16	3 19	2 26	9 2	—	16 2/12
High Mowthorpe	574	0 0	8 9	3 3	8 10	7 11	3 1	8 3	0 0	37 37
		—	5 27	5 2	3½ 17	4 10	½ 26	4 2	—	5 27/11

TABLE 2 (*continued*)

Number of days with snow falling and snow lying and maximum depth in inches, with date, during the month/season

Station	Height *feet*	Oct.	Nov.	Dec.	Jan.	Feb.	Mar.	Apr.	May	Season
Slaidburn	629	0 0	4 6	2 5	5 1	6 8	5 1	7 9	0 0	29 30
		—	16 30	16 1	—	2 9	1 26	15 2	—	16 30/11
Chelker	730	0 0	12 8	2 3	4 9	9 12	3 0	7 5	0 0	37 37
		—	12 29	12 1	2 15	4 9	—	4 2	—	12 29/11
Nelson	500	0 0	12 7	3 3	8 9	9 7	5 1	7 5	0 0	44 32
		—	4½ 30	4 1	1⅓ 16	½ 9	¼ 25	10½ 2	—	10½ 2/4
Belmont Spring	810	0 0	11 6	6 6	6 9	2 11	5 2	7 6	0 0	37 40
		—	4 28	4 1	¾ 16	½ 8	1 25	5 1	—	5 1/4
Cragg (Bacup)	1348	0 0	13 8	9 9	7 9	11 12	5 3	10 9	0 0	55 50
		—	8½ 29	8 1	1½ 17	1½ 9	—	7½ 2	—	8½ 29/11
Strinesdale	791	0 0	7 4	2 0	3 2	0 0	2 0	2 2	0 0	16 8
		—	4 29	—	2 14	—	—	4 1	—	4 29/11
Chew Res.	1600	0 0	15 6	9 9	9 13	9 12	4 3	12 13	0 0	58 56
		—	15 30	15 1	3 21	3 10	3 26	3 2	—	15 30/11
Hallbroom	1050	0 0	7 8	0 0	6 14	4 16		8 21	0 0	
		—	15 30	—	4 17	7 16		4 3	—	
Howden Dam	845	0 0	13 6	6 5	12 11	12 11	10 3	10 9	0 0	63 45
		—	6½ 29	—	2 17	2 9	2½ 26	6 2	—	6½ 29/11
Redmires	1100	0 0	8 8	2 2	7 7	7 12	5 6	8 12	0 0	37 47
		—	17¼ 29	8 1	3¼ 21	7¼ 15	3⅞ 26	8½ 2	—	17¼ 29/11
Grenoside	560	0 0	10 5	3 4	8 10	8 10	3 2	4 6	0 0	36 37
		—	7¾ 30	7½ 1	2½ 17	3½ 11	¼ 24	2½ 2	—	7¾ 30/11
Bamford	509	0 0	8 5	4 1	4 6	9 11	2 0	7 4	0 0	34 27
		—	1½ 29	1 28	2 17	1 8	—	2 3	—	2 17/1
Southrey	20	0 0	6 2	1 0	5 0	2 0	5 0	3 0	0 0	22 2
		—	⅜ 23	—	—	—	—	—	—	⅜ 23/11
Hunstanton	45	0 0	2 1	0 0	7 8	5 2	0 0	1 0	0 0	15 3
		—	½ 15	—	1⅞ 16	1¾ 1	—	—	—	1⅞ 16/1
Derby, Brayfield Rd	233	0 0	9 3	3 1	8 8	8 2	3 0	7 1	0 0	38 15
		—	½ 30	—	1¼ 21	1½ 10	—	1½ 15	—	1½ 10/2
Hednesford	770	0 0	3 1	0 0	2 1	2 2	1 0	3 3	0 0	11 7
		—	½ 28	—	¾ 17	3 10	—	1 20	—	3 10/2
Countesthorpe	300	0 0	6 0	0 0	7 8	3 4	2 0	4 0	0 0	22 12
		—	—	—	1 17	2 11	—	—	—	2 11/2
West Bromwich	543	0 0	5 0	0 0	10 2	4 3	1 0	1 1	0 0	21 6
		—	—	—	1½ 21	1½ 11	—	1 15	—	1½ 21/1
Market Harborough	300	0 0	4 2	0 1	7 7	3 5	2 1	2 1	0 0	18 17
		—	½ 23	½ 1	½ 16	3 11	½ 25	½ 14	—	3 11/2
Barnt Green	620	0 0	5 2	1 0	8 12	5 7	4 1	4 3	0 0	27 25
		—	¾ 30	—	1 16	—	—	—	—	1 16/1
Baunton	396	0 0	2 0	0 0	3 1	1 0	3 0	3 3	0 0	12 4
		—	—	—	2 10	—	—	1½ 14	—	2 10/1
Hinksey Hill	330	0 0	3 0	0 0	4 8	1 0	4 2	2 2	0 0	14 12
		—	—	—	1 16	—	—	—	—	1 16/1
Garston	256	0 0	4 1	1 0	6 7	1 0	3 0	2 0	0 0	17 8
		—	—	—	1 16	—	—	—	—	1 16/1
Little Chalfont	425	0 0	7 2	1 0	8 8	5 0	4 1	2 1	0 0	27 12
		—	¼ 23	—	1½ 21	—	¼ 26	2 15	—	2 15/4
Rayleigh	240	0 0	3 0	0 0	5 5	0 0	3 0	1 0	0 0	12 5
		—	—	—	1 17	—	—	—	—	1 17/1

TABLE 2 (*continued*)

Number of days with snow falling and snow lying and maximum depth in inches, with date, during the month/season

Station	Height *feet*	Oct.	Nov.	Dec.	Jan.	Feb.	Mar.	Apr.	May	Season
Eastcote	175	0 0	3 1	0 0	4 6	0 0	4 0	1 0	0 0	12 7
		—	½ 17	—	¾ 21	—	—	¾ 5	—	¾ 21/1
Southend	90	0 0	4 1	0 0	4 4	0 0	2 0	1 1	0 0	11 6
		—	1 22	—	2 13	—	—	2 14	—	2 13/1
Chisledon	510	0 0	7 1	0 0	9 13	5 1	6 2	4 3	0 0	31 20
		—	1 30	—	3 17	½ 15	1 25	4 15	—	4 15/4
Charlton Park	150	0 0	3 1	0 0	2 3	0 0	0 0	1 1	0 0	6 5
		—	½ 23	—	1 17	—	—	2 15	—	2 15/4
Camberley	216	0 0	5 0	0 0	9 8	2 0	3 0	2 1	0 0	21 9
		—	— —	—	¾ 11	—	—	1¼ 14	—	1¼ 14/4
Watchet	155	0 0	1 0	0 0	6 2	0 0	0 0	2 0	0 0	9 2
		—	— —	—	4 11	—	— —	— —	—	4 11/1
Biddenden	165	0 0	4 0	0 0	6 9	0 0	3 2	1 1	0 0	14 12
		—	— —	— —	8 16	—	—	1 14	—	8 16/1
Washington	170	0 0	5 1	0 0	7 11	0 0	3 0	1 2	0 0	16 14
		—	— —	—	1¼ 11	—	—	1½ 14	—	1½ 14/4
Stoke Abbot	350	0 0	2 0	0 0	1 3	1 0	3 0	1 1	0 0	8 4
		—	— —	— —	1 11	—	—	1 14	—	1 11/1
Batworthy	1250	0 0	4 2	3 0	2 1	0 0	1 1	2 1	0 0	12 5
		—	8½ 15	—	¾ 20	—	1 25	3 14	—	8½ 15/11
Plymouth	200	0 0	4 2	2 1	5 0	0 0	2 0	1 0	0 0	14 3
		—	¼ 22	—	—	—	—	—	—	¼ 22/11
Bwlch Tunnel	900	0 0	8 8	3 3	4 7	5 8	3 3	8 8	0 0	31 37
		—	10 29	—	3 14	3 10	½ 28	3 19	—	10 29/11
Mount Pleasant	523	0 0	5 3	2 2	2 9	5 4	3 4	4 2	0 0	21 24
		—	4 29	—	2 15	2 8	1 28	2 2	—	4 29/11
Capel Curig	640	0 0	7 1	3 0	7 4	3 3	2 2	4 2	0 0	26 12
		—	4 29	—	2 15	—	—	1½ 15	—	4 29/11
Alwen	1200	0 0	12 7	6 11	8 10	8 11	6 6	8 6	0 0	48 51
		—	5½ 30	4 1	2 15	1 11	¾ 29	1⅝ 20	—	5½ 30/11
Clawdd Newydd	998	0 0	8 5	1 7	3 13	4 4	5 0	8 7	0 0	29 36
		—	15 28	2 27	2 14	2 9	—	3 17	—	15 28/11
Gronwen, Cae Llwyd	1000	0 0	5 7	2 3	7 8	6 11	4 3	7 5	0 0	31 37
		—	10 29	8 1	¾ 17	3½ 12	1 25	2½ 16	—	10 29/11
Dolgellau	90	0 0	4 2	1 0	5 2	4 0	1 1	4 0	0 0	19 5
		—	¼ 21	—	½ 16	—	¼ 25	—	—	½ 16/1
Evancoyd	744	0 0	5 3	1 1	8 14	5 5	2 1	4 3	0 0	25 27
		—	2 30	1 1	3½ 10	3 10	—	4 15	—	4 15/4
Tairbull	660	0 0	3 3	1 0	4 1	2 0	2 0	2 0	0 0	14 4
		—	12 16	—	6 10	—	—	—	—	12 16/11
Llangynidr	1410	0 0	5 5	0 0	7 7	1 3	3 3	3 3	0 0	19 21
		—	4½ 29	—	6 10	1 15	2½ 25	5 14	—	6 10/1
Merthyr Tydfil	770	0 0	6 5	4 2	6 15	3 1	1 1	2 3	0 0	22 27
		—	2 30	1¼ 1	5¾ 11	—	—	4 14	—	5¾ 11/1
Maesteg	590	0 0	4 4	2 2	3 7	2 0	2 1	2 3	0 0	15 17
		—	1 29	—	1½ 10	—	¾ 25	6 14	—	6 14/4
Swansea	75	0 0	3 1	1 1	4 1	0 0	2 1	2 1	0 0	12 5
		—	—	—	¼ 21	—	¼ 25	1 14	—	1 14/4
Port St Mary	25	0 0	9 0	2 1	3 0	6 0	4 0	4 0	0 0	28 1
		—	—	—	—	—	—	—	—	—

TABLE 3

Number of days with snow lying in the British mountains, 1965–66

(*a*) Near summit. (*b*) At about 2500 feet. (*c*) At station level.

		Oct.	Nov.	Dec.	Jan.	Feb.	Mar.	Apr.	May	Season
A'Chailleach (3276 ft)	*a*	0	30	31	31	28	31	30	0	181
Station: Glackour (80 ft)	*b*	0	15	31	31	22	31	22	0	152
Ross-shire	*c*	0	10	17	4	1	6	0	0	38
Cairngorms (4296 ft)	*a*	4	30	31	31	27	31	26	31	211
Station: Achnagoichan (1000 ft)	*b*	3	23	31	31	27	24	26	4	169
Inverness-shire	*c*	0	15	20	13	14	7	3	0	72
Creag Meagaidh (3700 ft)	*a*	3	22	31	31	28	31	30	11	187
Station: Fersit (800 ft)	*b*	2	21	31	31	24	19	23	1	152
Inverness-shire	*c*	0	10	18	5	5	9	1	0	48
Ben Vane (3004 ft)	*a*	0	18	26	22	20	13	22	0	121
Station: Loch Arklet (480 ft)	*b*	0	18	26	22	20	13	22	0	121
Stirlingshire	*c*	0	7	15	3	15	4	4	0	48
Ben More (3170 ft)	*a*	0	18	8	1	10	6	6	0	49
Station: Rhuvaal (63 ft)	*b*	0	18	8	1	10	6	6	0	49
Argyll	*c*	0	0	0	0	1	0	0	0	1
Kells Range (2668 ft)	*a*	0	24	15	0	12	8	7	0	66
Station: Forrest Lodge (500 ft)	*b*	0	24	15	0	12	8	7	0	66
Kirkcudbrightshire	*c*	0	8	5	0	7	1	1	0	22
Cross Fell (2930 ft)	*a*	0	21	31	31	28	29	30	0	170
Station: Alston (1070 ft)	*b*	0	21	31	31	28	29	30	0	170
Cumberland	*c*	0	18	11	9	12	4	13	0	67
Helvellyn (3118 ft)	*a*	0	20	14	15	18	10	7	0	84
Station: Patterdale (520 ft)	*b*	0	19	13	13	17	7	6	0	75
Westmorland	*c*	0	7	1	4	13	1	3	0	29
Snowdonia (3560 ft)	*a*	0	15	7	6	4	3	8	0	43
Station: Capel Curig (640 ft)	*b*	0	14	6	6	3	3	7	0	39
Caernarvonshire	*c*	0	1	0	4	3	2	2	0	12
Cader Idris (2927 ft)	*a*	0	12	9	6	9	7	7	0	50
Station: Dolgellau (90 ft)	*b*	0	12	9	6	9	7	7	0	50
Merioneth	*c*	0	2	0	2	0	1	0	0	5
Brecon Beacons (2906 ft)	*a*	0	13	16	26	11	9	23	0	98
Station: Tairbull (660 ft)	*b*	0	13	16	26	11	9	23	0	98
Brecknockshire	*c*	0	3	0	1	0	0	0	0	4

100 feet = 30·48 metres.

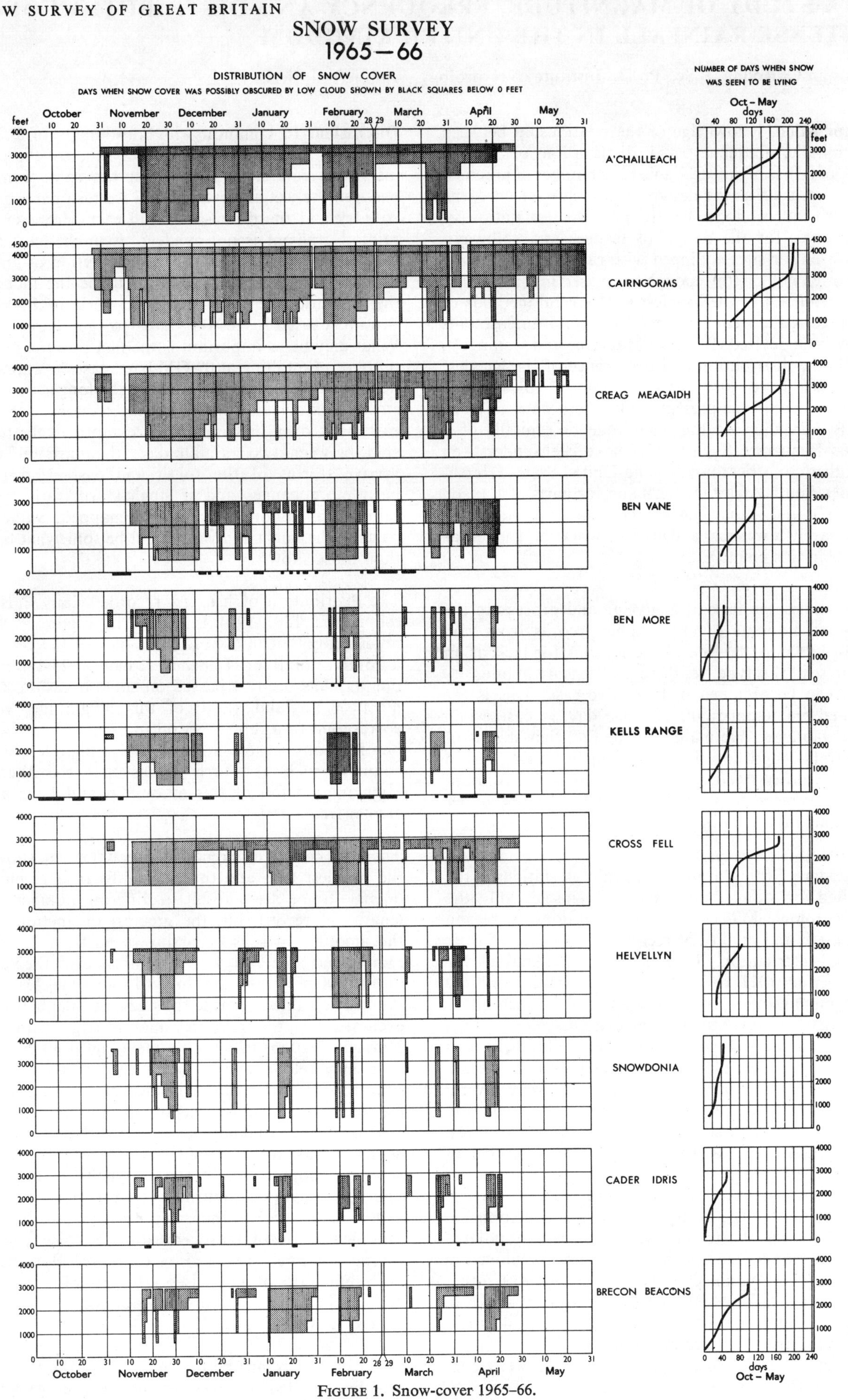

FIGURE 1. Snow-cover 1965–66.

3 A STUDY OF MAGNITUDE, FREQUENCY AND DISTRIBUTION OF INTENSE RAINFALL IN THE UNITED KINGDOM

by John C. Rodda, B.Sc., Ph.D., Institute of Hydrology, Wallingford, Berks.

Introduction. Knowledge of the relationship between magnitude, frequency and distribution of intense falls of rain is highly desirable for a number of practical purposes such as the design of drainage schemes, the assessment of flood hazard and the estimation of discharges for the design of bridges and spillways. But even for sites equipped with rain-gauges, records are often not lengthy enough to be used in establishing the frequency of the heaviest fall. Several methods of estimating the frequency and duration of intense falls have been devised (Bilham,[1] Maclean[2] and Collinge[3]), while the distribution and occurrence of the heaviest amounts have also been described (Bleasdale[4]). However, until the present study, there were no maps of Britain showing the magnitude of rainfalls of a given frequency. Such studies have been undertaken in other countries; notably the United States (Hershfield[5]), South Africa (Reich[6]) and for parts of Canada (Storr[7]) and Japan (Asai, Hatano and Hatano[8]), but rarely for such a dense network of gauges as exists in this country.

The wearisome task of analysing and processing the copious amounts of autographic records from rain recorders must be one of the reasons for lack of an exercise of this type, yet there is a measure of intensity that can be obtained with relative ease. This is the annual maximum rainfall in one day. It has certain disadvantages, due mainly to the fact that storms of arbitrary length are fitted to a single time-base, but on the other hand, the network of standard rain-gauges is more dense, better distributed and of longer standing than the rain-recorder network.

Procedure. From the 6000 or so rainfall stations in the United Kingdom, about 100 have been selected by the Meteorological Office as reference stations, by reason of length and quality of record and because of their spatial distribution. Records from these stations for the period 1911 to 1960 were scrutinized and the maximum rainfall recorded in one day for each of the 50 years was extracted. Then an extreme-value analysis of the Gumbel Type 1 distribution (Gumbel[9]) was carried out using the records of each station in turn. Even with this number of gauges, there were several gaps in the network, so records were obtained from stations within these areas to increase the coverage. In addition, records were included for a number of sites where exceptional daily falls have occurred. Not all these extra stations possessed 50 years of records, so that the magnitude–frequency relationship found for the shorter period had to be extrapolated in some cases. A list of the 121 stations employed in the study is contained in Table 1 along with other pertinent details, while their distribution is shown in Figure 1. From results of the analyses, maps illustrating the distribution and magnitude of falls of rain for 10, 20, 50 and 100-year return periods were constructed (Figures 2 to 5).

Discussion. In common with most maps of Britain showing the distribution of one or other of the parameters of rainfall, these exhibit the same contrast between the wetter northern and western parts of the country and the drier south and east. However, the rainfall gradient is not as steep as might have been anticipated on the basis of the distribution of annual rainfall. One reason for this might be the fact that the limitations of the gauge network in highland Britain caused excessive falls to go unrecorded. What may be as important is the fact that the probability of rain falling at 09 GMT increases towards the north and west (Figure 6). Hence the longer-duration storms associated with frontal systems moving eastwards over high ground are more likely to be split between two rainfall days than are the convective storms of the south and east. Had there been more gauges and if single-storm rather than rainfall-day amounts had been mapped, values in parts of the north and west might be somewhat larger than those shown.

The distribution of heaviest recorded daily falls for the period 1881–1964 is shown in Figure 7. It is noteworthy that in almost every part of the country, a total rainfall of 4 inches (1 inch = 25·4 mm) in one day has been surpassed on some occasion. Over most of England a fall of this magnitude would correspond to a return period in excess of 100 years, but for considerable areas in the remainder of the country, such as north Wales and the Western Highlands, a fall of this amount would have a far greater frequency.

There are certain differences between the map showing the heaviest falls and those for falls of a given frequency. Some are a result of differing numbers and lengths of record, but the greatest distinction is in the amounts of falls occurring in parts of Norfolk, Dorset and Somerset. For these areas even the values on the 100-year map are three to four times less than the heaviest falls, because the heaviest falls had to be excluded from the frequency analysis (e.g. Sprowston, Norfolk: 7·20 in) to obtain a meaningful magnitude–frequency relationship (Figures 8(*a*) and (*b*)). Such occurrences make for considerable difficulties in frequency analyses. An exceptional amount might be a portion of the same population but with an enormous return period or it could be part of an entirely separate population. However, neither of these alternatives can be confirmed to the exclusion of the other. In the case of the exceptional falls in Norfolk and the south-west, they were estimated to have return periods of one million years or more! Fortunately such difficulties do not arise in the case of records for stations in the north and west, even for sites with the largest falls (Figure 9).

The different return-period maps show the falls that will be equalled or exceeded on the average within

FIGURE 1

Stations used for Figures 2 – 5

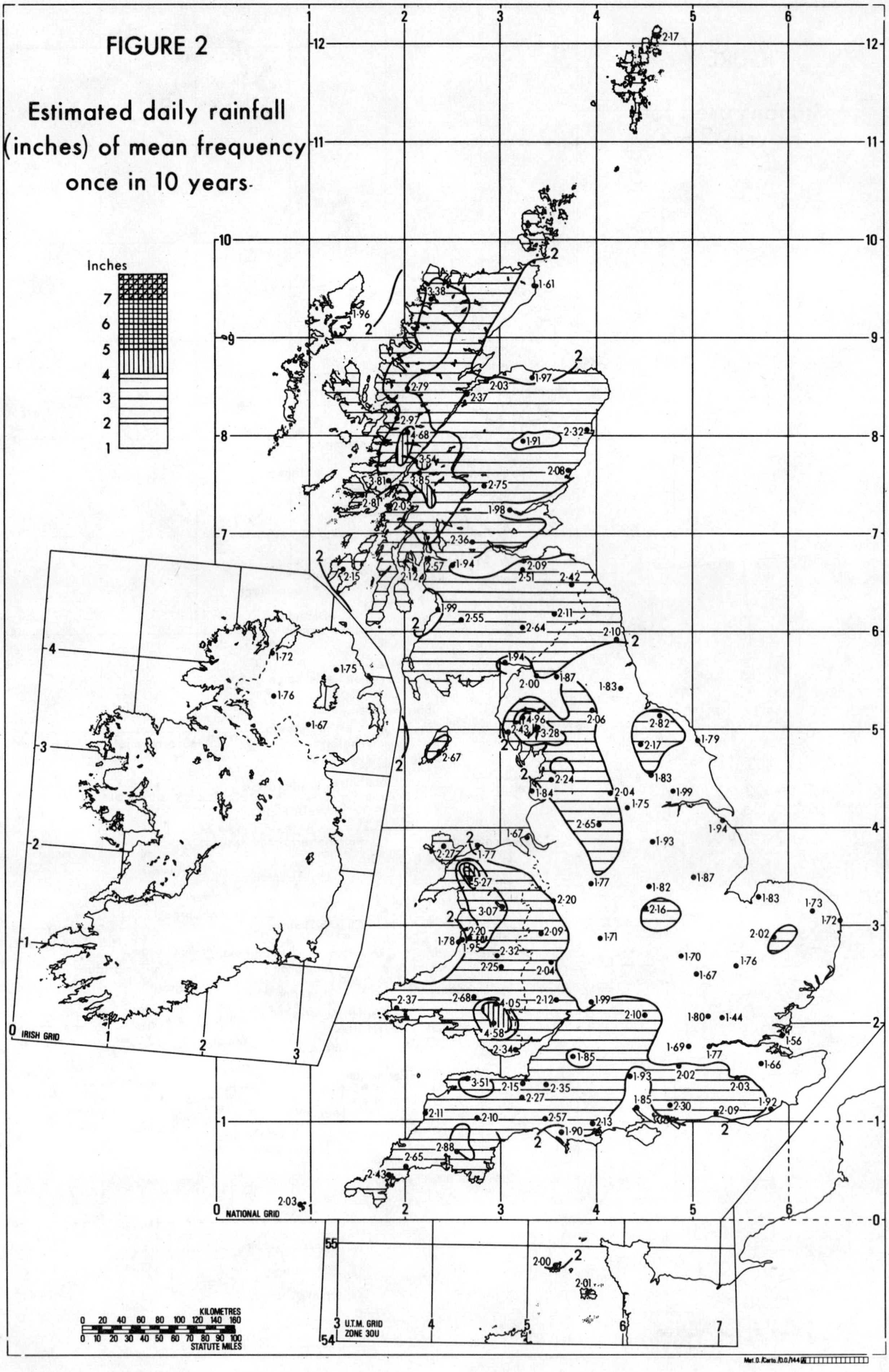
FIGURE 2
Estimated daily rainfall (inches) of mean frequency once in 10 years
Inches
7
6
5
4
3
2
1
IRISH GRID
NATIONAL GRID
U.T.M. GRID ZONE 30U
KILOMETRES
STATUTE MILES

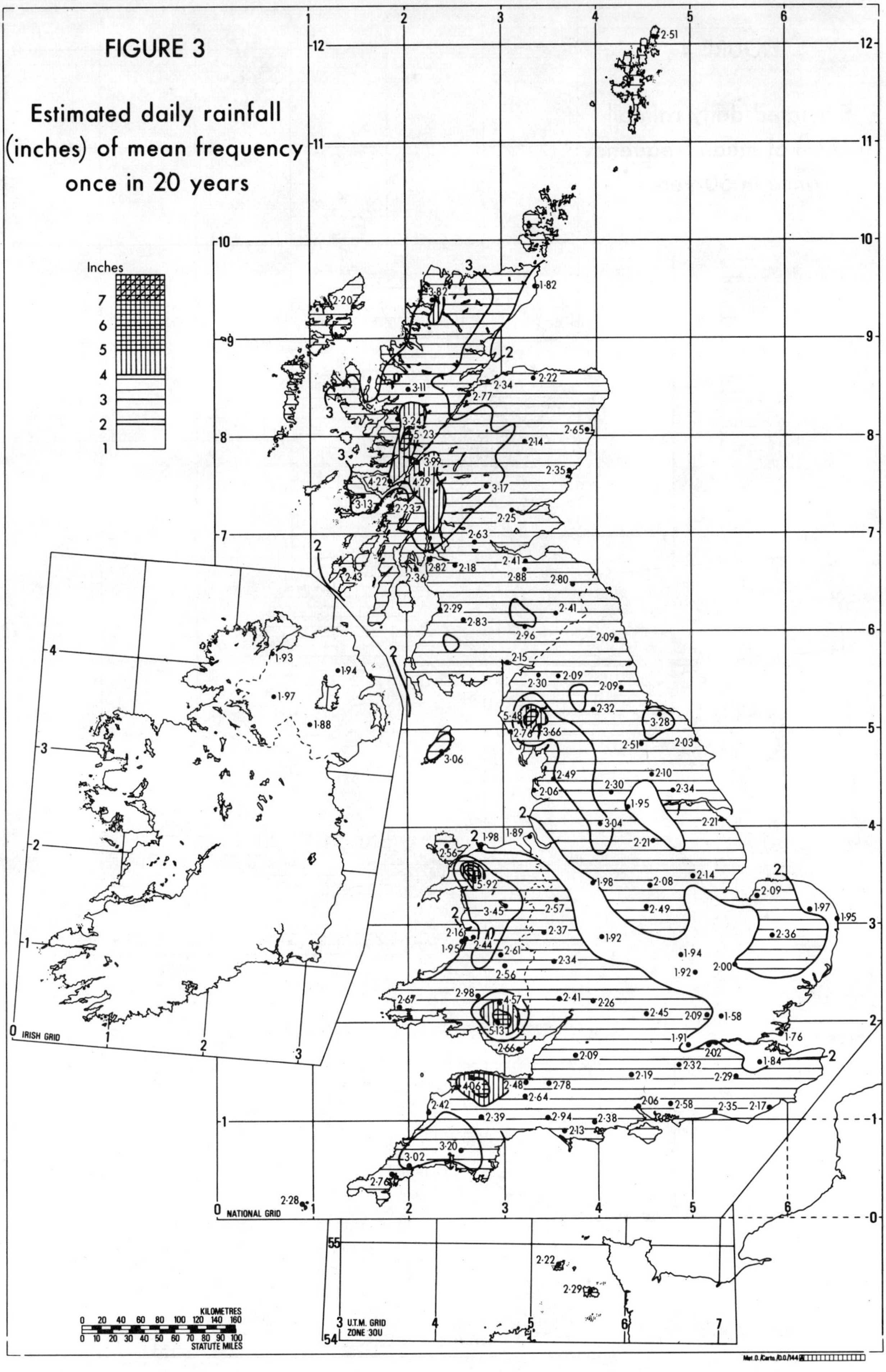

FIGURE 3

Estimated daily rainfall (inches) of mean frequency once in 20 years

Met.O./Carto./D.O./1449

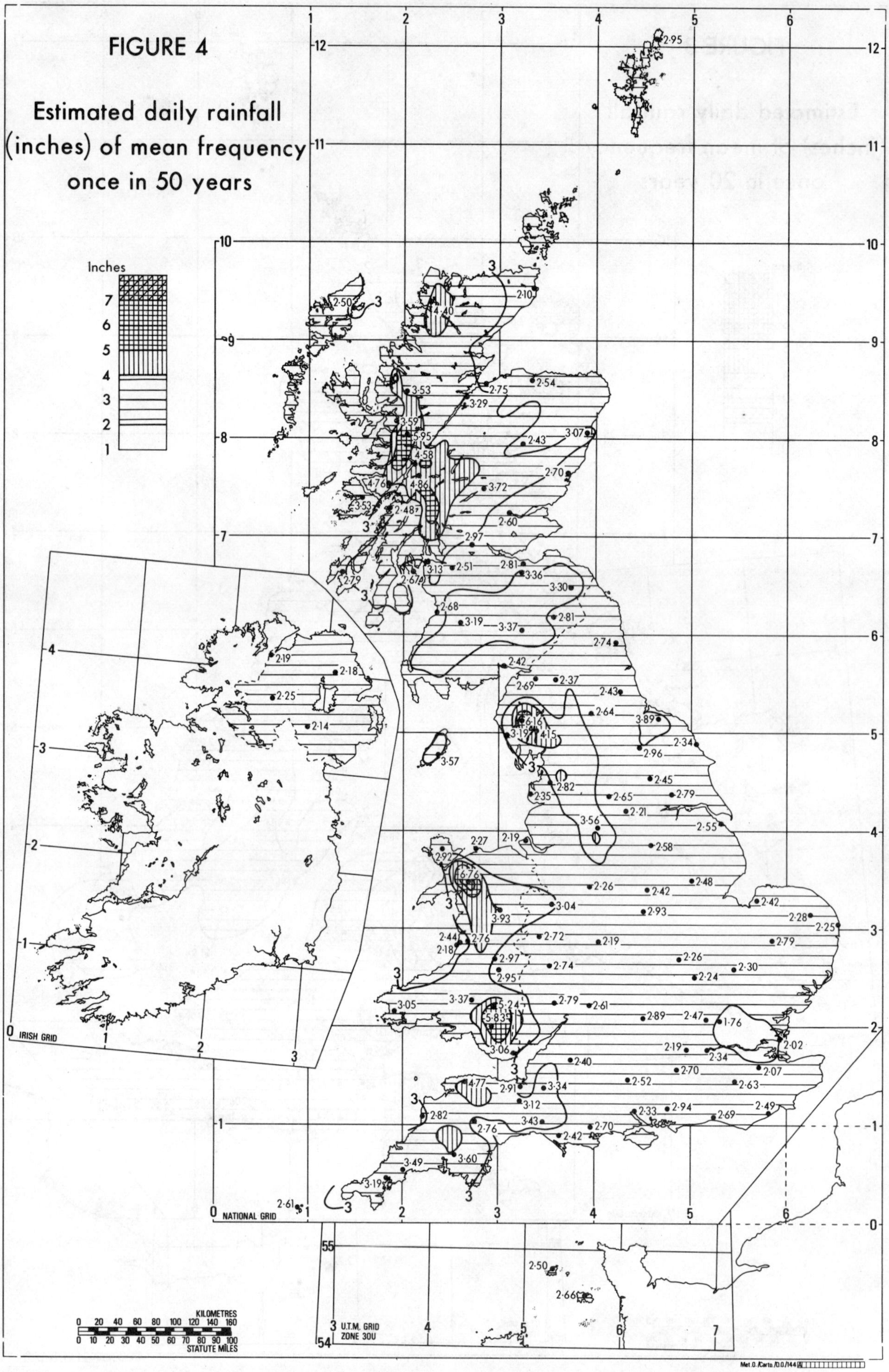

FIGURE 4

Estimated daily rainfall (inches) of mean frequency once in 50 years

FIGURE 5

Estimated daily rainfall (inches) of mean frequency once in 100 years

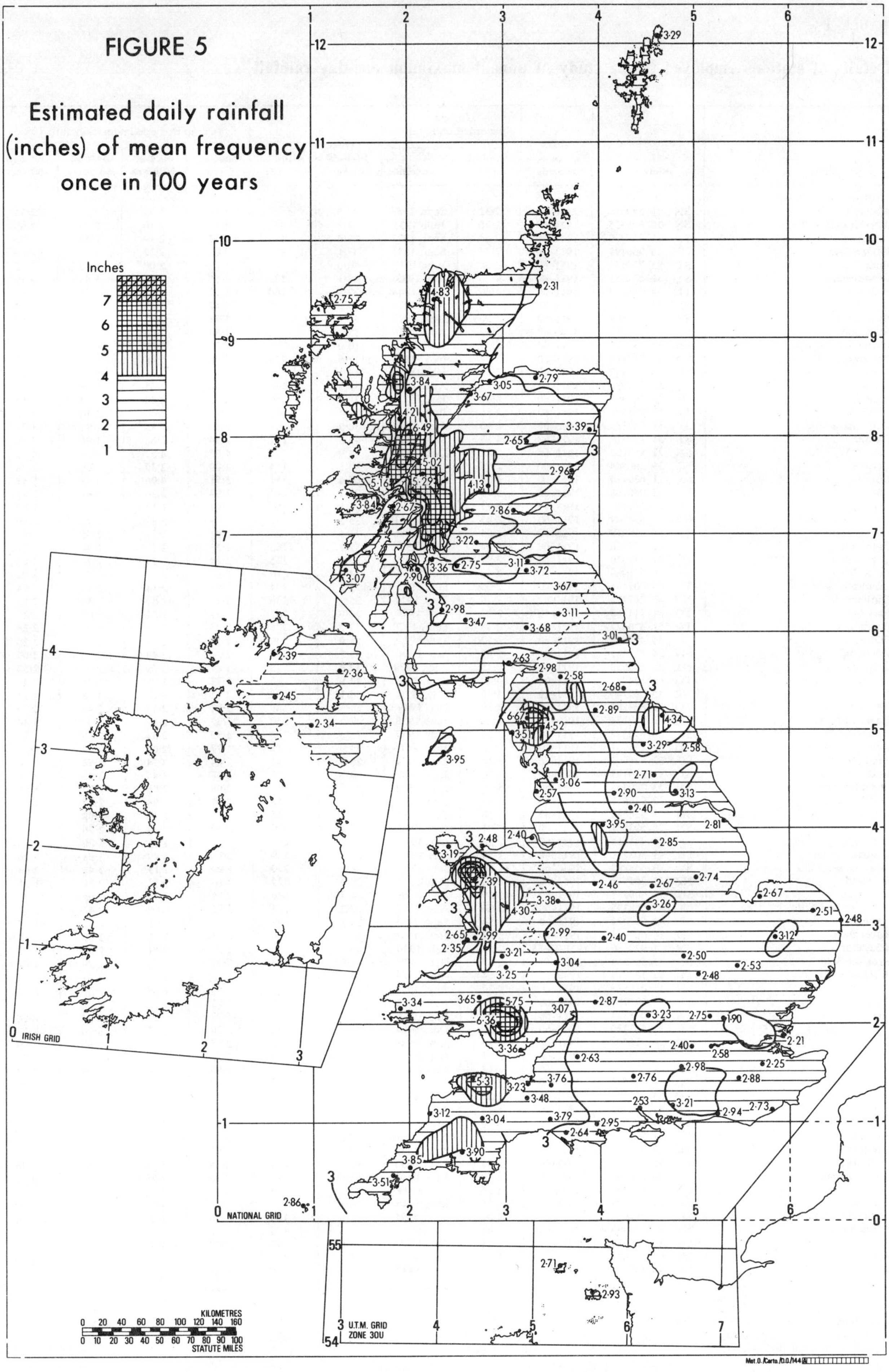

TABLE 1

Details of stations employed in the study of annual maximum one-day rainfall

	Grid reference	Years of records	Highest recorded daily fall		Average 1916–50 *inches*	Altitude *feet*	Computed maximum daily fall			
			Amount *inches*	Occurrence			once in 10 years	once in 20 years	once in 50 years	once in 100 years
ENGLAND AND WALES										
Aberystwyth	SN 22 580815	1910–60	2·32	Sept. 1958	37·5	12	1·78	1·95	2·18	2·35
Aberystwyth P.B.S.	SN 22 605825	1924–60	2·76	June 1935	41·6	452	1·95	2·16	2·44	2·65
Andover	SU 41 360450	1911–60	2·65	Sept. 1927	29·5	180	1·93	2·19	2·52	2·76
Baldersdale	NY 35 966193	1911–60	2·46	Sept. 1935	37·1	866	2·06	2·32	2·64	2·89
Bath	ST 31 754653	1911–60	2·24	June 1917	30·9	67	1·85	2·09	2·40	2·63
Beaminster	ST 31 482011	1911–60	4·53	July 1955	40·1	183	2·57	2·94	3·43	3·79
Bedford	TL 52 057498	1911–60	3·46	Aug. 1948	22·5	100	1·67	1·92	2·24	2·48
Birkenhead	SJ 33 287897	1911–60	2·36	Aug. 1954	29·0	198	1·67	1·89	2·19	2·40
Birmingham	SP 42 046864	1911–60	2·20	Aug. 1922	30·7	536	1·71	1·92	2·19	2·40
Blackpool	SD 34 324361	1940–60	1·75	Aug. 1950	35·1	65	1·84	2·06	2·35	2·57
Blaenau Ffestiniog	SH 23 699470	1911–60	7·75	June 1928	110·0	700	5·27	5·92	6·76	7·39
Borrowdale (1)	NY 35 237119	1928–60	5·24	Nov. 1931	131·6	423	4·96	5·48	6·16	6·67
Bradford	SE 44 144352	1911–60	2·32	Jan. 1944	34·1	531	2·04	2·30	2·65	2·90
Brighton W. Wks	TQ 51 243080	1911–60	2·78	Aug. 1946	31·5	166	2·09	2·35	2·69	2·94
Bude (2)	SS 21 208063	1911–60	3·42	Jan. 1927	34·2	50	2·11	2·42	2·82	3·12
Cambridge	TL 52 453572	1911–60	2·26	July 1959	21·7	41	1·76	2·00	2·30	2·53
Cannington	ST 31 237385	1920–60	9·40	Aug. 1924	30·9	102	2·15	2·48	2·91	3·23
Cardiff	ST 31 171729	1911–60	4·18	May 1931	41·3	121	2·34	2·66	3·06	3·36
Cardiff W. Wks	SN 22 987192	1911–60	5·21	Dec. 1960	83·9	1357	4·05	4·57	5·24	5·75
Carlisle	NY 35 390544	1911–60	2·95	Oct. 1949	32·1	114	2·00	2·30	2·69	2·98
Challacombe	SS 21 689410	1911–60	7·58	Aug. 1952	65·4	851	3·51	4·06	4·77	5·31
Cheltenham	SO 32 940209	1911–60	3·29	June 1930	27·3	210	1·99	2·26	2·61	2·87
Cheshunt	TL 52 311048	1911–60	1·75	July 1956	26·5	250	1·44	1·58	1·76	1·90
Church Stretton	SO 32 438909	1911–60	2·85	Aug. 1957	35·3	582	2·09	2·37	2·72	2·99
Cleethorpes	TA 54 314079	1925–60	1·84	Sept. 1935	23·4	23	1·94	2·21	2·55	2·81
Compton	SU 41 773150	1911–60	2·94	Sept. 1927	37·9	263	2·30	2·58	2·94	3·21
Copplestone	SS 21 771020	1911–60	2·97	Aug. 1952	36·0	325	2·10	2·39	2·76	3·04
Corfe Mullen	SY 30 976963	1911–60	2·41	July 1922	34·5	210	2·13	2·38	2·70	2·95
Cranwell	TF 53 003494	1911–60	2·65	Oct. 1913	23·5	204	1·87	2·14	2·48	2·74
Durham	NZ 45 267415	1911–60	2·76	May 1924	26·3	336	1·83	2·09	2·43	2·68
East Malling	TQ 51 711568	1911–60	2·23	Oct. 1955	26·9	132	1·66	1·84	2·07	2·25
Edenbridge	TQ 51 470427	1911–60	2·48	Sept. 1958	32·2	375	2·03	2·29	2·63	2·88
Felixkirk	SE 44 474848	1911–60	3·12	July 1932	28·9	540	2·17	2·51	2·96	3·29
Garstang	SD 34 527481	1911–60	2·82	Jan. 1954	48·7	616	2·24	2·49	2·82	3·06
Geltsdale	NY 35 590533	1911–60	2·15	Aug. 1928	38·0	778	1·87	2·09	2·37	2·58
Gorleston	TG 63 534037	1911–60	4·80	Aug. 1912	23·9	5	1·72	1·95	2·25	2·48
Guisborough	NZ 45 669140	1911–60	4·46	July 1930	34·1	632	2·82	3·28	3·89	4·34
Hastings	TQ 51 804102	1911–60	2·39	Oct. 1949	28·5	149	1·92	2·17	2·49	2·73
Haverfordwest	SM 12 919154	1930–60	3·24	June 1938	47·9	250	2·37	2·67	3·05	3·34
Holme-on-Spalding Moor	SE 44 813382	1911–60	3·58	May 1932	26·3	30	1·99	2·34	2·79	3·13
Kew	TQ 51 171757	1911–60	2·38	July 1956	23·9	18	1·77	2·02	2·34	2·58
Kimbolton	SO 32 526602	1911–60	3·40	May 1924	28·3	424	2·04	2·34	2·74	3·04
Lake Vyrnwy	SJ 33 019193	1911–60	3·70	July 1926	63·2	770	3·07	3·45	3·93	4·30
Llanddeusant	SN 22 730242	1911–60	3·98	Nov. 1931	58·5	323	2·68	2·98	3·37	3·65
Llandudno	SH 23 776818	1911–60	1·98	Sept. 1931	28·8	13	1·77	1·98	2·27	2·48
Llanerchymedd	SH 23 436843	1911–60	3·80	Oct. 1937	40.3	175	2·27	2·56	2·92	3·19
Lletty-Evan-hên	SN 22 693851	1936–60	3·01	July 1940	57·3	950	2·20	2·44	2·76	2·99
Longton	SJ 33 941418	1911–60	2·28	Oct. 1933	32·9	617	1·77	1·98	2·26	2·46
Loughborough	SK 43 508172	1911–60	3·24	July 1915	28·1	269	2·16	2·49	2·93	3·26
Martinstown (3)	SY 30 648888	1920–60	11·00	July 1955	38·5	270	1·90	2·13	2·42	2·64
Morpeth	NZ 45 201912	1911–60	2·84	May 1924	28·7	325	2·10	2·37	2·74	3·01
Newbridge	SO 32 009585	1911–60	3·50	July 1926	44·4	646	2·25	2·56	2·95	3·25
Nottingham Castle	SK 43 568395	1911–60	2·60	Aug. 1922	24·2	192	1·82	2·08	2·42	2·67
Oxford	SP 42 509072	1911–60	3·34	Sept. 1951	25·7	208	2·10	2·45	2·89	3·23
Ravenglass	SD 34 087965	1911–60	3·53	Nov. 1951	44·1	128	2·43	2·76	3·19	3·51
Rhayader	SN 22 972680	1921,25–55	2·97	Feb. 1946	49·1	700	2·32	2·61	2·97	3·21
Ross-on-Wye	SO 32 598233	1922–40	2·47	Oct. 1929	27·9	223	2·12	2·41	2·79	3·07
St Albans	TL 52 182076	1940–60	2·64	Aug. 1952	25·2	272	1·80	2·09	2·47	2·75
St Austell	SX 20 018528	1911–60	4·39	Aug. 1959	48·4	300	2·65	3·02	3·49	3·85
Sandringham	TF 53 695288	1911–60	2·91	Aug. 1912	27·2	121	1·83	2·09	2·42	2·67
Scarborough	TA 54 044884	1911–60	2·91	July 1940	25·6	118	1·79	2·03	2·34	2·58
Shawbury (4)	SJ 33 553220	1942–60	3·22	Dec. 1960	26·0	235	2·20	2·57	3·04	3·38
Sheepstor	SX 20 558682	1911–60	3·95	Aug. 1929	62·6	760	2·88	3·20	3·60	3·90
Shoeburyness	TQ 51 948857	1911–60	2·01	Oct. 1955	20·3	11	1·56	1·76	2·02	2·21
Southampton	SU 41 419119	1911–60	2·14	July 1922	31·6	65	1·85	2·06	2·33	2·53
South Farnborough	SU 41 867544	1940–60	2·43	July 1945	26·9	226	2·02	2·32	2·70	2·98
Sprowston	TG 63 251123	1911–60	7·20	Aug. 1912	25·4	93	1·73	1·97	2·28	2·51
Street	ST 31 492363	1911–60	5·13	June 1917	30·1	100	2·35	2·78	3·34	3·76
Swineshaw Moor	SE 44 020003	1911–60	4·50	May 1924	52·1	1190	2·65	3·04	3·56	3·95
Taunton	ST 31 228238	1911–60	4·33	June 1917	29·3	71	2·27	2·64	3·12	3·48
Thetford W. Wks	TL 52 864864	1911–60	3·53	Aug. 1912	24·6	165	2·02	2·36	2·79	3·12
Treherbert	SS 21 932994	1911–60	5·93	Dec. 1960	93·6	801	4·58	5·13	5·83	6·36
Tresco Abbey	SV 00 896143	1911–60	2·70	Feb. 1936	33·8	40	2·03	2·28	2·61	2·86
Truro	SW 10 827450	1911–60	3·26	Oct. 1924	42·9	65	2·43	2·76	3·19	3·51
Wakefield	SE 44 327199	1911–60	2·21	May 1945	26·4	115	1·75	1·95	2·21	2·40
Wellingborough	SP 42 894675	1911–60	3·05	Aug. 1922	24·5	169	1·70	1·94	2·26	2·50
Windermere	NY 35 389009	1911–60	4·03	July 1938	68·6	175	3·28	3·66	4·15	4·52
Windsor	SU 41 979754	1911–60	2·24	Oct. 1955	25·8	70	1·69	1·91	2·19	2·40
Worksop	SK 43 594835	1911–60	3·08	Aug. 1922	24·2	100	1·93	2·21	2·58	2·85
York	SE 44 598522	1911–60	2·77	May 1932	24·7	57	1·83	2·10	2·45	2·71

(1) 3 months missing 1953.
(2) June, July and August 1956 missing.
(3) Abbotsbury used prior to 1930.
(4) Ternhill used 1942–43.

TABLE 1 (*continued*)

Details of stations employed in the study of annual maximum one-day rainfall

	Grid reference	Years of records	Highest recorded daily fall: Amount *inches*	Highest recorded daily fall: Occurrence	Average 1916–50 *inches*	Altitude *feet*	Computed maximum daily fall: once in 10 years	once in 20 years	once in 50 years	once in 100 years
SCOTLAND										
Aberdeen (5)	NJ 38 915042	1911–60	3·01	Oct. 1941	32·9	171	2·32	2·65	3·07	3·39
Aberfeldy (6)	NN 27 852492	1911–60	4·25	June 1935	38·8	320	2·75	3·17	3·72	4·13
Achnashellach	NH 28 040491	1923–43	3·54	Mar. 1926	81·0	220	2·79	3·11	3·53	3·84
Ayr Cemetery (7)	NS 26 348211	1911–60	3·57	June 1919	35·9	40	1·99	2·29	2·68	2·98
Balmoral	NO 37 253949	1911–60	2·65	Jan. 1937	33·3	927	1·91	2·14	2·43	2·65
Baltasound	HP N42 609090	1911–60	4·28	July 1935	44·9	30	2·17	2·51	2·95	3·29
Denholm (8)	NT 36 587182	1911–60	3·06	Aug. 1948	31·1	425	2·11	2·41	2·81	3·11
Edinburgh	NT 36 259706	1911–60	3·00	Aug. 1920	27·5	441	2·09	2·41	2·81	3·11
Eskdalemuir	NT 36 235026	1911–60	4·18	June 1953	62·3	794	2·64	2·96	3·37	3·68
Fochabers	NJ 38 350595	1911–60	2·65	Sept. 1915	29·6	104	1·97	2·22	2·54	2·79
Fort William	NN 27 101735	1911–60	5·01	Dec. 1936	78·7	216	3·54	3·99	4·58	5·02
Glen Afton	NS 26 617101	1911–60	3·45	Dec. 1936	57·5	833	2·55	2·83	3·19	3·47
Glenleven	NN 27 187619	1911–60	4·80	Dec. 1957	94·1	140	3·85	4·29	4·86	5·29
Greenock W. Wks (9)	NS 26 256724	1911–60	2·85	Aug. 1920	70·4	643	2·57	2·82	3·13	3·36
Inverness (10)	NH 28 665412	1911–60	3·64	Sept. 1915	28·4	242	2·37	2·77	3·29	3·67
Islay	NR 26 335634	1911–60	3·17	Sept. 1939	53·1	68	2·15	2·43	2·79	3·07
Loch Duich	NG 18 918199	1926–59	3·07	Sept. 1952	80·4	12	2·97	3·24	3·59	4·21
Loch More (11)	NC 29 292396	1911–60	3·90	Apr. 1946	84·9	137	3·38	3·82	4·40	4·83
Loch Quoich	NH 28 028028	1911–54	6·28	Nov. 1947	113·3	569	4·68	5·23	5·95	6·49
Marchmont House	NT 36 743484	1911–60	4·00	Aug. 1948	30·7	498	2·42	2·80	3·30	3·67
Montrose	NO 37 707617	1911–60	2·72	Sept. 1927	28·7	200	2·08	2·35	2·70	2·96
Morvern	NM 17 836531	1911–60	4·92	Nov. 1941	105·1	40	3·81	4·22	4·76	5·16
Mull (12)	NM 17 552393	1911–60	3·27	Nov. 1917	79·8	100	2·81	3·13	3·53	3·84
Nairn	NH 28 868565	1911–60	3·29	Sept. 1915	26·4	59	2·03	2·34	2·75	3·05
Oban (13)	NM 17 852294	1911–60	2·41	June 1957	56·6	229	2·03	2·23	2·48	2·67
Pentland Hills (14)	NT 36 223636	1911–60	3·82	Sept. 1927	38·1	739	2·51	2·88	3·36	3·72
Perth	ND 37 101239	1911–60	3·42	July 1916	30·7	77	1·98	2·25	2·60	2·86
Renfrew	NS 26 508662	1911–60	3·20	Aug. 1920	40·7	30	1·94	2·18	2·51	2·75
Rothesay (15)	NS 26 106647	1911–60	2·51	Dec. 1952	56·2	200	2·12	2·36	2·67	2·90
Ruthwell	NY 35 078691	1911–60	2·25	Mar. 1950	36·9	95	1·94	2·15	2·41	2·63
Stirling	NS 26 733915	1911–60	3·12	Aug. 1920	52·8	759	2·36	2·63	2·97	3·22
Stornoway	NB 19 465332	1914–43	2·75	Aug. 1933	39·5	11	1·96	2·20	2·50	2·75
Wick	ND 39 366523	1911–60	2·27	Apr. 1934	30·0	119	1·61	1·82	2·10	2·31
ISLE OF MAN										
Douglas	SC 24 383776	1911–50, 54–57	4·67	Sept. 1918	45·2	284	2·67	3·06	3·57	3·95
NORTHERN IRELAND										
Armagh	IH 23 878457	1911–60	2·17	Oct. 1929	33·3	205	1·67	1·88	2·14	2·34
Ballymena	ID 34 108028	1911–60	2·08	Aug. 1952	43·4	150	1·75	1·94	2·18	2·36
Londonderry	IC 24 415173	1911–60	2·25	Aug. 1917	46·1	325	1·72	1·93	2·19	2·39
Omagh	IH 23 464718	1911–60	2·78	Oct. 1954	42·8	280	1·76	1·97	2·25	2·45
CHANNEL ISLANDS										
Guernsey	WV 554 311762	1911–39, 1950–60	2·50	Sept. 1919	34·5	340	2·00	2·22	2·50	2·71
Jersey	WV 554 632507	1911–60	2·95	Aug. 1931	34·3	53	2·01	2·29	2·66	2·93

(5) 1911 estimated.
(6) 1926 estimated.
(7) 1911 and 1912 estimated.
(8) 1911 to 1915 estimated.
(9) 1911, 1913 and 1915 estimated.
(10) 1922, 1957 and 1958 estimated.
(11) 1926, 1951 to 1960 estimated.
(12) 1959 and 1960 estimated.
(13) 1956 and 1957 estimated.
(14) 1911 and 1912 estimated.
(15) 1958 and 1959 estimated.

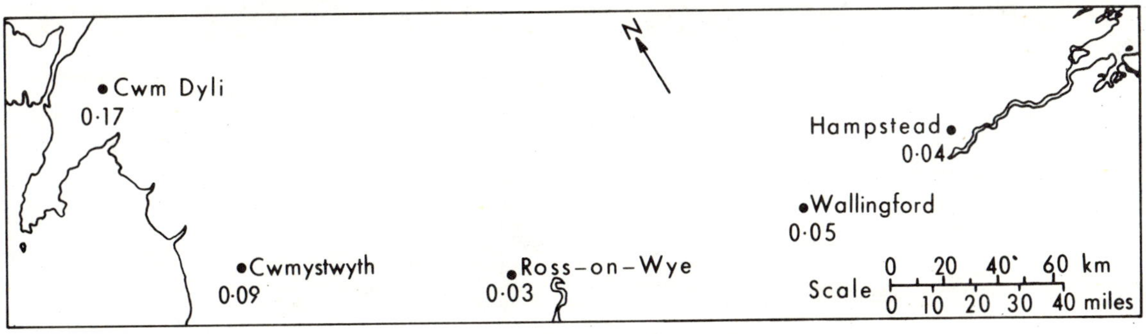

FIGURE 6. The probability of rain falling at 09 GMT

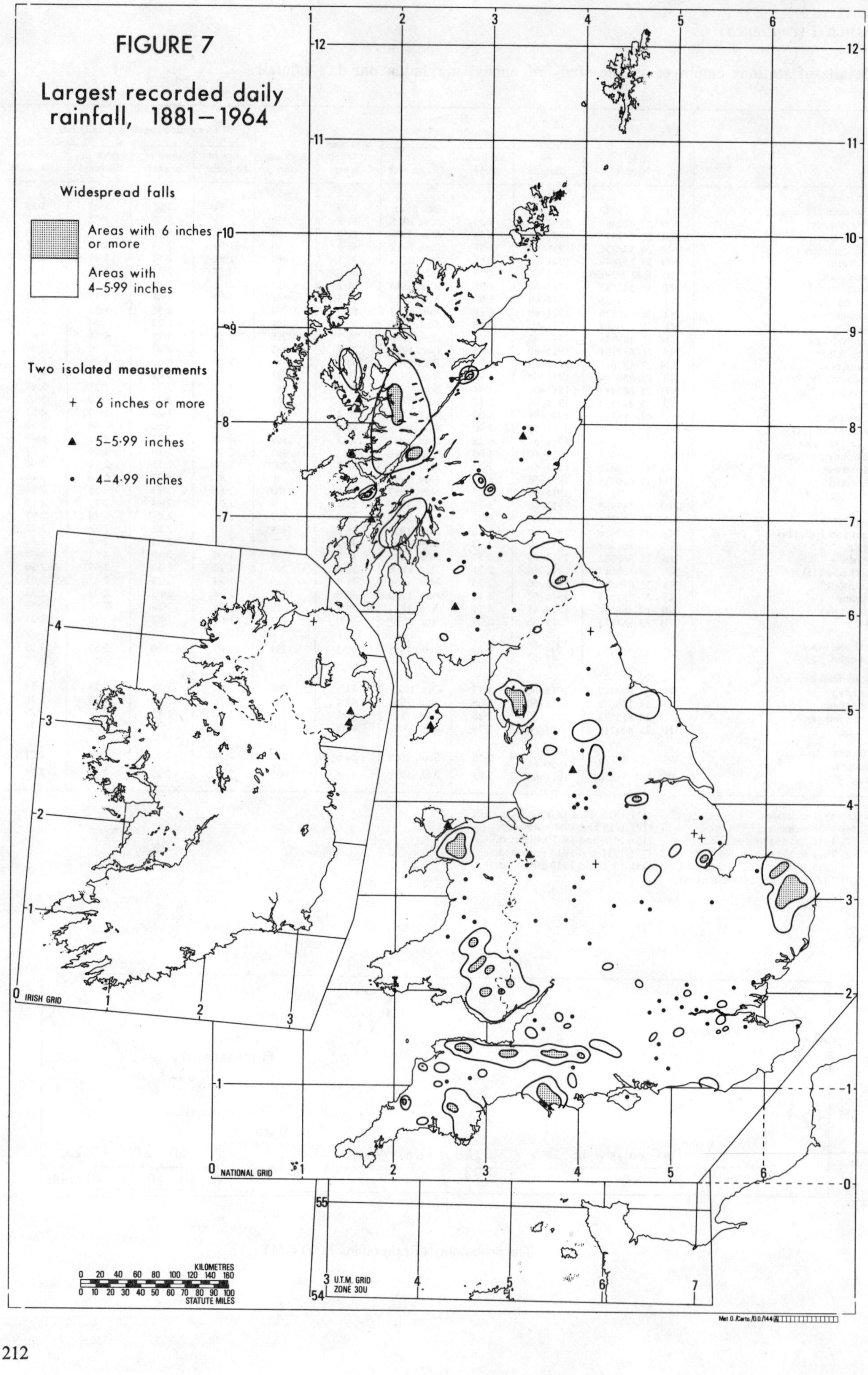
FIGURE 7
Largest recorded daily rainfall, 1881–1964
Widespread falls
Areas with 6 inches or more
Areas with 4–5·99 inches
Two isolated measurements
+ 6 inches or more
▲ 5–5·99 inches
• 4–4·99 inches
IRISH GRID
NATIONAL GRID
U.T.M. GRID ZONE 30U
KILOMETRES
0 20 40 60 80 100 120 140 160
0 10 20 30 40 50 60 70 80 90 100
STATUTE MILES
Met O./Carto./D.D./144

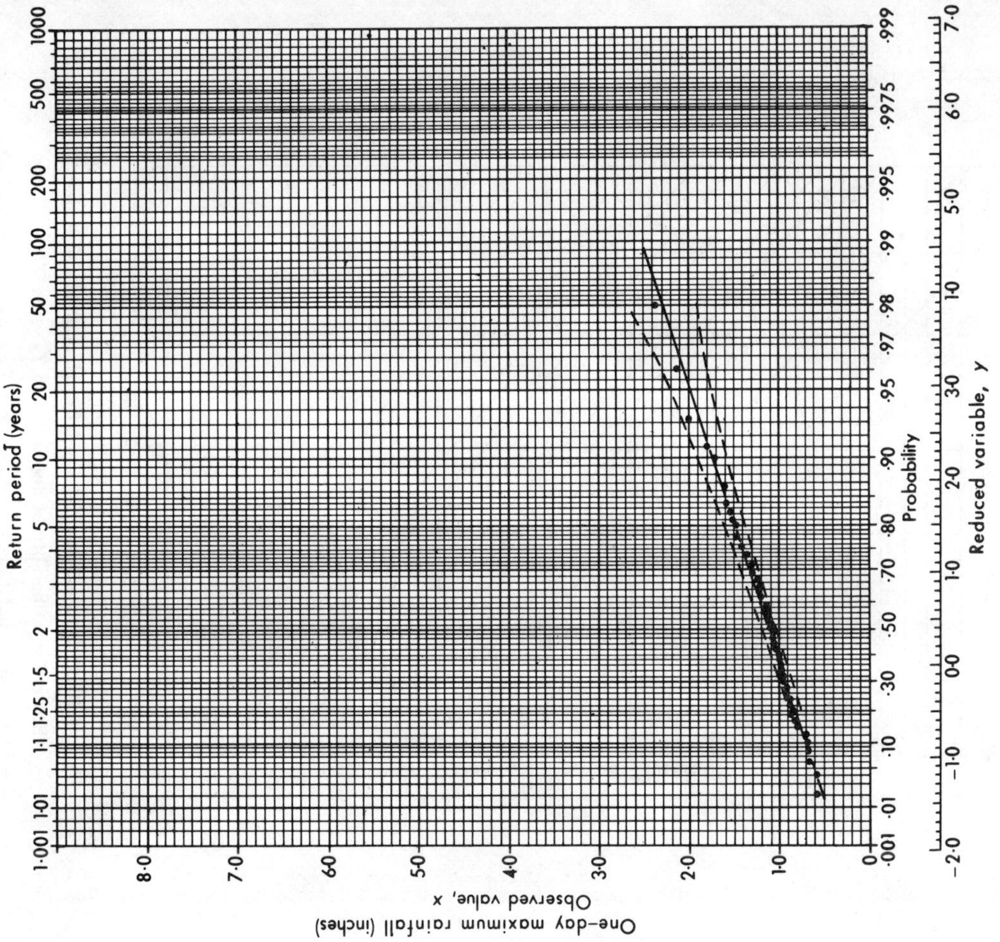

FIGURE 8(*b*). One-day rainfall magnitude–frequency relationship for Sprowston excluding the highest maximum

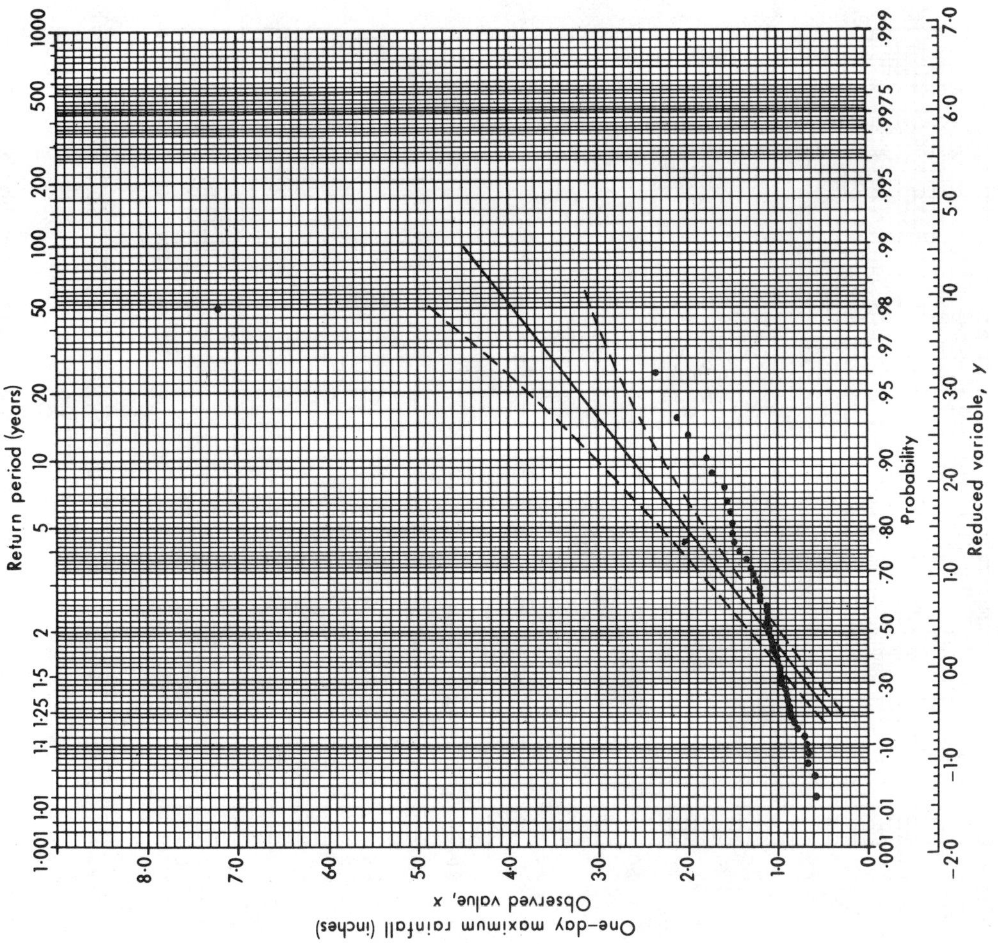

FIGURE 8(*a*). One-day rainfall magnitude–frequency relationship for Sprowston including the highest maximum

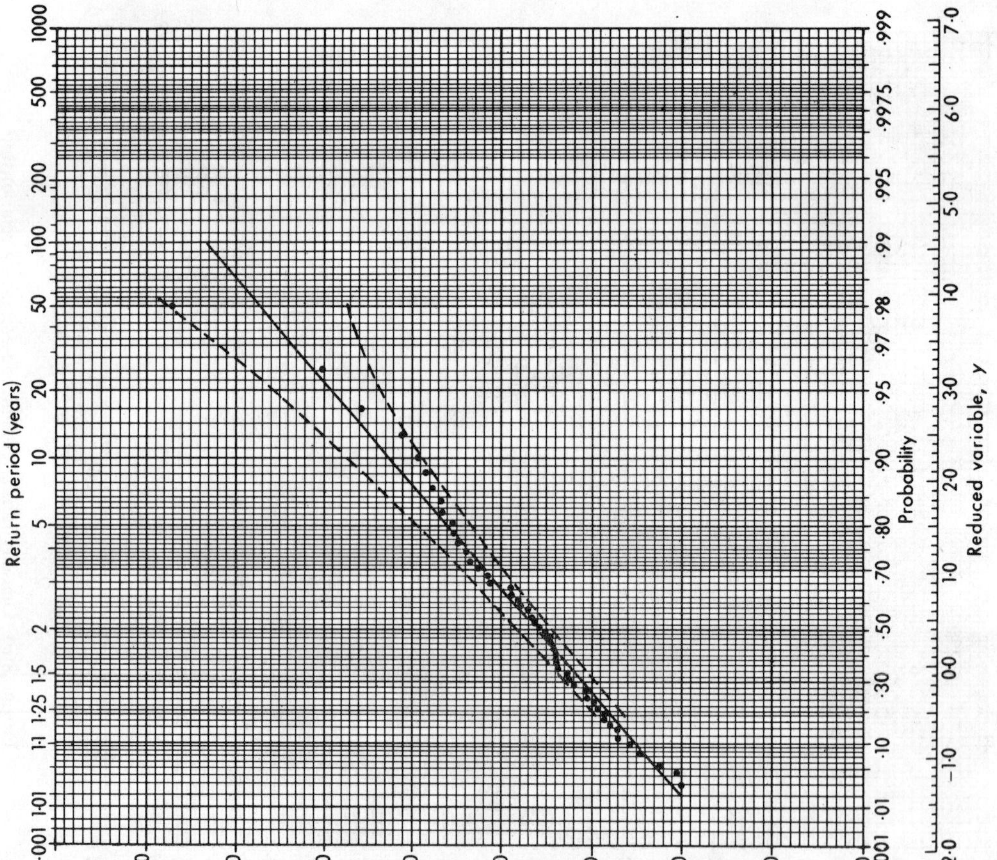

FIGURE 9. One-day rainfall magnitude–frequency relationship for Blaenau Ffestiniog

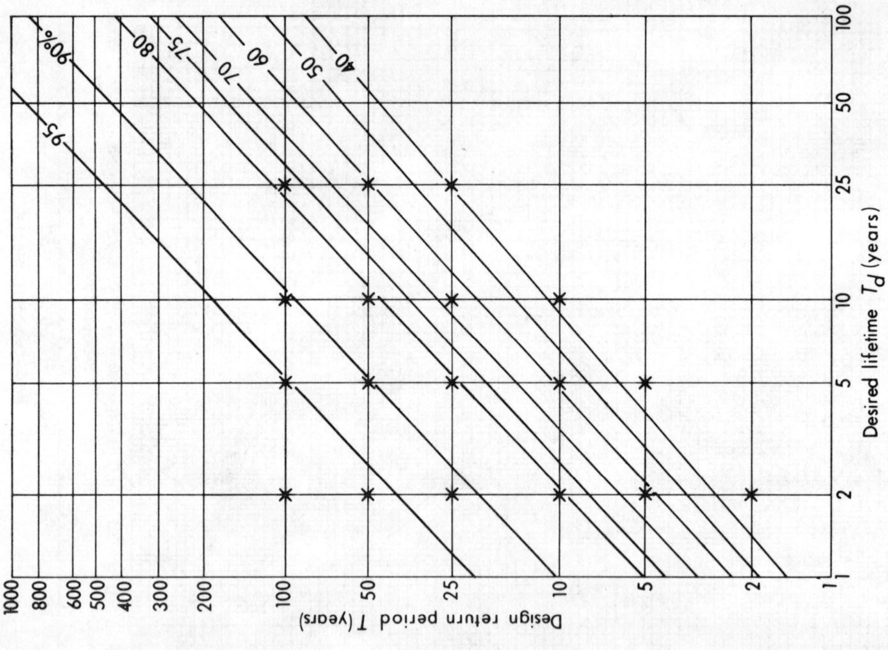

FIGURE 10. Calculated-risk diagram (after Hershfield and Kohler[10])

— Theoretical probability of not failing in T_d years.

× Observed probabilities of not failing in T_d years (25-year record).

those return periods. However, as an n-year rainfall could occur next year or the year after, knowledge of the return period of a given fall is more valuable for design purposes, if it is related to the probability of that fall occurring at least once before the next n years have elapsed. It is known that the probability P of the event occurring at least once in n years is:

$$P = 1 - \left(1 - \frac{1}{n}\right)^n \approx 1 - e^{-1} = 0{\cdot}63.$$

By using this formula, a calculated-risk diagram (Figure 10) has been produced (Hershfield and Kohler[10]) to show the probability of the event not occurring within any interval up to the design return period. Thus the probability of a structure designed to a return period of 100 years not failing in the first 10 years is given as 90 per cent.

Conclusion. The merits of this study depend upon two sets of factors, one concerned with the basic data and the other with the method of analysis. The fundamentals of this type of extreme-value analysis were neither tested, nor compared with others of a similar kind which might have given slightly different results. However, it is worth recognizing this point and the fact that errors could arise because of differences in lengths of record and climatic change. That the climate of Britain is declining from the optimum attained during the first half of this century is well documented (Lamb[11]), but it is not clear whether this change is affecting the frequency and magnitude of intense falls. These errors are likely to be less serious than those associated with the gauge itself (Rodda[12]) but gauge errors are in any case common to all studies of this country's rainfall. While such shortcomings must be considered, they should be insufficient to detract from the value of this type of approach to the study of intense rainfall.

References

1. BILHAM, E. G.; Classification of heavy falls of rain in short periods. *Brit Rainf* 1935, *London*, 1936, pp. 262–280.
2. MACLEAN, D. J.; Rainstorm data. *Survr Munic Cty Engr*, *London*, **104**, 1945, p. 36 and p. 56.
3. COLLINGE, V. K.; The frequency of heavy rainfalls in the British Isles. *Civ Engng Publ Wks Rev*, *London*, **56**, 1961, Part I pp. 341–344 and Part II pp. 497–500.
4. BLEASDALE, A.; The distribution of exceptionally heavy daily falls of rain in the United Kingdom, 1863 to 1960. *J Instn Wat Engrs*, *London*, **17**, 1963, pp. 45–55.
5. HERSHFIELD, D. M.; Rainfall frequency atlas of the United States for durations from 30 minutes to 24 hours and return periods 1 to 100 years. *Tech Pap U.S. Weath Bur*, *Washington*, No. 40, 1961.
6. REICH, B. M.; Short-duration rainfall-intensity estimates and other design aids for regions of sparse data. *J Hydrol*, *Amsterdam*, **1**, 1963, pp. 3–28.
7. STORR, D.; Maximum one-day rainfall frequencies in Alberta. *Circ Met Div Dep Transp*, *Toronto*, No. 3796, 1963.
8. ASAI, T., HATANO, T. and HATANO, M.; The return periods of daily heavy rainfall at 33 climatological stations in Northern Kyūshū, Japan. *Misc Rep Inst Nat Resour*, *Tokyo*, No. 61, 1963, pp. 1–6.
9. GUMBEL, E. J.; Statistics of extremes. New York, Columbia University Press, 1958.
10. HERSHFIELD, D. M. and KOHLER, M. A.; An empirical appraisal of the Gumbel extreme-value procedure. *J Geophys Res*, *Baltimore*, **65**, 1960, pp. 1737–1746.
11. LAMB, H. H.; Britain's changing climate. Johnson, C. G. and Smith, L. P. (editors). The biological significance of climatic changes in Britain (Proceedings of a Symposium held at the Royal Geographical Society, London, on 29 and 30 October 1964). London and New York, Academic Press, 1965, pp. 3–31.
12. RODDA, J. C.; The systematic error in rainfall measurement. *J Instn Wat Engrs*, *London*, **21**, 1967, pp. 173–177.

Dd 503133 K7 10/72